Geophysical Monograph Series

Including

IUGG Volumes
Maurice Ewing Volumes
Mineral Physics Volumes

Geophysical Monograph Series

107 **Rivers Over Rock: Fluvial Processes in Bedrock Channels** *Keith J. Tinkler and Ellen E. Wohl (Eds.)*

108 **Assessment of Non-Point Source Pollution in the Vadose Zone** *Dennis L. Corwin, Keith Loague, and Timothy R. Ellsworth (Eds.)*

109 **Sun-Earth Plasma Interactions** *J. L. Burch, R. L. Carovillano, and S. K. Antiochos (Eds.)*

110 **The Controlled Flood in Grand Canyon** *Robert H. Webb, John C. Schmidt, G. Richard Marzolf, and Richard A. Valdez (Eds.)*

111 **Magnetic Helicity in Space and Laboratory Plasmas** *Michael R. Brown, Richard C. Canfield, and Alexei A. Pevtsov (Eds.)*

112 **Mechanisms of Global Climate Change at Millennial Time Scales** *Peter U. Clark, Robert S. Webb, and Lloyd D. Keigwin (Eds.)*

113 **Faults and Subsurface Fluid Flow in the Shallow Crust** *William C. Haneberg, Peter S. Mozley, J. Casey Moore, and Laurel B. Goodwin (Eds.)*

114 **Inverse Methods in Global Biogeochemical Cycles** *Prasad Kasibhatla, Martin Heimann, Peter Rayner, Natalie Mahowald, Ronald G. Prinn, and Dana E. Hartley (Eds.)*

115 **Atlantic Rifts and Continental Margins** *Webster Mohriak and Manik Talwani (Eds.)*

116 **Remote Sensing of Active Volcanism** *Peter J. Mouginis-Mark, Joy A. Crisp, and Jonathan H. Fink (Eds.)*

117 **Earth's Deep Interior: Mineral Physics and Tomography From the Atomic to the Global Scale** *Shun-ichiro Karato, Alessandro Forte, Robert Liebermann, Guy Masters, and Lars Stixrude (Eds.)*

118 **Magnetospheric Current Systems** *Shin-ichi Ohtani, Ryoichi Fujii, Michael Hesse, and Robert L. Lysak (Eds.)*

119 **Radio Astronomy at Long Wavelengths** *Robert G. Stone, Kurt W. Weiler, Melvyn L. Goldstein, and Jean-Louis Bougeret (Eds.)*

120 **GeoComplexity and the Physics of Earthquakes** *John B. Rundle, Donald L. Turcotte, and William Klein (Eds.)*

121 **The History and Dynamics of Global Plate Motions** *Mark A. Richards, Richard G. Gordon, and Rob D. van der Hilst (Eds.)*

122 **Dynamics of Fluids in Fractured Rock** *Boris Faybishenko, Paul A. Witherspoon, and Sally M. Benson (Eds.)*

123 **Atmospheric Science Across the Stratopause** *David E. Siskind, Stephen D. Eckerman, and Michael E. Summers (Eds.)*

124 **Natural Gas Hydrates: Occurrence, Distribution, and Detection** *Charles K. Paull and Willam P. Dillon (Eds.)*

125 **Space Weather** *Paul Song, Howard J. Singer, and George L. Siscoe (Eds.)*

126 **The Oceans and Rapid Climate Change: Past, Present, and Future** *Dan Seidov, Bernd J. Haupt, and Mark Maslin (Eds.)*

127 **Gas Transfer at Water Surfaces** *M. A. Donelan, W. M. Drennan, E. S. Saltzman, and R. Wanninkhof (Eds.)*

128 **Hawaiian Volcanoes: Deep Underwater Perspectives** *Eiichi Takahashi, Peter W. Lipman, Michael O. Garcia, Jiro Naka, and Shigeo Aramaki (Eds.)*

129 **Environmental Mechanics: Water, Mass and Energy Transfer in the Biosphere** *Peter A.C. Raats, David Smiles, and Arthur W. Warrick (Eds.)*

130 **Atmospheres in the Solar System: Comparative Aeronomy** *Michael Mendillo, Andrew Nagy, and J. H. Waite (Eds.)*

131 **The Ostracoda: Applications in Quaternary Research** *Jonathan A. Holmes and Allan R. Chivas (Eds.)*

132 **Mountain Building in the Uralides Pangea to the Present** *Dennis Brown, Christopher Juhlin, and Victor Puchkov (Eds.)*

133 **Earth's Low-Latitude Boundary Layer** *Patrick T. Newell and Terry Onsage (Eds.)*

134 **The North Atlantic Oscillation: Climatic Significance and Environmental Impact** *James W. Hurrell, Yochanan Kushnir, Geir Ottersen, and Martin Visbeck (Eds.)*

135 **Prediction in Geomorphology** *Peter R. Wilcock and Richard M. Iverson (Eds.)*

136 **The Central Atlantic Magmatic Province: Insights from Fragments of Pangea** *W. Hames, J. G. McHone, P. Renne, and C. Ruppel (Eds.)*

137 **Earth's Climate and Orbital Eccentricity: The Marine Isotope Stage 11 Question** *André W. Droxler, Richard Z. Poore, and Lloyd H. Burckle (Eds.)*

138 **Inside the Subduction Factory** *John Eiler (Ed.)*

139 **Volcanism and the Earth's Atmosphere** *Alan Robock and Clive Oppenheimer (Eds.)*

140 **Explosive Subaqueous Volcanism** *(James D. L. White, John L. Smellie, and David A. Clague (Eds.)*

141 **Solar Variability and Its Effects on Climate** *(Judit M. Pap and Peter Fox (Eds.)*

Geophysical Monograph 142

Disturbances in Geospace
The Storm-Substorm Relationship

A. Surjalal Sharma
Yohsuke Kamide
Gurbax S. Lakhina
Editors

American Geophysical Union
Washington, DC

Published under the aegis of the AGU Books Board

Jean-Louis Bougeret, Chair; Gray E. Bebout, Carl T. Friedrichs, James L. Horwitz, Lisa A. Levin, W. Berry Lyons, Kenneth R. Minschwaner, Darrell Strobel, and William R. Young, members.

Library of Congress Cataloging-in-Publication Data
Disturbances in geospace : the storm-substorm relationship / A. Surjalal Sharma, Yohsuke Kamide, Gurbax S. Lakhina, editors.
      p. cm. - (Geophysical monograph ; 142)
      Includes bibliographical references.
      ISBN 0-87590-407-6
  1. Magnetic storms. 2. Magnetosphere. 3. Magnetospheric substorms. 4. Ionosphere.
  5. Space environment. I. Sharma, A. Surjalal, 1951- II. Kamide, Y. III. Lakhina, Gurbax S.
  IV. Series.

QC835.D57 2003
538'.76-dc22   2003066286

ISSN 0065-8448
ISBN 0-87590-407-6

Cover images
Large auroral image: Aurora is one of the manifestations of substorms. Photograph by Nori Sakamoto (http://www.auroraphoto.net/).

Inset: The inset shows an energetic neutral atom (ENA) image of the ring current in the late main phase (~22:00 UT) of the 4 October 2000 storm obtained in the 27-60 keV range by the ENA imager HENA on board the IMAGE satellite [*C:son Brandt et al.,* this volume]. The white circle outlines the Earth's limb and the dipole magnetic field lines are drawn for reference at dawn, noon, dusk, and midnight for L-shells 4 and 8. The image shows the ring current just shortly after a substorm injection.

Copyright 2003 by the American Geophysical Union
2000 Florida Avenue, N.W.
Washington, DC 20009

Figures, tables, and short excerpts may be reprinted in scientific books and journals if the source is properly cited.

Authorization to photocopy items for internal or personal use, or the internal or personal use of specific clients, is granted by the American Geophysical Union for libraries and other users registered with the Copyright Clearance Center (CCC) Transactional Reporting Service, provided that the base fee of $1.50 per copy plus $0.35 per page is paid directly to CCC, 222 Rosewood Dr., Danvers, MA 01923. 0065-8448/03/$01.50+0.35.

This consent does not extend to other kinds of copying, such as copying for creating new collective works or for resale. The reproduction of multiple copies and the use of full articles or the use of extracts, including figures and tables, for commercial purposes requires permission from the American Geophysical Union.

Printed in the United States of America.

CONTENTS

Preface
A. Surjalal Sharma, Yohsuke Kamide, and Gurbax S. Lakhina vii

Panel Overview

Storm-Substorm Relationship: Current Understanding and Outlook
*A. S. Sharma, D. N. Baker, M. Grande, Y. Kamide, G. S. Lakhina, R. M. McPherron, G. D. Reeves,
G. Rostoker, R. Vondrak, and L. Zelenyi* 1

How Storms and Substorms Respond to Solar Wind Driver

Effects of Solar Wind Density on the Auroral Electrojets and Global Auroras During Geomagnetic Storms
Y. Kamide, J.-H. Shue, and M. Brittnacher 15

A Lack of Substorm Expansion Phases During Magnetic Storms Induced by Magnetic Clouds
B. T. Tsurutani, X.-Y. Zhou, and W. D. Gonzalez 23

Magnetic Storms in the Magnetotail

Storm-time and Quiet-time Substorms in the Magnetotail
A. A. Petrukovich 37

High Time Resolution Observations of Magnetospheric Disturbances During Auroral Activity
M. O. Fillingim, G. K. Parks, R. P. Lin, M. McCarthy, and A. Szabo 45

Substorms, Storms, and the Storm-Time Plasma Sheet
Wolfgang Baumjohann, Rumi Nakamura, Rainer Schödel, and Kai Dierschke 55

Storm-Substorm Relationships in the Inner Magnetosphere

O+ Transport into the Ring Current: Storm versus Substorm
A. Korth, R. H. W. Friedel, M. G. Henderson, F. Frutos-Alfaro, and C. G. Mouikis 59

What is the Effect of Substorms on the Ring Current Ion Population During a Geomagnetic Storm?
M. Grande, C. H. Perry, A. Hall, J. Fennell, R. Nakamura, and Y. Kamide 75

IMAGE, POLAR, and Geosynchronous Observations of Substorm and Ring Current Ion Injection
*G. D. Reeves, M. C. Henderson, R. M. Skoug, M. F. Thomsen, J. E. Borovsky, H. O. Funsten,
P. C:son Brandt, D. J. Mitchell, J.-M. Jahn, C. J. Pollock, D. J. McComas, and S. B. Mende* 91

Storm-Substorm Relationships During the 4 October, 2000 Storm. IMAGE Global ENA Imaging Results
*Pontus C:son Brandt, Donald G. Mitchell, Robert Demajistre, Edmond C. Roelof, Shin Ohtani, Jörg-Micha Jahn,
Craig Pollock, and Geoff Reeves* 103

CONTENTS

The Role of Substorms in Storm-time Particle Acceleration
Ioannis A. Daglis and Yohsuke Kamide 119

Modeling the Magnetosphere

Role of Plasma Instabilities Driven by Oxygen Ions During Magnetic Storms and Substorms
G. S. Lakhina and S. V. Singh 131

The Relationship of Storms and Substorms Determined From Mid-latitude Ground-based Magnetic Maps
C. Robert Clauer, Michael W. Liemohn, Janet U. Kozyra, and Michelle L. Reno 143

A Wavelet Analysis of Storm-substorm Relationships
W. B. Cade III, J. J. Sojka, L. Zhu, and Y. Kamide 159

Response of Ionosphere and Atmosphere to Storm-time Substorms

Energetics of Isolated and Storm Time Substorms
N. Østgaard and E. Tanskanen 169

Equatorial Ionosphere-Thermosphere System During Geomagnetic Storms
J. Hanumath. Sastri, R. Sridharan, and Tarun Kumar Pant 185

Structure of Turbulent Irregularities in High-Latitude Plasma Patches—3D Nonlinear Simulations
N. A. Gondarenko and P. N. Guzdar 205

Space Weather and Related Issues

Relativistic Electron Flux Enhancements During Strong Geomagnetic Activity
D. N. Baker and X. Li 217

Modeling the Magnetosphere Using Time Series Data
A. S. Sharma, A. Y. Ukhorskiy, M. I. Sitnov, and J. A. Valdivia 231

Comments on Some Long-Standing Problems in Storm/Substorm Studies
S.-I. Akasofu, W. Sun, and B.-H. Ahn 243

Measurement Strategies for Future Missions to Understand Geospace Dynamics
R. Vondrak, J. Slavin, L. Zelenyi, M. Guhathakurta, S. Curtis, and B. Tsurutani 255

PREFACE

Advancing our knowledge of the Sun-Earth connection and our capabilities to predict conditions in near-Earth geospace has captured the attention of geospace, solar and other scientists, prompting initiatives in many countries. These advances rely heavily on our understanding of the coupling processes between the solar wind and magnetosphere, such as geomagnetic storms and substorms. In this regard, the science of the storm-substorm relationship is not an end in itself, but a critical step in unveiling the Sun-Earth connection.

Sidney Chapman (1889-1970) was a pioneer in the study of the influence of the Sun on geospace. In fact, it was he who first coined the term "substorms" to describe the many intense disturbances that occur during a magnetic storm. Chapman also provided us with a comprehensive view of the relationship between storms and substorms. Since Chapman's initial work, our understanding has undergone vast changes. The classical notion of substorms as building blocks of storms has evolved, allowing us to note complex synergies between storms, substorms, and convection enhancements in the magnetosphere.

Recent progress has accelerated our understanding here. In particular, scientists now recognize that the ring current during the main phase of magnetic storms is asymmetric and driven mainly by plasma convection in the magnetosphere. This result, obtained from analysis of ground based magnetometers, and confirmed by the global images of the ring current from the IMAGE spacecraft, is regarded as one of the most important new results in magnetospheric physics.

The present volume thus reviews several important aspects of the storm-substorm relationship as well as what returns scientists expect from forthcoming exploratory missions in the sun-earth system. These missions, which should provide us with unprecedented data on magnetic storms and substorms, have the potential of resolving several long-standing issues.

We have organized the papers in the volume to represent the key issues of the storm-substorm relationship, including their importance to associated areas of research. The processes that start on the Sun, such as coronal mass ejections, lead to strong disturbances propagating in the solar wind, which in turn become the drivers for geomagnetic activity extending to the ionosphere and upper atmosphere. Although the plethora of events in the sun-earth connection is often viewed as a single chain, it is more like a matrix with many cross connections among the different elements. The multiple facets of the storm-substorm relationships exemplify such a view.

The introduction to the book provides a broad overview of the storm-substorm relationship, and acts as a guide to the papers that follow - a progressive view of the processes starting from the Sun and culminating in the Earth's atmosphere. The concluding group of papers deals with the issues of space weather, future missions, and some of the open questions that require our attention.

A recent claim that storms and substorms are unrelated is an example of how scientists often claim the resolution of long-standing problems by calling unrelated two phenomena previously found inter-related. Such a divorce can serve to clarify certain aspects of both, while leaving other significant issues unresolved. While our new understanding - the ring current during the main phase of the storm to be asymmetric and driven mainly by convection - is an important advance, it is still not clear how the different aspects of storms and substorms are unrelated or related. Certainly questions remain, including whether the inner magnetosphere is populated mainly by convective processes or through impulsive injections driven by substorm-associated inductive electric fields. As substorms are almost always accompanied by the injection of energetic particles, we would expect them to contribute to the storm-time ring current. In view of the electrodynamic nature of the interaction between different regions of the magnetosphere and the dominantly global nature of its dynamics, we would not expect these two manifestations of geomagnetic activity merely to co-exist. Yet how they influence each other is still unclear.

These and other questions prompted us to convene a Chapman Conference on the storm-substorm relationship in Lonavala, India, March 2001. This volume derives from the proceedings of the conference, which brought together scientists involved in storm and substorm studies. Our purpose was to comprehensively pursue and clarify the specific ways in which storms and substroms influence each other from both sides of the issue, within the context of most recent developments.

Even in a volume of this size it has not been able to do justice to many aspects of the subject, which claim as much right to attention as those that have received their due here. Nonetheless, the editors hope that the topics selected represent the key issues that will be most actively pursued in the near future. Among the topics not included here are the formation and development of the ring current, which is influenced by substorms in some form not fully understood; global MHD modeling of the solar wind-magnetosphere coupling, which when coupled to inner magnetospheric models is likely to yield a comprehensive picture; and models of the inner magnetosphere with descriptions of the ring current during storms.

Our colleagues who reviewed many papers in the volume have contributed significantly to its development. We wish to thank the reviewers for their cooperation and assistance in generous measure: T. Araki, P. K. Chaturvedi, C. R. Clauer, I. A. Daglis, A. C. Das, Y. Dimant, M. C. Fok, M. Grande, P. N. Guzdar, T. Hada, B. Hultqvist, T. Iyemori, A. Korth, J. U. Kozyra, D. Laxmi, M. Liemohn, G. Lu, R. M. McPherron, G. M. Milikh, T. Mukai, R. Nakamura, A. Nishida, N. Nishitani, T. Obara, S. I. Ohtani, T. P. O'Brien, A. Petrukovich, B. M. Reddy, G. Reeves, G. Rostoker, X. Shao, V. A. Sergeev, M. I. Sitnov, W. Sun, J. A. Valdivia and P. Yoon. Finally we thank A. Graubard and B. Matsko of AGU Books for their efforts in making this volume possible.

A. Surjalal Sharma
Department of Astronomy
University of Maryland
College Park, Maryland

Yohsuke Kamide
Solar Terrestrial Environment Laboratory
Nagoya University
Toyokawa, Japan

Gurbax S. Lakhina
Indian Institute of Geomagnetism
Colaba, Mumbai, India

The Storm-Substorm Relationship: Current Understanding and Outlook

A. S. Sharma[1], D. N. Baker[2], M. Grande[3], Y. Kamide[4], G. S. Lakhina[5], R. M. McPherron[6], G. D. Reeves[7], G. Rostoker[8], R. Vondrak[9], and L. Zelenyi[10]

The intensification of the ring current during a geospace storm has been one of the key issues in space physics. Substorms have been considered responsible for bringing in particles from the magnetotail, which get trapped on closed drift paths to form a symmetrical ring current. It is now recognized that the ring current develops dominantly from a sustained enhancement of the convection electric field. The magnetic perturbations observed during a storm main phase are then due to a partial ring current, which closes in part through the ionosphere and in part through the magnetopause. An enhanced convection electric field moves the plasma Earth-ward, thus energizing it, and when this field is reduced, the particles become trapped and a symmetric ring current is formed. Substorms, however, are always accompanied by the injection of energetic particles and their contribution to the storm-time ring current is a matter of current debate. Considering the electrodynamic nature of the interaction between different regions of the magnetosphere and the dominantly global nature of its dynamics, storms and substorms are not expected to just co-exist, but the ways in which they influence each other are not clear yet. The Chapman Conference at Lonavala (2001) saw the cementing of a new paradigm for the ring current and the storm-substorm relationship. The accumulating evidence against the substorms being the main constituents of storm main phase and the recognition of the dominant role of partial ring current led to a consensus (Lonavala consensus) marking a turning point in the understanding of the storm-substorm relationship.

---

[1]Department of Astronomy, University of Maryland, College Park, Maryland.
[2]Laboratory for Atmospheric and Space Physics, University of Colorado, Boulder, Colorado.
[3]Rutherford Appleton Laboratory, Chilton, Didcot, Oxon, U. K.
[4]Solar Terrestrial Environment Laboratory, Nagoya University, Toyokawa, Japan.
[5]Indian Institute of Geomagnetism, Colaba, Mumbai, India.
[6]Institute of Geophysics and Planetary Physics, University of California, Los Angeles, California.
[7]Los Alamos National Laboratory, Los Alamos, New Mexico.
[8]Department of Physics, University of Alberta, Edmonton, Alberta, Canada.
[9]Laboratory for Extraterrestrial Physics, Goddard Space Flight Center, Greenbelt, Maryland.
[10]Space Research Institute, Moscow, Russia.

Disturbances in Geospace: The Storm-Substorm Relationship
Geophysical Monograph 142
Copyright 2003 by the American Geophysical Union
10.1029/142GM01

1. INTRODUCTION

The relationship between storms and substorms is a key problem in solar wind-magnetosphere-ionosphere coupling and one of the long-standing questions has been what causes the development of the ring current, leading to the main phase of a magnetic storm. The relative importance of the substorm occurrence and sustained enhancements of magnetospheric convention in the development of magnetic storms has been an open issue. It is widely believed that the main phase of a magnetic storm is the interval in which many intense substorms must take place successively. Among the questions that have been raised in this context [*Gonzalez et al.*, 1994; *Kamide et al.*, 1998] are: Is this a necessary condition for the occurrence of magnetic storm? How many is "many"? How intense is "intense"? How successive is "successive"? These are the type of practical questions which have not been settled in storm-substorm relationship studies.

In terms of physical processes in the magnetosphere, the two major questions are: (1) What is the physical difference between storm-time substorms and non-storm-time substorms, if any? (2) Is the main phase of magnetic storms a result of (a) the impact of southward interplanetary magnetic field (IMF) which also relates to substorm activity, or (b) the successive occurrence of substorms which also have a direct relationship with southward IMF? Numerous studies in the past have not been conclusive and only recently is a coherent picture emerging.

The Chapman Conference at Lonavala [*Sharma et al.*, 2001] was held at a critical juncture, and it saw the cementing of a new paradigm for the ring current and the storm-substorm relationship. The change has been summarized by *Clauer* et al. [this volume]. In the past, the conventional idea has been that the ring current results from the accumulation of many elementary disturbances, viz. substorms. Each substorm would produce an enhanced westward electric field near the outer ring current boundary and bring particles in from the plasma sheet, a so-called "substorm injection". These particles become trapped on closed azimuthal drift paths, and form a symmetrical ring current. The substorm would promptly enhance the plasma sheet population with ionospheric particles which would also then contribute to the ring current. The new view is that the ring current development results directly from a sustained enhancement of the convection electric field. Most of the ring current magnetic perturbations are due to a partial ring current, which closes in part through the ionosphere and in part through the magnetopause. The energetic neutral analyzers aboard IMAGE spacecraft observe that the ring current is not symmetrical during storm onset [*C:son Brandt et al.*, *Reeves et al.*, this volume]. The enhanced cross-magnetospheric electric field moves the Alfven layers inward thus energizing the plasma and also moving the ring current closer to the Earth. Only after the enhanced field is reduced do particles find themselves on closed drift paths and the ring current becomes symmetric. Thus the classical two-stage decay of the ring current during recovery is seen as a prompt initial decay due to plasma on open paths convecting out of the system, followed by a slow decay due to charge exchange within the trapped population. Substorms can play a role in storm development in different ways, e.g., in initiating ionospheric ion out flow [*Grande et al.*, this volume; *McFadden et al.*, 2001], or injection of energetic ions into the inner magnetosphere [*Reeves et al.*, this volume; *Daglis and Kamide*, this volume]. The emerging picture of storms consists of a dominant role of convection which brings in the plasma into the inner magnetosphere thus increasing its energy due partly to the conservation of the adiabatic invariants, and substorms which may contribute to the energization of the particles and facilitate their trapping in the ring current region.

The debate on the role of substorms in the storm development can now be viewed from a new perspective of determining the relative roles of substorms and convection. At this stage the role of convection is clearer than that of substorms. The plasma and fields in the different regions of the magnetosphere are coupled strongly and efficiently due to the electrodynamic nature of interaction of the plasma in the anchor dipole field of the Earth [*Sharma*, 1995]. This coupled with the close proximity of the ring current region, (4–6) R_E, and the region of substorm initiation, which could be as close or closer to Earth than $8R_E$, make the development of storms without any influence from the substorms very unlikely. Moreover the substorm injections during storms are ubiquitous during storms [*Reeves et al.*, this volume].

Our understanding of the storm-substorm relationship has been derived mainly from the global characteristics of the magnetosphere. For example, from the observational point of view, the global indices such as Dst and AL have been used extensively to study storms and substorms, respectively. On the theoretical side, energy balance conditions such as the virial theorem have been used to evaluate the partition of stored magnetic energy into the kinetic energy of the ring current and the dissipation in the ionosphere [*Siscoe and Petschek*, 1997]. This use of the global quantities to describe many essentially local processes is perhaps the main cause of many of the previous misunderstandings. The study of substorm injection during a storm main phase using indices is one such example. Neither the auroral indices such as AL representing substorms, nor the Dst representing the ring current, can yield spatial dependences of a substorm injection, which is a local process. The study of the global as well as the local processes is needed to understand the interrelationship between convection, substorms and storms. Many studies presented in this volume use data from spatial extended locations [e.g., *Clauer et al.*, *C:son Brandt et al.*, *Filligim et al.*, *Reeves et al.*, this volume].

2. THE RING CURRENT: A NEW UNDERSTANDING

Our understanding of the ring current has been derived mainly from the Dst index and its correlation with the solar wind. Another part is obtained from an interpretation of the Dst index in terms of simple models including

Chapman-Ferraro theory of the magnetopause, single particle drift in a dipole field, magnetospheric convection, substorm collapse of the tail field and charge exchange loss of ring current protons. Only a small part of our understanding is derived directly from in-situ observation or remote sensing.

A simple model of the ring current, represented by the Dst index, is the *Burton et al.* [1975] equation, which can reproduce nearly all the variance in hourly values of Dst [*McPherron*, 1997; *McPherron and O'Brien*, 2001]. The root-mean-square prediction residual is of order 8 nT, comparable to most estimates of the noise in the solar wind and Dst data. This equation uses only solar wind data to make the prediction, completely ignoring substorms, while the standard paradigm for storm development [*Gonzalez et al.*, 1994] requires a substorm expansion to inject ions and accelerate them to the energies observed in the ring current. This indicates that something is lacking in the standard paradigm since it requires substorms.

In its simplest form the *Burton et al* [1975] equation states that the rate of change of dynamic pressure-corrected Dst is a balance of injection controlled by the solar wind electric field, and decay controlled by charge exchange. For data of hourly resolution, the injection filter is a delta function at lag zero, i.e., it is a simple constant of proportionality equal to 4.4 nT/hr per mV/m and the decay time is generally taken to be 8 hours. The interpretation of this empirical relation is that the rectified solar wind electric field drives magnetospheric convection that injects particles into the ring current that then decay away by charge exchange. This alternative model however has a problem as there is no known way for a steady electric field to produce a symmetric ring current. A steady convection electric field drives convection from the tail along drift paths that are open to the magnetopause and thus the particles do not form a symmetric ring current.

It might be thought that this is the role of the substorm expansion. Although the solar wind electric field is steady, the internal convection electric field is not. Each fluctuation in the internal field traps some of the drifting particles onto closed drift paths. Unfortunately the data do not support this conjecture. *Iyemori and Rao* [1996] have shown that the rate of decrease of Dst becomes smaller after a substorm expansion, not larger. Also *Fay et al.* [1986] showed that the high time-resolution injection filter is a Gaussian pulse peaked at about 20 minutes delay, a time scale much shorter than the average time for an expansion phase to develop after a southward IMF turning.

The Chapman Conference on Storm-Substorm Relationship [*Sharma et al.*, 2001] coincided with the emergence of a new paradigm for the ring current and the storm-substorm relationship. The change has been summarized by *Clauer* et al. [this volume] and *C:son Brandt et al.* [this volume] and presents the ring current and Dst index in a new light. In the standard view the Dst is caused by the symmetric part of the ring current, and asymmetry by a separate partial ring current. However the new results show that the ring current is not symmetric during the storm main phase. In this view ions simply drift Sunward in the convection electric field until gradient and curvature drifts take over close to the Earth. On the night side the ions drift across equipotentials of the convection electric field gaining energy and increasing their density as they approach the Earth. This produces an effective westward drift current that increases across the night side all the way to dusk. As the ions move into the day sector they drift opposite to the convection electric field, lose energy, move outward, and the current decreases. Of course, the divergence of the westward current must be connected to the ionosphere with an outward current on the nightside and an inward current on the dayside.

In this picture there is no symmetric ring current, only a combination of tail current, partial ring current, and dayside continuation to the magnetopause. However, if the convection electric field begins to decrease slowly, some of the open drift paths will be converted to closed drift paths and a symmetric ring current should begin to develop. A sudden northward turning of the IMF should convert all open drift paths to closed drift paths and eventually the ring current should become completely symmetric. It is easy to demonstrate that this never happens. A plot of the asymmetry index versus the Dst index reveals that the two vary together, and that asymmetry is never zero. In fact Dst can explain over half of the variance in asymmetry. It appears that a substantial fraction of asymmetry is produced by the same current as Dst. The fact that asymmetry is never zero might be explained by the existence of a convection electric field driven by the viscous interaction. Such a field would maintain a background convection electric field and some open drift paths that lead to asymmetry by the mechanism described above.

This model may also explain a perplexing observation that the initial ring current recovery rate after VB_S reaches its minimum value is much faster than any possible charge exchange lifetime of observed ring current ions [*O'Brien and McPherron*, 2000]. For very strong solar wind electric fields (E>10 mV/m) the recovery rate is of order of four hours. This time scale is close to the travel time of particles from dusk to the magnetopause, a loss mechanism that has been named the "flow-out effect" [*Takahashi et al.*, 1990]. Thus the observed dependence of the ring current recovery rate on the solar wind electric field may be a consequence of a gradual transition from the flow-out effect during strong convection to charge exchange further and further from the Earth as the convection field diminishes.

Many papers in this volume emphasize and elaborate different aspects of the point of view that substorms are not essential to the ring current development during the storm main phase. *Tsurutani et al.* [this volume] have shown that storms are possible without substorms. *Korth et al.* [this volume] showed that the cycling of O^+ to the tail was independent of substorm activity. FAST data showed the action of a partial ring current during stormtime [*McFadden et al.*, 2002]. *Baker and Li* [this volume] showed that the same prediction algorithms relate Dst to solar wind parameters at solar maximum and at solar minimum. It is interesting to consider that if this is so, then ionospheric material, which is largely absent at solar minimum, can not be directly mediating the process. *Grande et al.* [this volume] showed that on the timescale of order one hour or less there is no major change in the average behavior of the Dst index itself, when ordered by substorm onset, which implies that there is no prompt connection between the injection of energetic particles at substorm onset and changes in the energetic particle population of the ring current.

Overall, it is becoming clear that whereas in the past there was a tendency to think of substorms as the "quanta" of storms, a new view is emerging with a whole range or "family" of coexisting magnetospheric disturbances, including storms, pressure pulse events, enhanced convection interludes, multiple and single onset substorms, pseudo-breakups and enhanced flows. All of these enable the magnetosphere to respond to solar wind drivers, and dissipate energy. The question of to what extent, and under what circumstances, these responses are associated, will form one of the important areas for the field in future. A further major question is the need to understand the build-up of relativistic electrons during storm-time [*Baker and Li*, this volume]. While the association is clear, the precise correlations are certainly not. It seems clear [*Grande et al.*, this volume] that substorm activity does not directly provide these electrons. However, it may provide the seed population, as in the idea that substorms "precondition" the storm-time plasma.

3. ROLE OF OXYGEN ION AND WAVE-PARTICLE INTERACTIONS

The role of energetic O^+ ions of ionospheric origin in the development of magnetic storms is among the key unsolved problems [*Daglis and Kamide*, *Grande* et al., *Korth et al.*, this volume]. The storm triggers ionospheric upflow, and if it persists long enough, this material finds its way into the plasma sheet, and eventually injected into the ring current. Simulations of ionospheric processes show the formation of patches at altitudes where the oxygen concentration is the highest, thus indicating their potential role in the outflow of oxygen ions [*Gondarenko and Guzdar*, this volume]. Because of its high mass, the oxygen ions contribute a large part of the energy density, which can be represented in terms of the Dst index. However, the role of the oxygen ions is not quite clear. The case when it does not play any special role in the process is presented in *Grande et al.* [this volume]. The other view in which they play a significant role is presented in *Daglis and Kamide* [this volume]. An important issue is how the thermal oxygen ions are extracted from the ionosphere and then energized to ring current energies of~ a few keV to hundreds of keV. As the particles convect into the region of stronger magnetic field closer to Earth they gain energy due to conservation of the first adiabatic invariant. Studies of single-particle dynamics in models of magnetic field dipolarizations indicate that low-energy ionospheric-origin O^+ ions can be accelerated up to a few hundreds of keV and injected earthward during substorm-related dipolarization events. However this does not yield the required energies and different types of fluctuating fields have been considered as sources of ion energization. Intense plasma waves could provide an efficient mechanism for energy transfer between different ion species and may prove important for selectively heating and accelerating thermal heavy ions [*Thorne and Horne*, 1994]. In fact, assessing the integrated effect on storm-time ring current losses due to the scattering of ions by waves is one of the unsolved problems concerning magnetic storms [*Lakhina and Singh*, this volume]. The fact that the occurrence of particular plasma modes is usually limited in time or confined to localized regions casts doubts on the ability of the wave-scattering processes to affect significantly the energy balance of the ring current globally.

Interaction between the ring current ions and modes such as electromagnetic ion cyclotron (EMIC) waves is considered as an important ring current loss process. The global impact of ion cyclotron waves on the ring current have been modeled by using the Ring Current-Atmosphere Interaction Model (RAM) which follows the evolution of three major ring current ion species (H^+, He^+, and O^+) considering adiabatic drift motion, Coulomb collisions, charge exchange, and pitch-angle scattering of protons in the field of EMIC waves [*Jordanova et al.*, 1996; *Kozyra et al.*, 1997]. The model produced order-of-magnitude enhancements in the ion precipitation as a result of diffusion in the ion cyclotron waves within the unstable region. However, no significant impact of the wave losses was seen in the global energy balance even though the waves reduced the anisotropy in the proton pitch angle distributions locally. An improved scheme has been used for estimating the global distribution and amplitude of EMIC activity using a warm plasma ray tracing code in the RAM model [*Kozyra et al.*, 1997]. The comparison of the storm development from this model with that from the maps of magnetic field disturbances obtained from ground-based measurements has contributed to a new view of the storm main phase dominated by partial ring current [*Clauer et al.*, this volume].

The O^+ ions of ionospheric origin can also excite the electromagnetic helicon mode in the near-Earth plasma sheet region [*Lakhina and Tsurutani*, 1997, 1998; *Singh et al.*, 2002]. This instability may play an important role by facilitating the excitation of tearing instability, which is important during substorm onset. During storms, plasma sheet oxygen ions can be accelerated by the helicon mode waves and injected earthwards and become part of the ring current. In addition to the EMIC and helicon mode waves, quasi-electrostatic instabilities driven by loss-cone distributions of the ring current ions can also occur during magnetic storms [*Lakhina and Singh*, this volume]. The role of the wave-particle interaction in storm development can be best assessed when reliable models of the magnetic field in the inner magnetosphere are available.

4. SUBSTORM AND MAGNETOTAIL EFFECTS IN THE INNER MAGNETOSPHERE

The injection of energetic particles from the near-Earth plasmasheet into the inner magnetosphere is one of the key processes during storms and substorms. Injections are so commonly observed in association with the dipolarization of the magnetic field at substorm onset that if a substorm is documented without observing an injection it is often assumed that there was an injection at another location but that there may not have been a suitably-located observing spacecraft. Recent results and understanding of the role of injections during substorms have been reviewed by *Reeves* [1998].

Among the key issues in storm-substorm relationship is the differences and similarities of storm-time and isolated substorms [*Baumjohann et al., Petrukovich*, this volume]. "Isolated" substorm injections are produced by localized, inductive electric fields with little or no change in the large-scale, externally-imposed, "convection" electric field. Therefore the substorm injection serves to move particles from open or untrapped drift trajectories to closed, trapped drift orbits. During storms there are changes in both the large-scale convection electric field and superimposed fluctuations of the more localized inductive electric fields (which are often hard to separate observationally). Either process or, more commonly, both together serve to move energetic particles from the magnetotail to the inner magnetosphere. Whether those particles are eventually trapped on closed drift trajectories or lost to the dayside magnetopause depends on the precise time history of the electric fields experienced by the particles, but, whatever their fate, those particles (particularly the ions) are the particles that carry the storm-time ring current which is the defining feature of a geomagnetic storm.

With the advent of global Energetic Neutral Atom (ENA) imaging the limitations due to the inability to obtain a global picture of the injection and transport of the actual current-carriers through the inner magnetosphere is beginning to be overcome. Energetic neutral atoms are produced by the charge exchange of magnetospheric ions with tenuous, cold, exospheric neutrals. Since the first application of ENA observations to geomagnetic storms [*Roelof*, 1987] this technique has led to many new advances. Among these is a direct relationship between ENA flux and the Dst index during the recovery phase of storms. This is expected because the charge-exchange of ring current ions which produces the ENA fluxes is also a direct loss process for ring current ions. If charge exchange is the only loss process then the time rate of change of Dst is proportional to Dst. While the ENA emission during the recovery phase is generally proportional to Dst it is not the only loss process [*Jorgensen et al.*, 2001]. In a two-phase recovery, charge exchange accounts for roughly 75% of the decay of the ring current but in the early, rapid recovery phase it accounts for only a small proportion of the loss. The rapid recovery of Dst is produced primarily by loss of ring current particles to the magnetopause—what is often called the "partial ring current" but might better be called the "untrapped ring current". *Liemohn et al.* [2001] recently provided compelling observational and model-based evidence for the importance of magnetopause loss in rapid recovery and *Reeves et al.* [this volume] show ENA evidence that Dst (and SYM-H) can reach values of –100 nT without any substantial flux of ring current ions extending past noon local time.

The injection of energetic particles during storms and substorms need to be studied using simultaneous data from many different observations. *Reeves and Henderson* [2001] undertook a study which compared 7 isolated substorm injections with 7 storm-time injections using POLAR ENA observations and in situ geosynchronous fluxes in order to gain better understanding of the storm-substorm relationship. For the geomagnetic storms they used the first injection in order to have a clear and unambiguous timing signature. One conclusion was that while main-phase substorms can be difficult to identify, essentially all storms began with a clear substorm and a clear substorm injection. Further they found that the storm-time injections were essentially identical to isolated substorm injections. They were neither larger (in flux or local time) or more intense (e.g., in spectral hardness). What distinguished storm-times from isolated events was (a) continued injection activity for a period of hours following the initial injection, (b) a spreading of the local time extent of ion injection toward dawn - opposite to the direction of ion drift, and (c) an immediate response in Dst for the storm-time injections compared to no measurable response for the isolated events. This study led to the conclusion that it was the presence of large-scale, externally-imposed, "convection" electric field superimposed on localized, inductive electric fields which differentiated storm-time particle injections from typical substorm injections. More recently *Lui et al.* [2001] reached a similar conclusion using Geotail ENA observations of the ring current and SuperDARN radar observations of polar cap convection. Two papers in this volume [*C:son Brandt et al.*, and *Reeves et al.*] present the first IMAGE observations of storm-time substorms and ring current evolution with unprecedented spatial and temporal resolution of the injection and transport of inner magnetospheric ions.

One of the unique features of ENA observations is that they remotely sense the dominant current carriers in the inner magnetosphere and yet are completely insensitive to other magnetic perturbations such as magnetopause currents, substorm current wedges, or ionospheric electrojets all of which have added ambiguity to understanding the storm-substorm relationship. With the pace of recent developments it is not difficult to imagine the day when the most popular proxy for ring current intensity, Dst, is superceded by actual, global, time-dependent measurements of the ring current ions themselves.

5. RELATIVISTIC ELECTRONS DURING STORMS

As noted by *Baker and Li* [this volume], most major geomagnetic storms give rise to relativistic electron enhancements in the Earth's outer radiation belt. However, some large storms do not show such electron enhancements [*Reeves*, 1998]. Thus, it is an area of active research to try to understand in detail how high-energy electron acceleration occurs in the magnetosphere during the course of strong geomagnetic activity. Long-term studies of relativistic electrons in the magnetosphere have shown many of the occurrence characteristics. A very obvious role is played by solar wind speed in producing subsequent relativistic electron enhancements. In fact, the solar wind speed is the single biggest determinant of electron enhancement. However, there is also a key role played by the north-south component of the IMF. There typically must be a significant interval of southward IMF along with a period of high (V_{SW} = 845 km/s) solar wind speed. Thus, it is generally thought that enhancement in geomagnetic activity (e.g., magnetospheric substorms) is a key first step in the acceleration of magnetospheric electrons to high energies. A second step is then thought to be a period of powerful low-frequency waves that is closely related to high values of V_{SW}. In this picture, substorms provide a "seed" population, while high-speed solar wind drives the acceleration to relativistic energies in a two-step geomagnetic storm scenario. This picture seems to apply to most storms examined whether associated with high-speed streams or with CME-related events.

The terrestrial magnetosphere clearly is an efficient accelerator and effective trapping device for energetic particles. The acceleration and transport processes for energetic electrons remain primary issues in magnetospheric physics even four decades after the discovery of the radiation belts. High energy electrons hold special interest because of their continuous presence in the magnetosphere and their effect on human technology. Present-day spacecraft missions have given a remarkable view of energetic particle phenomena. Long-term measurements have unveiled many interesting features of relativistic electrons and have also presented a great variety of new challenges in understanding the dynamics of these particles in the Earth's magnetosphere. Examination of the 10-year record of SAMPEX data [*Baker and Li*, this volume] shows that the highest electron fluxes were seen in late 1993 and in 1994. The 1993–94 period was a time of very prominent high-speed solar wind streams and was also the period of most extreme relativistic electron radiation in the past solar cycle. There was a clear and prominent 27-day periodicity in the electron flux enhancements. This was well associated with solar wind velocity enhancements. Thus, during the approach to sunspot minimum, high-energy electrons are at their highest levels throughout the outer radiation belt and this population is well associated with recurrent geomagnetic storms. Many authors have studied mechanisms that might account for acceleration of electrons to relativistic energies during geomagnetic storms. An important correlation has been found between electron flux enhancements and ULF waved power in the magnetosphere [*Baker et al.*, 1998a]. Data show increases from quiet day wave power by as much as a factor of 1000 in the frequency range ~1.0 to 20 mHz. Based on correlation studies, it is argued that these ULF waves can play an active role in electron acceleration [e.g., *Rostoker et al.*, 1998; *Hudson et al.*, 2000]. Another important point, however, is that there needs to be a "seed population" of electrons available on which the ULF waves (or other agents) act [e.g., *Baker et al.*, 1998a]. Using plasma wave and particle data from the CRRES satellite, *Meredith et al.* [2002] suggested that the gradual acceleration of electrons to relativistic energies during geomagnetic storms can be effective only when there are periods of prolonged substorm activity following the main phase of the geomagnetic storm. Thus, magnetospheric substorms are essential to providing the seed population [*Baker et al.*, 1998b]. *Baker and Li* [this volume] suggest that magnetospheric substorms and geomagnetic storms are closely related to one another when it comes to energetic electron phenomena. They note that it would be remarkable if a southward turning of the IMF that opens the magnetosphere to energy input would lead to two totally separate and unrelated phenomena. The original view that storms are merely a superposition of substorms was clearly too limited. But, on the other hand, it seems unlikely that storms could occur without substorms.

Baker and Li [this volume] espouse the belief that substorms are an important step along the way to geomagnetic storms. The magnetosphere crosses many thresholds in its progression of development and it begins to admit many new forms of energy dissipation as it is driven harder and harder by the solar wind. Substorms are an elementary (and essential) component in this progression. Substorms have many important properties like nonlinearity, complexity, self-organization, and even criticality [*Sharma et al.*, this volume]. As the magnetosphere progresses toward major storms, however, the external driver (the strong flow of the solar wind energy) overwhelms and drives the magnetosphere into a mode of powerful direct response. This strong driving of magnetospheric convection, in turn, produces the conditions that, very frequently at least, produces highly relativistic electrons during geomagnetic storms.

6. GROUND BASED DATA: THE NEED FOR A PROPER INTERPRETATION

The community of researchers working on the storm-substorm relationship has, for years, tried to define the global behavior of particles and fields in the magnetosphere that lead up to substorm expansive phase and to explore the global response of the system as it goes through the development of the expansive phase and the ensuing recovery. The primary reason that these issues have not yet been successfully resolved has been the lack of observation points in the vast volume of space in which the expansive phase develops. The research community is almost always in the position of being unable to separate spatial and temporal effects when trying to analyse satellite data, simply because the magnetosphere is not a homogenous medium in which all perturbations detected can be identified as temporal changes. In an attempt to alleviate this problem, researchers use ground-based data obtained from instruments such as magnetometers and auroral imagers. These can provide continuous two-dimensional imaging of the auroral ionosphere and the input required by models to infer the three dimensional structure of the electric current distribution and particle populations. Ground-based magnetometer data, in particular, have been used to identify substorm onsets and to track the evolution of the phases of substorm activity. However, these data are rarely used to the full extent possible with the consequence that some conclusions reached are not warranted. Two such situations are identified below to elucidate how treatment of only a limited part of the ground based magnetometer data available may lead to incorrect conclusions.

The first case is on the use of the auroral electrojet indices for individual event studies. It is very common to see the state of activity in the magnetosphere quantified by a plot of the AE or AL index, and sometimes onset times will be determined from sudden increases in the value of either of the aforementioned indices. In terms of the storm-substorm relationship, the problem arises because the indices contain contributions of two different types of activity— directly driven and storage-release (cf. *Rostoker et al.*, 1987 for a review of these processes). It is well to remember that the AL index does not reflect the disturbance at any particular point in space. It is derived by superposing all available records of the north-south (H) component of the magnetic perturbations from a specific set of stations (normally 12) at average auroral zone latitudes distributed as uniformly as possible around the world. The maximum negative value from all these records is chosen and AL is assigned that value. The problem that arises is best seen referring to Figure 1. In that figure, the slowly changing perturbation represents a single H-component magnetogram featuring a rise and fall of the directly driven system, whose peak negative magnetic perturbation at any time is normally found in the dawn sector. A second trace represents a single H-component magnetogram showing a substorm expansive phase onset, which normally occurs in the midnight sector. The third trace is what would appear as the time series of AL values for this event. It is immediately evident that the expansive phase onset would only be a small short-lived increase in AL occurring some time after the start of the increase in the index value. Even more important to note is that, if the

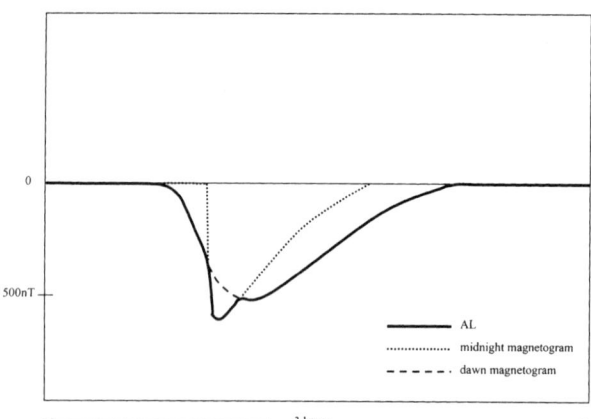

Figure 1. Schematic diagram showing midnight sector and dawn sector H-component magnetograms during a substorm disturbance, together with the AL index that would be formed if these two magnetograms provided the largest disturbance at any time during the three hour interval portrayed. Similar scenario was also discussed in *Kamide and Kokubun* [1996].

peak magnetic perturbation of the midnight sector magnetogram had been less than or equal to the perturbation due to the directly driven activity, the expansive phase onset could not have been identified from the time series of the index. It is clear that for substorm expansive phase identification, AL is unsuitable for either statistical or individual event studies.

The second case deals with the source region of substorm-like disturbances. One point is that there are actually two regions in the auroral oval in which substorm-like perturbations can be found, namely the near the equatorward edge of the oval (expansive phase onsets and pseudo-breakups) and at the poleward edge of the oval (poleward border intensifications or PBIs). In establishing the storm-substorm relationship, it is important to first understand the origin of the belief that the development of the storm main phase is always accompanied by substorm activity.

Figure 2 gives some indication of the nature of this problem. For the event shown, there are two clear disturbances that look very similar in terms of their signatures in the north-south component of the magnetic field. It is only when one examines the evolution of this episode of substorm activity, that one recognizes that the disturbance around 0430 UT is initiated at the equatorward edge of the oval (i.e. is a normal expansive phase onset) while the disturbance at ~ 0630 UT is at the poleward edge of the oval (i.e. is a PBI). The reader is referred to *Rostoker* [2002] for a full description of this event. Figure 3 shows latitude profiles for the 0630 UT event which illustrate this point unam-

biguously. Panel a shows a latitude profile near local midnight just before the onset of the disturbance in question. Panel b shows a differential profile using 0630 UT as the baseline from which the perturbations are measured. Prior to the intensification there is a strong broad westward electrojet centered at ~ 66.5° N (PACE) with a poleward border at ~ 67.5° N. The intensification is centered north of the poleward border of the pre-existing westward electrojet with the poleward edge of the latter coinciding with the equatorward edge of the new electrojet (marked by a vertical line in the figure). The weak perturbations (< 100 nT) in the latitude regime of that pre-existing electrojet seen in Panel b clearly show the new system to be an independent entity, with the old system not responding significantly at the time of the ~0633 UT intensification. Clearly this event

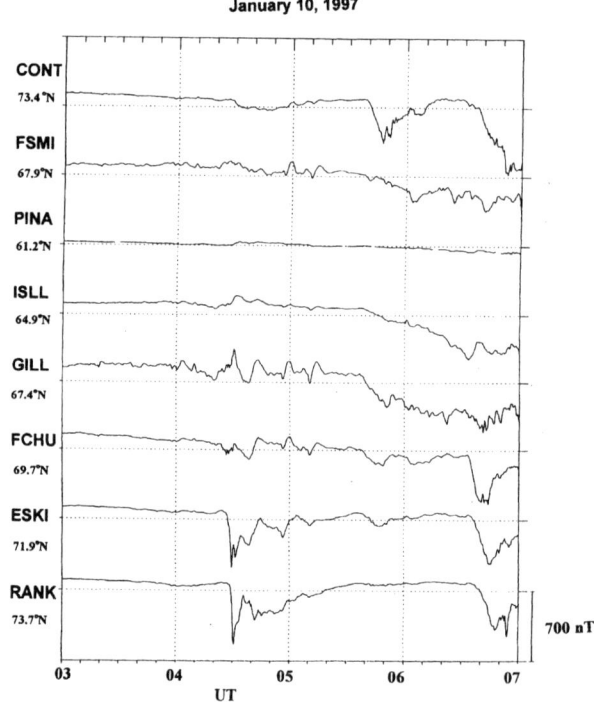

Figure 2. North-south component magnetograms from selected CANOPUS stations for a four hour interval on January 10, 1997 (after Rostoker, 2001). The four bottom traces are from the Churchill meridian line from PINA in the south to RANK in the north, while CONT and FSMI are auroral oval stations approximately two hours in local time west of the Churchill line. The attention of the reader is drawn to the ESKI and RANK disturbances at ~ 0430 UT and ~ 0630 UT, which look quite similar to one another and might be identifed as substorm expansive phase onsets. The ~0430 UT event is, in fact, an expansive phase disturbance.

was a poleward border intensification, with little or no response in the equatorward portion of the electrojet (which maps close to the earth near the inner edge of the plasma sheet).

We now can see why the claim that the main phase of a magnetic storm is accompanied by substorm activity must be treated with care. PBIs, as seen in normal magnetograms, look very much like expansive phase onsets. Therefore, it is quite possible that the strong magnetic perturbations seen at average auroral zone locations during storm main phase development are really PBIs. In fact, since the polar cap expands during the development of a magnetic storm, one would expect the poleward edge of the oval to move to latitudes normally occupied by the equatorward edge of the oval. Thus it is entirely possible that the large magnetic perturbations detected at average auroral zone stations do not reflect substorm expansive phase onsets.

The magnetic field variations measured on the ground are driven by the solar wind and in specific cases the relevant solar wind variable can be identified. In the case of January 1997 magnetic cloud event the ground magnetic variations are found to be strongly correlated with the solar wind plasma density [Kamide et al., this volume].

7. FORECASTING STORMS AND SUBSTORMS: ROLE IN SPACE WEATHER

Storms and substorms are the major geomagnetic events of interest for space weather and arise from an efficient coupling of the solar wind energy and momentum to the magnetosphere and ionosphere. They can cause severe damage to our technological systems in space as well as on Earth, and forecasting storms and substorms is essential for protecting these systems. Nonlinear dynamical models of the magnetosphere derived from observational time series data using phase space reconstruction techniques have yielded new advances in the understanding of its dynamics [Sharma, 1995]. In particular, it is now recognized that the dynamics has a strong global component which leads to overall coherence in the magnetosphere. This forms a basis for the predictability of the magnetospheric behavior and space weather. These techniques have been successful in developing nonlinear models for predicting the global dynamics in terms of the geomagnetic indices, such as the auroral electrojet index AL, or the storm time disturbance index Dst, from the spacecraft data of the incoming solar wind.

The importance of nonlinear dynamical studies to space weather arises from its ability to reconstruct the dynamics from the observational data of a limited number of variables. The reconstructed phase space of the system yields a dynamical description independent of particular modeling assumptions, and embodies the information in the past observational data [Sharma et al., this volume].

Predictability is a natural consequence of the low dimensionality, which represents the global aspects of the magnetospheric dynamics. The input-output nature of the solar wind-magnetosphere interaction is incorporated into the local-linear techniques by including the solar wind input along with the geomagnetic response. In input-output studies, the local linear technique has been successful in yielding simple predictive models of the global magnetospheric dynamics by using the main features of the system. In the

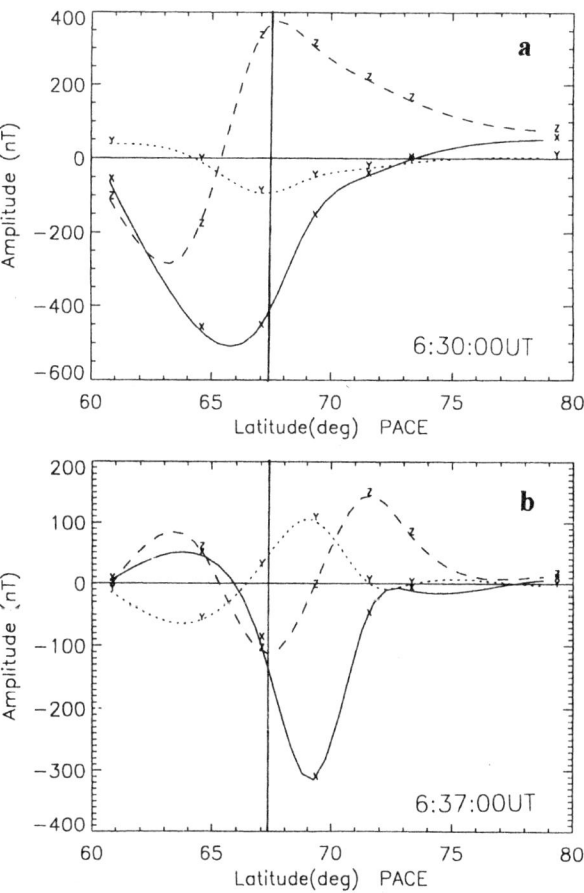

Figure 3. Latitude profiles composed from data taken along the Churchill line just prior to (Panel a) and shortly after (Panel b) a substorm intensification that occurred at ~ 0633 UT. The profile in Panel a shows the disturbance with respect to a baseline before the ~0430 UT onset while Panel b is a differential profile showing the disturbance with respect to a baseline at 0630 UT, just before the intensification in question. The intensification is, in fact, a PBI originating further back in the magnetotail than where expansive phase onsets take place.

study of storm-substorm relationship, such a model was developed from the high resolution data set for 1979 using AL as the input and Dst as the input. This model [*Kamide et al.*, 1998] showed that AL can predict Dst well. Another issue that was examined using the input-output models is the relative timing between storms and substorms from AL and Dst indices data. The linear prediction filters relating AL to Dst for data averaged over 10–50 min showed no delay between AL and Dst indices [*Sharma et al.*, 1998]. This indicates that there is no relative delay between substorms and storms, consistent with the conclusion that solar wind, rather than the substorms, is the main driver for storms. The nonlinear dynamical techniques thus yield predictive models and in the same time provide insights on the relationships between different processes. Many techniques are now used to study the solar wind—magnetosphere coupling and among these is the wavelet analysis [*Cade et al.*, this volume].

The effects of geomagnetic storms on the equatorial ionosphere are enhanced during active periods and this has important implications for space weather forecasting [*Sastri et al.*, this volume].

8. SUN-EARTH CONNECTION

Although the role of storms and substorms in the basic structure and dynamics of the magnetosphere has been identified, our knowledge and understanding is still incomplete. Because storms and substorms are temporal reconfigurations, not objects, a successful integration of knowledge requires a paradigm that identifies the sequence of significant events. Some specific unresolved questions are as follows.

The primary question is the initiation of storms and substorms. Geomagnetic storms result from changes in the solar wind that modify the dynamics of the magnetosphere. However, the relative effectiveness of solar wind pressure gradients and interplanetary magnetic field variations is not well established. Similarly, substorms may result from solar wind drivers or they may be initiated within the magnetosphere. So, important progress will be made when we understand the entire sequence of events that initiates a substorm or a storm. Furthermore, there is insufficient understanding of the detailed processes that trigger a substorm, or even how a substorm might be connected to a storm [*Reeves et al.*, this volume].

The energy budget of the magnetosphere during the course of a storm or substorm is another important issue. Although the solar wind is the primary energy source of both storms and substorms, we do not understand how that solar wind energy is coupled to the magnetosphere and how it is dissipated, e.g., by means of the ring current, high-latitude Joule Heating and particle energy input, acceleration of energetic trapped particles, magnetopauselosses, etc. [*Ostgaard and Tanskanen*, this volume]. We need to establish the efficiency of energy transfer from the solar wind into the magnetosphere for a variety of conditions. Understanding this transfer will enable predictive models of the response of the magnetosphere to extreme solar wind conditions.

The understanding of the entire Sun-Earth coupled system is the main issue. Ultimately we need a predictive capability for space weather that allows us to go from observations of the solar surface to ultimate consequences at Earth. Current predictive ability is mainly in solar wind-magnetosphere coupling [*Baker and Li, Sharma et al.*, this volume].

The solutions to these problems require many ingredients. However two key elements can be identified. The first is the development of a comprehensive paradigm of magnetospheric dynamics in which the sequence of events that occur during a geomagnetic storm and substorm are represented. This paradigm should be a synthesized picture that contains the essential features and processes. It would be a modern, contemporary version of the auroral substorm sequence developed by Syun Akasofu in the early 1960's [*Akasofu et al.*, this volume]. His paradigm was based only on ground-based observations, yet it revolutionized our understanding of the auroral current systems and of magnetospheric dynamics. Such an integrated picture need not include all the details, but should be sufficiently comprehensive so as to describe the significant events, the controlling processes, and the most important consequences.

This difficulty in reconstructing an accurate, integrated picture of the essential characteristics of storms and substorms is reminiscent of the classic story of the several blind men who encounter an elephant. Each provides an accurate description of a single part of the elephant, but their combined descriptions are misleading and confusing.In a similar way we have developed very precise characterizations of the various pieces of storms and substorms, but we have yet to develop a satisfactory representation of the entire storm/substorm elephant that lives within the magnetosphere. In fact, there are almost certainly several types of magnetospheric elephants, which only compounds our difficulty in reconstructing the true characteristics of each kind.

Perhaps the first thing would be to provide a better parameterization of the stormtime current as observed on the ground. This could be done by combining simple models for the magnetopause current, the tail current, the symmetric ring current, the partial ring current, and the substorm current wedge. Roughly 15–20 parameters might be determined as a function of time, and used for further analysis. A more sophisticated inversion of these currents could be done using the Assimilative Mapping of Ionospheric Electrodynamics (AMIE) technique [*Richmond and Kamide*, 1988]. These models should be motivated by more detailed statistical surveys of satellite magnetometer and plasma data during various phases of a storm. The newly emerging technique of Energetic Neutral Atom imaging should be used to develop a picture of where ions are charge exchanging with atmospheric neutrals as a function of storm phase. In conjunction with these experimental measurements there should be continued improvement of the numerical simulations. In particular these simulations need to be made self consistent so that the effects of the storm time currents are fed back to alter the magnetic field in which particles drift. The inner magnetosphere should also be linked to global MHD simulations so that their outer boundary conditions are more realistically defined.

The absence of a dynamical storm-time magnetic field model is perhaps the greatest impediment to progress in the field of research. This is desperately needed as a framework in which to interpret experimental results.

The second key element consists of measurements targeted at important questions [*Vondrak et al.*, this volume]. Improved measurements are needed to reveal the processes that couple the magnetospheric regions and to establish the relationships between cause and effect. The present data are too sparse, with inadequate spatial and temporal resolution. Furthermore our present models and simulations have more detail than can be measured. We cannot validate some aspects of existing models, nor can we judge which model is correct when their predictions differ. In general, the current data sets are not adequate to answer the science questions!

The ambiguity in the observations results from the fact that processes occur on scales smaller than our measurement resolution. For example, the coherence length of the plasma sheet has been observed to be about one Earth radius, based on the autocorrelation of measurements by a single spacecraft. This scale size is much finer than the spacing of most multi-point measurements that are available today. To attain such a fine resolution over a large volume, we need improved methods such as imaging systems, and networks of multi-spacecraft constellations.

Future measurements need to be targeted at key questions. They should not provide only additional phenomenological data that simply characterize the magnetospheric environment. The measurements should be aimed at identification of physical processes, rather than just data collection. For example, they should resolve the roles of the convection electric field versus pressure gradients during the initiation of magnetospheric disturbances. Similarly, they should provide comprehensive measurements of the energy budget and its partition, as well as other significant physical parameters.

Eventually there must be more measurements in the ring current region by multiple spacecraft. Several spacecraft in each of several orbits separated in local time are needed to track the growth and decay of the ring current. Measurements spanning the inner magnetosphere in the equatorial plane, as well as stereoscopic measurements in energetic neutrals, are needed. However, overall, it is apparent that a new consensus has emerged in storm-substorm relationship, which will form the basis for future considerations of the field.

It is of general interest to examine the cause-effect relationships in a broader context of the interaction of solar wind with planetary environments. In the absence of adequate data in this area, we can use a Gedanken experiment to reach reasonable conclusions. There are three main components which control the solar-planetary relations: solar wind supersonic flow with frozen in interplanetary magnetic field B_{IMF}, internal magnetic field of the planet, B_g, and planetary atmosphere and the ionosphere, I. We can then examine the effect of switching on and off one or two of these factors. The more familiar cases are those of Venus (developed ionosphere and almost absent internal magnetic field) and Mercury (significant internal B_g and almost absent ionosphere). Also, one could imagine that the interaction of solar wind with these planets occurs during the (exceptional) intervals when the intensity of IMF field is very low ($B_{IMF} \rightarrow 0$). The question is what types of dynamical phenomena in planetary magnetospheres one might expect during such hypothetical (although partially realistic) interactions? Would the resulting response be storms, substorms, both or something else? Do we expect, for example, that storm–type and or substorm-type phenomena might occur in the Hermean or Venusian magnetospheres? A list of the possible combinations is shown in Table 1 in terms of possible "Yes" or "No" answers. However, it is not so easy to reach conclusive answers and this Table is meant to stimulate further discussion.

9. SUMMARY

The prevalent view in storm-substorm relationship has been that substorms are the main building blocks of storms [*Chapman*, 1962]. This view was questioned, based mainly

Table 1. Combinations of factors that can lead to substorms and storms in magnetospheres with different characteristics

Combination of Factors	Storms	Substorms
1. B_{IMF}, B_g, I Magnetosphere of Earth & Giant planets	Yes	Yes
2. $B_{IMF}, B_g, I = 0$ Hermean magnetosphere	Yes?	No
3. $B_{IMF}, B_g = 0, I$ Venus magnetosphere	No	Yes
4. $B_{IMF}, B_g = 0, I = 0$ Moon like interactions	No	No
5. $B_{IMF} = 0, B_g, I$	No (weak driving)	Yes (viscous interaction)
6. $B_{IMF} = 0, B_g, I = 0$	No	Yes
7. $B_{IMF} = 0, B_g = 0, I$	No	Yes
8. $B_{IMF} = 0, B_6 = 0, I = 0$	No	No

on the important role of the solar wind as the driver of geomagnetic activity [*Kamide*, 1992]. Advances in modeling and measurements have intensified the pace of development in this arena [*Siscoe*, 1997]. A new understanding of the ring current, responsible for the magnetic disturbances during storms, was reached at the Chapman Conference on Storm-Substorm Relationship [*Sharma et al.*, 2001]. During the storm main phase the magnetic disturbances are caused by a partial ring current driven mainly by convection, and the fast decay of this current is due to the losses to the ionosphere and the magnetopause. The ring current becomes symmetric as the recovery phase develops and its decay is governed mainly by charge exchange processes. The accumulated results from recent modeling studies and the agreement with the ENA imaging observations have led to this consensus.

However there many issues that need to be settled and some of these are: What role, then, do substorms play? What does substorm injections do or accomplish? In the broader context, the more fundamental than the storm-substorm relationship is: How do two different processes in the magnetosphere with the same origin, the solar wind, and occurring in the same spatial volume (in a highly coupled system) evolve and how they affect each other?

Acknowledgments. The authors gratefully acknowledge many constructive discussions with many colleagues and grant support by many funding agencies. The supporting grants at the University of Maryland are NSF grants ATM-0001676 and ATM-0119196.

REFERENCES

Baker, D.N., T.I. Pulkkinen, X. Li, S.G. Kanekal, J.B. Blake, R.S. Selesnick, M.G.Henderson, G.D. Reeves, H.E. Spence, and G. Rostoker, Coronal mass ejections, magnetic clouds, and relativistic magnetospheric electron events: ISTP, *J. Geophys. Res.*, *103*, 17,279, 1998a.

Baker, D.N., X. Li, J.B. Blake, and S. Kanekal, Strong electron acceleration in the Earth's magnetosphere, *Adv. Space Res.*, *21(4)* 609, 1998b.

Baker, D.N., and X. Li, Relativistic electron flux enhancements during strong geomagnetic activity, this volume, 2003.

Burton R.K., R.L. McPherron, and C.T. Russell, An empirical relationship between interplanetary conditions and Dst, *J. Geophys. Res.*, *80*, 4204-4214, 1975.

C:son Brandt, P., D. G. Mitchell, R. Demajistre, E. C. Roel of, S. Ohtani, J.-M. Jahn, C. Pollock and G. Reeves, Storm-substorm relationships during the 4 October 2000 storm: IMAGE global ENA imaging results, this volume, 2003.

Chapman, S., Earth storms: Retrospect and prospect, *J. Phys. Soc. Japan, 6,* Suppl. A-I, 17, 1962.

Clauer, C. R., M. W. Liemohn, J. U. Kozyra, M. L. Reno,The relationasihp of storms and substorms determined from mid-latitude ground-based magnetic maps, this volume, 2002.

Daglis, I. A., and Y. Kamide, The role of substorms in storm-time particle acceleration, this volume, 2003.

Fay, R.A., C.R. Garrity, R.L. McPherron, and L.F. Bargatze, Prediction filters for the Dst index and the polar cap potential, in *Solar Wind Magnetosphere Coupling*, eds. Y. Kamide, and J.A. Slavin, pp. 111-117, Terra Sci. Pub., Tokyo, Japan, 1986

Gonzalez W.D., J.A. Joselyn, Y. Kamide, H.W. Kroehl, G. Rostoker, B.T. Tsurutani, and V.M. Vasyliunas, What is a geomagnetic storm?, *J.Geophys. Res., 99,* 5771, 1994.

Grande, M., C. H. Perry, A. Hall, J. Fennell, R. Nakamura and Y. Kamide, What is the Effect of Substorms on the Ring Current ion Population During a Geomagnetic Storm?, this volume, 2003.

Hudson, M.K., S.R. Elkington, J.G. Lyon, and C.C. Goodrich, Increase in relativistic electron flux in the inner magnetosphere: ULF wave move structure, *Adv. Space Res., 25,* 2327, 2000.

Iyemori, T., and D.R.K. Rao, Decay of the Dst field of geomagnetic disturbances after substorm onset and its implication to substorm relation, *Annales Geophysicae,* 14, 6087, 1996.

Jordanova, V. K., L. M. Kistler, J. U. Kozyra, G. V. Khazanov, and A. F. Nagy, Collisional losses of ring current ions, *J. Geophys. Res., 101,* 111, 1996.

Jorgensen, A. M., M. G. Henderson, E. C. Roelof, G. D. Reeves, and H. E. Spence, The charge-exchange contribution to the decay of the ring current measured by Energetic Neutral Atoms (ENAs), *J. Geophys. Res., 106,* 1931, 2001.

Kamide, Y., Is substorm occurrences a necessary condition for a magnetic storm?, *J. Geomag. Geoelectr., 44,* 109, 1992.

Kamide Y., and S. Kokubun, Two-component electrojet: Importance for substorm studies, *J. Geophys.Res., 101,* 13,027,1996.

Kamide Y.,W. Baumjohann, I. A. Daglis, W. D. Gonzalez, M. Grande, J. A. Joselyn, R. L. McPherron, J. L. Phillips, E. G. D. Reeves, G. Rostoker, A. S. Sharma, H. J. Singer, B. T. Tsurutani and V. M. Vasyliunas,Current Understanding of Magnetic Storms: Storm-substorm Relationships, *J. Geophys.Res., 103,* 17,705, 1998.

Korth, A., R. Friedel and F. Frutos-Alfaro, Storm-time Contribution to Oxygen/Hydrogen Ratio by Substorms: Storm-Substorm Relationship, this volume 2003.

Kozyra, J. U., V. K. Jordanova, R. B. Horne, and R. M. Thorne, Modeling of the contribution of electromagnetic ion cyclotron (EMIC) waves to stormtime ring current erosion, in *Magnetic Storms*, Geophys. Monogr. Ser., vol. 98, edited by B. T. Tsurutani, W. D. Gonzalez, Y. Kamide, and J. K. Arballo, pp. 187-202, Amer. Geophys. Union, Washington, DC, 1997.

Lakhina, G. S., and B. T. Tsurutani, Helicon modes driven by ionospheric O^+ ions in the plasma sheet region, *Geophys. Res. Lett., 24,* 1463-1466, 1997.

Lakhina, G. S., and B. T. Tsurutani, Role of Helicon Modes in Substorm Processes, *Substorms-4,* eds. S. Kokubun and Y. Kamide, Terra Scientific, pp 511-516, 1998.

Lakhina, G. S., and S. Singh, Role of plasma instabilities driven by oxygen ions during magnetic storms and substorms, this volume, 2003.

Liemohn, M. W., J. U. Kozyra, M. F. Thomsen, J. L. Roeder, G. Lu, J. E. Borovsky, and T. E. Cayton, The dominant role of the asymmetric ring current in producing the stormtime Dst*, *J. Geophys. Res., 106,* 10,883, 2001.

Lui A. T. Y., R. W. McEntire, and K. B. Baker, A New Insight on the Cause of Magnetic Storms A New Insight on the Cause of Magnetic Storms, *Geophys. Res. Lett., 28,* 3413, 2001.

McFadden, J. P., Y. K. Tung, C. W. carlson, R. J. Strangeway, E. Mobius and L. M. Kistler, FAST observations of ion outflow associated with magnetic storms, in *Space Weather*, edited by P. Song, H. J. Singer and G. L. Siscoe, pp. 413-421, Amer. Geophys. Union, Washington, DC, 2001

McPherron, R.L., Role of substorms in the development of magnetic storms, in *Magnetic Storms*, edited by B.T. Tsurutani, W.D. Gonzalez, Y. Kamide, and J.K. Arballo, pp. 131-147, Amer. Geophys. Union, Washington, DC, 1997.

McPherron, R.L., and T.P. O'Brien, Predicting Geomagnetic Activity: The Dst Index, in *Space Weather,* edited by P. Song, G.L. Siscoe, andH. Singer, pp. 339-345, Amer. Geophys. Union, 2001.

Meredith, N.P., R.B. Horne, R.H. Iles, R.M. Thorne, D. Heynderickx, and R.R. Anderson, Outer zone relativistic electron acceleration associated with substorm enhanced whistler mode chorus, *J. Geophys. Res.*, in press, 2002

O'Brien, T.P., and R.L. McPherron, An empirical phase-space analysis of ring current dynamics: solar wind control of injection and decay, *J. Geophys. Res., 105,* 7707, 2000.

Ostgaard, N., and E. Tanskanen, Energetics of Isolated and Storm-time Substorms, this volume, 2003.

Pulkkinen, T. I., and A. S. Sharma Storm-substorm Relationship (Session summary),in *Substorms-5: Proc. 5th Internl. Conf. on Substorms,* ESA Pub. SP-443, 2000.

Reeves, G. D., New perspectives on substorm injections, *Substorms-4, Proc. ICS-4*, Hamanako, Japan, S. Kokubun and Y. Kamide, eds., Kluwer Academic Publishers, Boston, p. 785, 1998.

Reeves, G.D., Relativistic electrons and magnetic storms: 1992-1995, *Geophys. Res. Lett., 25,* 1817, 1998.

Reeves, G. D., and M. G. Henderson, The storm-substorm relationship: Ion injections in geosynchronous measurements and composite energetic neutral atom images, *J. Geophys. Res., 106,* 5833-5844, 2001.

Reeves, G., M. G. Henderson, R. M. Skoug, M. F. Thomsen, J. E. Borovsky, H. O. Funsten, P. C. Brandt and D. J. Mitchell, IMAGE, POLAR, and Geosynchronous Observations of Substorm and Ring Current Ion Injection, this volume, 2003.

Richmond, A. D., and Y. Kamide, Mapping electrodynamic features of the high latitude ionosphere from localized observations: technique, *J. Geophys. Res., 93,* 5741, 1988.

Roelof, E. C., Energetic neutral atom image of a storm-time ring current, *Geophys. Res. Lett., 14,* 652, 1987.

Rostoker, G., Identification of substorm expansive phase onsets, *J. Geophys. Res.*, in press, 2002.

Rostoker, G., S.-I. Akasofu, W. Baumjohann, Y. Kamide and R.L. McPherron, The roles of direct input of energy from the solar wind and unloading of stored magnetotail energy in driving magnetospheric substorms, *Space Sci. Rev., 46 ,* 93 , 1987.

Rostoker , G., S. Skone, and D.N. Baker, On the origin of relativistic electrons in the magnetosphere associated with some geomagnetic storms, Geophys. Res. Lett., 25, 3701, 1998.

Sharma, A. S., Assessing the magnetosphere's nonlinear behavior: Its dimension is low, its predictability, high (US National Report to IUGG, 1991-1994), *Rev. Geophys, 33(. Supple.),* 645, 1995.

Sharma, A. S., J. A. Valdivia and Y. Kamide, Dynamical Relationship Between Storms and Substorms, in *Substorms-4: Proc. 4th Internl. Conf. on Substorms,* edited by S. Kokubun and Y. Kamide, Terra Sci., Tokyo, pp. 737-740, 1998.

Sharma, A. S., G.S. Lakhina and Y. Kamide, Conference explores relationship between geomagnetic storms and substorms, *Eos, Trans. AGU, 82(49),* 609-610, 2001.

Siscoe, G., Big storms make little storms, *Nature, 390,* 448, 1997.

Siscoe, G. L., and H. E. Petschek, On storm weaking during substorm expansion phase, *Ann. Geophys., 15,* 211, 1997.

Singh, S. V., R. V. Reddy, and G. S. Lakhina, Low-frequency instabilities due to energetic oxygen ions in the ring current, *Ann. Geophys.*, in press, 2003.

Takahashi, K., T. Iyemori, and M. Takeda, A simulation of the storm-time ring current, *Planetary Space Science, 38 (9),* 1133-1141, 1990.

Thorne, R.M. and R.B. Horne, Energy transfer between ring current H$^+$ and O$^+$ by electromagnetic ion cyclotron waves, *J. Geophys. Res., 99*, 17275, 1994.

Vondrak, R.,J. Slavin, L. Zelenyi, L. Guhathakurta, S. Curtis and B. Tsurutani, Measurement Strategies for Future Missions to Understand Geospace Dynamics, this volume, 2003

D. N. Baker, Laboratory for Atmospheric and Space Physics, University of Colorado, Boulder, CO 80309, USA. (baker@lasp.colorado.edu)

M. Grande, Rutherford Appleton Laboratory, Chilton, Didcot, Oxon OX11 0QX, United Kingdom. (M.Grande@rl.ac.uk)

Y. Kamide, Solar-Terrestrial Environment Laboratory, Nagoya University, Toyokawa, 442, Japan. (kamide@stelab.nagoya-u.ac.jp)

G. S. Lakhina, Indian Institute of Geomagnetism, Colaba, Mumbai 400 005, India. (lakhina@iig.iigm.res.in)

R. M. McPherron, Institute of Geophysics and Planetary Physics, University of California, Los angeles, CA 90024. (rmcpherron@igpp.ucla.edu)

G. D. Reeves, Los Alamos National Laboratory, Los Alamos, NM 87545, USA. (reeves@lanl.gov)

G. Rostoker, Department of Physics, University of Alberta, Edmonton, Alberta, T6G 2J1, Canada. (rostoker@space.ualberta.edu)

A. S. Sharma, Department of Astronomy, University of Maryland, College Park, MD 20742, USA. (ssh@astro.umd.edu)

R. Vondrak, Laboratory for Extraterrestrial Physics, Goddard Space Flight Center, Greenbelt, MD 20771, USA. (vondrak@gsfc.nasa.gov)

L. Zelenyi, Space Research Institute, RAS, Profsoyuznaya Street, 84/32, 117810 Moscow, Russia. (lzelenyi@iki.rssi.ru)

Effects of Solar Wind Density on the Auroral Electrojets and Global Auroras During Geomagnetic Storms

Y. Kamide

Solar-Terrestrial Environment Laboratory, Nagoya University, Toyokawa, Aichi, Japan

J.-H. Shue

Applied Physics Laboratory, The Johns Hopkins University, Laurel, Maryland

M. Brittnacher

Geophysical Program, University of Washington, Seattle, Washington

It was shown statistically, some thirty years ago, that substorms occurring during the initial phase of geomagnetic storms and those during the main phase are different in character. Substorms during geomagnetic storms, regardless of the phases in which they occur, were also shown to be different from normal isolated substorms in terms of the relative strength between the eastward and westward auroral electrojets. Based on the intensities of the auroral electrojets estimated from ground magnetometer data and on the distribution of large-scale auroras seen in Polar auroral images, the present study shows that the solar wind density does in fact control the intensity of the auroral electrojets and the associated auroral activity, but the efficiency of the control depends strongly on the polarity of interplanetary magnetic field (IMF). The earlier statistical results can consistently be accounted for by considering that the initial phase of geomagnetic storms is caused by the high dynamic pressure (or density) of the solar wind, while southward IMF dominates the main phase.

1. INTRODUCTION

Although magnetospheric substorms are known to take place when a certain condition (yet unknown) in the magnetosphere-ionosphere system and the magnetotail is met, it has been found that a subset of substorms are triggered "externally" by sudden changes in the solar wind [e.g., *Burch*, 1972; *Lyons*, 1995; *Zesta et al.*, 2000; *Chua et al.*, 2001). Those sudden changes include sudden northward turnings of the interplanetary magnetic field (IMF) and pressure pulses in the solar wind. Pressure enhancements in the solar wind are usually associated with interplanetary shocks [*Akasofu and Chao*, 1980], generating sudden changes in the geomagnetic field, such as sudden impulses (SIs) and storm sudden commencements (SSCs): see *Araki* [1977]. It has also been shown that substorms are activated by SIs and SSCs under a precondition of southward IMF: see *Burch* [1972], *Iijima* [1973], *Kokubun et al.* [1977], and *Zhou and Tsurutani* [2001].

Disturbances in Geospace: The Storm-Substorm Relationship
Geophysical Monograph 142
Copyright 2003 by the American Geophysical Union
10.1029/142GM02

The purpose of this brief paper is to report that solar wind density controls the intensity of substorms and auroral activity. The efficiency of the control is shown to depend on the polarity of the IMF.

2. THE AURORAL ELECTROJETS DURING THE INITIAL PHASE AND THE MAIN PHASE OF GEOMAGNETIC STORMS

It is quite reasonable to assume tacitly that characteristics of isolated substorms are the same as those of storm-time substorms, in the sense that we should begin our empirical study with the simplest assumption unless some fundamental difficulty regarding the assumption is initially encountered. There is no reason, however, to believe that there must be only one type of substorm even though solar wind conditions associated with major solar disturbances are known to be considerably different from those during relatively quiet times, during which normal isolated substorms occur. What about substorms occurring during the state of the compressed magnetosphere (for example, at the initial phase of geomagnetic storms) and substorms occurring during inflated state of the magnetosphere (at the main phase)? Are those substorms the same in character in spite of the fact that the magnetosphere clearly achieves different states depending on the different conditions in the interplanetary medium?

Figure 1 shows the result of a statistical study in which substorms that occurred during the initial phase of geomagnetic storms when solar wind pressure was enhanced, are distinguished from those during the main phase [*Kamide*, 1980]. In this diagram, the eastward and westward auroral electrojets (represented by AU and AL, respectively) for substorms observed during all the major geomagnetic storms ($Dst < -100$ nT) during the 1957–1964 (solar maximum to minimum) period are shown, by classifying all the cases into two subsets according to the two storm phases. The corresponding Dst traces were used to identify the storm phases. A significant difference can be noticed between those two subsets of substorms in terms of the AU/AL ratio. That is, the eastward electrojet is, on average, more intense than the westward electrojet during the initial phase, while the trend is reversed during the main phase during which the eastward electrojet often disappears. Note that the line $AU = |AL|/3$ is located in between: this line indicates the average relationship between AU and AL for isolated substorms that take place without being associated with major geomagnetic storms [*Kamide and Fukushima*, 1972].

One may argue that the small AU values during the main phase are biased by the equatorward shift of the eastward electrojet region, so that AU during the main phase, monitored by the standard AE observatories, cannot be very reliable. However, regarding the initial phase, we even see cases where the eastward electrojet reaches at times very high values, such as 3000 nT during the August 1972 storm, whereas the westward electrojet, which is supposed to be three times more intense than the eastward electrojet during normal substorms, is rather weak, such that the dependence of the AU/AL ratio on the storm phases does truly exist. We contend that this clear difference between the initial phase-substorms and the main phase-substorms results from the difference in the state of the magnetosphere at two different phases, although we admit that detailed mechanisms have not still been identified.

3. EFFECTS OF SOLAR WIND DENSITY ON THE AURORAL ELECTROJETS

In most of the early studies, examinations of the effects of the solar wind density on the auroral electrojets have been made for cases in which not only the solar wind density but also other variables changed considerably, so that it has often been difficult to isolate the effects of the solar wind density. *Shue and Kamide* [2001] have recently shown cases of the period of January 10–11, 1997, during which all available solar wind parameters, except for the density, were fortuitously nearly constant [*Burlaga et al.*, 1998; *Fox et al.*, 1998] demonstrating that the solar wind density has the ability of enhancing the auroral electrojets. It has also been contended, however, that its efficiency varies considerably, depending on the IMF polarity. Shue and Kamide have presented two

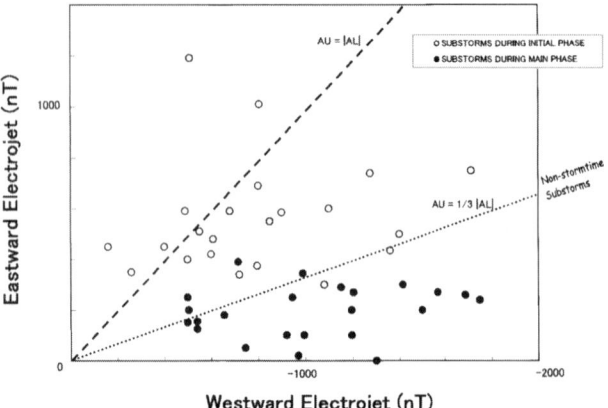

Figure 1. Comparison between the AU and AL indices for substorms occurred during major geomagnetic storms. Cases are grouped into two subsets; substorms during the initial phase and those observed during the main phase.

periods of auroral electrojet enhancements during which only the solar wind density changed significantly. One was the 0600–1200 UT, January 10, 1997 event, which occurred when the IMF was directed steadily southward: see their Figure 1. The other was the 0030–0300 UT, January 11, 1997 event in the interplanetary medium, which occurred under a northward IMF condition: see their Figure 3. Data of the solar wind plasma and the magnetic field during these events were obtained from the Wind satellite.

Figure 2a shows the relationship between AL (58) values and the solar wind density for periods during southward IMF. AL was derived from the superposed X_m component traces of 58 high-latitude ground-based magnetometer observatories. The scattered points in Figure 2a are fitted to a straight line, as indicated by a solid line, using a least-squares method. The slope of the line is more than 100 nT cm^3, meaning that the solar wind density significantly affects the westward electrojet intensity when the IMF is directed southward.

Figure 2b, on the other hand, is a scattered plot of AU (58) values against the solar wind density for periods during which the IMF was directed nearly steadily northward. *Shue and Kamide* [2001] noted that during periods of northward IMF, the eastward electrojet is best correlated with the solar wind density. The slope determined by a least squares method is, however, only 0.9 nT cm^3, much smaller than the slope in Figure 2a. The two sets of scattered plots demonstrate undoubtedly that the solar wind density does affect the auroral electrojets regardless of the IMF orientation, but its efficiency of the enhancement depends strongly on whether the IMF is directed northward or southward. The local time where the solar wind effect is greatest appears to vary, according to the IMF polarity.

We feel that to claim that the density effect is most effective in the westward (or eastward) electrojet when the IMF is directed southward (or northward), it is essential to show that the dependence of the westward (or eastward) electrojet on the number density is not dominant when the IMF is northward-directed (or southward-directed). Figure A1 in the APPENDIX presents that this is really the case.

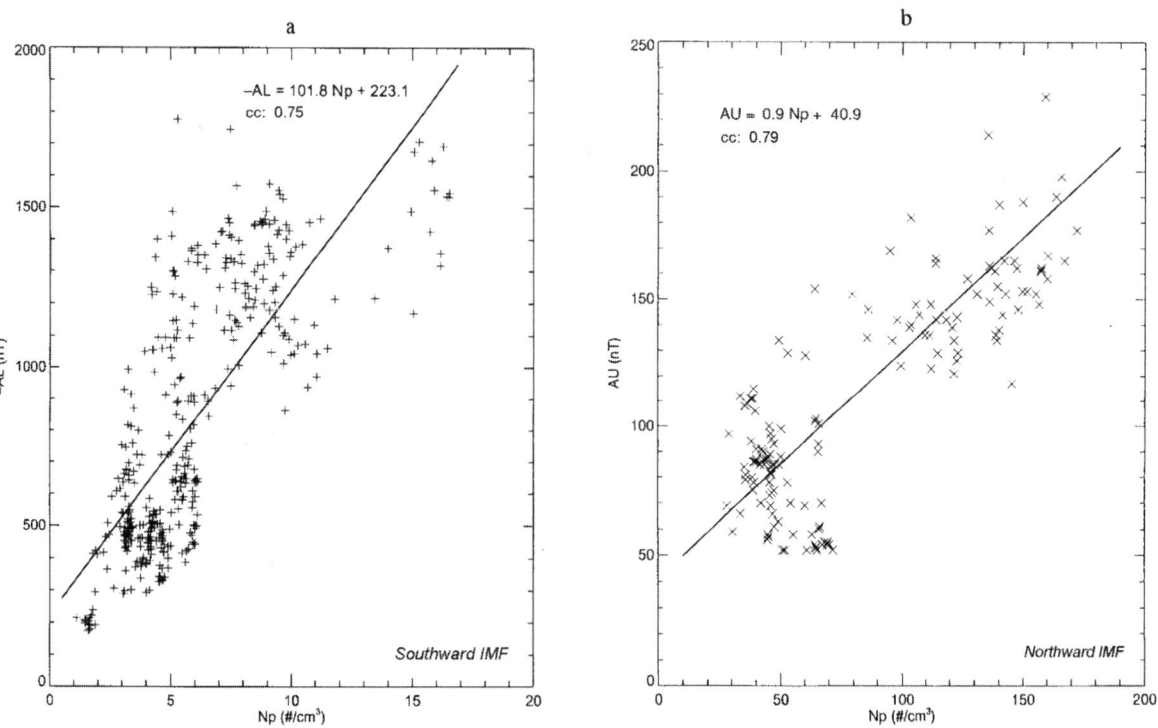

Figure 2. (a) Scatter plot for the AL index versus solar wind density (N_p) for the period 0600–1200 UT, January 10, 1997. The data points are fitted to a straight line by a linear least-squares method. (b) Same as Fig. 2a, except for the AU index for the period 0030–0300 UT, January 11, 1997.

4. AURORAL DISTRIBUTION

In this section, we will examine the effect of the solar wind pressure on auroral activity. A series of auroral photographs taken from the Ultraviolet Imager on the Polar satellite for a period of high solar wind pressure are shown in Figure 3. On this day, i.e., April 9, 1996, the number density and the bulk speed of the solar wind showed a very interesting feature. In the first half of the day, the number density was quite high, as high as 20 /cm^3, whereas in the second half of the day, the solar wind speed hiked from 350 km/s to 500 km/s as the number density dropped to a normal level, 5–10 /cm^3. There was no major substorm activity and *Dst* was between +10 and –20 nT throughout the interval. It is seen in Figure 3 that relatively intense auroral activity waxed and waned along the auroral oval but it was separated into two parts, one on the day side and the other on the night side with somewhat weak auroral activity in between on the dawn and dusk sectors. This set of observations suggests strongly that two separate sources and mechanisms operate at the same time, generating the two separate regions of auroral displays. Although Figure 3 demonstrates global auroral displays only for the second half of this day, the features for the first half is basically the same, showing two separate auroral regions.

Figure 4 shows two individual Polar images in which it is possible to identify the effect of an enhancement of the solar wind pressure on global auroral distributions. These two auroral distributions on October 1, 1997 were separated each other by merely three minutes, during which an intense interplanetary shock hit the magnetosphere. In the data processing stage, dayglow has been removed, although some filtering noise might still have remained. Figure 5 shows magnetograms from two mid-latitude magnetic observatories, one on the day side and the other on the night side. As indicated by arrows, an evident effect of an interplanetary shock, i.e., an SSC, is seen at the two observatories at 0058:48 UT. Note that this is followed by the main phase of a medium-sized geomagnetic storm, reaching –100 nT.

According to IMP 8 measurements of the solar wind, the interplanetary shock was registered as a rapid rise in the dynamic pressure from 4 nPa to 15 nPa within a few minutes [*Brittnacher et al.*, 2000]. This increase in the dynamic pressure was due to an increase of about 10% in the solar wind speed and nearly a doubling of the density. The B_z component of the IMF had a nearly zero average value with some fluctuations.

In Figure 4a, recorded one minute before the shock's arrival, a rather weak display of the auroral oval with a

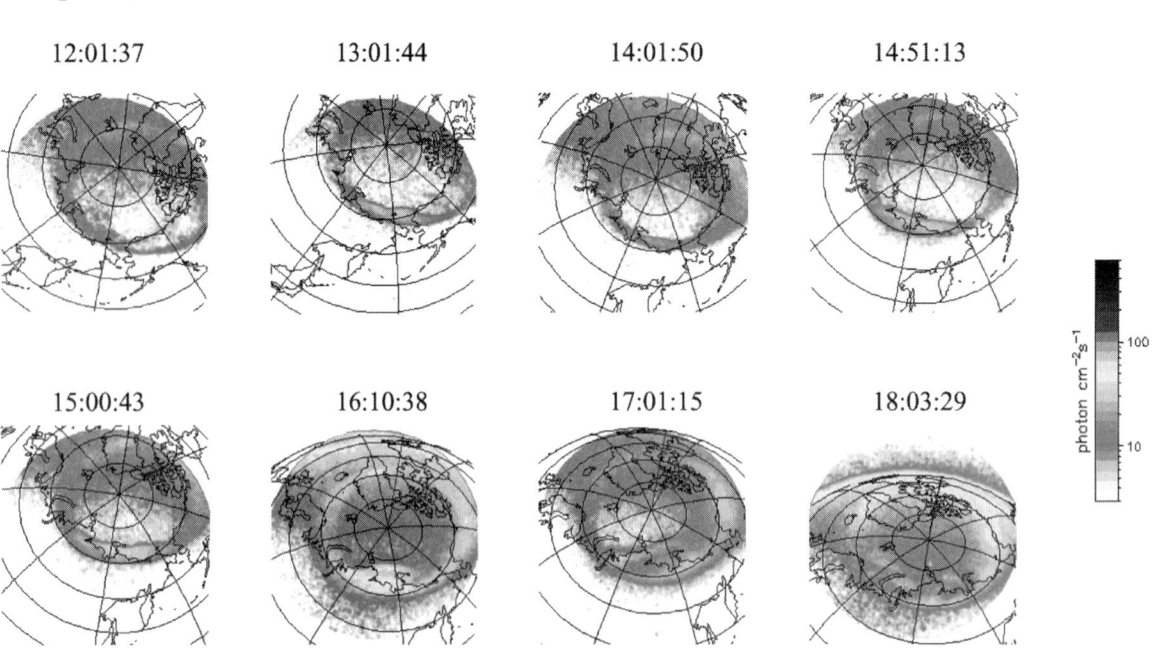

Figure 3. A series of global auroral images when solar wind pressure was continuously high: the dayside at the top. Two separate auroral displays on the day side and on the night side are seen.

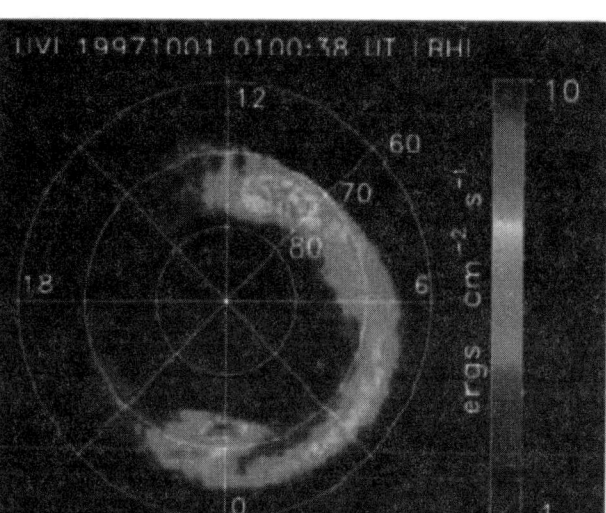

Figure 4. UVI auroras from Polar: (a) just before the arrival of an interplanetary shock, and (b) just after the arrival of the interplanetary shock.

small bright spot in the midnight sector is noticed. After the arrival of the shock, however, a very bright spot was created along with an enhancement of the entire auroral belt as seen in Figure 4b. These spots, on the night side and day side, soon became "seeds" for a further development of auroral activity on a global scale. The auroral oval continued to expand both latitudinally and longitudinally.

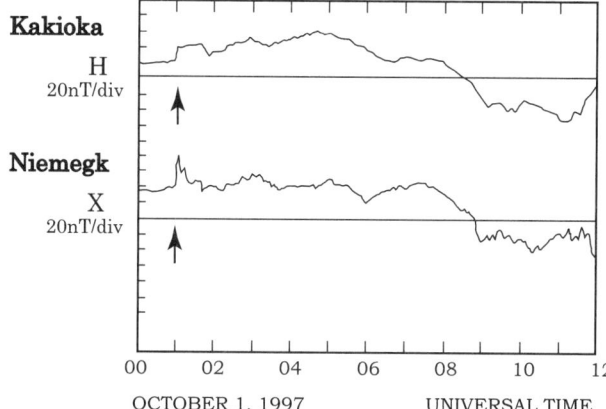

Figure 5. Magnetograms from Kakioka and Niemegk, where a clear indication of the arrival of the interplanetary shock at 0058 UT of October 1, 1997 is seen.

5. DISCUSSION

This brief report has shown two types of observations, i.e., auroral activity and the corresponding auroral electrojet current, that demonstrate the importance of solar wind density (or pressure) in determining the intensity of the auroral electrojets. It has been shown that the electrojet intensity as well as the auroral distribution are well correlated with an increase/decrease in the solar wind density. The efficiency of the correlation has been shown to depend strongly on the orientation of the IMF. That is, the correlation itself is found to be lower for northward IMF than for southward IMF (see Figures 2a and 2b), and moreover, when the IMF is directed southward, the density effect is dominant in the westward electrojet, while it is most effective in the eastward electrojet when the IMF is directed northward. Plausible mechanisms of such density effects are not known at present, although it seems likely that the density of the plasma sheet, which is strongly correlated with the density of the solar wind, plays some key role in enhancing the auroral electrojets [*Borovsky et al.*, 1998]. *Shue and Kamide* [2001] cautioned, however, that the time lag between the solar wind density and the plasma sheet density is more than 2 hours, whereas the solar wind density affects the auroral electrojet nearly simultaneously, as reported in Figure 2 in the present paper.

We have to note that unless fortuitous periods, such as those used for Figure 2, are found, it is generally difficult to separate the effects of the solar wind density and those of IMF on the auroral electrojets. It is well known that when the IMF is directed northward for several hours, the magnetosphere reaches the so-called ground state, meaning that both AU and AL approach basically zero during such a prolonged period of northward IMF. This in turn implies that an AU enhancement during the initial phase of geomagnetic storms when the IMF is directed northward, as shown in Figure 1, can in fact be the result of an enhancement in the solar wind pressure. On the contrary, during the main phase during which southward IMF prevails, everything including AU and AL is so dynamic that it is impossible to isolate the effects of only one solar wind parameter in the corresponding electrojet activity. Perhaps, the effects of the number density of the solar wind and that of IMF are coupled in a nonlinear way.

Our finding has at least one important implication. That is, the effects of the solar wind density are dominant on the night side of the earth when the IMF is strongly southward directed, whereas it is most effective on the day side when the IMF is directed northward. This local-time dependence is confirmed directly with auroral observations in which the dayside portions of the auroral oval responded nearly instantly to an interplanetary shock: see Figures 3 and 4. *Brittnacher et al.* [2000] have also recently shown that dayside bright spots in the aurora appear often in conjunction with a sudden increase in the solar wind density. Note, however, that our preliminary study on the basis of a number of auroral images has indicated that sudden intensifications of dayside auroras are not simply related to solar wind conditions but to previous auroral activity on the night side, which relates to the polarity of the IMF.

We have found that the observations shown in this paper are consistent with some of the earlier observations where only indirect information on the solar wind was available. Figure 1 has evidently shown a clear difference in the relative importance of the westward and eastward electrojets between the initial phase and the main phase of geomagnetic storms, which correspond presumably to periods of high and low solar wind pressure, respectively. One may argue that since the initial phase is generally associated with northward IMF, it may be difficult to distinguish the effects of high pressure of the solar wind with those of northward IMF. *Shue and Kamide* [2001] have shown, however, that even when IMF is steadily northward, the eastward electrojet on the day side increases in harmony with an increase in the solar wind density. It is also true that no studies have ever shown that there is one-to-one correspondence between AU increases and an increase (or decrease) in the intensity of northward IMF.

It is speculated that associated with an abrupt increase in the density and the dynamic pressure of the solar wind, the dayside aurora can dynamically be enhanced and the area of the dynamic aurora expands latitudinally, accompanied by an intensification of the eastward electrojet. In terms of phenomenology, this nature is quite similar to the well-established feature of the auroral substorm in the midnight sector, except for the direction of the corresponding electrojet current. Although the detailed time sequence of the dayside phenomena has not been studied yet, this characteristic disturbance may be called "the dayside substorm." Time scales between the dayside substorm and the nightside substorm (or normal substorm) are quite different, by perhaps a factor of 10 or more. The area involved in the dayside substorm is, also, much smaller than that of the normal substorm. It is interesting to infer that the dayside substorm on the night side is caused by suddenly dumping energy by the solar wind, whereas the normal substorm is provided by an excess amount of energy which has been stored for some time during the so-called growth phase.

Figure 6 summarizes our observations reported in this paper. For simplicity, this diagram shows only four practical choices as a combination of the variability in the IMF and solar wind density. When the solar wind is relatively low or normal and the IMF is steadily northward, there would practically be no possibility of substorm occurrence [*Kamide et al.*, 1977]. On the contrary, under a "stormy" condition of high solar wind density and southward IMF, intense substorms characterized by the intense westward electrojet occur subsequently, as typically seen during the main phase of strong geomagnetic storms. Between these two extreme conditions, there can, of course, be a variety of combinations

Solar Wind Density	IMF Northward	IMF Southward
High	Large-AU substorms	Large-AL substorms
Low	No substorm	Normal-AU/AL substorms

Figure 6. Summary of the observations for the characteristics of substorms, shown for four possible combinations of the solar wind density level and the IMF polarity.

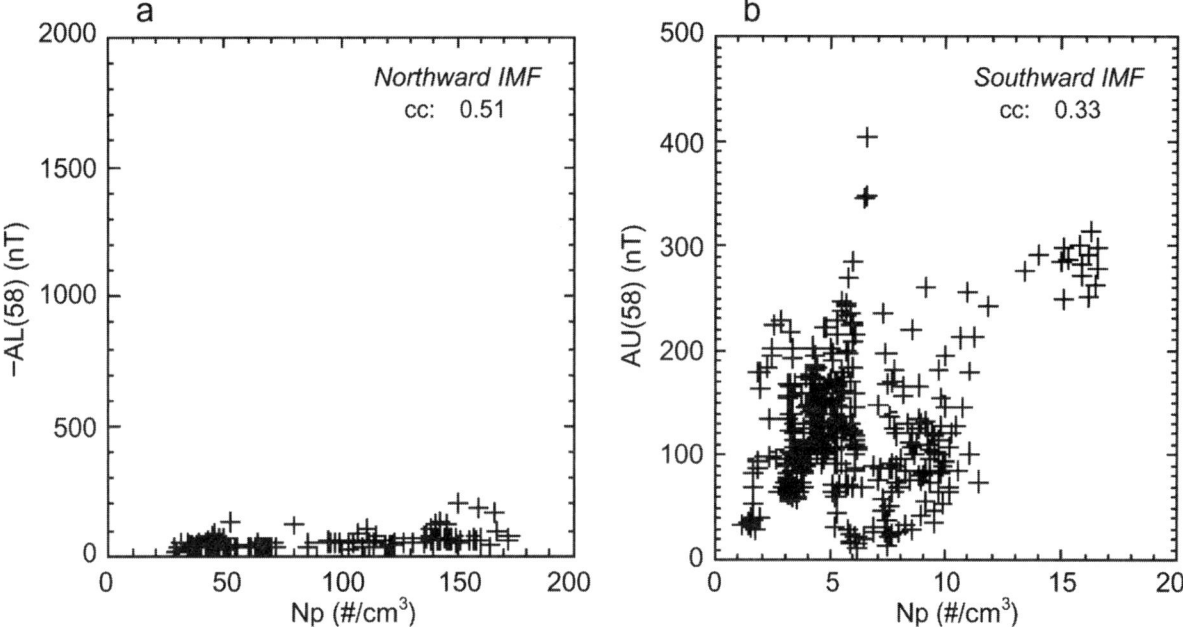

Figure A1. (a) Scatter plot for the *AL* index versus solar wind density (N_p) for the period of northward IMF. (b) Same as Fig. A1a, except for the *AU* index for the period of southward IMF.

of the solar wind density and the IMF and their time changes, making observations quite complicated.

6. APPENDIX

Figure A1, which is very similar to Figure 2, shows two diagrams where the correlation is quite low between the *AL* (58), or *AU* (58), and the number density of the solar wind when the IMF is directed northward (or southward). Figures A1a and A1b correspond to northward and southward IMF, respectively. It is evident in Figure A1a that the number density of the solar wind does not show any systematic control on *AL*. It is particularly important to point out that the *AL* values stay mostly below 100 nT during periods of northward IMF, regardless of a wide dynamic range in the variability of the number density. On the other hand, Figure A1b indicates that *AU* can change significantly during periods of southward IMF, but again no systematic correlation with the number density is seen.

Acknowledgments. We are grateful to A. J. Lazarus of MIT and R. P. Lepping of NASA/GSFC for making available the Wind SWE and MFI records. We also thank the World Data Center-C2 at Kyoto University for providing magnetogram data used in Figure 5 and the *Dst* index which we referred to for discussions relating to Figure 3.

REFERENCES

Akasofu, S.-I., and J. K. Chao, Interplanetary shock waves and magnetospheric substorms, *Planet. Space Sci.*, *28*, 381, 1980.

Araki, T., Global structure of geomagnetic sudden commencement, *Planet. Space Sci.*, *25*, 373, 1977.

Borovsky, J. E., M. F. Thomsen, and R. C. Elphic, The driving of the plasma sheet by the solar wind, *J. Geophys. Res., 103*, 17,617, 1998.

Brittnacher, M., M. Wilber, M. Fillingim, D. Chua, G. Parks, J. Spann, and G. Germany, Global auroral response to a solar wind pressure, *Adv. Space Res.*, *25*, 1377, 2000.

Burch, J. L., Preconditions for the triggering of polar magnetic substorms by storm sudden commencements, *J. Geophys. Res.*, *77*, 5629, 1972.

Burlaga, L., R. Fitzenreiter, R. Lepping, K. Ogilvie, A. Szabo, A. Lazarus, J. Steinberg, G. Gloeckler, R. Howard, D. Michels, C. Farrugia, R. P. Lin, and D. E. Larson, A magnetic cloud containing prominence material: January, 1997, *J. Geophys. Res.*, *103*, 277, 1998.

Chua, D., G. Parks, M. Brittnacher, W. Peria, G. Germany, J. Spann, and C. Carlson, Energy characteristics of auroral electron precipitation: A comparison of substorms and pressure pulse related auroral activity, *J. Geophys. Res.*, *106*, 5945, 2001.

Fox, N. J., M. Peredo, and B. J. Thompson, Cradle to grave tracking of the January 6–11, 1997, Sun-Earth connection event, *Geophys. Res. Lett.*, *25*, 2461, 1998.

Iijima, T., Interplanetary and ground magnetic conditions preceding SSC-triggered substorms, *Rep. Ionos. Space Res. Japan*, *27*, 205, 1973.

Kamide, Y., Relationship between storms and substorms, in *Dynamics of the Magnetosphere*, edited by S.-I. Akasofu, p. 425, Reidel Pub., 1980.

Kamide, Y., and N. Fukushima, Positive geomagnetic bays in evening high-latitudes and their possible connection with partial ring current, *Rep. Ionos. Space Res. Japan*, *26*, 79, 1972.

Kamide, Y., P. D. Perreault, S.-I. Akasofu, and J. D. Winningham, Dependence of substorm occurrence probability on the interplanetary magnetic field and on the size of the auroral oval, *J. Geophys. Res.*, *82*, 5521, 1977.

Kokubun, S., R. L. McPherron, and C. T. Russell, Triggering of substorms by solar wind discontinuities, *J. Geophys. Res.*, *82*, 74, 1977.

Lyons, L. R., A new theory for magnetospheric substorms, *J. Geophys. Res.*, *100*, 19,069, 1995.

Shue, J.-H., and Y. Kamide, Effects of solar wind density on auroral electrojets, *Geophys. Res. Lett.*, *28*, 2181, 2001.

Zesta, E., H. J. Singer, D. Lummerzheim, C. T. Russell, L. R. Lyons, and M. J. Brittnacher, The effect of the January 10, 1997 pressure pulse on the magnetosphere-ionosphere current system, in Magnetospheric Current Systems, *Geophys. Monogr. Ser.*, Vol. 118, p. 217, edited by S.-I. Ohtani, R. Fujii, M. Hesse, and R. L. Lysak, AGU, Washington, DC, 2000.

Zhou, X., and B. T. Tsurutani, Interplanetary shock triggering of nightside geomagnetic activity: Substorms, pseudobreakups, and quiescent events, *J. Geophys. Res.*, *106*, 18,957, 2001.

M. Brittnacher, Geophysical Program, University of Washington, Seattle, WA 98195-1650, USA.

Y. Kamide, Solar-Terrestrial Environment Laboratory, Nagoya University, Toyokawa 422-8507, Japan.

J.-H. Shue, Applied Physics Laboratory, The Johns Hopkins University, Laurel, MD 20723-6099, USA.

A Lack of Substorm Expansion Phases During Magnetic Storms Induced by Magnetic Clouds

B. T. Tsurutani, X.-Y. Zhou and W. D. Gonzalez

Jet Propulsion Laboratory, California Institute of Technology, Pasadena, California
Instituto Nacional Pesquisas de Espaciais (INPE), Sao Paulo, Brazil

Eleven magnetic storms induced by magnetic clouds (MCs) are studied using Wind interplanetary data and Polar nightside auroral UV images. MCs were specifically chosen because the magnetic field directionality typically varies smoothly with time. Substorms were detected in only ~half of the storm main phases (where UV images were available). In the other 5 events, the storm main phase auroras were characterized by longitudinally broad, moderate-intensity displays distributed over the entire nightside oval. The auroras in the 19 to 24 LT sector were centered at ~62°–65° MLAT and often with a second band at higher, ~70°–72° MLAT. Dawn and dusk auroras were typically more intense than midnight sector auroras. The most prominent auroral forms were north-south aligned patches. They extended ~1 hr in LT by ~5° in MLAT, and had durations of ~3 to 6 min. The patches occurred at all local times in the nightside sector and often attached the two auroral bands (double oval) in the premidnight sector. Researchers have ascribed these patches to be the ionospheric manifestations of bursty bulk flows occurring in the plasmasheet. Two of the storms without apparent substorms were caused by interplanetary electric fields > 5 mV/m with durations > 3 hr, but had anomalously low D_{ST} values (D_{ST} > –100 nT). We suggest 3 possible mechanisms to explain the weak storm intensities: 1) a lack of substorms lead to lower O^+ ion production, 2) a lack of substorm dipolarizations lead to lower plasmasheet temperatures, and 3) the low density magnetic cloud plasma enters the near-Earth plasmasheet rapidly. Finally we present a testable scenario for substorms occurring during a magnetic cloud induced magnetic storm.

1. INTRODUCTION

The solar and interplanetary causes of magnetic storms are now well understood. Magnetic reconnection (*Dungey*, 1961) between intense, southwardly-directed interplanetary magnetic fields (B_S) and the Earth's dayside magnetospheric fields lead to solar wind energy and momentum transfer to the magnetosphere. *Gonzalez and Tsurutani* (1987) have empirically established a one-to-one relationship between interplanetary electric fields and magnetic storm intensities. Every interplanetary event with E > 5mV/m and τ >3 hr created a storm with intensity $D_{ST} \leq$ –100 nT, and vice versa, large major storms with $D_{ST} \leq$ –100 nT had the above stated interplanetary electric field criterion. This empirical criterion which was established during the 1978–1979 solar maximum era also held true for the following 11-year solar cycle.

What is the origin of the intense interplanetary magnetic fields? During the solar maximum phase of the solar cycle and shortly thereafter, fast (>400 km/s) interplanetary coronal mass ejections (ICMEs) and their effects (upstream shocks and sheaths) caused the predominance of large magnetic storms (*Gonzalez and Tsurutani*, 1987; *Gosling et al.*, 1991; *Gonzalez et al.*, 1994). The two interplanetary regions with intense magnetic fields are the sheaths behind fast forward shocks and the following driver gas proper (*Tsurutani et al.*, 1988; *Huttunen et al.*, 2001). If both of these regions have southwardly directed magnetic fields, then "double storms" resulted (*Kamide et al.*, 1998a). During the declining phase of the solar cycle, another solar/interplanetary phenomena is the dominant cause of magnetic storms (*Tsurutani et al.*, 1995). Collisions between high-speed solar wind streams (emanating from solar coronal holes) and slower speed streams, form compressed interplanetary magnetic field regions called corotating interaction regions or CIRs (*Smith and Wolf*, 1976; *Forsyth and Gosling*, 2001). However, because the B_Z component within CIRs is typically highly fluctuating, the consequential magnetic storms at Earth are smaller in magnitude ($D_{ST} > -100$ nT) than those caused by ICMEs and their upstream sheaths (*Tsurutani et al.*, 1995). For a review of the interplanetary and solar causes of magnetic storms, we refer the reader to *Gonzalez et al.* (1994) and *Tsurutani* (2001).

In sharp contrast to our general understanding of magnetic storms, the role of substorms within magnetic storms is not well understood. The original substorm definition paper by *Akasofu* (1964) implicitly assumes (see also *Akasofu*, 1968) that storms are composed of less intense events in the form of substorms. *Kamide et al.* (1998b) recently reviewed the situation, and stated that all storms studied in the past contain substorms. On the other hand, it has been shown that intense and continuous AE activity (High Intensity Long-Duration Continuous AE Activity or HILDCAAs) can occur without significant ring current development (*Tsurutani and Gonzalez*, 1987). The importance of substorms within storms is currently being strongly debated (*McPherron et al.*, 1997; *Rostoker et al.*, 1997). *Iyemori and Rao* (1996) argue that the SYM-H index decreases after substorm onsets. Contrarily, *Rostoker* (2000) find that D_{ST} is positively correlated with substorm expansive phase activity. *Sergeev et al.* (1998; 1999) have demonstrated that during continuous magnetospheric activity periods, many substorm-like features (recurrent magnetic field dipolarizations and bursty bulk flows) are present, but without traditional large scale plasma injections (a lack of substorm expansion phases). *Kamide* (2001) through a "gedanken" experiment has predicted that steady southward interplanetary magnetic fields would result in a geomagnetic storm with a lack of substorm expansion phases.

The purpose of this paper is to examine the substorm-magnetic storm relationship with a specific subset of all magnetic storms. Of the many ICMEs occurring during solar maximum, we will study only (the effects of) interplanetary magnetic cloud events (not all ICMEs contain magnetic clouds). Magnetic clouds (*Klein and Burlaga*, 1982) have the properties of low plasma beta, smooth north-south (or south-north) magnetic field rotations and a general absence of Alfvén waves and discontinuities (*Tsurutani et al.*, 1988; *Tsurutani and Gonzalez*, 1997; *Farrugia et al.*, 1997). It is these latter qualities that are of importance here. Smooth IMF B_Z rotations with a lack of abrupt and strong B_Z variations should not provide rapid interplanetary/magnetospheric electric field changes that are believed to trigger substorm expansion phases (*Tsurutani and Meng*, 1972; *Meng et al.*, 1973, *Caan et al.*, 1975; *Iyemori*, 1980; *Samson and Yeung*, 1986; *Rostoker et al.*, 1982; *McPherron et al.*, 1986; *Lyons et al.*, 1997). To perform this study, we will use the Wind 1997 interplanetary field and plasma data and corresponding Polar UV auroral images. We will also address the issue of whether magnetic storms with a lack of substorms are different from other storms. Finally, based on previously reported substorm triggering mechanisms, we will present a scenario (based on numerous models) of substorm occurrence as a function of the phase of the magnetic cloud passage. This scenario can easily be tested using space plasma data.

2. METHOD OF ANALYSES

All 11 (fast) magnetic clouds that occurred in 1997 are used in the study. These events were identified by R.P. Lepping (personal comm., 2001; http://lepmfi.gsfc.nasa.gov/mfi/mag_cloud_pub1.html) and are indicated in Table 1. The Wind plasma and magnetic field data have been used to calculate the solar wind ram pressure and plasma beta. The Wind magnetometer is described in *Lepping et al.* (1995) and the plasma instrument in *Ogilvie et al.* (1995). The magnetic cloud intervals are identified from the plasma beta and the magnetic field characteristics. Low beta values (typically < 0.1) and smooth magnetic fields are characteristics of clouds, but we note that there is no one unique set of criteria (see *Zwickl et al.*, 1983). The magnetic field magnitude |B| and the solar wind speed V_{SW} shown in Table 2 are values at the centers of the magnetic clouds. The average IMF B_S values and the duration for the B_S events (τ) are given in the 6th and 7th columns. The resultant magnetic storm D_{ST} values are given in the right-hand column.

Table 1. 1997 Magnetic Cloud Events

Event	Date (1997)	Cloud Interval (UT)	|B| (nT)	V_{sw} (km/s)	B_s (nT)	τ for B_s (hr)	$D_{ST\ min}$ (nT)
1	Jan 10–11	0440–2040	15	450	–12	4.5	–78
2	Feb 10	0245–1900	9	500	–7	7.0	–68
3	Apr 21–23	1205(4/21)–0700(4/23)	15	390	–10	>0.2	–107[1]
4	May 15–16	0945(5/15)–0100(5/16)	25	450	–25	4.0	–115
5	Jun 09	0200–2300	12	360	–8	2.2	–84[2]
6	Jul 15–16	0615(7/15)–0100(7/16)	13	350	–11	5.8	–45
7	Aug 03–04	1350(8/3)–0100(8/4)	15	450	–13	3.0	–49
8	Oct 01–02	1600(10/1)–2300(10/2)	10	460	–10+	>4.0+	–98
9	Oct 10–12	2300(10/10)–0000(10/12)	13	440	–10	4.0	–130[3]
10	Nov 07–08	0530(11/7)–1200(11/8)	15	450	–13	0.5	–110[4]
11	Nov 22–23	1850(11/22)–1800(11/23)	25	500	–12	3.0	–106

[1]Primed by IMF B_z = –10 nT for 2 hr.
[2]Primed by IMF B_z = –8 nT for 6.1 hr.
[3]Primed by IMF B_z = –11 nT for 2 hr.
[4]Primed by IMF B_z = –11 nT for >1.5 hr.
+There was a data gap.

The auroral forms were taken by the Polar UV imager using the LBH Long wavelength filter. The highest time resolution used was ~3 min, more than adequate to resolve the question of the presence or absence of substorm expansion phases. Due to Polar orbital restrictions, the entire main phases of each of the magnetic storms were not observed. Only images for parts of each event were available. However, in most cases substantial lengths of time (hours) were available, so definitive statements about the detection or lack of detection of classical substorm expansion phases during magnetic storm main phases can be made.

Classical substorm expansion phases have been described by *Akasofu* (1964). Some of the salient features are: 1) a brightening of the equatorward-most arc in 0–5 min; 2) poleward, westward and eastward expansion of auroral forms in 5–10 min; 3) the maximum auroral expansion ends after 10–30 min; and 4) arcs reform and drift back to their quiescent latitudes in 30 min – 1 hr. During intervals of strong activity, the substorm expansion phases (steps 1 through 3) can take place in 5 to 10 min (*Akasofu*, 1964). It will be this "classic" substorm expansion phase sequence that will be used to identify substorms.

2.1. Substorm Expansion Phase Examples

An example of two substorm expansion phases that occurred during a magnetic storm that was associated with a magnetic cloud passage are shown in Plate 1. These events occurred on November 22–23 (storm event 11 in Table 1). The temporal resolution is ~6 min. In each image, noon is at the top and dawn at the right. The first substorm expansion phase onset event is visible at 0839:42 UT. An arrow is placed between that image and the previous image indicating that the onset occurred somewhere between the two. In the 0839:42 UT image there is a sudden brightening (red color) from 23 to 03 MLT (magnetic local time) at latitudes between 60° and 68° MLAT (magnetic latitude). In the next image, at 0845:50 UT, the intense aurora has expanded westward (to 2230 MLT), poleward (to 71° MLAT), and eastward (to 0330 MLT). The expansion phase is completed by 0851:58 UT, giving a duration of ~12 min.

The second substorm expansion phase onset occurs between 0922:38 and 0928:46 UT. An arrow again indicates the onset time. At 0959:26 UT, the auroral forms reached a maximum in spatial extent, and spanned from ~21 to 05 MLT and from 60° to 75° MLAT. The duration of this expansion phase was ~30 min.

The interplanetary magnetic field and plasma data and the D_{ST} and AE indices for the substorm events are given in Plate 2. The Polar UV imaging data interval shown in Plate 1 is shaded in Plate 2 for reference. The solar wind data (top 7 panels) have been time shifted (by 34 min) to subtract the convection time from the upstream Wind spacecraft to the nose of the Earth's magnetopause. The UT at Wind is shown at the top. The AE and D_{ST} indices are given in the bottom two panels. Ground UT is given at the bottom. The substorm expansion phase onsets are indicated by arrows in the Plate (the times are ground UT). Examining the second panel, the first substorm expansion phase which began between 0833:34 and 0839:45 UT (labeled ~0836 UT in the Plate) could have been triggered either by the sharp northward turning of the IMF B_z (at 0802 UT at Wind) or by

Table 2. The 5 magnetic storms without observed substorm expansion phases

Event	V_{sw} (km/s)	B_s (nT)	E_{sw} (mV/m)	τ (hr)	N (cm^{-3})	G-T Intp. Criteria	D_{ST} (nT)	D^*_{ST} (nT)
1	450	11–15	>5.4	4.5	7	Y	–78	–82
2	425	7	3.0	7.0	<1	N	–68	–57
5+	360	8	3.1	3.3	3	N	–84	–78
6	360	11	4.0	5.8	8	N	–45	–45
7	470	13	6.1	3.0	5	Y	–49	–50

+Precursor IMF B_S for this storm.

a sharp southward turning of B_z and a simultaneous ram pressure decrease (top panel) that occurred ten minutes later. The second substorm expansion phase onset at ~0925 UT has no obvious interplanetary trigger.

The first substorm occurred after the IMF B_z had slowly increased from –15 nT to ~0 nT (from ~0100 UT to 0836 UT). The corresponding D_{ST} value was increasing, indicating the storm recovery phase (*Gonzalez et al.* 1994 noted an hour lag of D_{ST}, apparent here as well). The southward turning of B_Z (at 0812 UT at Wind) just after the substorm was a somewhat unusual feature of the magnetic cloud. This rotation in the field towards a southwardly direction altered the storm recovery, and led to a second D_{ST} intensification with a delay time of ~1 hr (shown in bottom panel). The second substorm occurred after the IMF southward turning.

2.2. Magnetic Storms Without Substorm Expansion Phases

An example of a magnetic storm without classical substorm expansion phases is shown in Figure 1 (event 6 in Table 1). The top 7 panels have been shifted by 23 min to take out the solar wind convection delay. The magnetic cloud is indicated in panel 2 by the horizontal bar. The available UV imaging data interval is indicated by shading. The bottom two panels give the AE and the D_{ST} indices. Continuous imaging is available during 6 hr of the storm's main phase.

The interval with UV imaging available (~1310 to 1937 UT) occurs during the magnetic storm main phase. D_{ST} (bottom panel) decreased from ~ –12 to –42 nT during the 6 hr interval. The interplanetary magnetic field B_Z component was relatively steady, in a range of –9 to –11 nT except at the very end of the interval where the field increased to B_Z ~ –4 nT.

The AE index was ~650 nT at 1310 UT, rose to a peak value of ~1300 nT at ~1445 UT (corresponding to the peak IMF B_S), and then smoothly decreased back down to ~500 nT at the end of the interval. Individual ground magnetograms exhibited similar (smooth) features (not shown). There were no ground stations that recorded strong H component time variations, as one would expect for substorm expansions.

Plate 3 shows auroral images for the entire 6 hr interval at a ~12 min cadence. Although 3 min time resolution images were available, only every fourth image is shown for brevity (the ~3 min resolution images were examined throughout the entire ~6 hr interval to search for evidence of substorm expansion phases. None were found). The images from 1332:15 to 1522:02 UT (at the IMF B_s peak) show a double band of aurora similar to those reported by Elphinstone (1995a, b). In the 19 to 24 MLT sector, one band was centered at ~62°–65° MLAT and the other centered at ~70°–72° MLAT. A good example of these double bands is found in the image of 1511:00 UT. The double bands exist primarily in the premidnight sector. The bands are not continuous, but are composed of "patches" that are typically smaller than ~1 hr MLT and have a poleward extent of ~5° MLAT. The patches last for ~3–6 min. When a patch from the equatorward band extends into the poleward band, these auroras have the form of "torches", noted earlier by *Akasofu* (1974).

Similar double auroral bands during substorm recovery phases have been reported by *Elphinstone* (1995a, b). *Sergeev et al.* (1999) using Polar UV images and *Sandholt and Farrugia* (2001) using high spatial and temporal resolution ground based optical observations, have noted that these "auroral streamers" (same as the *Akasofu* [1974] "torches") first appear as a bright spot in the poleward part of the double oval and then rapidly propagate down towards the lower branch. Both sets of authors have correlated these events to energetic electron injections and argue that the auroral patches are ionospheric manifestations of bursty bulk flows in the plasma sheet (*Angelopoulos et al.*, 1992, 1997).

From 1522:02 UT onward, the aurora became quieter and primarily single-banded. From 1544:07 to 1754:42 UT the post-midnight aurora was brighter than the premidnight aurora. There was a general brightness minimum at local midnight (see the 1713:40 UT image). An auroral loop is found in the postmidnight sector at 1735:44 UT.

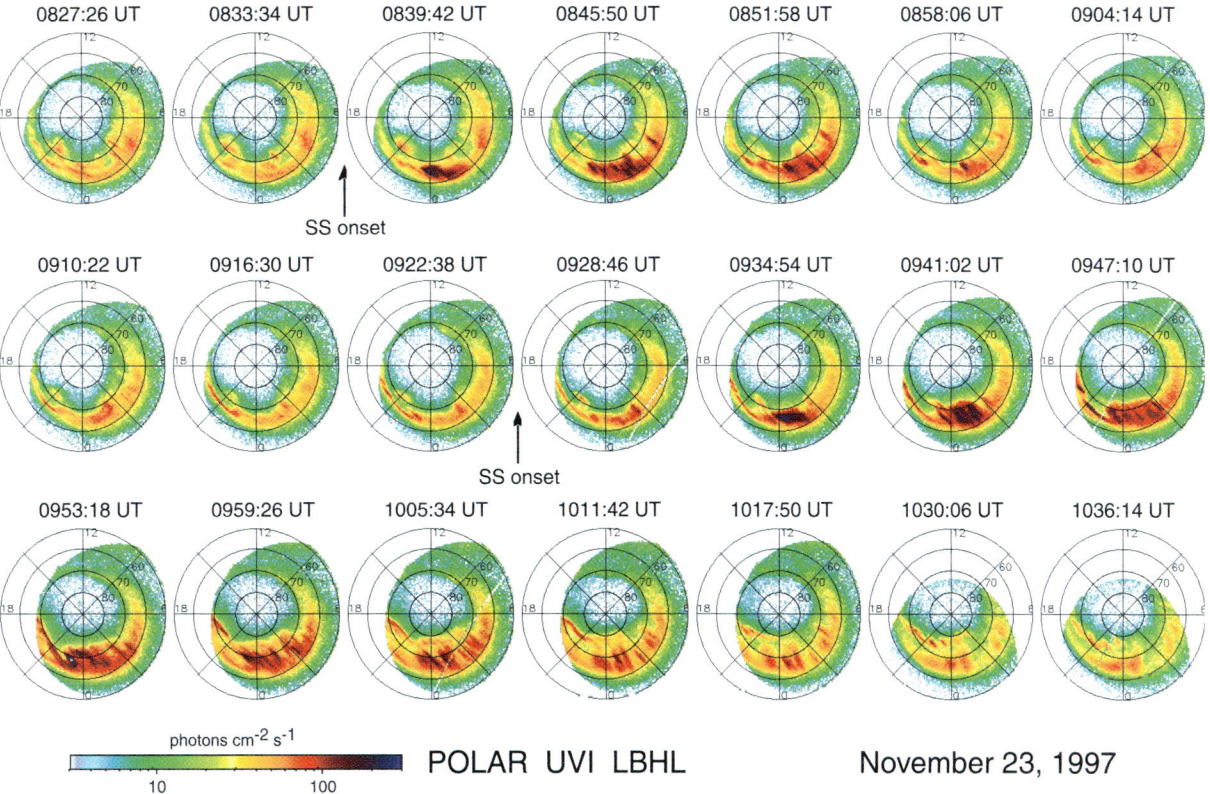

Plate 1. Two substorm expansion phases detected in the Polar UV global images. The events occurred on November 23, 1997. The substorms follow the Akasofu (1964) expansion sequence. The event onsets are indicated by black arrows.

28 A LACK OF SUBSTORM EXPANSION PHASES

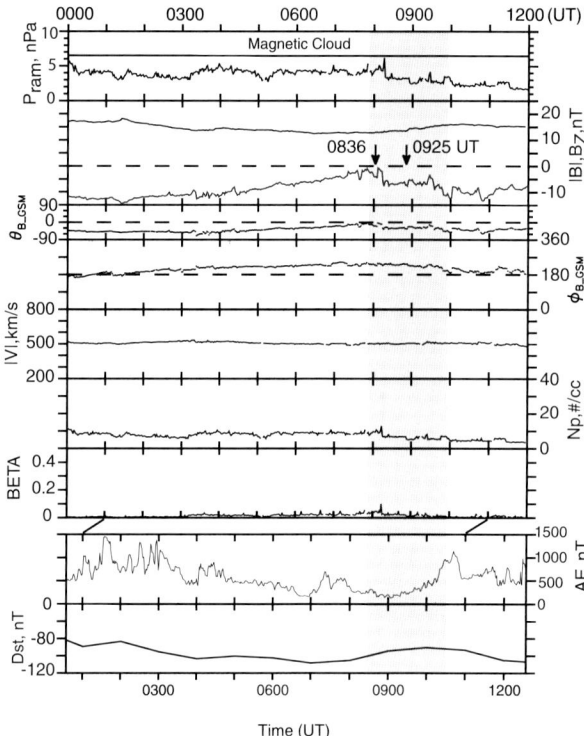

Plate 2. Interplanetary parameters (taken from Wind) and resultant geomagnetic activity. The times of two substorms shown in Plate 1 are indicated. The substorms cannot be identified in the AE index even though the ground coverage was "adequate" (10 stations AE indices with station coverage in the local time sector).

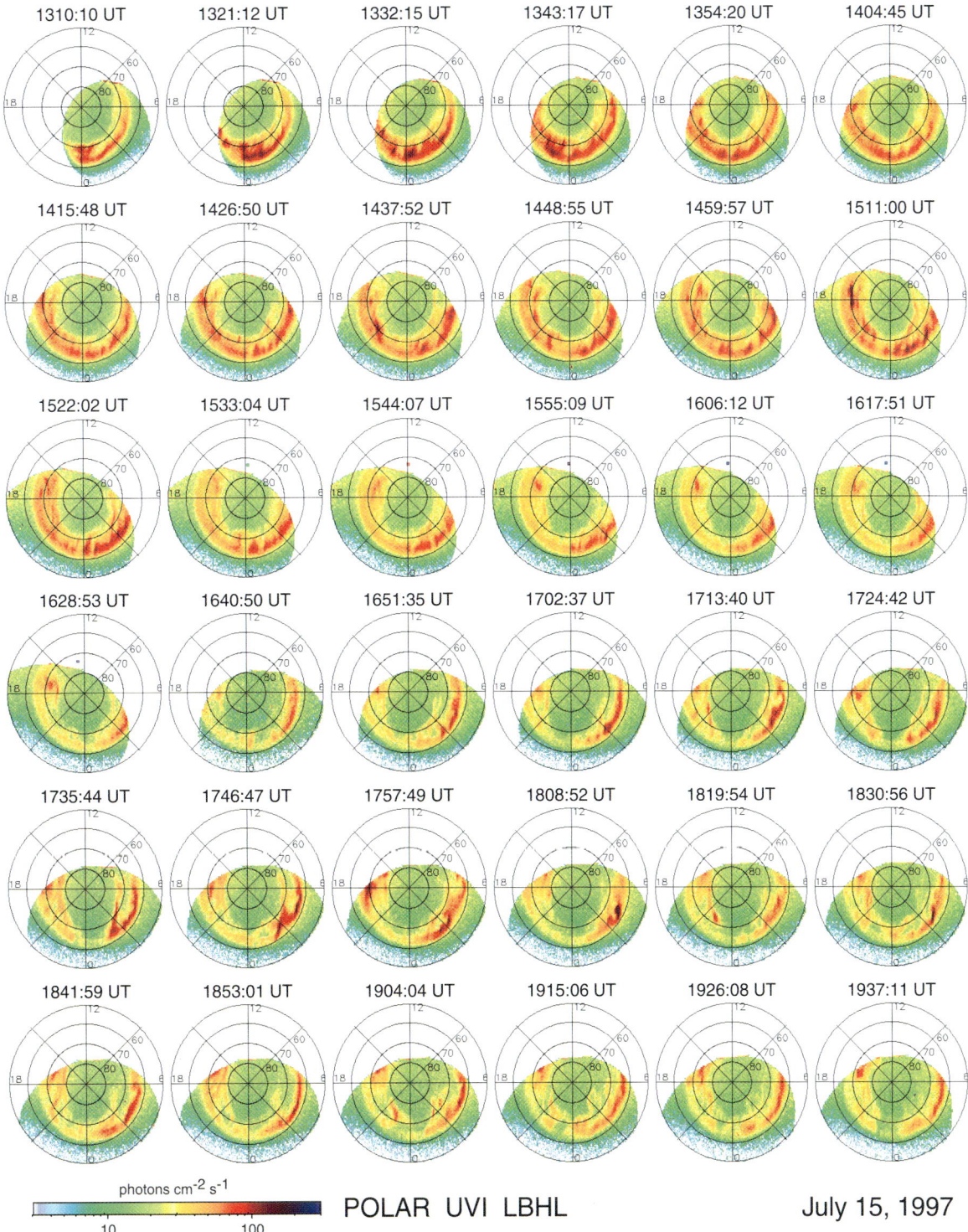

Plate 3. Imaging data for the shaded region of Figure 1. The cadence is ~12 min. There are no identifiable substorm expansion phases during the ~6 hr interval.

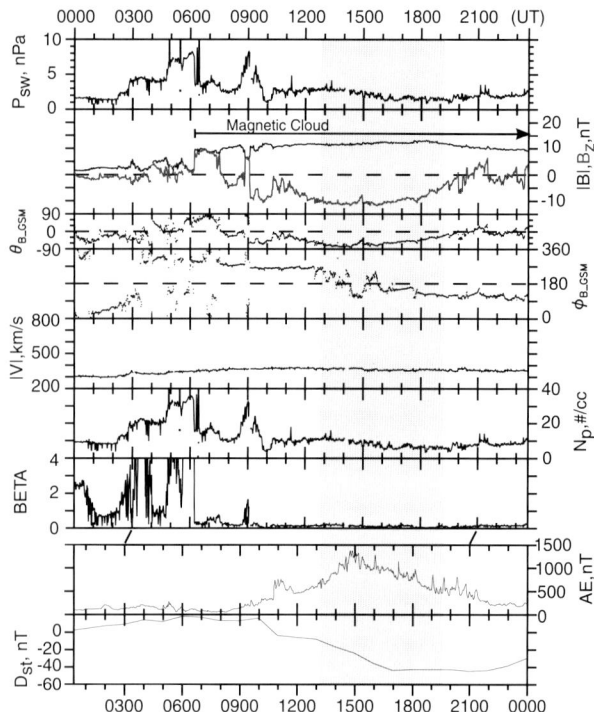

Figure 1. A magnetic cloud event causing a magnetic storm on July 15–16, 1997. The interval when Polar UV images are available is designated by shading. This interval is the majority of the storm main phase. The AE index gradually increases and decreases during the storm main phase. There are no substorm expansion phases within the ~ 6 hr of imaging data.

Figure 2 is another example of a magnetic storm caused by magnetic cloud B_s fields. This event occurred on February 10, 1997 (event 2 in Table 1). The solar wind data (top 7 panels) have been time-shifted by 40 min to remove the convection delay. The available Polar UV imaging data is again indicated by the shading. Imaging is available for 5 hr during the storm main phase (and the start of the recovery phase). The peak interplanetary magnetic field strength of ~8 nT and peak B_s (of 8 nT) occurred at ~1100 UT, close to the interval of peak D_{ST} (1100–1130 UT).

In the shaded interval, the magnetic storm D_{ST} intensity decreased from ~ –55 to –73 nT. The IMF B_Z was in a range from ~ –5 to –8 nT. The IMF B_Z profile was smooth with very little variations. The AE profile was relatively smooth. Individual ground magnetograms were studied for the interval (not shown). Substorm expansion phases were not detected.

Auroral images for the entire shaded interval of Figure 2 are shown in Plate 4. Images every ~6 min are given (3 min data are available, but not shown to save space). At 0722:28 UT, the brightest UV aurora occurred at premidnight local times at ~70° MLAT from 17 to 23 MLT. At 0753:08 UT, an intense bright spot appeared at ~70° MLAT and ~21 MLT. With time, the spot expanded equatorward until it spanned from 65° to 72° MLAT at 0829:56 UT. The spot disappeared by 0848:20 UT and there were patchy auroras from 18 to 05 MLT at ~65° MLAT. At 0909:48 UT, the aurora again changed in a subtle way. A new "spot" formed at ~70° MLAT at ~19 MLT. By 1023:24 UT the aurora again became similar to that at the beginning of this sequence: bright, high latitude aurora in the premidnight sector (~15 – 20 MLT at ~70° MLAT) and a fainter aurora from ~21 MLT through ~06 MLT at lower latitudes (~60°–70° MLAT). The "break" between these two types of aurora forms occurred at ~21 MLT. This break is clearly visible in the 1023:24 UT image.

Poleward-aligned, small patchy forms (torches, auroral streamers) are again found in the midnight sector. Examples can be seen in the images at 0854:28 UT and 1203:59 UT. These are similar with (but less apparent than) the features shown in Plate 3.

2.3. Polar Wobble Effects

The Polar spacecraft wobbles ~0.5° in the direction perpendicular to the orbit plane (J. Spann, personal comm., 2001). For the July 15, 1997 event (Plate 3), from 1300 to 1900 UT, the distance between the Polar spacecraft and the ionosphere was ~4.5 to 8 R_E, respectively. The orbit plane was approximately in the GSM Y-Z plane. Thus the maximum smearing error would be ~±220 km (~2.2°) along the noon-midnight direction when Polar was near apogee at ~1900 UT. The wobble error was significantly less, ~±120 km at ~ 1300 UT. Thus the wobble effect is too small to cause elongation of the aurora in the noon-midnight direction for the double-banded aurora (separated by ~5°). However, the wobble effect can be noted in the 1904:04 UT image (in Plate 3). In the area near ~03 MLT and ~72° MLAT, there are small spots that have been elongated into small bars. For the Feb 10, 1997 event (Plate 4), the smearing error values were approximately the same, but the smearing direction was along the 09–21 MLT direction.

The north-south patches spanning the double bands of aurora shown in Plates 3 and 4 are real features and cannot be due to spacecraft wobble effects. The extent of these features (~5° or ~500 km) are much larger than any effects that wobbling may produce. Additionally, the double banded aurora would be totally smeared out by such an effect.

2.4. Interplanetary Parameters and Magnetic Storm Intensities

Table 1 gives the interplanetary parameters and the peak D_{ST} values for the 11 storms studied. The principal quantities for the magnetic merging process are the interplanetary B_S and V_{SW} values (*Dungey*, 1961). The duration τ of the B_S event is also listed. These 1997 magnetic storms can be tested against the *Gonzalez and Tsurutani* (1987) criteria (hereafter called G-T) which was empirically developed during another solar cycle.

Of the 5 magnetic storms with D_{ST} peak values less than –100 nT (events 3, 4, 9, 10 and 11), events 4, 10 and 11 satisfy the G-T interplanetary criteria: $V_{SW} B_S > 5$ mV/m and $\tau > 3$ hr. Event 3 had a data gap (not all of the interplanetary data for the event were available). Event 9 is in clear violation of the G-T criteria, being more intense than expected. The event is noted to have precursor IMF B_S "priming". This may lead to the greater storm intensity (but is beyond the scope of the present paper).

In Table 2, we show the 5 magnetic storms that did not exhibit substorm expansion phases for the intervals where UV imaging data were available. The value $V_{SW} B_S$ (E_{SW}) is given in the 4th column. The density values which are used in the calculations for the pressure corrections are shown in the 6th column. The (ram) pressure corrected D_{ST} values, D^*_{ST}, is given in the right-hand column. The D^*_{ST} values are not significantly different from the D_{ST} values. Two events, events 1 and 7 have large E_{SW} values (>5.4 and 6.1 mV/m, respectively) and large τ values (4.5 and 3.0 hr, respectively), satisfying interplanetary conditions for the G-T criteria. However the corresponding magnetic storm intensities ($D_{ST}^* = –82$ and $–50$ nT, respectively), were abnormally small.

3. DISCUSSION AND CONCLUSIONS

We have shown examples of long stretches of magnetic storm main phases where there was an absence of "classical" substorm expansion phases. These intervals occurred when the interplanetary magnetic field B_S component was either steady or changing slowly. This result has shown for the first time that there is a lack of substorm expansion phases during steady IMF B_S intervals (predicted by *Kamide*, 2001).

Based on previous reports of substorm expansion phase triggerings due to sudden IMF B_Z turnings and ram pressure pulses (see earlier references), we have constructed an aggregate model of substorm occurrence as a function of magnetic cloud features. Figure 3 shows a magnetic cloud field profile (top three panels) and the magnetospheric/ionospheric response (bottom two panels). In this paper we have only explored region 2 where the IMF B_Z is relatively steady or its change with time (dB_Z/dt) is small. We show a lack of substorms in this region (even though AE values could be high). The remainder of substorm expansion phases in this Figure are based on arguments presented in the substorm literature.

The interplanetary field magnitude, B_Z component and ram pressure changes in magnetic clouds are shown the top three panels (this particular example is similar to an event published by *Lepping et al.*, 1997). The sheath fields antisunward of the cloud (from the far left of the Figure to point 1) are compressed by the interplanetary shock (they are more intense than the upstream quiet fields) and are characterized by large amplitude fluctuations in B_Z (Alfvén waves). At the cloud boundary (point 1) there is a tangential discontinuity (*Tsurutani et al.*, 1988) when there is an abrupt increase in $|B|$, sharp change in B_Z and also a substantial ram pressure decrease. The dB_Z/dt value is the largest in this region. Region 2 of the cloud is characterized by slowly changing B_Z values, typical of the two events

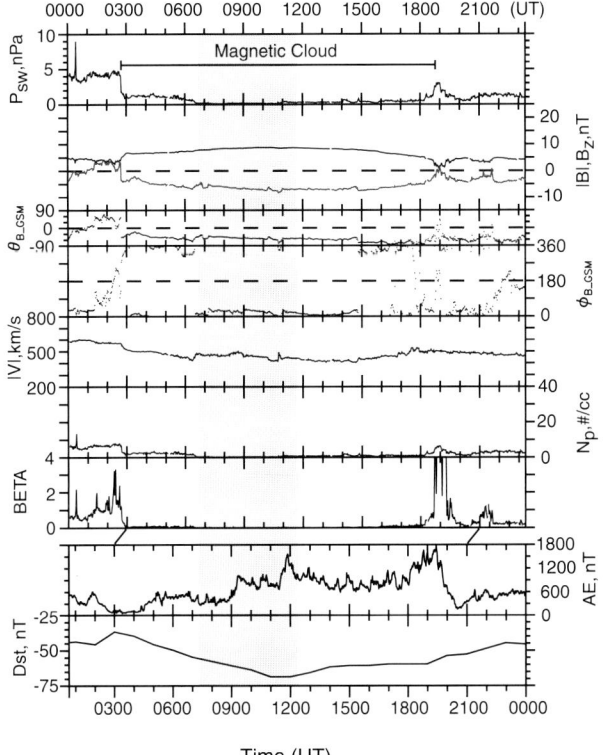

Figure 2. A storm event on February 10, 1997. This figure uses the same format as Figure 1.

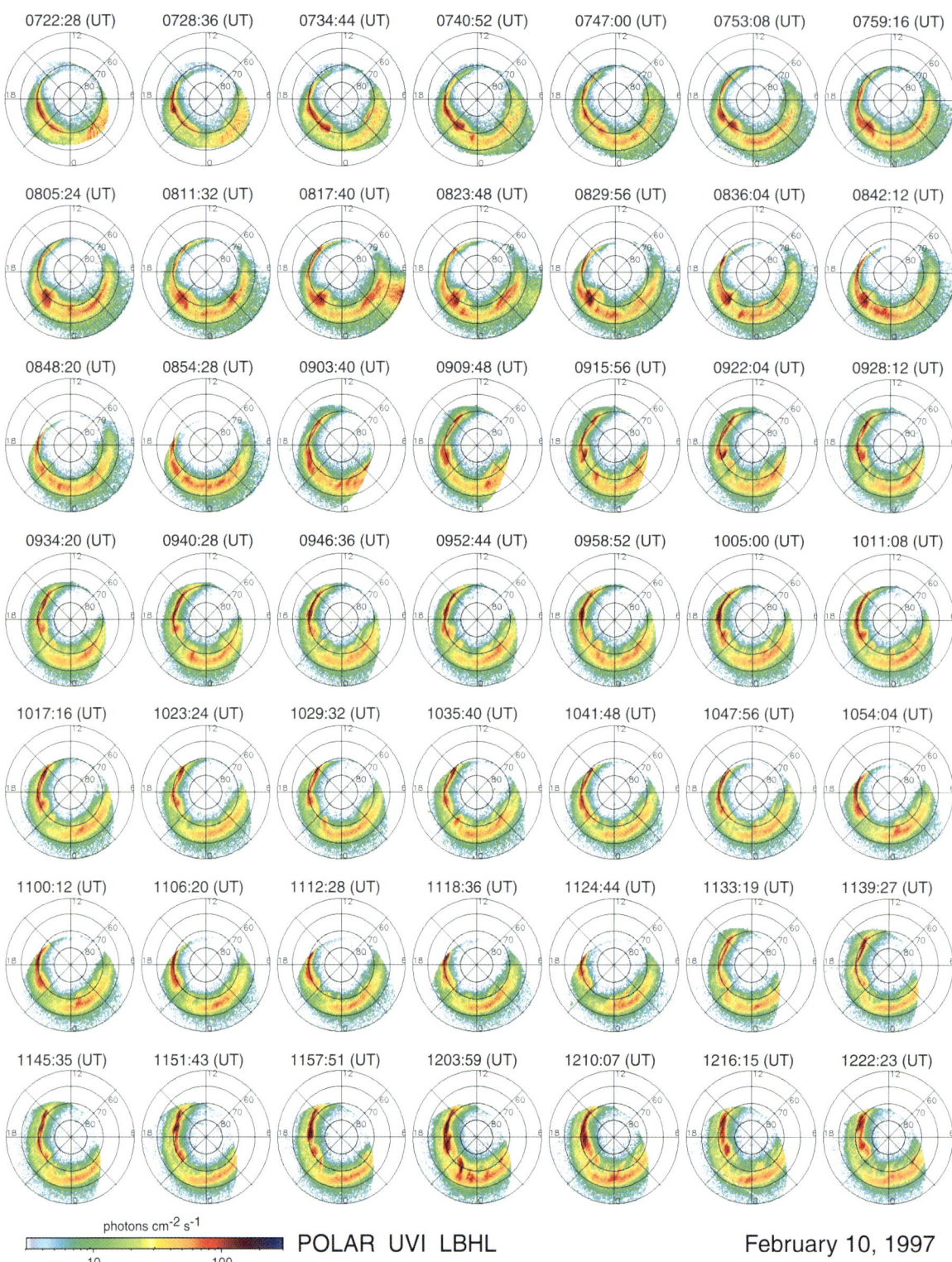

Plate 4. Polar UVI data for the shaded region of Figure 2. There are no identifiable substorm expansion phases during this ~6 hr interval.

shown in this paper. Region 3 is where the IMF rotates from south-to-north (or north-to-south). Region 4 is similar to region 2, but with the opposite polarity of magnetic fields. Region 5 can have sharp discontinuities like region 1 or it can evolve smoothly into the high speed medium. There is also an increase in the solar wind ram pressure at this location.

The magnetospheric/ionospheric responses to the various parts of the magnetic cloud are predicted to be as follows: the intense, highly fluctuating sheath B_S values will lead to many substorm expansion phases. The sheath will prime the magnetosphere/ionosphere plasma sheet system. At the outset of the magnetic cloud where there is a sharp IMF B_S (or B_N) turning (point 1), intense substorm expansion phases occur. Substorm expansion phases will not occur in region 2 where $dB_S/dt \sim 0$, but there may still be high AE indices (results of this paper) due to large amplitude convection bays. In region 3 when the IMF rotates from south-to-north, expansion phases would again be expected if the rate of rotation is "sufficiently fast". In region 4, magnetic quiet is expected when the IMF B_Z is northward and steady (*Tsurutani and Gonzalez,* 1995). In region 5 at the end of the magnetic cloud, the "predictions" are less clear. Because of the depletion of stored plasmasheet energy from the prior (IMF B_n) time interval, the IMF B_Z rotation from northward to zero may not lead to substorms. The details of this general scenario can and should be tested using multiple magnetic cloud events. There will be ample number of events in the 2001–2003 era.

4. FINAL COMMENTS

North-south elongated patches (torches and streamers) were a common features of the magnetic storm auroras studied here. The polewardly aligned patches had characteristics of: ~1 hr in MLT, ~5° in MLAT, with durations of 3 to 5 min. *Nakamura et al.* (1993) have discussed similar north-south structures during active periods. *Sergeev et al.* (1999) have shown that bright spots in the poleward part of the double oval migrate equatorward and sometimes connect to the equatorward oval. Finite spatial extent (< 1 hr in MLT), transient (1–2 min) ~100 keV particle injections have been detected at geosynchronous orbit in relation to the auroral spots (*Sergeev et al.,* 1999). These injections have been interpreted as bursty bulk flows (*Baumjohann et al,* 1990; *Angelopoulos et al.,* 1992; 1995). *Sandholt and Farrugia* (2001) have demonstrated a one-to-one correspondence between 5–20 keV electron injections and auroral events detected in the Polar UV images. The patches lasted ~2 min. Bursty bulk flows may be the main process of plasma injection during substantial portions of magnetic storms that are driven by magnetic clouds.

For the storm main phases without substorm expansive phases, ground-based magnetograms were characterized by intense, long duration magnetic convection bays. There were no signatures of substorm expansion phases within the bays. One interpretation of these cloud-driven storm events are that they are extremely intense convection bays (*Pytte et al.,* 1978; *Sergeev et al.,* 1996). *Pytte et al.* (1978) examined a weaker (than the cases presented here) interplanetary B_Z event ($B_S \approx 3$–4 nT) that was relatively constant for ~5 hr. The authors determined that during intense (~500 to 1000 nT) bays, weak and infrequent substorm onset signatures occurred. Our present results are similar with the exception that the cases presented here have much stronger interplanetary magnetic fields (B_S) and the resultant geomagnetic AE and D_{ST} indices are also more intense.

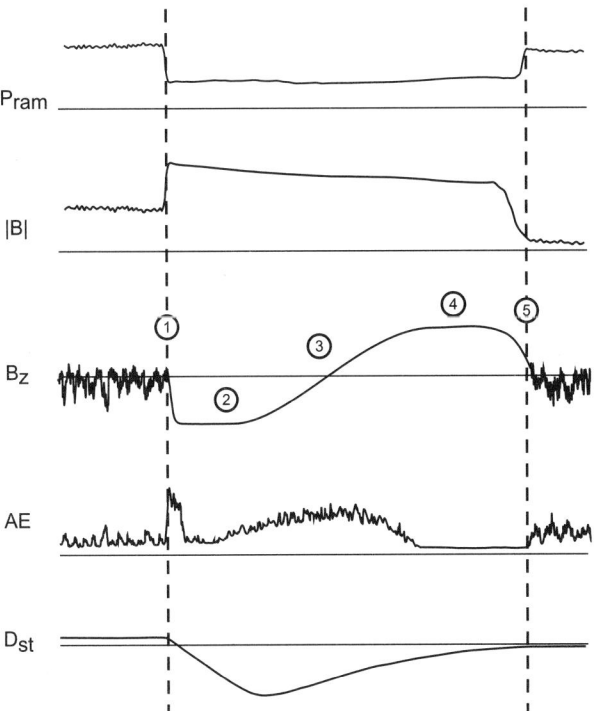

Figure 3. A schematic of a magnetic cloud and expected geomagnetic activity. Intervals where dB_Z/dt are sufficiently large are predicated to trigger substorm expansion phases (indicated by bursty AE indices). The purpose of this schematic is to provide a model to be used as a test for substorm triggering ideas.

Another result of this study is that magnetic storms that did not have detectable substorm expansion phases had substantially lower (less negative) intensities than those that did. Two events that met the G-T interplanetary criteria had peak D_{ST} values of only −82 and −50 nT. Three possible mechanisms for the low D_{ST} values come to mind. *Daglis et al.* (1994) have statistically shown a strong correlation between substorm expansion phase AE indices and near-Earth magnetotail O^+ ion energy densities. *Daglis and Axford* (1996), reviewing many ionospheric and magnetospheric ion observations, have concluded that the auroral ionosphere is capable of responding to solar wind energy input on substorm time scales, loading the near-Earth magnetotail with O^+ ions. Thus our scenario 1) is that a lack of substorm expansion phases leads to a lowering of oxygen ion densities in the plasmasheet and therefore a lowering of oxygen densities in the storm-time ring-current as well.

Scenario 2 emphasizes the role that substorm expansion phase "dipolarizations" play. Dipolarizations may cause significant plasma compression (*Delcourt*, 2002), leading to "preheating" of the plasma being injected into the ring current. Scenario 3 assumes the solar wind plasma is able to enter the near-Earth plasmasheet on rapid time scales (*Borovsky et al.*, 1998). If this occurs, the low plasma beta within magnetic clouds will lead to a low density plasmasheet and hence a lower density ring-current. One or more of these possibilities may be operative leading to the abnormally low magnetic storm peak D_{ST} values. We hope to investigate these possibilities in the near future.

Acknowledgements: We thank J.U. Kozyra for stimulating scientific conversations. Portions of this research were performed at the Jet Propulsion Laboratory, California Institute of Technology, under contract with National Aeronautics and Space Administration.

REFERENCES

Akasofu, S.-I., The development of the auroral substorm, *Planet. Space Sci., 12*, 273, 1964.

Akasofu, S.-I., *Polar and Magnetospheric Substorms*, D. Reidel, Dordrecht, 1968.

Akasofu, S.-I., A study of auroral display photographed from the DMSP-2 satellite and from the Alaska meridian chain of stations, *Space Sci. Rev., 16*, 617, 1974.

Angelopoulos, V., W. Baumjohanm, C.F. Kennel, et al., Bursty bulk flows in the inner central plasma sheet, *J. Geophys. Res., 97*, 4027, 1992.

Angelopoulos, V., T.D. Phan, D.E. Larson, et al., Magnetotail flow bursts: Association to global magnetospheric circulation, relationship to ionospheric activity and direct evidence for localization, *Geophys. Res. Lett., 24*, 2271, 1997.

Baumjohann, W., G. Paschmann, and H. Luhr, Characteristics of high-speed flows in the plasma sheet, *J. Geophys. Res., 95*, 3801, 1990.

Borovsky, J.E., M.F. Thomsen, and R.C. Elphic, The driving of the plasma sheet by the solar wind, *J. Geophys. Res., 103*, 17617, 1998.

Caan, M.N., R.L. McPherron, and C.T. Russell, Substorm and interplanetary magnetic field effects on the geomagnetic tail lobes, *J. Geophys. Res., 80*, 191, 1975.

Daglis, I.A., S. Livi, E.T. Sarris, B. Wilken, Energy density of ionospheric and solar wind origin ions in the near-Earth magnetotail during substorms, *J. Geophys. Res., 99*, 5691, 1994.

Daglis, I.A., and W.I. Axford, Fast ionospheric response to enhanced activity in geospace: Ion feeding of the inner magnetotail, *J. Geophys. Res., 101*, 5047, 1996.

Daglis, I.A., Space storms, ring current and space-atmosphere coupling—Critical elements of space weather, in *Space storms and space weather hazards*, edited by I.A. Daglis, Kluwer Academic Publishers, Dordrecht, 2001.

Delcourt, D.C., Particle acceleration by inductive electric fields in the inner magnetosphere, *J. Atmos. Sol. Terr. Phys.*, in press 2002.

Elphinstone, R.D,., J.S. Murphree, D.J. Hearn, L.L. Cogger, I. Sandahl, P.T. Newell, D.M. Klumper, S. Ohtani, J.A. Sauvaud, T.A. Potemra, K. Mursula, A. Wright and M. Shapshak, The double oval UV auroral distribution 1. Implications for the mapping of auroras, *J. Geophys. Res., 100*, 12075, 1995a.

Elphinstone, R.D., D.J. Hearn, L.L. Cogger, J.S. Murphree, A. Wright, I. Sandahl, S. Ohtani, P.T. Newell, D.M. Klumpar, M. Shapshak, T.A. Potemra, K. Mursula, and J.A. Sauvaud, The double oval UV auroral distribution 2. The most poleward arc system and the dynamics of the magnetotail, *J. Geophys. Res.*, 12093, 1995b.

Farrugia, C.J., L.F. Burlaga and R.P. Lepping, Magnetic clouds and the quiet-storm effect at Earth, in *Magnetic Storms*, edited by B.T. Tsurutani, W.D. Gonzalez, Y. Kamide and J.K. Arballo, Amer. Geophys. U., Washington D.C., *98*, 91, 1997.

Forsyth, R.J. and J.T. Gosling, Corotating and transient structures in the heliosphere, in *The Heliosphere Near Solar Minimum: The Ulysses Perspective,* edited by A. Balogh, R.G. Marsden and E.J. Smith, *Springer, 107*, 2001.

Gonzalez, W.D., and B.T. Tsurutani, Criteria of interplanetary parameters causing intense magnetic storms ($D_{ST} < -100$ nT), *Planet. Space Sci., 35,* 1101, 1987.

Gonzalez, W.D., J.A. Joselyn, Y. Kamide, H.W. Kroehl, G. Rostoker, B.T. Tsurutani and V.M. Vasyliunas, What is a geomagnetic storm?, *J. Geophys. Res., 99*, 5771, 1994.

Gosling, J.T., D.J. McComas, J.L. Phillips, and S.J. Bame, Geomagnetic activity associated with Earth passage of interplanetary shock disturbances and coronal mass ejections, *J. Geophys. Res., 96*, 7831, 1991.

Huttunen, K.E.J., H.E.J. Koskinen and R.S. Schwenn, Variability of magnetospheric storms driven by different solar wind perturbations, to appear in *J. Geophys. Res.*, 2002.

Iyemori, T., Time delay of the substorm onset from the IMF southward turning, *J. Geomag. Geoelectr., 32*, 267, 1980.

Iyemori, T. and D.R.K. Rao, Decay of the D_{ST} field of geomagnetic disturbance after substorm onset and its implications for storm-substorm relation, *Annales Geophysicae, 19*, 608, 1996.

Kamide, Y., N. Yokoyama, W. Gonzalez, B.T. Tsurutani, I.A. Daglis, A. Brekke and S. Masuda, Two-step development of geomagnetic storms, *J. Geophys. Res., 103*, 6917, 1998a.

Kamide, Y., W. Baumjohann, I.A. Daglis, W.D. Gonzalez, M. Grande, J.A. Joselyn, R.L. McPherron, J.L. Phillips, E.G.D. Reeves, G. Rostoker, A. Sharma, H. Singer, B.T. Tsurutani, and V.M. Vasyliunas, Current understanding of magnetic storms: Storm-substorm relationships, *J. Geophys. Res., 103*, 17705, 1998b.

Kamide, Y., Interplanetary and magnetospheric electric fields during geomagnetic storms: what is more important, steady-static fields or fluctuating fields?, *J. Atmos. Sol.-Terr. Phys., 63*, 413, 2001.

Klein, L.W. and L.F. Burlaga, Interplanetary magnetic clouds at 1 AU, *J. Geophys. Res., 87*, 613, 1982.

Lepping, R.P., M.H., Acuna, L.F. Burlaga, W.M. Farrell, J.A. Slavin, K.H. Schatten, F. Mariani, N.F. Ness, F.M. Neubauer, Y.C. Whang, J.B. Byrnes, R.S. Kennon, P.V. Panetta, J. Scheiffle, and E.M. Worley, The WIND magnetic field investigation, *Space Sci. Rev., 71*, 207, 1995.

Lepping, R.P., L.F. Burlaga, A. Szabo, K.W. Ogilvie, W.H. Mish, D. Vassiliadis, A.J. Lazarus, J.T. Steinberg, C.J. Farrugia, L. Janoo, and F. Mariani, The Wind magnetic cloud and events of October 18-20, 1995: Interplanetary properties and as triggers for geomagnetic activity, *J. Geophys. Res., 102*, 14,049, 1997.

Lyons, L.R., G.T. Blanchard, J.C. Samson, R.P. Lepping, T. Yamamoto, and T. Moretto, Coordinated observations demonstrating external substorm triggering, *J. Geophys. Res., 102*, 27039, 1997.

McPherron, R.L., T. Terasawa and A. Nishida, Solar wind triggering of substorm onset, *J. Geomag. Geoelectr., 38*, 1089, 1986.

McPherron, R.L., The role of substorms in the generation of magnetic storms, in *Magnetic Storm*, edited by B. Tsurutani et al., AGU Monograph, *98*, 131, 1997.

Meng, C.I., B.T. Tsurutani, K. Kawasaki, and S.-I. Akasofu, Cross-correlation analysis of the AE index and interplanetary magnetic field B_Z component, *J. Geophys. Res., 78*, 617, 1973.

Nakamura, R., T. Oguti, T. Yamamoto, and S. Kokubun, Equatorward and poleward expansion of the auroras during auroral substorms, *J. Geophys. Res., 98*, 5743, 1993.

Ogilvie, K.W., D.J. Chornay, R.J. Fritzenreiter, F. Hunsaker, J. Keller, J. Lobell, G. Miller, J.D. Scudder, E.C. Sittler, Jr., R.B. Torbert, D. Bodet, G. Needell, A.J. Lazarus, J.T. Steinberg, J.H. Tappan, A. Mavertic, and E. Gergin, SWE, A comprehensive plasma instrument for the WIND spacecraft, *Space Sci. Rev., 71*, 55, 1995.

Pytte, T., R.L. McPherron, E.W. Hones, Jr., and H.I. West, Jr., Multiple-satellite studies of magnetospheric substorms: Distinction between polar magnetic substorms and convection driven negative bays, *J. Geophys. Res., 83*, 663, 1978.

Richardson, I.G., E.W. Cliver, and H.V. Cane, Sources of geomagnetic storms for solar minimum and maximum conditions during 1972-2000, *Geophys. Res., Lett., 28*, 2569, 2001.

Rostoker, G., M. Mareschal, and J.C. Samson, Response of dayside net downward field-aligned current to changes in the interplanetary magnetic field and to substorm perturbation, *J. Geophys. Res., 87*, 3489, 1982.

Rostoker, G., E. Friedrich, and M. Dobbs, Physics of magnetic storms in *Magnetic Storm*, edited by B. Tsurutani et al., AGU Monograph, *98*, 149, 1997.

Rostoker, G., Effects of substorms on the stormtime ring-current index D_{ST}, *Annales Geophysicae, 18*, 1390, 2000.

Samson, J.C. and K.L. Yeung, Some generalizations on the method of superposed epoch analysis, *Planet. Space. Sci., 34*, 1133, 1986.

Sandholt, P.E., and C.I. Farrugia, Multipoint observations of substorm intensifications: The high-latitude aurora and electron injections in the inner equatorial plasmasheet, *Geophys. Res. Lett., 28*, 483, 2001.

Sergeev, V.A., R.J. Pellinen and T.I. Pulkkinen, Steady magnetospheric convection: A review of recent results, *Space Sci. Rev., 75*, 551, 1996.

Sergeev, V.A., L.I. Vagina, K. Kauristie, H. Koskinen, A. Huuskonen, A. Pajunpaa, R. Pellinen, T. Phan, V. Angelopoulos, R.P. Lin, R.P. Lepping, G.D. Reeves, and P.T. Newell, Continuous activity and substorm activations during a weak magnetic storm (WIND tail passage), SUBSTORMS-4, edited by S. Kokubun and Y. Kamide, *Terra. Sci. Publ. Co., 681*, 1998.

Sergeev, V.A., K. Liou, C.-I. Meng, P.T. Newell, M. Brittnacher, G. Parks, and G.D. Reeves., Development of auroral streamers in association with localized injections to the inner magnetotail, *Geophys. Res., Lett., 26*, 417, 1999.

Smith, E.J. and J.H. Wolfe, Observations of interaction regions and corotating shocks between one and five AU: Pioneers 10 and 11, *Geophys. Res. Letts., 2*, 137, 1976.

Tsurutani, B.T. and C.-I. Meng, Interplanetary magnetic-field variations and substorm activity, *J. Geophys. Res., 77*, 2964, 1972.

Tsurutani, B.T. and W.D. Gonzalez, The cause of high intensity long-duration continuous AE activity (HILDCAAs): Interplanetary Alfvén trains, *Planet. Space Sci., 35*, 405, 1987.

Tsurutani, B.T., and W.D. Gonzalez, The efficiency of "viscous interaction" between the solar wind and the magnetosphere during intense northward IMF events, *Geophys. Res. Letts., 22*, 663, 1995.

Tsurutani, B.T., W.D. Gonzalez, F. Tang, S.-I. Akasofu, and E.J. Smith, Origin of interplanetary southward magnetic fields responsible for major magnetic storms near solar maximum (1978–1979), *J. Geophys. Res., 93*, 8519, 1988.

Tsurutani, B.T., W.D. Gonzalez, A.L.C. Gonzalez, F. Tang, J.K. Arballo and M. Okada, Interplanetary origin of geomagnetic activity in the declining phase of the solar cycle, *J. Geophys. Res., 100*, 21717, 1995.

Tsurutani, B.T. and W.D. Gonzalez, The interplanetary causes of magnetic storms: A review, in *Magnetic Storm*, edited by B. Tsurutani et al., AGU Monograph, *98*, 77, 1997.

Tsurutani, B.T., The interplanetary causes of magnetic storms, substorms and geomagnetic quiet, in *Space storms and space weather hazards*, edited by I.A. Daglis, 103, Kluwer Academic Publishers, Dordrecht, 2001.

Zwickl, R.D., J.R. Ashbridge, S.J. Bame, W.C. Feldman, J.T. Gosling, and E.J. Smith, Plasma properties of driver gas following interplanetary shocks observed by ISEE-3, in *Solar Wind Five*, NASA Conf. Publ. CP-2280, 711, 1983.

Jet Propulsion Laboratory, 4800 Oak Grove Drive, Pasadena, CA 91109.

INPE-CEA/DGE, CAIXA POSTAL 515, 12201-970 S.J.DOS CAMPOS, SP, Brazil.

Storm-time and Quiet-time Substorms in the Magnetotail

A. A. Petrukovich

Space Research Institute, Russian Academy of Sciences, Moscow, Russia

Analysis of the storm-time magnetotail data reveals distinct signatures of the substorm loading-unloading sequence similar to that of quiet-time substorms, except larger amplitudes of the lobe magnetic pressure increase and the nominal solar wind electric field input. Substorm intervals are intermittent with episodes of quasi-stationary enhanced convection. The difference between these large substorms and smaller quiet-time events measured in terms of accumulated magnetic flux can reach an order of magnitude. Therefore, a substorm onset likely cannot be directly associated with some universal critical level of magnetic flux or energy input. The SYM-H index, which due to higher than D_{st} time resolution is capable to resolve the effect of substorm phases on the ring current, actually readily reacts to IMF changes only. Only short-living and small partial recoveries of SYM-H are observed at substorm expansions. These findings suggest primarily convective mechanism of the ring current control.

1. INTRODUCTION

Since the first detailed descriptions of storm and substorm phenomena [*McPherron et al.*, 1973; *Akasofu et al.*, 1973; *Perreault and Akasofu*, 1978] the problem of storm-substorm relationship emerged on the agenda of magnetospheric physics, being inspired partially by the verbal form of the term "substorm" itself. It is frequently stated that the storm effect on the magnetosphere can be described as a superposition of a series of substorm effects, which complement to the ring current (measured as the geomagnetic index D_{st} or SYM-H) by injections of plasma sheet particles. The alternative point of view emphasizes the role of the IMF-driven convection in the ring current formation, while the magnetotail substorm activity is considered as a mere coincidence. This topic is described in more detail by *Gonzalez et al.* [1994] and *Kamide et al.* [1998].

It should be noted that currently the term "geomagnetic storm" is associated with extreme global disturbances of the magnetosphere and has very general sense, while the "geomagnetic substorm" notion is associated with the specific type of polar geomagnetic and magnetotail patterns. Namely, substorms exhibit the clear cyclic behavior with growth, expansion and recovery phases, which are driven by loading and unloading of solar magnetic energy [*Baker et al.*, 1996]. Substorms are distinct from other frequent types of polar activity such as quasi-stationary enhanced convection [*Sergeev et al.*, 2001] and pseudo-breakup.

To understand the proper role of substorms during a geomagnetic storm it is important to find what makes storm-time substorms more effective in generation of the ring current (inner magnetosphere) changes. In some papers magnetotail pressure changes and, hence, open magnetic flux reconnection were attributed only to storm-time substorms [*Kamide et al.*, 1998]. Another key to the problem of the ring current driver is the accurate timing of magnetotail and ring current phenomena. Now researchers have the advantage to use the 1-min resolution SYM-H index, which might help to resolve differences between growth (when convection effects dominate) and expansion (when substorm injection dominates) substorm phases. The hourly D_{st}, routinely used for such studies previously, allows one to study only the overall substorm effect.

Investigation of storm-time substorms might be also important for the substorm science proper. Analysis of the magnetotail dynamics under extreme solar wind loading might be used to reveal plasma sheet stability criteria and discriminate between the models of a substorm onset.

2. THE DATA SET

With the formation of comprehensive ISTP and associated databases including a wealth of data taken in the key magnetospheric regions and on the ground, detailed studies of many interesting problems became feasible. The current investigation is an extension of the previous analyses of substorms during contracted oval [*Petrukovich et al.*, 2000] and normal conditions [*Petrukovich*, 2000] with the addition of the storm-time substorm statistics. In such an approach the magnetic flux accumulation during a growth phase and its disposal during an expansion phase (observed as the lobe magnetic pressure maximum and the plasma sheet thinning [*McPherron*, 1972; *McPherron et al.*, 1973; *Baker et al.*, 1996]) are emphasized as general magnetotail signatures, common to substorms of any magnitude. Our statistics includes 66 contracted-oval or small substorms, 23 normal-size substorms, 11 large substorms and 8 storm-time substorms. Here, the small events were selected as cases with the auroral zone magnetic bays within –300 nT and the large events were selected as cases with the IMF B_z less than –5 nT (hereafter GSM frame of reference always used). Storm-time events were selected when D_{st} was below –50 nT. From the five-year 1995–1999 period only sufficiently isolated substorms with sufficient ground and magnetotail data coverage were chosen.

During contracted-oval conditions magnetotail pressure increase was often marginal [*Petrukovich et al.*, 2000], so that it was easier to locate a substorm initially with the use of ground and auroral data. For larger-size geomagnetic disturbances, the magnetotail pressure profile (with some important caveats [*Petrukovich et al.*, 1999]) was already a reliable signature of an ongoing substorm, clearly discriminating it from, e.g., enhanced convection periods, even if ground magnetic or optical data were absent or insufficient. During storms, a substorm expansion phase onset in ground magnetograms was often obscured by intense pseudo-breakups or convection bays often reaching some hundred nT in magnitude. Nevertheless, it turned out that individual substorms can be reliably identified as the magnetic pressure maxima in the lobe. However, it should be noted that almost all 1995–1999 geomagnetic storms with D_{st} lower than -100 nT happened to be not covered by reliable magnetotail measurements, so that analyzed storm-time substorms commenced during rather mild conditions.

3. EVENT DESCRIPTIONS

3.1. Event 24 November, 1996

We start with the illustration of a typical substorm, the famous event of 24 November, 1996 (Figure 1) (see also *Petrukovich et al.* [1999]). The Interball-Tail satellite in the tail lobe (coordinates were -26.4, 1.3, 9.3 R_E) detected gradual growth of the magnetic pressure starting with the IMF southward turning at 21:07 UT. Hereafter Wind (or ACE) IMF and solar wind data were shifted forward in time to allow for the flow propagation time from the satellite position to Earth. Such changes of lobe magnetic field are indicative of the accumulation of the open magnetic flux in the Earth's tail during a substorm growth phase. The concurrent plasma sheet thinning (decrease of the plasma sheet B_z) was observed by Geotail (at –25.2, –7.9, –3.2 R_E). The onset of the expansion phase according to ground magnetograms happened at about 22:27 UT, almost simultaneously with the magnetotail lobe pressure maximum and the start of the pressure decrease. The tail returned to the pre-substorm state 30 minutes after the onset.

While the more detailed descriptions of this event can be found elsewhere, here we would like to compare these data with the profile of the ring current SYM-H index (bottom panel of Figure 1). The SYM-H clearly reacted on the switch-on of the southward IMF and reached its minimum (–24 nT) at 22:45 UT, i.e., with the northward IMF turn. No significant changes of SYM-H were observed with the substorm onset at 22:25–22:30 UT. It is interesting to note the ~15 min delay between the moment of IMF southward turning (as observed by a distant spacecraft with the appropriate delay) and the start of SYM-H decrease and lobe magnetic field increase. However, discussion of this phenomenon is beyond the limits of this report.

3.2. Event 7 November, 1999

Probably optimal conditions for a detailed storm-time substorm study occur during a first substorm after southward IMF turn. Such a case was well documented by several satellites on 7 November, 1999. In Figure 2 the lobe magnetic pressure and local B_z measured by Interball-Tail (at -22, –5.5, 1.6 R_E), IMF B_z, auroral zone magnetogram and SYM-H index are presented. This was an isolated mild

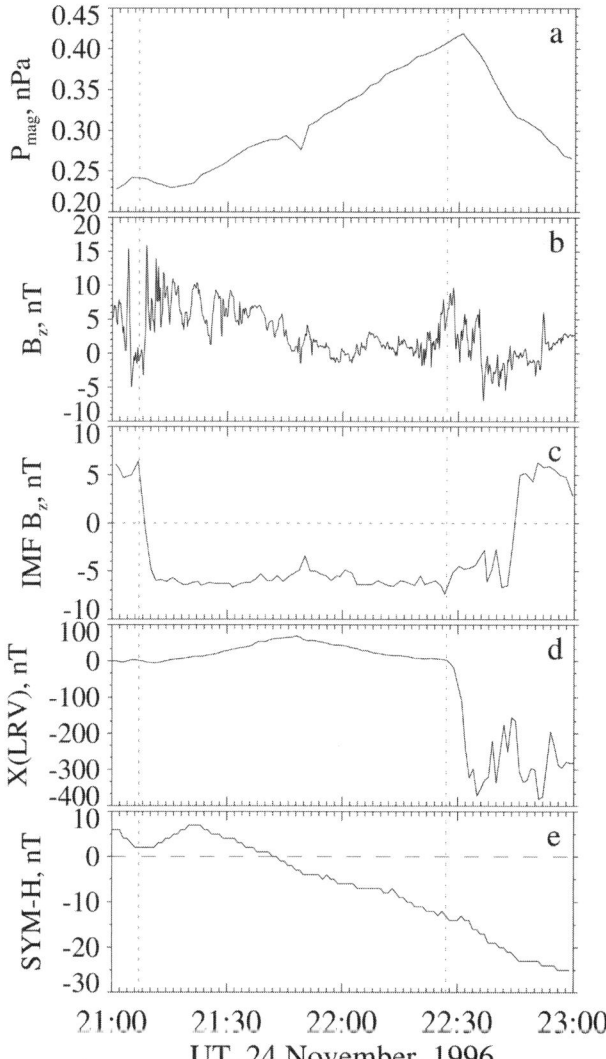

Figure 1. Substorm event of 24 November 1996. (a) Interball-Tail lobe magnetic pressure. (b) Geotail B_z in the plasma sheet. (c) Solar wind IMF B_z (Wind spacecraft). (d) Ground night-side auroral zone magnetogram from the Leirvogur station. (e) SYM-H index. Vertical lines show instants of the growth phase start and the onset.

storm period with the minimum SYM-H = -60 nT (D_{st} = -50 nT, Kp=5-) caused by the 2-hour long interval of intense southward IMF. During it only one substorm commenced and, in a certain sense, here the substorm and the storm are one and the same event.

The lobe magnetic data exhibit the same standard sequence of accumulation and discharge of the open magnetic flux during growth and expansion phases as for previous example. The onset of the expansion phase was at 05:25 UT. On the upper panel of Figure 2, magnetic pressure from the first substorm example is also shown by a shaded line. For the purpose of easier comparison, it was normalized by the solar wind dynamic pressure ratio and the spacecraft downtail distance ratio according to the tail magnetic field model of *Fairfield and Jones* [1996]. While the initial and final pressure values for two events are close, in the November 7th event the extra accumulated magnetic energy is almost twice as large.

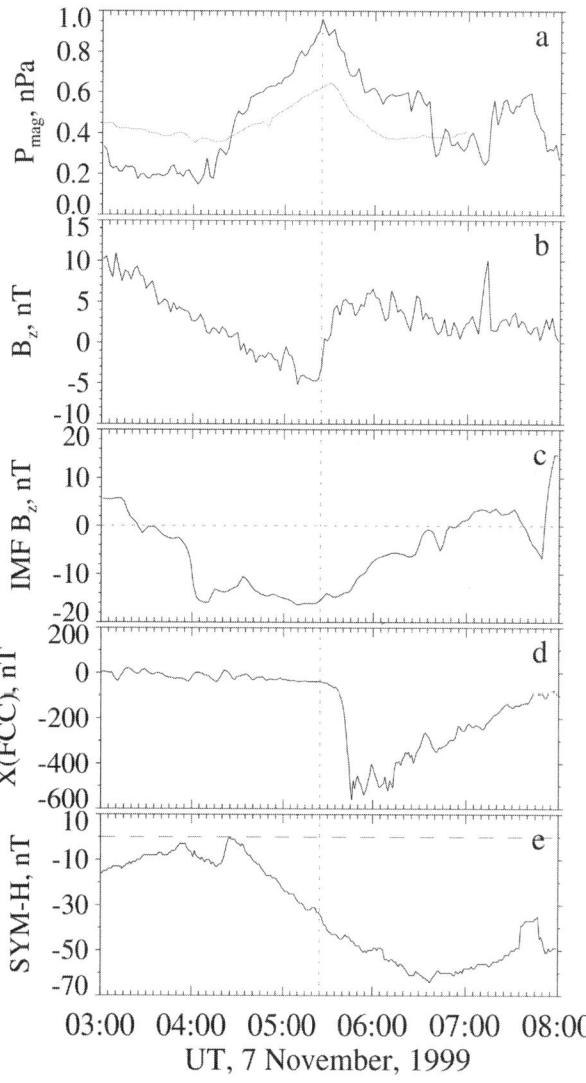

Figure 2. Substorm event of 7 November 1999. (a) Interball-Tail lobe magnetic pressure. Shaded line shows corrected pressure from Figure 1 (see text for details). (b) Interball-Tail lobe flaring (B_z). (c) Solar wind IMF B_z (ACE spacecraft). (d) Ground nightside auroral zone magnetogram from the Fort Churchill station. (e) SYM-H index. Vertical line shows the instant of the onset.

The SYM-H index in the bottom panel of Figure 2 closely follows the IMF profile irrespective of the substorm phases. However, it is saturated at the minimal value only at 06:30 UT, that is 30–50 minutes later than the IMF gradual turn to north (at least according to ACE data measured at the L1 point).

3.3. Event 22–23 November, 1997

Another example of a substorm sequence during the mild storm of 22–23 November 1997 is in Figure 3. Here the lobe magnetic field was measured by IMP-8 satellite. During the interval of the steady IMF driving from 20:00 UT to 08:00 UT one can pick up several clear magnetic pressure maxima corresponding to substorm onsets (marked by vertical lines in Figure 3). Distinct growth phases preceding them are visible as the magnetic pressure and flaring increases. It's interesting, that in between the end of each pressure decrease (of the previous substorm) and the start of the new growth phase, there exist intervals of low steady pressure and high B_z (approximately 23:30–24:00, 02:30–03:30, 05:00–06:00 UT). According to classification of *Sergeev et al.* [2001], these are enhanced convection events.

The instants of the substorm onsets are traced down to SYM-H and auroral magnetograms panels. Intensifications of auroral electrojets to -1000 nT levels accompany the onsets, while between the minima negative magnetic bays are 300–400 nT deep. The SYM-H index essentially follows the IMF profile with transient ~20 nT recoveries at the substorm onset moments.

3.4. Statistical Analysis

In Figure 4 growth phase duration and estimate of the magnetic flux input for all 108 analyzed substorms are presented with respect to intensity of the solar wind electric field $E_{sw} = V_{sw} \cdot \sqrt{B_z^2 + B_y^2 / 2} \cdot \sin^4(\theta/2)$. The estimate of the magnetic flux input on the lower panel is just an integral of E_{sw} over the growth phase interval. Similar statistics (without storm-time events) was discussed by *Petrukovich* [2000].

Focusing on the high-intensity input part of the plots (on the right) one can see that while the inverse proportion of growth phase duration vs. input intensity (upper panel) exists for moderate substorms (E_{sw} = 1–3 mV/m), growth phases of large substorms are 0.5–1.5 hours long irrespective of the IMF input. The addition of the storm-time cases (marked by triangles) only reinforces this conclusion. The resulting difference of the solar wind input between between small and large substorms is more than an order of magnitude. That is, the amount of magnetic flux loaded in

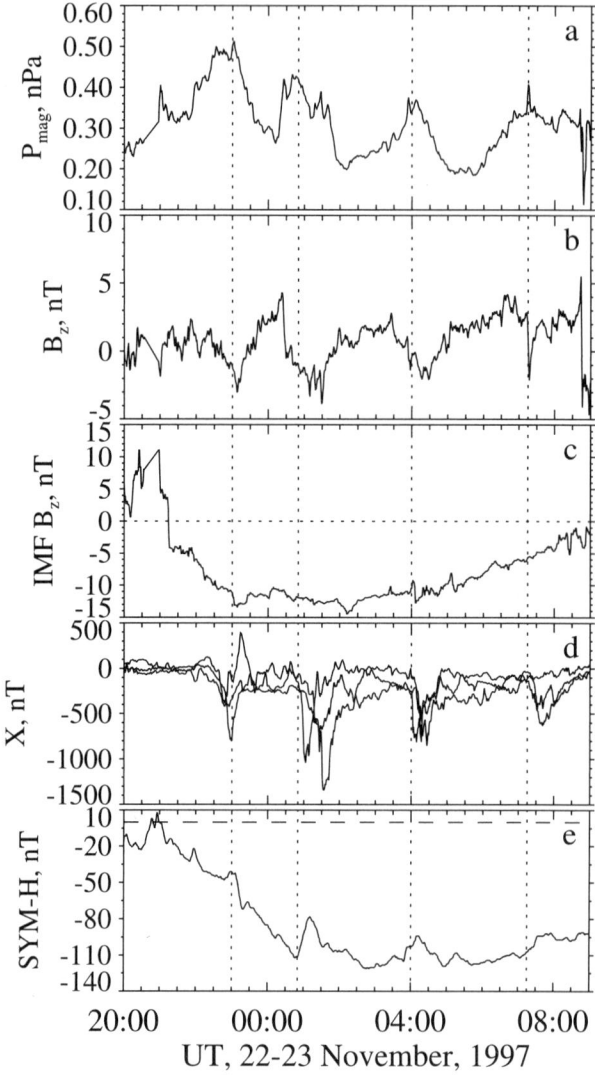

Figure 3. Substorms of 22–23 November 1997. (a) IMP-8 lobe magnetic pressure. (b) IMP-8 lobe flaring (B_z). (c) Solar wind IMF B_z (Wind spacecraft). (d) Ground night-side auroral zone magnetograms from Leirvogur, Poste de la Belin, Fort Churchill station. (e) SYM-H index. Vertical lines show instants of the onsets.

the tail before a substorm increases with the increasing intensity of solar wind electric field. The external trigger effect [*Lyons*, 1996] might be responsible for the vertical scatter of duration times for close electric field values.

4. DISCUSSION

4.1. Comparison of Substorms

Substorm examples presented in this report and in previous studies demonstrate essentially very similar patterns of magnetotail measurements and in the ring current index. However, *Kamide et al.* [1998] stated basing on the AMPTE-IRM substorm statistics, that only storm-time substorms exhibit noticeable changes in the lobe magnetic field, while during quiet-time events magnetic pressure remains flat. The origin of this discrepancy is likely related with the orbit parameters of the AMPTE-IRM satellite. All magnetic lobe data for our statistics were taken at the downtail distances 20–30 R_E, while AMPTE-IRM apogee was 19 R_E, so that most of data were taken around 15 R_E. Due to significant downtail pressure gradient at these distances, relatively small magnetic pressure variations during quiet-time events might be less visible at the AMPTE positions.

The magnitude of substorm signatures depends on the intensity of the solar wind input during the growth phase: the larger is the solar wind electric field E_{sw}, the stronger is the substorm (with the natural exception of substorms triggered in the early growth phase). Continuing during the expansion phase solar wind input also makes the substorm larger [*Kallio et al.*, 2000], but discussion of these details is beyond the current topic. It is important that the difference in substorm magnitude does exist not only as estimated by the nominal solar wind input (Figure 4), but also is really observed in the tail magnetic pressure profiles (Figure 2). The magnetotail pressure value depends on the tail cross-section (related to solar wind dynamic pressure), down-tail distance, and amount of open magnetic flux in the lobes. If the influence of dynamic pressure and distance is corrected in some approximation [*Fairfield and Jones*, 1996], the remaining difference in magnetic pressure might be interpreted as the open magnetic flux difference. Therefore, according to our examples, larger substorms also accumulate larger amount of the open magnetic flux.

4.2. Substorm Instability

Though it is intuitively clear that substorms do not have the same strength (in fact, we speak about contracted oval—small events and about storm-time large events) it is still necessary to comment on several issues. *Dmitrieva and Sergeev* [1985] analyzed several spontaneous substorms and found that growth phase duration was inversely proportional to VB_s, that is, there exists some critical value of accumulated magnetic flux or amount of solar wind input. Unfortunately, their statistics did not include contracted-oval and very strong substorms, for which this conclusion appears not to be valid (see Figure 4 and relevant description above [*Petrukovich*, 2000]). Since a ground state of the magnetotail is supposed to exist, in the frame of *Dmitrieva and Sergeev* [1985] interpretation all "hills" (excluding externally triggered substorms) above this ground level have the same height.

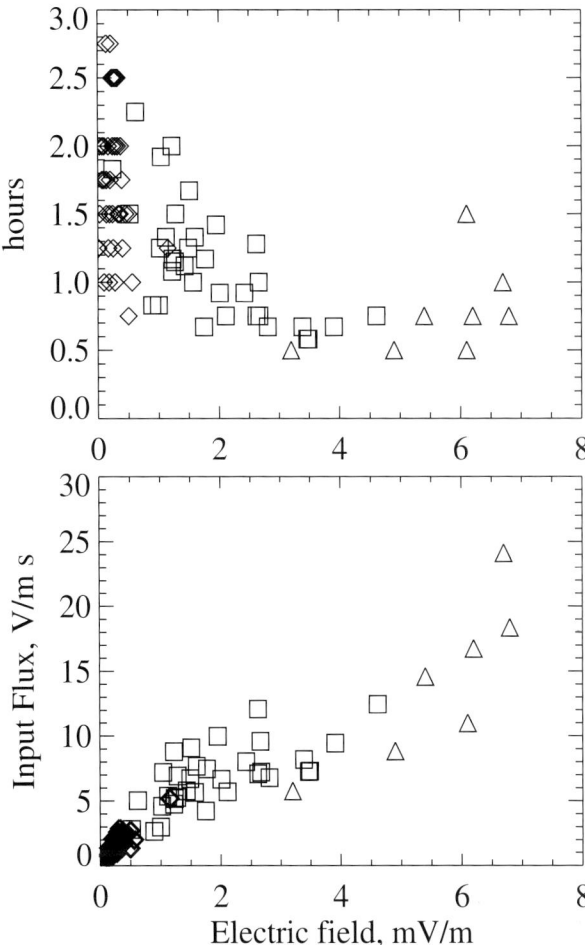

Figure 4. Substorm growth phase statistics. Growth phase duration (upper panel) and solar wind input (lower panel) vs solar wind electric field (see text for details) Diamonds denote contracted-oval cases. Squares denote normal-size cases. Triangles denote storm-time cases.

According to the model of *Freeman and Farrugia* [1999], which was built on the analysis of a substorm sequence during one magnetic cloud encounter, the substorm expansion starts, when some universal critical amount of the magnetic flux in the tail is achieved. Substorms are allowed to have variable magnitude, depending on the intensity of solar wind input immediately before the onset. The growth phase of the substorm, next in a sequence, starts from the low flux state determined by the previous substorm and ends up with the critical flux value. Following the same landscape analogy, all hills have the same height, but there is no ground level—valleys are allowed to have any depth.

The existence of such universal critical flux level is in contradiction with our results. There exist also other simple arguments. Let us consider again a contracted-oval substorm and a storm-time substorm. The former commences on the contracted oval, that is, with the small polar cap area. The latter is usually associated with the extended auroral zone, that is, with the large polar cap area. Since the polar cap area is believed to be closely related to the open magnetic flux amount, small substorms at the moment of the onset should have much smaller flux in the tail lobes, than storm-time substorms. While, according to *Freeman and Farrugia* [1999], the amount should be the same. It is also unclear in the frame of their model what is the magnetic flux state of the magnetotail during long periods of magnetic quietness.

Existence of some critical or threshold value of open magnetic flux (or of the solar wind input) is an important part of models describing the magnetotail stability and emphasizing internal reasons for a substorm onset. Local plasma instabilities such as tearing or cross-field current instability were considered as possible onset drivers, when certain thresholds in the thickness of the current sheet or current density are reached [*Baker et al.*, 1996; *Lui*, 1996]. Our findings, on the contrary, claim the absence of any straightforward threshold of the magnetotail loading. On our landscape, hills can be of any height. Moreover, the apparent existence of the typical growth phase duration for large and very large substorms suggests that the magnetotail system has a temporal limit rather than the loading limit.

4.3. Inner Magnetosphere Effect

Considering substorms as individual phenomena, we did not reveal any principal difference between substorm size classes. Storm-time substorms are stronger than their quiet-time counterparts, but this can be naturally explained by the higher solar wind input during storm intervals. In a certain sense, the division on storm and non-storm substorms appears mainly due to introduction of somewhat artificial storm threshold, e.g., $D_{st} = -50$ nT. The effect of individual substorms might be reinforced, when a series of strong substorms commences in the course of a storm.

Increases of the ring current intensity (measured as the SYM-H or D_{st} index) appear to be associated with the significant solar wind input and follow IMF changes rather than substorm phases. SYM-H decrease starts together with the substorm growth phase well before the expansion phase onset. It is true not only for storm-time events, but also for smaller substorms, though SYM-H changes then are smaller and are below accepted storm thresholds.

Therefore, at least on the initial stage, ring current changes are likely directly produced by the IMF-driven convection, while the influence of the substorm injections might gradually appear later, for example, in sustaining the achieved level of ring current after the IMF northward turning. The cross-tail current, intensifying in the near tail during the growth phase, likely also contributes to SYM-H index. However, this effect should vanish after the substorm onset due to cross-tail current decay and dipolarization. Observations of the transient recoveries of the index observed at onsets of strong substorms of 22–23 November, 1997 (Event 3) are consistent with the hypothesis of the cross-tail current influence.

5. CONCLUSIONS

Currently available new geomagnetic indices and extensive sets of spacecraft and ground data help to investigate important details of the solar-terrestrial relationship. As a result of such studies, both the complexity of the magnetosphere and the generality of principal governing mechanisms appear simultaneously. One can also notice nowadays a certain renaissance of the convection-dominated models of the magnetospheric behavior, which are returning after the period of preference to spontaneous and/or internal instability approaches to magnetotail stability and global dynamics.

In particular, this investigation shows that substorms have no any straightforward loading threshold, usually implied when internal instabilities are emphasized. Instead, the duration of growth phase appears to be stable for a very broad range of substorm magnitudes indicating existence of some temporal (convective) limit. During geomagnetic storms the SYM-H geomagnetic index follows IMF changes rather than associated substorm growth-expansion-recovery cycles, which suggests dominantly convective nature of the ring current formation.

Nevertheless, a lot of work still needs to be done in order to reach the proper understanding level of key problems. The detailed analysis of the ring current index components, or even of individual station magnetic records, might be necessary to separate full and partial ring current contributions and reveal the effect of substorm injections in certain time sectors. New magnetotail observations of the Cluster mission might add important data for super-storm periods.

Acknowledgments. We are grateful to Interball, Geotail, IMP-8, and ground magnetic experiments for their data available online and to WDC and CDAWeb data centers. A.A.P. is thankful to RFBR travel grant and the Conference organizing committee for their financial support. The research was funded by INTAS grant 99–0078.

REFERENCES

Akasofu, S.-I., P.D. Perreault, F. Yasuhara, and C.-I. Meng, Auroral substorms and the interplanetary magnetic field, *J. Geophys. Res.*, *78*, 7490–7508, 1973.

Baker, D.N., T.I. Pulkkinen, V. Angelopoulos, W. Baumjohann, and R.L. McPherron, Neutral line model of substorms: Past results and present view, *J. Geophys. Res., 101*, 12,975–13,010, 1996.

Caan, M., R.L. McPherron, and C.T. Russell, substorm and interplanetary magnetic field effects on the geomagnetic tail lobes, *J. Geophys. Res., 80*, 191–194, 1975.

Dmitrieva, N.P. and V.A. Sergeev, *Geomagn. Aeron., Engl. Transl. 25*, 352, 1985.

Fairfield, D.H., and J. Jones, Variability of the tail lobe field strength, *J. Geophys. Res., 101*, 7785–7791, 1996.

Freeman, M.P. and C.J. Farrugia, Solar wind input between substorm onsets during and after the October 18–20, 1995, magnetic cloud, *J. Geophys. Res., 104*, 22,729–22,744, 1999.

Gonzalez, W.D., J.A. Joselyn, Y. Kamide, H.W. Kroehl, G. Rostoker, B.T. Tsurutani, and V.M. Vasyliunas, What is a geomagnetic storm? *J. Geophys. Res., 99*, 5771, 1994.

Kallio, E.I., T.I. Pulkkinen, H.E.J. Koskinen, A. Viljanen, J.A. Slavin, and K. Ogilvie, Loading-unloading processes in the nightside ionosphere, *Geophys. Res. Lett., 27*, 1627–1630, 2000.

Kamide, Y., W. Baumjohann, I.A. Daglis, W.D. Gonzalez, M. Grande, J.A. Joselyn, R.L. McPherron, J.L. Phillips, E.G.D. Reeves, G. Rostoker, A.S. Sharma, H.J. Singer, B.T. Tsurutani, V.M. Vasyliunas, Current understanding of magnetic storms: Storm-substorm relationships, *J. Geophys. Res., 103*, 17,705–17,728, 1998.

Lui, A. T. Y., Current disruption in the Earth's magnetosphere: Observations and models, *J. Geophys. Res., 101*, 13,067, 1996.

Lyons, L.R., Substorms: Fundamental observational features, distinction from other disturbances, and external triggering, *J. Geophys. Res., 101*, 13,011, 1996

McPherron, R.L., Substorms related changes in the geomagnetic tail: The growth phase, *Planet. Space Sci., 20*, 1521–1539, 1972.

McPherron, R.L., C.T. Russell, and M.P. Aubry, Satellite studies of magnetospheric substorms on August 15, 1968, *J. Geophys. Res., 78*, 3131, 1973.

Perreault, P., and S.-I. Akasofu, A study of geomagnetic storms, *Geophys. J. R. Astron. Soc., 54*, 547, 1978.

Petrukovich, A.A., T. Mukai, S. Kokubun, S.A. Romanov, Y. Saito, T. Yamamoto, and L.M. Zelenyi, Substorm-associated pressure variations in the magnetotail plasma sheet and lobe, *J. Geophys. Res., 104*, 4501–4514, 1999.

Petrukovich, A.A., The growth phase: comparison of small and large substorms, In *Proceedings of the Fifth International Conference on Substorms*, ESA SP-443, 9–14, 2000.

Petrukovich, A.A., W. Baumjohann, R. Nakamura, T. Mukai, and O.A. Troshichev. Small substorms: solar wind input and magnetotail dynamics. *J. Geophys. Res., 105*, 21,109–21,118, 2000.

Sergeev, V.A., M.V. Kubyshkina, K. Liou, P.T. Newell, G. Parks, R. Nakamura, and T. Mukai, Substorm and convection bay compared: Auroral and magnetotail dynamics during convection bay, *J. Geophys. Res., 106*, 18,843, 2001.

A. A. Petrukovich, Space Research Institute, Russian Academy of Sciences, 84/32 Profsoyuznaya St., 117997, Moscow, Russia. (apetruko@iki.rssi.ru)

High Time Resolution Observations of Magnetospheric Disturbances During Auroral Activity

M. O. Fillingim[1], G. K. Parks[1], R. P. Lin[1], M. McCarthy[2], and A. Szabo[3]

We present high time resolution plasma and magnetic field data from the near-Earth plasma sheet during times of active aurorae. The plasma sheet disturbance associated with the auroral activity is composed of Earthward traveling ions with large mean velocities ($\langle \mathbf{v} \rangle = \int \mathbf{v}\, f(\mathbf{v}) d^3v$) and large amplitude, high frequency magnetic field fluctuations. The ion $\langle \mathbf{v} \rangle$ can change substantially (by up to 100%) on time scales comparable to the local proton gyroperiod. The magnetic fluctuations lead to large, rapidly varying induced electric fields. Power spectral analysis of the magnetic field data shows a significant amount of wave power present at frequencies up to and greater than the local proton cyclotron frequency. Examination of the three-dimensional ion distribution functions indicates that the distributions are complex and nongyrotropic, with large gradients and anisotropies, and dynamic, with considerable changes in the phase space features within one gyroperiod. These results illustrate that kinetic physics controls the plasma behavior during times of plasma sheet disturbances associated with auroral activity. This conclusion is in contrast to the usual interpretation that the large ion velocity moments observed during plasma sheet disturbances are convective in nature (i.e., bursty bulk flows). Additionally, we show that these kinetic effects are important in the near-Earth plasma sheet over a wide range of geomagnetic activity, from pseudobreakups to substorms. Therefore, we suggest that these kinetic processes operate during all types of geomagnetic disturbances and can occur throughout large regions of the magnetotail during substorms and storms.

[1]Space Sciences Laboratory, University of California, Berkeley, California, USA.
[2]Department of Earth and Space Sciences, University of Washington, Seattle, Washington, USA.
[3]Laboratory for Extraterrestrial Physics, NASA Goddard Space Flight Center, Greenbelt, Maryland, USA.

Disturbances in Geospace: The Storm-Substorm Relationship
Geophysical Monograph 142
Copyright 2003 by the American Geophysical Union
10.1029/142GM05

1. INTRODUCTION

Aurorae come in a large variety of spatial sizes and intensities, from the smallest pseudobreakup to large scale substorms to storm-time aurora which may extend down to mid-latitudes. Three examples of different levels of auroral activity are shown in Figure 1. The image on the far left illustrates a pseudobreakup. The region of intense auroral emission is limited to only a few degrees in latitude and a few hours of local time. Additionally, the duration of the emission is very short, on the order of a few minutes. In this case the strength of the geomagnetic disturbance was quite weak. The AE index was < 100 nT and the K_P index was 1⁻.

The middle image shows a moderate substorm near the end of the expansive phase when the area of the active auro-

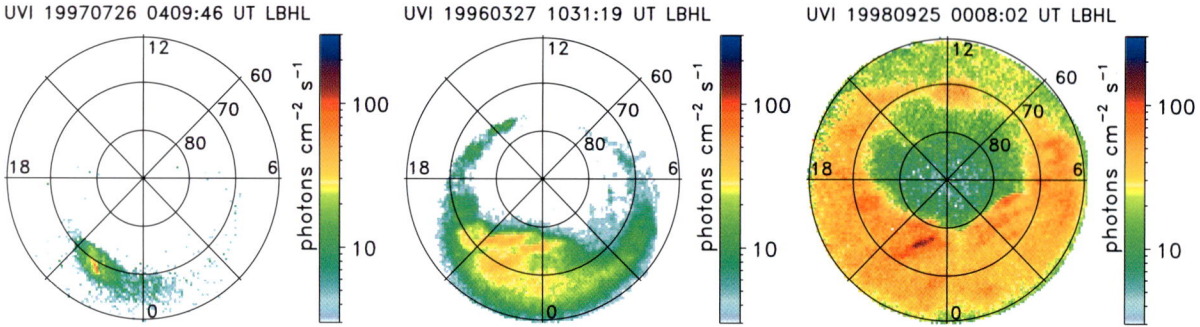

Figure 1. Examples of different levels of auroral activity. From left to right: pseudobreakup, substorm, storm.

ra reaches its maximum size. The region of intense emission covers nearly the entire midnight sector, spanning 20° in latitude and over 6 hours of local time. The expansive phase of such a substorm can last an hour or more. For this particular substorm, the AE index reached ~ 400 nT and K_p was 3⁻.

On the far right, the image shows the extent of energetic electron precipitation during a magnetospheric storm. The auroral emission extends equatorward of 60° at all local times except near local noon and reaches poleward of 80° near midnight. Intense aurora associated with storms can last several hours to a significant fraction of a day. Storms produce strong geomagnetic disturbances. During the early phase of the storm when this image was taken, AE was ~ 2000 nT and K_p was high at 8⁻. Several hours later D_{st} reached its minimum value of about –200 nT

All of these auroral events are linked magnetically to the magnetosphere. Because of this, auroral images can be used to get a global view of magnetospheric disturbances. It is also invaluable to use *in situ* plasma and magnetic field data in the magnetosphere to get a local, detailed view of the processes involved. Several of our previous studies have combined global auroral images from POLAR/UVI [*Torr et al.*, 1995] and in situ plasma and magnetic field measurements from perigee passes of the WIND spacecraft through the near-Earth plasma sheet to form a coherent picture of how magnetospheric disturbances are related to auroral activity [*Fillingim et al.*, 2000, 2001; *Parks et al.*, 2001, 2002].

Both *Fillingim et al.* [2001] and *Parks et al.* [2002] have recently shown that the plasma dynamics in the plasma sheet appear the same during pseudobreakups and substorms. The left hand side of Figure 2 shows four hours of auroral and plasma sheet data during a series of pseudobrekaups observed on 26 July 1997. The right hand side shows data from two large substorms observed over seven hours on 27 March 1997. (See Figure 1, left and middle images.) Both panels show, from top to bottom, a keogram (auroral intensity as a function of latitude and Universal Time) centered on local midnight constructed from UVI images with the latitudinal position of the WIND footprint marked; the energy deposition rate computed from UVI images in units of gigawatts carried by precipitating electrons into the nightside ionosphere; the plasma sheet magnetic field measured by WIND/MFI [*Lepping et al.*, 1995] in GSM coordinates; the ion velocity moment computed from the three-dimensional distributions measured by WIND/3DP [*Lin et al.*, 1995] in GSM coordinates; the ion velocity in magnetic field aligned coordinates; and the omnidirectional ion flux in units of $(cm^2\text{-sec-ster-keV})^{-1}$ at four different energies from 10 keV to 1 MeV. During the four hour period on 26 July 1997, the WIND footprint moved from a local time of 21 to midnight at a radius of 11 R_E. For the interval shown from 27 March 1996, the WIND footprint moved from a local time of 20:30 to 22:30 at a radius of about 17 R_E.

An important feature to note is that when the footprint of WIND maps to the region of intense auroral emission, several repeatable phenomena are seen. These include large ion velocity moments ($\langle \mathbf{v} \rangle = \int \mathbf{v}\, f(\mathbf{v}) d^3 v$) typically in the Earthward direction at WIND perigee distances of –10 to –20 R_E with, at times, a sizable component directed parallel to the magnetic field; large amplitude, high frequency fluctuations in all three components of the magnetic field; and increases in the energetic ion (electron) fluxes up to MeV (several hundred keV) energies. These phenomena appear quite similar in the plasma sheet even though the auroral activity is quite different.

Many of the same features in the plasma sheet have been noted by *Angelopoulos et al.* [1997] and *Fairfield et al.* [1999] in conjunction with substorm onsets and intensifications identified using global auroral images. The typical interpretation for the large ion velocity moments is that they are caused by an enhanced, quasistatic electric field resulting in a large $\mathbf{E} \times \mathbf{B}$ drift. That is, they are due to enhanced convection and have been termed burst bulk flows (BBFs)

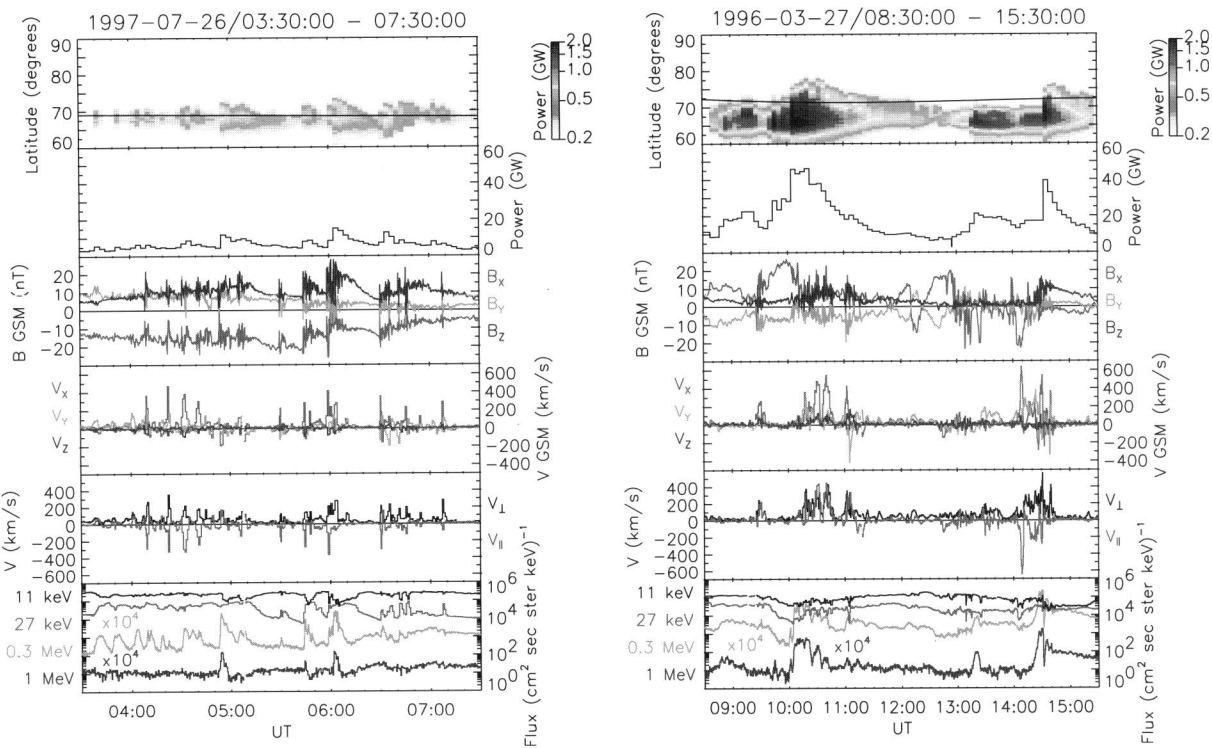

Figure 2. Auroral and plasma sheet data from a series of pseudobreakups (left) and a sequence of substorms (right). From top to bottom: keogram, energy deposition rate in the nightside ionosphere (both computed from UVI images), GSM components of the plasma sheet magnetic field, GSM components of the ion velocity moment, ion $\langle\mathbf{v}\rangle$ in magnetic field aligned coordinates, and omnidirectional ion flux (bottom four panels measured by WIND).

[*Angelopoulos et al.*, 1992]. The electric field that would be necessary to produce the observed $\langle\mathbf{v}\rangle$ is about 5 to 10 mV/m ($\mathbf{E} = -\langle\mathbf{v}\rangle \times \mathbf{B}$), roughly an order of magnitude larger than typical estimates for the cross tail convection electric field. BBFs have been interpreted to be the result of reconnection in the tail and have been invoked to explain the bulk transport of mass, energy, and flux from the tail to the near-Earth region.

However, close examination of the particle distribution functions during both pseudobreakups and substorms shows that the distributions can be quite complex and indicates that the large $\langle\mathbf{v}\rangle$ are mainly due to highly anisotropic energetic ions. Figure 3 shows some representative ion distribution functions. Two-dimensional slices of the three-dimensional distribution functions in the $\mathbf{B}$–$\mathbf{V}_\perp$ plane in the spacecraft frame of reference are shown along with one-dimensional cuts in the directions parallel (red asterisks) and perpendicular (blue diamonds) to the averaged magnetic field. The horizontal axis in the contour plots is velocity parallel to $\mathbf{B}$ while the vertical axis is velocity perpendicular. The numbers in the upper left, upper right, and lower right of the contour plots are the elevation angle, azimuth angle, and magnitude of the magnetic field averaged over the integration time (50 seconds, except for the lower right distribution which is integrated over 25 seconds). The plus signs in each contour plot indicates the direction and magnitude of $\langle\mathbf{v}\rangle$. The horizontal axis in the 1-D cuts is velocity; the vertical axis is the phase space density in units of $s^3 km^{-6}$. The solid red and blue lines represent the one-count level in that particular direction. Since the detector has a variable angular resolution depending upon look direction, the one-count level depends upon look direction. The top three contour plots and 1-D cuts show data from pseudobreakups observed on 26 July 1997 while the bottom plots contain data obtained during substorms on 27 March 1996.

The first column shows the ion distributions when WIND does not map to the region of intense auroral emission. The distributions appear relatively isotropic, though there are clearly two populations present. The two component nature of the distributions is most visible in the bottom most plots where the break in the spectrum occurs at ~ 5 keV or about 1000 km/s (assuming the ions are protons).

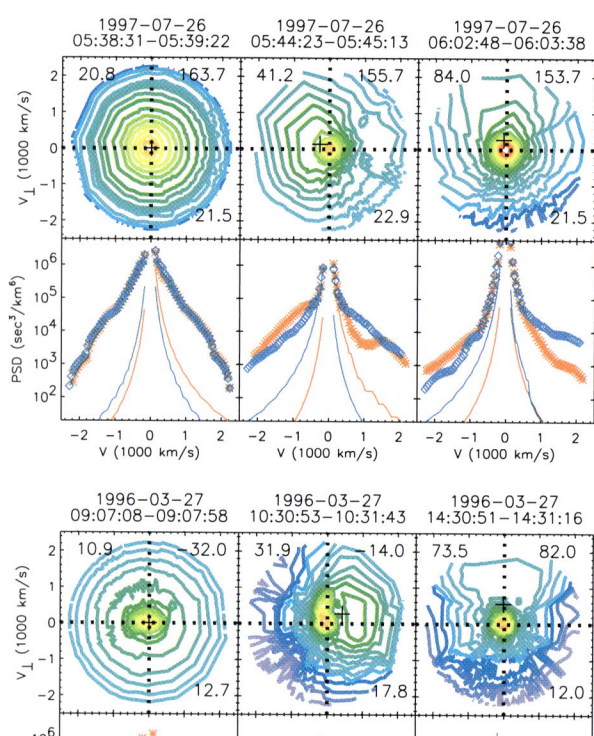

Figure 3. Ion phase space velocity distribution functions before and during large ⟨**v**⟩ associated with pseudobreakups (top row) and substorms (bottom row). The contour plots show 2-D slices of the 3-D distribution in the **B**–**V**⊥ plane in the spacecraft frame. The line plots show 1-D cuts parallel (red asterisks) and perpendicular (blue diamonds) to **B**.

The plots in the middle column again illustrate the multi-component nature of the ion distributions. Additionally, they show that the different components may behave dynamically differently. In both cases, there is a parallel beam present along with a core population. In the top example, a parallel beam with an energy of about 10 keV is present along with a relatively energetic ion population that is shifted anti-parallel and somewhat perpendicular to the magnetic field (upper left quadrant) as indicated by ⟨**v**⟩. The beam does not appear shifted in the perpendicular direction. The bottom distribution in the middle column similarly contains a wide parallel beam with an energy of about 5 keV. A very low energy, anisotropic population is also present. The low energy population is shifted slightly in the anti-parallel direction. Both the core and the beam are shifted by a few hundred km/s in the perpendicular direction suggesting that the plasma is being convected by an electric field.

The distributions in the last column are highly anisotropic and appear nongyrotropic at energies greater than a few keV. The energetic populations are shifted perpendicular to **B** relative to the spacecraft, and steep phase space density gradients are present. The 1-D cuts show plateaus extending to the highest energy channels of the detector (> 30 keV). *Chen et al.* [2000b] have shown that the anisotropy can persist up to MeV energies.

These examples show how the single component fluid assumption [*Nakamura et al.*, 1991] can breakdown in the plasma sheet during active intervals. The computed velocity moments in these cases do not accurately represent the plasma behavior. Multi-component distributions in which the different populations are behaving dynamically differently cannot be adequately explained solely by convection. Other processes must be operating to create these complex and dynamic distributions with multiple populations, beams, and strong phase space density gradients. It is for this reason that we do not refer to these events as BBFs. The ⟨**v**⟩ are not due to the bulk motion of a single component plasma and are not flows in the sense that motion is only perpendicular to **B**.

The similarity of the phase space features present in the distributions during pseudobreakups and substorms leads us to conclude that the same, as yet unidentified, physical processes are operating in the plasma sheet during both types of auroral activity. *Parks et al.* [2002] extended this line of reasoning and suggested that just as there is a continuum of scale sizes and intensities of auroral activity, there is a continuum of scale sizes and intensities of plasma sheet disturbances. During a pseudobreakup only a small region of the plasma sheet is affected. A much larger region of the plasma sheet is disturbed during substorms, but the same physical mechanisms are operating. If we extrapolate this idea to storms, then the same processes may be active throughout nearly the entire nightside plasma sheet.

These previous conclusions were reached using plasma data integrated over several to several tens of ion gyroperiods which give only an average picture of the ion dynamics. In the next section, we show high resolution plasma data with a temporal resolution on the order of the proton gyroperiod. These data strongly demonstrate that the processes involved in plasma sheet disturbances associated with auroral activity are kinetic and nonlinear. Our observations stress the importance of using high time resolution measurements to get a clear picture of the dynamics that are active in the plasma sheet during pseudobreakups, substorms, and storms.

2. OBSERVATIONS

2.1. Pseudobreakup: 26 July 1997

On 26 July 1997, WIND was in the near-Earth plasma sheet at a radial distance of about 11 R_E while POLAR/UVI was imaging the entire northern auroral region. As shown in the left hand side of Figure 2, between 03:00 and 08:00 UT the WIND plasma and magnetic field instruments recorded a series of plasma sheet disturbances each composed of large Earthward ion $\langle v \rangle$, large amplitude, high frequency magnetic field fluctuations, and increases in the flux of energetic particles. At the same time, POLAR/UVI observed a series of minor auroral brightenings interpreted as pseudobreakups coincident with the plasma sheet disturbances. *Fillingim et al.* [2000] gives an overview of these observations. Here we focus on high resolution data during one event near 06:30 UT.

Figure 4 shows a detailed look at the ion $\langle v \rangle$ and magnetic field variations during the seven minute period from 06:29 to 06:36 UT. At this time, WIND was at a GSM position of [-11, 2, 0.5] R_E which translates to a magnetic local time of 23:30. The figure shows, from top to bottom, the components of $\langle v \rangle$ in GSM coordinates (plus $|\langle v \rangle|$), $\langle v \rangle$ in magnetic field aligned coordinates, three components of the

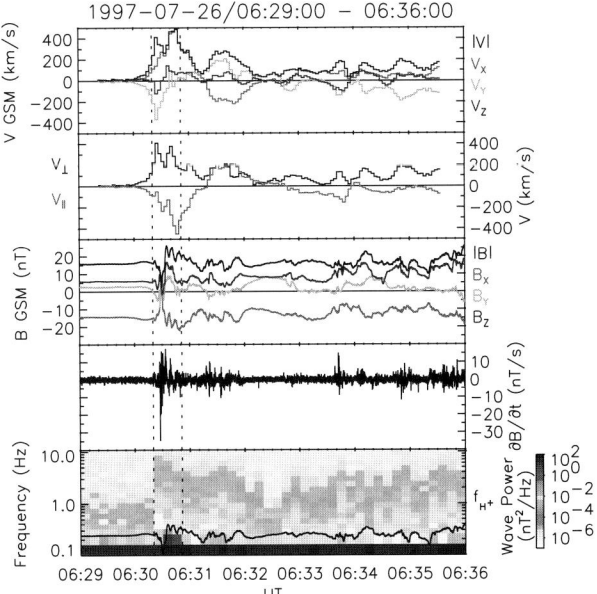

Figure 4. Plasma sheet ion and magnetic field data. From top to bottom: ion $\langle v \rangle$ in GSM coordinates (plus $|\langle v \rangle|$), ion $\langle v \rangle$ in magnetic coordinates, GSM components of **B** (plus $|\mathbf{B}|$), $\partial|\mathbf{B}|/\partial t$, dynamic frequency spectrogram.

magnetic field in GSM coordinates, the time derivative of the magnitude of the magnetic field, and a dynamic frequency spectrogram of the magnetic field.

The $\langle v \rangle$ are calculated from the three-dimensional ion distribution functions integrated over 3.1 seconds (one spacecraft rotation) which were continuously sampled for the entire nearly seven minute interval. This integration time is comparable to the local proton gyroperiod. The magnetic field data has a time resolution of 0.046 seconds. The dynamic frequency spectrogram was created by computing the Fast Fourier Transform for every 10 second segment of magnetic field data. The wave power in nT^2/Hz is plotted as a function of frequency from about 11 Hz (the Nyquist frequency) to 0.1 Hz and UT. The solid black line represents the local proton gyrofrequency. Note that there is a 3 second sinusoidal variation in **B** before 06:30 UT due to the spacecraft rotation. As a result, there is a horizontal band centered on 0.3 Hz and a weaker band centered on 0.6 Hz before 06:30 UT in the dynamic frequency spectrogram.

An important feature in this figure is the large changes seen in $\langle v \rangle$ on time scales comparable to the proton gyroperiod. Before 06:30:10 $\langle v \rangle$ increases very gradually, primarily in the $-Y$-direction. At 06:30:25, the component of the velocity perpendicular to **B** (mostly in the $-Y$-direction) increases by almost 100% in 3.1 seconds. The X-component of $\langle v \rangle$ increases by 50% near 06:30:40. Two peaks are also present in $|\langle v \rangle|$ within a few gyroperiods of each other. The first peak is centered about 06:30:25 UT and is directed mainly in the $-Y$-direction with a smaller $-Z$-component. This leads to a velocity perpendicular to the ambient magnetic field. The second peak is centered around 06:30:45 UT and is directed largely in the X-direction. During this second peak, as the direction of **B** changes, the velocity is first directed perpendicular then parallel to the magnetic field. In this case, the direction of $\langle v \rangle$ does not appear to be constrained by the direction of **B**.

The rapid, large amplitude variations in the magnetic field on time scales faster than the proton gyroperiod can be clearly seen throughout this interval. Starting around 06:30:20 UT, small oscillations with periods less than one second are seen in all three components of the magnetic field. As the magnitude of **B** decreases, the oscillations continue, and, in the Z-component, the amplitude of the oscillations grow. Near 06:30:30 UT $|\mathbf{B}|$ reaches its minimum as $\mathbf{B}_x$ briefly changes sign indicating that WIND may have crossed a current sheet or encountered strong local turbulence. There is also a brief excursion of B_z to negative values just after the B_x sign change during the period of Earthward dominated $\langle v \rangle$. As $|\mathbf{B}|$ increases, mainly due to increases in B_x and B_z, short period magnetic oscillations are present. The amplitude of the change in $|\mathbf{B}|$ is greater than 20 nT, larger than the original magnetic field strength. Since $\Delta B/B$ is greater than unity, this suggests that the instability is nonlinear.

$\partial B/\partial t$ can get quite large with a peak near –35 nT/s. More typical values are on the order of 10 nT/s. This translates to induced electric field strengths of several to 10 mV/m at scale lengths of the thermal proton gyroradius (several hundred km).

The dynamic frequency spectrogram shows significant wave power up to and exceeding the local proton gyrofrequency. The sharp decrease in $|\mathbf{B}|$ at 06:30:30 UT results in a broadband frequency response; however, short period oscillations are seen before, during, and after this spike. Wave power exists at frequencies up to and exceeding the proton gyrofrequency for several minutes. Since the fastest proton gyroperiod during this interval is about 2.5 seconds, and the more typical value is 4 seconds, the ions cannot be behaving adiabatically during such large magnetic field changes over such short time scales.

Phase space velocity distribution functions during the interval of large $\langle\mathbf{v}\rangle$ are presented in Figure 5. Ten two-dimensional slices of the 3-D distribution functions in the $\mathbf{B}-\mathbf{V}_\perp$ plane in the spacecraft frame of reference are again shown along with 1-D cuts. The format is the same as that in Figure 3, except that the distributions are integrated over 3.1 seconds. The vertical dashed lines in Figure 4 show the interval in which these distributions are taken. The second contour plot corresponds to the first peak in $|\langle\mathbf{v}\rangle|$ and $v_\perp$ and the negative peak in $\langle\mathbf{v}\rangle_Y$ in Figure 4.

The first distribution in this figure appears nearly isotropic in the plasma frame. There is a shift in the distribution toward the upper left quadrant resulting in a $\langle\mathbf{v}\rangle$ of just over 200 km/s, but the contours are nearly circular and concentric. It appears as though two components are present. There is a "ledge" in the distribution near 1000 km/s. It is also apparent in the 1-D cuts as a break in the spectrum. This feature is seen in earlier distributions as well.

From the first to the second distribution, the $|\langle\mathbf{v}\rangle|$ nearly doubles to over 400 km/s. The second distribution is much

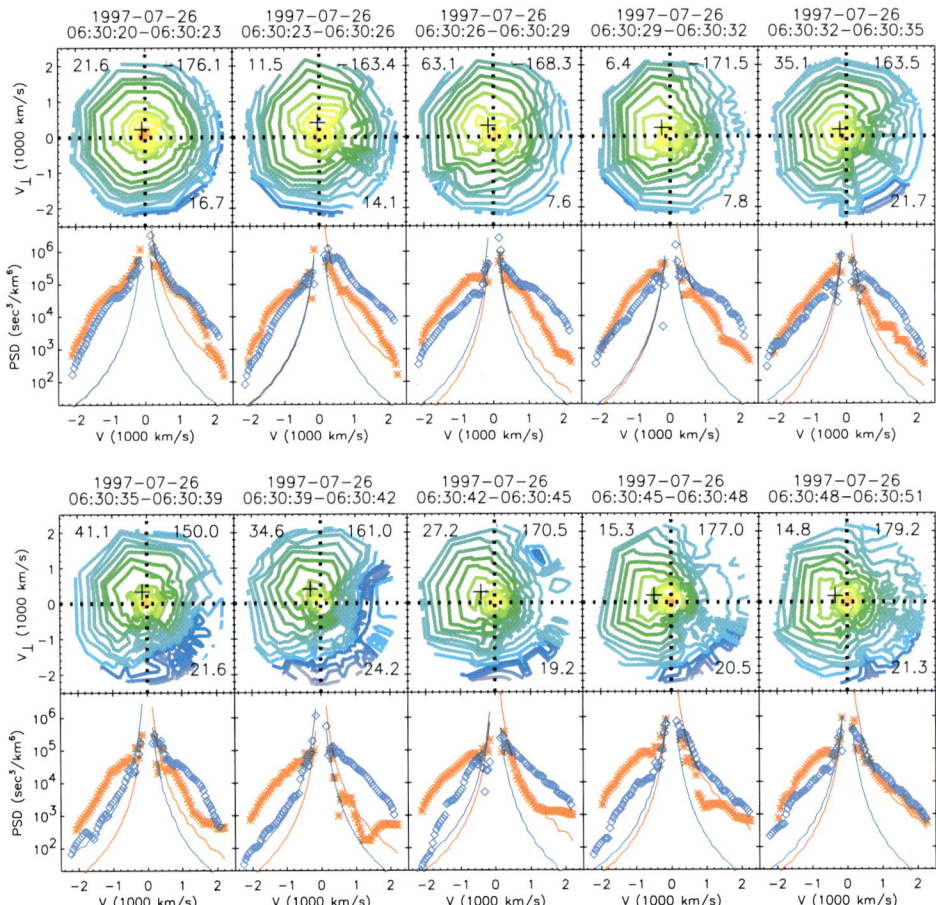

Figure 5. Ion phase space velocity distribution functions during the interval of large $\langle\mathbf{v}\rangle$.

more complex and anisotropic; sharp gradients in the phase space density are present. There is an enhancement of high energy particles perpendicular to the magnetic field which gives rise to the first peak in $\langle \mathbf{v} \rangle$. A Galilean transformation of an isotropic distribution cannot reproduce the features seen here. However, the magnetic field is changing during the 3.1 second integration time, so some time aliasing may be occurring.

Notice that from the fifth to the sixth distribution, there is a large decrease in the phase space density of high energy particles in the lower right quadrant, which corresponds to particles traveling in the tailward direction. The phase space density decreases by an order of magnitude at 10 keV in this direction in one gyroperiod. A careful analysis shows that the decrease in this direction is not as extreme at higher and lower energies.

In the last five distributions, note how the direction of the enhancement in energetic particles and $\langle \mathbf{v} \rangle$ as indicated by the plus sign changes from mostly perpendicular to $\mathbf{B}$ to nearly parallel to $\mathbf{B}$. At the same time, the elevation angle of $\mathbf{B}$ changes from about 40° to 14° while the azimuth angle remains fairly constant. However, note that, as shown in the top panel of Figure 4, $\langle \mathbf{v} \rangle$ is directed almost entirely in the X-direction throughout this time interval. (The seventh, eighth, and ninth distributions represent the second, broad peak in $\langle \mathbf{v} \rangle$ and the peak in the X-component of $\langle \mathbf{v} \rangle$.) Therefore, the direction of $\langle \mathbf{v} \rangle$ is not changing while the direction of $\mathbf{B}$ is changing. It appears that the ion dynamics at this point are not controlled by the local magnetic field.

2.2. Substorm: 30 September 1997

For completeness we also present high resolution data obtained during a substorm observed on 30 September 1997. At 04 UT on 30 September 1997, WIND was located at a GSM position of $[-14, 5, -2]$ R_E, a magnetic local time of about 22:30. When UVI begins observing the aurora shortly before 04 UT, a substorm is already in progress. Ground magnetograms suggest that the substorm onset (and a subsequent intensification) may have occurred between 03 and 04 UT. Figure 6 shows high resolution plasma sheet ion and magnetic field data from a time when WIND mapped to the region of active aurora during this substorm. The format of Figure 6 is identical to that of Figure 4. Again the ion data have a resolution of 3.1 seconds (one spacecraft rotation) and the magnetic field data have a resolution of 0.046 seconds.

There is some interesting behavior seen in the ion velocity moments. At the beginning of this interval, the ion $\langle \mathbf{v} \rangle$ is already elevated (> 100 km/s), pointed mostly in the Y-direction, and gradually increasing. The $\langle \mathbf{v} \rangle$ is directed mostly perpendicular to $\mathbf{B}$. As the velocity peaks in the Y-direc-

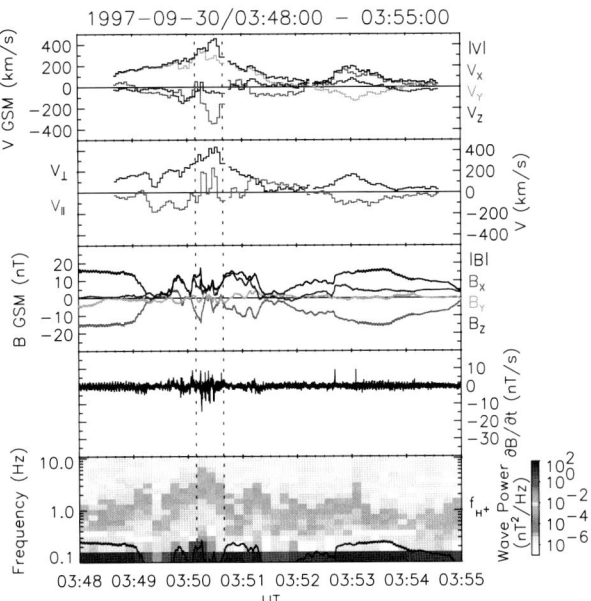

Figure 6. Same as Figure 4.

tion, the $-X$-component of $\langle \mathbf{v} \rangle$ rapidly increases and peaks at over 300 km/s in about 10 seconds. $|\mathbf{v}|$ and $v_\perp$ also peak at this time. Subsequently, the $-\langle \mathbf{v} \rangle_X$ component quickly decreases. Since $\mathbf{B}$ is between 5 and 10 nT, the duration of the $-\langle \mathbf{v} \rangle_X$ increase and decrease is just a few gyroperiods.

The bottom three panels show that the magnetic field is highly variable during this interval. B_X changes sign several times between 03:49 and 03:52 UT, indicating either multiple crossings of the current sheet or strong local turbulence. The minimum value of $|\mathbf{B}|$ is about 1 nT, indicating that WIND is near the neutral sheet. $\partial\mathbf{B}/\partial t$ routinely reaches values on the order of 10 nT/s. While smaller than the peak values observed on July 26, 1997, the induced electric field can still be ~ 10 mV/m at length scales of the gyroradii of the thermal protons. Similar to July 26, 1997, the dynamic frequency spectrogram shows that there is wave power at and above the local proton gyrofrequency throughout this interval.

Ten three-second resolution ion distribution functions are shown in Figure 7. These distributions cover the time interval of the $-X$-component of $\langle \mathbf{v} \rangle$. Again the dashed vertical lines in Figure 6 show the interval from which these distributions are taken.

The first two distributions are quite energetic with very few particles with energies below about 1 keV (~ 500 km/s) present. The distributions are anisotropic in the direction perpendicular to the magnetic field. In the third distribution, the phase space density of ions with energies around a few keV increases noticeably. This increase is most clearly seen in the upper right quadrant. This lower energy but anisotrop-

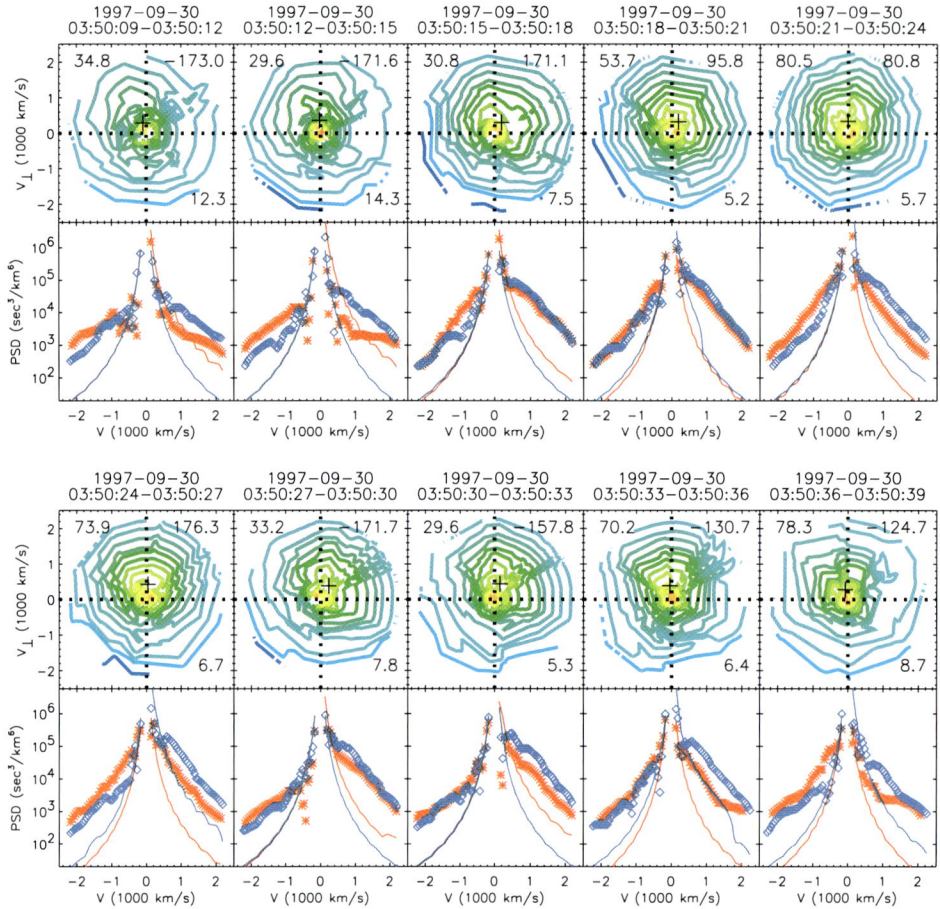

Figure 7. Same as Figure 5.

ic component which is responsible for the $-\langle\mathbf{v}\rangle_x$ persists for the next six distributions. Throughout these seven distributions, the direction of the $\langle\mathbf{v}\rangle$ does not change; it remains pointed duskward and tailward. However, the direction of $\langle\mathbf{v}\rangle$ as indicated by the small crosses in the contour plots does change as the magnetic field changes direction. In the final distribution, the medium energy component disappears leaving the anisotropic energetic component and a $\langle\mathbf{v}\rangle_Y > 200$ km/s behind.

3. DISCUSSION

The processes occurring during high ion $\langle\mathbf{v}\rangle$ events in the near-Earth plasma sheet are dynamic and possibly nonlinear. Here and in our previous work we have shown that large $\langle\mathbf{v}\rangle$ events are associated with anisotropic, complex ion distribution functions, increases in the flux of energetic particles up to MeV energies, large amplitude, high frequency magnetic field fluctuations, and intense aurora in the ionosphere [*Chen et al.*, 2000a,b; *Fillingim et al.*, 2000, 2001; *Parks et al.*, 2001, 2002]. Here we have shown high time resolution ion and magnetic field data from large ion $\langle\mathbf{v}\rangle$ events in the near-Earth plasma sheet associated with auroral activity to demonstrate that additional features are present. For example, large ion $\langle\mathbf{v}\rangle$ can contain significant structure, that is, multiple peaks in $|\langle\mathbf{v}\rangle|$ separated in time by a few proton gyroperiods and abrupt changes (up to 100%) in both the direction and magnitude of $\langle\mathbf{v}\rangle$ on time scales comparable to the local proton gyroperiod.

At the same time, large amplitude, high frequency magnetic field fluctuations are seen with $|\partial\mathbf{B}/\partial t|$ typically reaching ~ 10 nT/s. These magnetic field fluctuations will give rise to strong, rapidly changing induced electric fields. At scale lengths on the order of the gyroradius of a 1 keV proton (several hundred km in a 10 to 20 nT field), the induced electric field strength will be on the order of 10 mV/m. This is at least an order of magnitude larger than typical expected dawn-to-dusk electric fields of a few tenths of mV/m and is similar in magnitude to $\langle\mathbf{v}\rangle \times \mathbf{B}$ (500 km/s × 20 nT = 10 mV/m), which is usually interpreted as an enhanced convective electric field. However, the induced electric fields, rather than being quasi-static in the *Y*-direction for Earthward transport, are rapidly changing in direction and magnitude as **B** varies on time scales faster than the proton gyroperiod.

These magnetic fluctuations have significant wave power present at frequencies up to and exceeding the local proton gyrofrequency. *Lui et al.* [1992] saw similar broadband increases in wave power at similar locations and interpreted their observations as a signature of a cross field current instability giving rise to current disruption. *Perraut et al.* [2000] have also noted enhanced wave power at and above the proton gyrofrequency at geosynchronous altitudes which may be related to a parallel current-driven instability.

The ion distribution functions contain complex features that include large gradients and anisotropies in phase space. These features cannot be reproduced by a transformation of an isotropic distribution. Additionally, some of the features present appear nongyrotropic. Significant changes also occur in these distribution functions on time scales comparable to the local proton gyroperiod indicating that the plasma is very dynamic. It is clear from the rapid fluctuations in **B** and from the changes in the ion distribution functions that the ions cannot be behaving adiabatically. Furthermore, the direction of $\langle \mathbf{v} \rangle$ at times is not affected by changes in the direction of **B**, suggesting that the ions also may not be magnetized.

All of these observations taken together strongly indicate that kinetic processes are operating during periods of large $\langle \mathbf{v} \rangle$. *Chen et al.* [2000a, b] and *Parks et al.* [2001] have also argued that kinetic effects are important during large ion $\langle \mathbf{v} \rangle$ in the near-Earth plasma sheet. They found that the ion distributions contained multiple species, including those from the ionosphere. The measured $\langle \mathbf{v} \rangle$ were not always in agreement with the $\mathbf{E} \times \mathbf{B}$ convection velocity determined by analyzing the distributions. The exact kinetic processes operating have not yet been identified, but this is an area of continuing work.

The evidence presented here suggests that the large ion velocities may not be convective in nature as usually assumed. The interpretation that large ion $\langle \mathbf{v} \rangle$ are BBFs is based on lower resolution moment data [*Angelopoulos et al.*, 1992], not on gyroperiod resolution distribution functions. If large $\langle \mathbf{v} \rangle$ are not due to enhanced convection, then the role of BBFs as a transport mechanism of mass, energy, and flux from the tail to the near-Earth region will have to be reexamined.

With measurements from a single spacecraft, it is impossible to distinguish between spatial and temporal variations. However, the dynamic nature of the distributions, the large $\partial \mathbf{B}/\partial t$, and the close association with time varying phenomena such as the aurora as shown by *Fillingim et al.* [2000], lead us to interpret these observations as temporal variations in the properties of the local plasma. Multipoint measurements by multiple spacecraft missions such as Cluster may help resolve this ambiguity.

Our previous work suggests that the microphysical processes occurring in the near-Earth plasma sheet on minute time scales during pseudobreakups and substorms appear to be the same [*Fillingim et al.*, 2001; *Parks et al.*, 2002]. We suggest that the processes occurring in the near-Earth plasma sheet on gyroperiod and sub-gyroperiod time scales are also a characteristic of all types of auroral activity. Analogous to the aurora, which is observed with a continuum of spatial and temporal scales, plasma sheet disturbances occur with a similar continuum of scales [*Parks et al.*, 2002]. The longer scales are used to describe the average behavior of the dynamics by means of MHD physics. But in smaller regions, MHD physics is not valid as the faster and small scale features become important. In these regions the physics must be characterized by kinetic processes. The kinetic processes, while still not fully understood, are important since they could be regulating and controlling many of the large-scale current and transport generation mechanisms. We have demonstrated that these kinetic processes occur in the near-Earth plasma sheet during pseudobreakups and substorms. We also propose that the same processes operate throughout much of the magnetotail during storms.

Acknowledgments. The Wind magnetometer data is courtesy of Ronald P. Lepping. This work was supported in part by NASA grants NAG5-3170 and NAG5-26580.

REFERENCES

Angelopoulos, V., W. Baumjohann, C. F. Kennel, F. V. Coroniti, M. G. Kivelson, R. Pellat, R. J. Walker, H. Lühr, and G. Paschmann, Bursty bulk flows in the inner central plasma sheet, *J. Geophys. Res., 97*, 4027, 1992.

Angelopoulos, V., T. D. Phan, D. E. Larson, F. S. Mozer, R. P. Lin, K. Tsuruda, H. Hayakawa, T. Mukai, S. Kokubun, T. Yamamoto, D. J. Williams, R. W. McEntire, R. P. Lepping, G. K. Parks, M. Brittnacher, G. Germany, J. Spann, H. J. Singer, and K. Yumoto, Magnetotail flow bursts: association to global magnetospheric circulation, relationship to ionospheric activity and direct evidence for localization, *Geophys. Res. Lett, 24*, 2271, 1997.

Chen, L., D. Larson, R. P. Lin, M. McCarthy, and G. Parks, Multicomponent plasma distributions in the tail current sheet associated with substorms, *Geophys. Res. Lett., 27*, 843, 2000a.

Chen, L-J., G. K. Parks, M. McCarthy, D. Larson, and R. P. Lin, Kinetic properties of bursty bulk flow events, *Geophys. Res. Lett., 27*, 1847, 2000b.

Fairfield, D. H., T. Mukai, M. Brittnacher, G. D. Reeves, S. Kokubun, G. K. Parks, T. Nagai, H. Matsumoto, K. Hashimoto, D. A. Gurnett, and T. Yamamoto, Earthward flow bursts in the inner magnetotail and their relation to auroral brightenings, AKR intensifications, geosynchronous particle injections and magnetic activity, *J. Geophys. Res., 104*, 355, 1999.

Fillingim, M. O., G. K. Parks, L. J. Chen, M. Brittnacher, G. A. Germany, J. F. Spann, D. Larson, and R. P. Lin, Coincident POLAR/UVI and WIND observations of pseudobreakups, *Geophys. Res. Lett., 27*, 1379, 2000.

Fillingim, M. O., G. K. Parks, L. J. Chen, M. McCarthy, J. F. Spann, and R. P. Lin, Comparison of plasma sheet dynamics during pseudobreakups and expansive aurora, *Phys. Plasmas, 8*, 1127, 2001.

Lepping, R. P., M. H. Acuña, L. F. Burlaga, W. M. Farrell, J. A. Slavin, K. H. Schatten, F. Mariani, N. F. Ness, F. M. Neubauer, Y. C Whang, J. B. Byrnes, R. S. Kennon, P. V. Panetta, J. Scheifele, and E. M. Worley, The WIND magnetic field field investigation, *Space Sci. Rev., 71*, 207, 1995.

Lin, R. P., K. A. Anderson, S. Ashford, C. Carlson, D. Curtis, R. Ergun, D. Larson, J. McFadden, M. McCarthy, G. K. Parks, H. Rème, J. M. Bosqued, J. Coutelier, F. Cotin, C. D'uston, K.-P. Wenzel, T. R. Sanderson, J. Henrion, J. C. Ronnet, and G. Paschmann, A three-dimensional plasma and energetic particle investigation for the WIND spacecraft, *Space Sci. Rev., 71*, 125, 1995.

Lui, A. T. Y. R. E. Lopez, B. J. Anderson, K. Takahashi, L. J. Zanetti, R. W. McEntire, T. A. Potemra, D. M. Klumpar, E. M. Greene, and R. Strangeway, Current disruptions in the near-earth neutral sheet region, *J. Geophys. Res., 97*, 1461, 1992.

Nakamura, M., G. Paschmann, W. Baumjohann, and N. Sckopke Ion distributions and flows near the neutral sheet, *J. Geophys. Res., 96*, 5631, 1991.

Parks, G. K., L. J. Chen, M. Fillingim, and M. McCarthy, Kinetic characterization of plasma sheet dynamics, *Space Sci. Rev., 95*, 237, 2001.

Parks, G. K., L. J. Chen, M. Fillingim, R. P. Lin, D. Larson, and M. McCarthy, A new framework for studying the relationship of aurora and plasma sheet dynamics, *J. Atmos. Terr. Phys., 64*, 115, 2002.

Perraut, S., O. Le Contel, A. Roux, and A. Pedersen, Current-driven electromagnetic ion cyclotron instability at substorm onset, *J. Geophys. Res., 105*, 21,097, 2000.

Torr, M. R., D. G. Torr, M. Zukic, R. B. Johnson, J. Ajello, P. Banks, K. Clark, K. Cole, C. Keffer, G. Parks, B. Tsurutani, and J. Spann, A far ultraviolet imager for the international solar-terrestrial physics mission, *Space Sci. Rev., 71*, 329, 1995.

M. McCarthy, Department of Earth and Space Sciences, University of Washington, Seattle, Washington 98195, USA.

M. O. Fillingim, R. P. Lin, G. K. Parks, Space Sciences Laboratory, University of California, Berkeley, California 94720, USA.

A. Szabo, Laboratory for Extraterrestrial Physics, NASA Goddard Space Flight Center, Greenbelt, Maryland, USA.

Substorms, Storms, and the Storm-Time Plasma Sheet

Wolfgang Baumjohann and Rumi Nakamura

Space Research Institute, Austrian Academy of Sciences, Graz, Austria

Rainer Schödel and Kai Dierschke

Max-Planck-Institut für extraterrestrische Physik, Garching, Germany

Some substorms are associated with strong particle injection into the Earth's ring current, while others are not. The former type of substorms occurs during magnetic storm episodes. Those associated with strong injections are found during the main phase of magnetic storms and feature much more pronounced magnetic field dipolarization in the near-Earth tail than non-storm substorms. The more prolonged duration of southward IMF during storm times and the associated rapid recurrence of several substorms may explain the higher temperature and the much increased entropy in the storm-time plasma sheet. The stronger dipolarization and the higher temperature of the plasma sheet during storm-time substorms lead to a more effective injection of more energetic particles into the ring current and may, hence, explain the storm-time decrease in the Dst index.

1. INTRODUCTION

Data such as presented in Figure 1 clearly show that there is no one-to-one correlation between substorm activity, indicated by the multitude of AE index excursions, and magnetic storms, indicated by the negative excursion of the Dst index on 28 August 1978. On the other hand, there is a clear correlation between the southward component of the interplanetary magnetic field, B_z, and Dst. This has led to the idea that the decrease of the Dst index, and thus the injection of energetic particles into the ring current, is merely the result of an enhanced level of near-steady convection due to the enhanced solar wind-magnetosphere coupling caused by prolonged southward IMF.

However, the Earth's magnetotail behaves intrinsically in a non-steady fashion, there being only rare intervals of near-steady convection. The transport of magnetic flux, mass, momentum, and energy from the tail is governed by unsteady convection, ranging from the high-speed flow bursts, which have time scales of some minutes, to the classical substorm with its 1–2 hour duration (see, e.g., *Baumjohann*, 1996). In the present paper, it will be shown that a magnetic storm does not change this behavior. But during magnetic storms the dipolarization during substorm onset is more pronounced. Furthermore, the recurrence of several substorms in rapid succession produces a much hotter plasma sheet particle population.

2. MAGNETIC FLUX TRANSFER

Baumjohann et al. [1996] did a superposed epoch study to see possible differences in the behavior of the near-Earth tail around substorm onsets that occurred during the main phase of a magnetic storm and those that were not accompanied by magnetic storm activity ($Dst > -25$ nT). Figure 2 shows that the average behavior of the near-Earth tail magnetic field at radial distances between 10 and 20 R_E is significantly different during the two types of substorms.

56 STORM-TIME PLASMASHEET

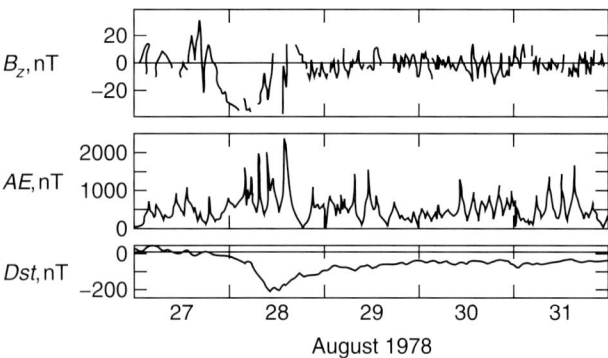

Figure 1. Interplanetary magnetic B_z component and *AE* and *Dst* indices during a magnetic storm (on 28 August) and a subsequent interval of large-amplitude interplanetary Alfvénic fluctuations (after *Gonzalez et al*, 1994).

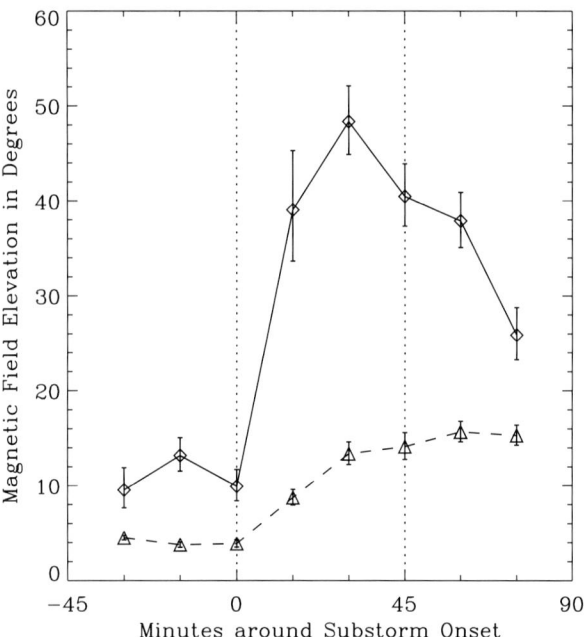

Figure 2. Superposed traces of the magnetic field elevation angle in the central plasma sheet (*left-hand diagram*) during storm-time (*solid traces*) and non-storm (*dashed traces*) substorms. The traces were constructed by averaging the measured values in 15-min bins with respect to substorm onset, separately for 7 substorms that occurred during the main phase of a magnetic storm and 35 substorms where the *Dst* index was above –25 nT. The dashed vertical lines mark substorm onset and the approximate start of the recovery phase (from *Baumjohann et al.*, 1996).

The difference between the two types of substorms lies mainly in the development of the magnetic field elevation angle. During substorms that are not accompanied by magnetic storm activity, the magnetic field dipolarization appears to be very gradual, reaching its highest elevation angles only during the recovery phase. Moreover, the dipolarization is not very pronounced, with an average maximum elevation angle of only 15°. On the other hand, for substorms which occur during the storm main phase, the magnetic field in the central plasma sheet starts to become rather dipolar immediately after substorm onset and the maximum field elevation reaches nearly 50°.

3. PLASMA SHEET HEATING

Figure 3, again from *Baumjohann et al.* [1996], shows the average ion temperature in the central plasma sheet, separately for storm-time and non-storm substorms. The figure illustrates that heating of the ion population in the central plasma sheet occurs during the substorm expansion phase. The temperature increase from substorm onset to the beginning of the recovery phase is about the same for both types of substorms, of the order of 2.5×10^7 K, or roughly 2 keV. However, the average ion temperature level before onset is clearly different for storm-time and non-storm substorms. In addition, the heating seems to occur more rapidly during storm-time expansion phases, resulting in an average ion energy of 8 keV only 15–30 min after the onset of a storm-time substorm, while the typical central plasma sheet ion

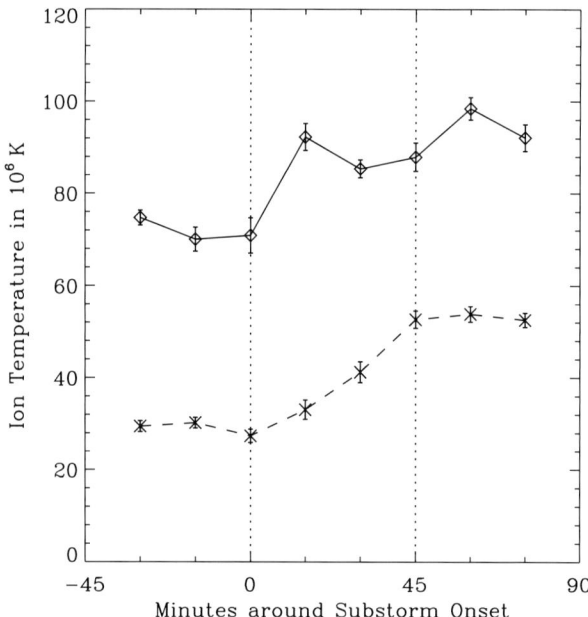

Figure 3. Average variation of the ion temperature in the central plasma sheet for substorms occurring during a storm main phase (*Dst* < –25 nT; *solid line*) and for non-storm substorms (*Dst* < –25 nT; *dashed line*). The vertical error bars give the errors of the mean values. The dashed vertical lines mark substorm onset and the average start of the recovery phase (from *Baumjohann et al.*, 1996).

has only 3–4 keV during the expansion phase of non-storm substorms.

The difference in the plasma sheet ion temperature between non-storm and storm times also is apparent in Figure 4, which shows the average radial variation in the central plasma sheet based on five years of Geotail data (courtesy of T. Mukai) during non-storm intervals ($Dst > -25$ nT) and during magnetic storms ($Dst < -25$ nT). Especially tailward of about 15 R_E the tail plasma is much more energetic during magnetic storms.

The strongly heated storm-time tail plasma is likely a result of ongoing substorm activity due to the sustained southward interplanetary magnetic field typical for the storm main phase. Actually, *Liu and Rostoker* [1995] have shown that the tail plasma will gain energy during a series of substorms if it is recirculated back and forth between the tail and the inner magnetosphere. During the growth phase, the field is stretched and gets weaker. Hence, the ions will experience Fermi deceleration (0°-particles), which is proportional to the square of the ratio of the field line length before and after stretching, and betatron (90°-particles) deceleration, which is proportional to the ratio of the field strength before and after. In the highly-stretched current sheet typical for the last stage of the growth phase, the ions will behave non-adiabatically and undergo pitch angle scattering. The subsequent dipolarization will accelerate the

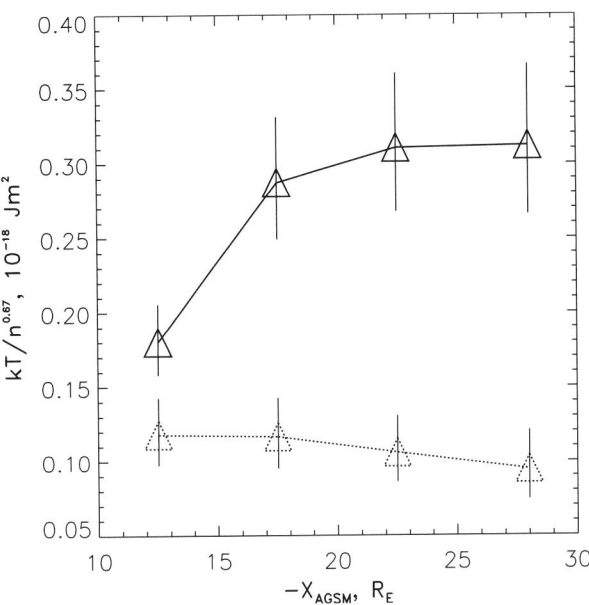

Figure 5. Average radial variation of specific entropy in the central plasma sheet during magnetic storms ($Dst < -25$ nT; *solid line*) and for non-storm intervals ($Dst < -25$ nT; *dashed line*). The vertical error bars give 0.2 s (Geotail data courtesy T. Mukai).

ions adiabatically by the Fermi and betatron process, since the field now gets stronger and the field lines shorter.

At first guess, one might think that there will be no net gain in energy during this process. However, due to non-adiabatic scattering, the Fermi-decelerated ions are betatron-accelerated and vice versa. *Liu and Rostoker* [1995] showed that a Maxwellian distribution developed a high-energy tail during subsequent recirculations. That the strong heating of the tail plasma during magnetic storms is indeed caused by non-adiabatic processes is apparent in Figure 5, which is based on the same Geotail data as Figure 4. During storm-time intervals the entropy level, again especially tailward of 15 R_E, is about a factor of three higher than during non-storm times.

4. RING CURRENT INJECTION

The *Dst* variation a magnetic disturbance caused by energetic particles encircling the Earth due to the combined effect of the gradient and curvature drift in a near-dipolar field. Hence, what is needed to create a notable *Dst* variation, is to bring energetic particles from the tail close enough to the Earth so that they experience strong enough gradient and curvature forces to perform complete or partial orbits around the Earth. Since the magnetic drift forces increase with increasing particle energy, more energetic particles will experience a stronger azimuthal drift for the same

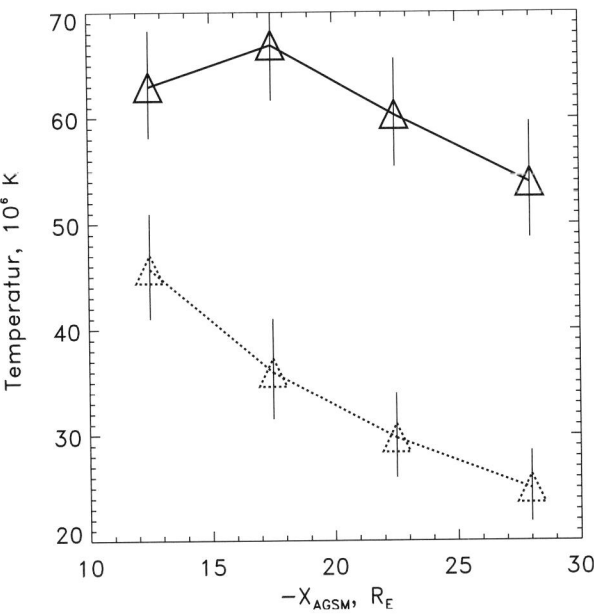

Figure 4. Average radial variation of the ion temperature in the central plasma sheet during magnetic storms ($Dst < -25$ nT; *solid line*) and for non-storm intervals ($Dst < -25$ nT; *dashed line*). The vertical error bars give 0.2s (Geotail data courtesy T. Mukai).

magnetic gradient and curvature. In addition, they will cause a larger *Dst* index, since the latter depends on the energy of the ring current particles. Apparently, the much stronger dipolarization during the storm-time substorms will bring the heated tail plasma closer to the Earth. Moreover, the more energetic particles will cause a stronger field depression.

5. CONCLUSIONS

The material presented above strongly supports the notion that substorms are, indeed, the building blocks of magnetic storms. However, there is another equally important ingredient to a magnetic storm, i.e., a prolonged interval of strongly southward interplanetary magnetic field, leading to continuous and strong solar wind-magnetosphere coupling and thus to a long interval of strong magnetospheric convection. Only then the basic mode of magnetic flux and particle transport from the tail into the inner magnetosphere, the magnetospheric substorm, is strong enough and has such a rapid recurrence that it can effectively enhance the energy content of the ring current.

REFERENCES

Gonzalez, W. D., J. A. Joselyn, Y. Kamide, H. W. Kroehl, G. Rostoker, B. T. Tsurutani, and V. M. Vasyliunas, What is a geomagnetic storm? *J. Geophys. Res.*, *99*, 5771–5792, 1994.

Baumjohann, W., Near-Earth plasma sheet dynamics, *Adv. Space Res.*, *18(8)*, 27–33, 1996.

Baumjohann, W., Y. Kamide, and R. Nakamura, Substorms, storms, and the near-Earth tail, *J. Geomag. Geoelec.*, *48*, 177–185, 1996.

Liu, W. W., and G. Rostoker, Energetic ring current particles generated by recurring substorm cycles, *J. Geophys. Res.*, *100*, 21897–21910, 1995.

W. Baumjohann and R. Nakamura, Space Research Institute, Austrian Academy of Sciences, Schmiedlstr. 6, A-8042 Graz, Austria.(e-mail: baumjohann@oeaw.ac.at)

R. Schödel and K. Dierschke, MPI f. extraterr. Physik, Postfach 1312, D-85741 Garching, Germany.

O^+ Transport into the Ring Current: Storm versus Substorm

A. Korth[1], R. H. W. Friedel[2], M. G. Henderson[2], F. Frutos-Alfaro[3], and C. G. Mouikis[4]

The importance of ionospheric oxygen ions during storms has been well documented. Substorms occurring during the main phase of storms have a pronounced effect on the storm-time buildup of O^+ in the ring current, being well correlated with the steepest decline in D_{st} during the main phase. The oxygen to hydrogen ratio (O^+/H^+) is investigated during six large storms of solar cycle 22, at solar-maximum. For this study we use CRRES particle and electric field data as well as indicators for storm (D_{st}) and substorm (AE) activity. In general, storm-time substorms are much more effective in transporting O^+ into the ring current region. We investigate the relative contribution of storms versus substorms to the observed buildup of the O^+/H^+ ratio during these six storms by developing a *measure* of the storm (ΣD_{st}/hour) and substorm (ΣAE/hour) contribution to O^+ transport. We find a virtually linear dependence of the O^+/H^+ ratio on the strength of a storm, with no saturation effects, indicating a limitless reservoir of O^+ in the near-Earth plasma sheet, while substorms only seem effective in adding to the O^+/H^+ ratio for small to medium sized storms. Much of the effectiveness of substorms occurring during storms can be explained by the ion outflow composition which favors O^+ over H^+ for most levels of AE activity, and the fact that ionospheric O^+ is preferentially transported closer to the Earth than ionospheric H^+, where it becomes available for subsequent substorm energization and transport.

[1]Max Planck-Institut für Aeronomie, Katlenburg-Lindau, Germany
[2]Los Alamos National Laboratory, Los Alamos, New Mexico
[3]University of Costa Rica, Physics Department, San José, Costa Rica (now at Max-Planck-Institut für Aeronomie, Katlenburg-Lindau, Germany)
[4]Space Science Center, Universtiy of New Hampshire, Durham, New Hampshire

Disturbances in Geospace: The Storm-Substorm Relationship
Geophysical Monograph 142
Copyright 2003 by the American Geophysical Union
10.1029/142GM07

1. INTRODUCTION

The role of ions, in particular ionospheric oxygen O^+ in the ring current dynamics during storms and substorms, have received considerable attention in the recent years and in the resurgent discussion of storm versus substorm relationships. While previous work [*McPherron*, 1997; *Korth et al.*, 1998, *Grande et al.*, 1998] has shown that the effect of substorms on D_{st} is at best minor, *Kamide et al.* [1998] speculates that the importance of ionospheric O^+ in storm-time ring current buildups is controlled by substorms.

Most of the body of observational data on the ion composition in the ring current during storms and substorms comes from AMPTE CCE CHEM [*Daglis et al.*, 1993] and CRRES, the last equatorial geosynchronous transfer orbit mission flown in this field. Using CRRES data *Daglis* [1997] showed that the abundance of O^+ ions are greatly enhanced during the main phase of storms, and that larger storms (as measured by D_{st}) generally have a higher abundance than moderate or weak storms. For the *superstorm* of March 1991, O$^+$ ions eventually dominated, contributing more than 70% of the ion energy density near storm maximum [*Daglis et al.*, 1999]. CRRES data has also been used to investigate the topic of this paper: *Daglis* [1998] showed that substorm-associated compositional changes are generally larger for substorms occurring during storms compared to non-storm substorms. This result was confirmed by *Korth et al.* [2001] who also showed that the O$^+$/H$^+$ ratio was enhanced during storm-time substorms compared to non-storm substorms.

While the CRRES observations firmly established an important role for the ion composition (and the O$^+$ ion abundance in particular) in storm and substorm dynamics, the possible reasons for this O$^+$ ion abundance increase during such disturbed times were left at the speculative stage. The mere fact that O$^+$ ions are involved pointed to a terrestrial ionospheric source [*Daglis et al.*, 1994], who speculated that "compositional changes ... are presumably due to the effective source geometry of the polar ionosphere and the transport paths of ionospheric oxygen."

In this paper we attempt to characterize the differences between the storm and non-storm substorm O$^+$ increases in a more quantitative way, and we further attempt to explain these differences in terms of the underlying ion transport in the presence of storm-time and substorm-time convection electric fields. This work both summarizes and extends the CRRES observations by *Daglis* and co-workers, by also offering an explanation for the observed O$^+$ dynamics during storms and substorms.

Work by *Korth et al.* [2000] has suggested that the dynamical access of O$^+$ ions to the ring current during storms is consistent with convective transport from the near-Earth plasma sheet. There is a direct response of the ionosphere to magnetic storm drivers in the solar wind such as CME's. Polar wind flux dominated by O$^+$ plasma is associated with these drivers [*Moore et al.*, 1999]. While the high-latitude ionosphere can act as a source for the mid- to far-Earth magnetotail, there is also evidence that the inner regions can be fed by more direct ionospheric access from the auroral regions [*You and Whalen*, 1992] for plasma sheet energies (~ several keV). There have also been some observations of outflowing ionospheric ions at ring current energies [*Kaye et al.*, 1981, *Lundin and Hultqvist*, 1989], but their quantitative significance is not established. We have given evidence that substorms are more effective in transporting O$^+$ into the inner magnetosphere during storms; but what both the *Daglis et al.* [1998] and the *Korth et al.* [2001] study left open was the question of which *agent* is the most significant for oxygen transport into the inner magnetosphere—large-scale convection in response to large scale storm-time electric fields, or localized, intense convection in response to substorm-induced electric fields. To complete the picture we also need to consider the ionospheric source / outflow rates for both O$^+$ and H$^+$, and some details for the different transport characteristics for these ions that bring them into the central plasma sheet. It is the combination of all these factors (outflow rates, transport to the plasma sheet, and storm-time / substorm-time dynamics transport) that will determine the observed O$^+$/H$^+$ ratio in the outer radiation belt.

In the context of this paper *transport* is used as a synonym for *acceleration*, since any transport that moves particles from regions of low to high magnetic field strengths automatically leads to particle energization as long as the first adiabatic invariant is preserved; which is the case for both storm and substorm transport.

In this study we investigate six magnetic storms observed during the CRRES period from January to July 1991, which includes the much studied *superstorm* of March, 1991. We first expand here on the results of *Daglis et al.* [1998] and *Korth et al.* [2001] and then attempt to distinguish between the convective transport effects of storms versus the convective transport effects of substorms.

We show here that the same results as obtained by *Daglis et al.* [1998] and *Korth et al.* [2001] apply to this expanded study of six storms, in that the O$^+$ energy density injected during non-storm-time substorms, in general, is less than that during storm-time substorms. We further show that the maximum O$^+$ ratio obtained during these storms is dependent on both the size of the storm and the frequency/strength of substorms during the storm. While the former dependence is almost linear, the dependence on substorms is double valued—a given O$^+$ ratio can be achieved for both low and high substorm activity. The two values refer to a small and a very large storm, which seems to indicate that substorms can contribute to the O$^+$ ratio up to some threshold main storm strength, when the contribution to the O$^+$ ratio from the storm-time electric fields is so large that the contribution from additional substorms becomes insignificant. However, since these results are based on a sample of six storms only, they must be treated as preliminary.

We also show that based on the measured ion outflow rate in response to storms and substorms [*Yau et al.*, 1988] substorms seem to be more effective in producing ion outflow than storms, and based on the fact that the slower outflow speeds of heavier ions lead to them being deposited into the central plasma sheet closer to Earth. Ionospheric O$^+$ is preferentially transported closer to the Earth compared to ionospheric H+ leading to an O+ rich source region for subsequent substorm energization and transport.

2. MISSION AND INSTRUMENTATION

This investigation uses particle and electric field measurements on the Combined Release and Radiation Effects Satellite (CRRES). CRRES had an elliptical, 18.1 degrees inclination orbit, an orbital period of ~ 10 hours, a perigee of 350 km, and an apogee of 33.580 km. CRRES covers the inner magnetosphere up to $L \sim 7.5$ with a time resolution of 5.5 hours per pass. The apogee of the orbit covers magnetic local times from 08:00 through midnight to 14:00.

We use particle data from the Magnetospheric Ion Composition Spectrometer (MICS) [*Wilken et al.*, 1992], the Medium Electrons B spectrometer (MEB) [*Korth et al.*, 1992], and the electric field probe [*Wygant et al.*, 1992] on CRRES.

In Figure 1 we show schematically the local time position of the CRRES orbits from which the data for the six storms in this study are taken. All the orbits are in the midnight to dusk sector, and lie in the direction of convective flow for ions from the near-Earth tail.

3. OBSERVATIONS

Here we present overview and detailed observation of the six storms used in this study. For completeness we also present some detailed observations of individual substorms during both storm and non-storm times.

3.1. Overview

The data presented in Plate 1 give an overview of 527 orbits from January 12 to August 20, 1991 of CRRES during the solar maximum of the solar cycle 22.

This is the most disturbed period during the lifetime of CRRES. The CRRES data are plotted in a L versus universal time presentation. Data are binned in L every half orbit (~ 5.5 hours). The spacecraft is traversing from the inner radiation belt through the slot region into the ring current and outer radiation belt.

Panel 1 gives a comprehensive view of the variations of the dawn-dusk (E_y) electric field component in a frame of reference corotating with the Earth. Panels 2–4 display the variation of the energy density of the ion species O+ and H+ (2 L ranges) which are the most abundant ions of the ring current. For each species the data are integrated over the full energy range. The energy ranges are representative for the main part of the ring current density (~ 30 to ~ 300 keV),

and are used here to show the importance of the ion composition in contributing to the ring current. All three panels are displayed in the same energy densities range from 0.1 to 100 keV/cm³. Panels 5 and 6 show the AE index and the D_{st} index, respectively. AE is an indicator for substorm activity and D_{st} is a measure of the variation of the horizontal component of the magnetic field, essentially caused by the ring current.

Several outstanding features can be determined from Plate 1 directly: Numerous rapid flux increases are observed in the ions in the L range between 2 and 6 throughout the whole period. The enhancements are correlated with strong decreases in D_{st}, indicating an increase in the ring current. Often these periods are associated with high substorm activity (AE index), but AE activity may also depress D_{st} during non-storm times or inhibit recovery of D_{st} during prolonged recovery phases. The local measure of the dawn-dusk field in Panel 1 also has good correlation with D_{st} showing that for larger D_{st} storms the electric field measured at CRRES is generally stronger, lasts longer and penetrates to lower L-values.

The width of the slot region depends on the level of disturbance and tends to be broader during quiet times. For strong magnetic storms with $D_{st} \lesssim -100$ nT the flux maximum moves into the slot region earthward and during the recovery phase away from the Earth. The decay time for O+ ions at ~ 100 keV is ten times faster than that for protons,

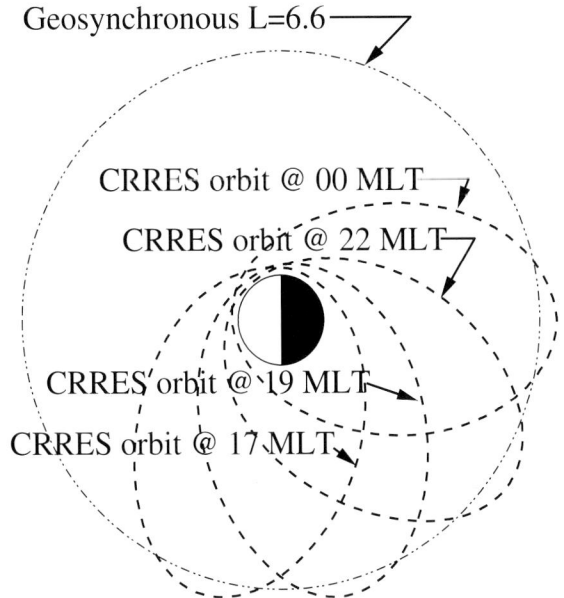

Figure 1. Schematic of the position of the CRRES orbits in the equatorial plane used for this study. Indicated in the plot are the apogees MLT value for the orbits.

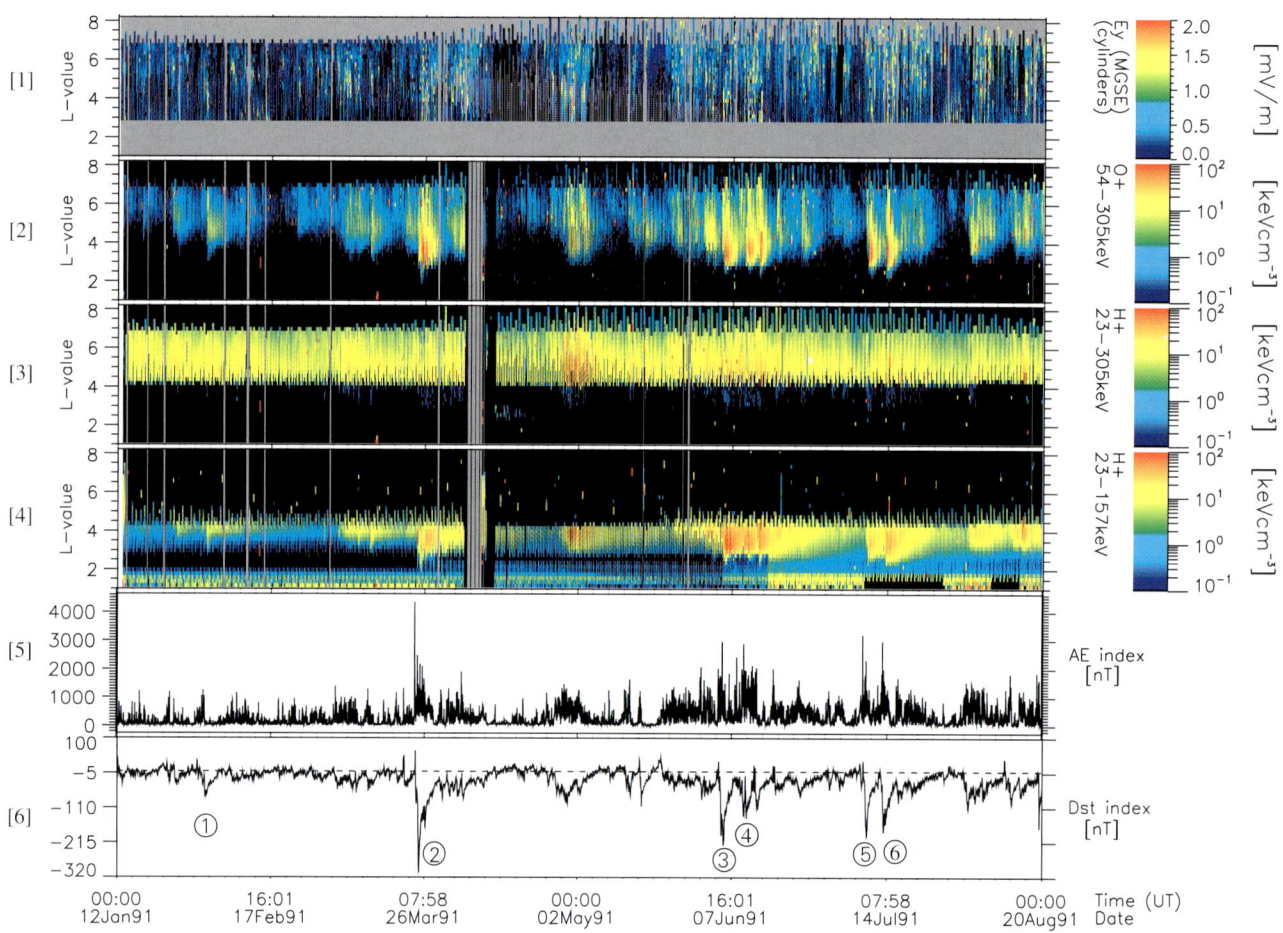

Plate 1. L sorted data for 527 CRRES orbits from January 12, 1991 to August 20, 1991.

which have lifetimes of days to weeks. The dawn-dusk electric field in Panel 1 is strongest during the main phase of magnetic storms and can penetrate to L values as low as three. The enhancements of the dawn-dusk electric field are associated with global enhanced convection. The numbers indicated in the D_{st} panel are the six storms to be investigated.

3.2. Six Storms—Details

The data presented in Figure 2 are the six magnetic storms marked in Panel 6 of Plate 1.

During this period the spacecraft's apogee was in the midnight to afternoon sector (see Figure 1) and data were available from both the MICS and the MEB instruments. We chose storms in this MLT sector because this is the sector through which all of the new ion populations convected in from the near-Earth plasma sheet must pass. We are thus always sampling in the main body of the full or partial ring current, avoiding the main phase asymmetries of the ring current observed at dawn. We know that during the main phase a partial ring current dominates D_{st} [*Korth et al.*, 2000], and in this way we are sure that we are also sampling the in situ particle populations contributing to the partial ring current.

For each magnetic storm we plotted three panels. Panels 3 show the D_{st} index as a measure of the ring current (the dashed horizontal lines indicate zero D_{st}), Panels 2 the AE index as a measure of substorm activity, and Panels 1 the O^+/H^+ energy density ratio in the L-range from 4.5 to 8.5 (apogee of CRRES). Each line segment in Panels 3 represents one CRRES orbit, and while some orbital dependence (radial gradients) is apparent at times (such as in Panel 1 of event ① before onset), during the main phase and recovery of the storm activity is strong enough so that radial effects are not apparent. The magnetic local time is given for each event, and the time base for all is 3 days.

In calculating the energy density ratio for O^+/H^+ care was taken to use the same energy range for each species. This assures that changes in the O^+/H^+ ratio are not due to changes in the average energy of the ring current. MICS had two proton counters that operated at different L-ranges; a digital counter restricted to $L < 4$ and an analog counter restricted to $L < 4$ (see Figure 2). Since we are restricting ourselves to the L-range from 4.5 to 8.5, only the digital counter is used. The overlapping energy range used for the two species is ~ 54 to ~ 305 keV.

There is a wide variety in these six storms when looking at their AE and D_{st} signatures. Events ② (March 25, 1991) and ⑤ (July 9, 1991) are the most *classical*: the event starts with an increase of D_{st} due to a sudden magnetopause compression, followed by a main phase of roughly 12 hours, and a slow recovery. There is strong AE activity right at the compression phase, which is probably *not* due to substorms. Noticeable in both these events is the slow decrease in D_{st} at the beginning of the main phase, turning steeply towards the minimum D_{st} for the event just as AE activity picks up. Both events show fairly continuous AE activity throughout the recovery phase. Events ① (February 2, 1991) and ⑥ (July 13, 1991) are similar minus the sudden magnetopause compression, both show an accelerated decrease of D_{st} when AE activity increases. Events ③ (June 5, 1991) and ④ (June 10, 1991) are unusual in that event ③ exhibits a long main phase and gradual decrease to minimum D_{st}, and both ③ and ④ show a double minimum in D_{st}.

In spite of these differences between the storms the most consistent signature throughout these events is that of the O^+/H^+ ratio. For all events the O^+/H^+ ratio increases throughout the main phase, reaching its maximum around the time of D_{st} minimum. The maximum ratio can be as high as 10 depending on the strength of the storm. Before the beginning of the magnetic storm or at the end of the recovery phase the O^+/H^+ ratio is as low as 0.1 to 0.5.

3.3. Storm Versus Non-Storm Substorms

In the data presented in Figure 2 it is difficult to see the effects of individual substorms, and we use the substorm activity proxy AE to indicate substorm activity. To convince ourselves that we really are seeing substorm events during the main phase of these storms (and that these storms *are* indicated by AE) we investigated several of the storms in more detail. As an example we picked the storm main phase of the July 13, 1991 magnetic storm which is event ⑥ of Figure 2. Figure 3 shows detailed particle data from two CRRES orbits, one just before and one just after storm onset in the storm main phase.

Panel 1 and 2 show the energy density of the oxygen content for ring current energies (~ 50 to 300 keV) and the O^+/H^+ energy density ratio (as described in Figure 2) from the MICS instrument. Panel 3 and 4 show the AE and D_{st} indices for this period. Panel 5 and 6 of Figure 3 display ion fluxes in the energy range from ~ 35 to 3200 keV in 12 channels and electron fluxes from ~ 20 to 300 keV in 14 channels perpendicular to the magnetic field from the MEB instrument. The highest flux is observed in the lowest energy channels. In both panels the storm-time substorms (injections) from the ions and from the electrons are observed between $L = 5$ and $L = 7.6$ during the main

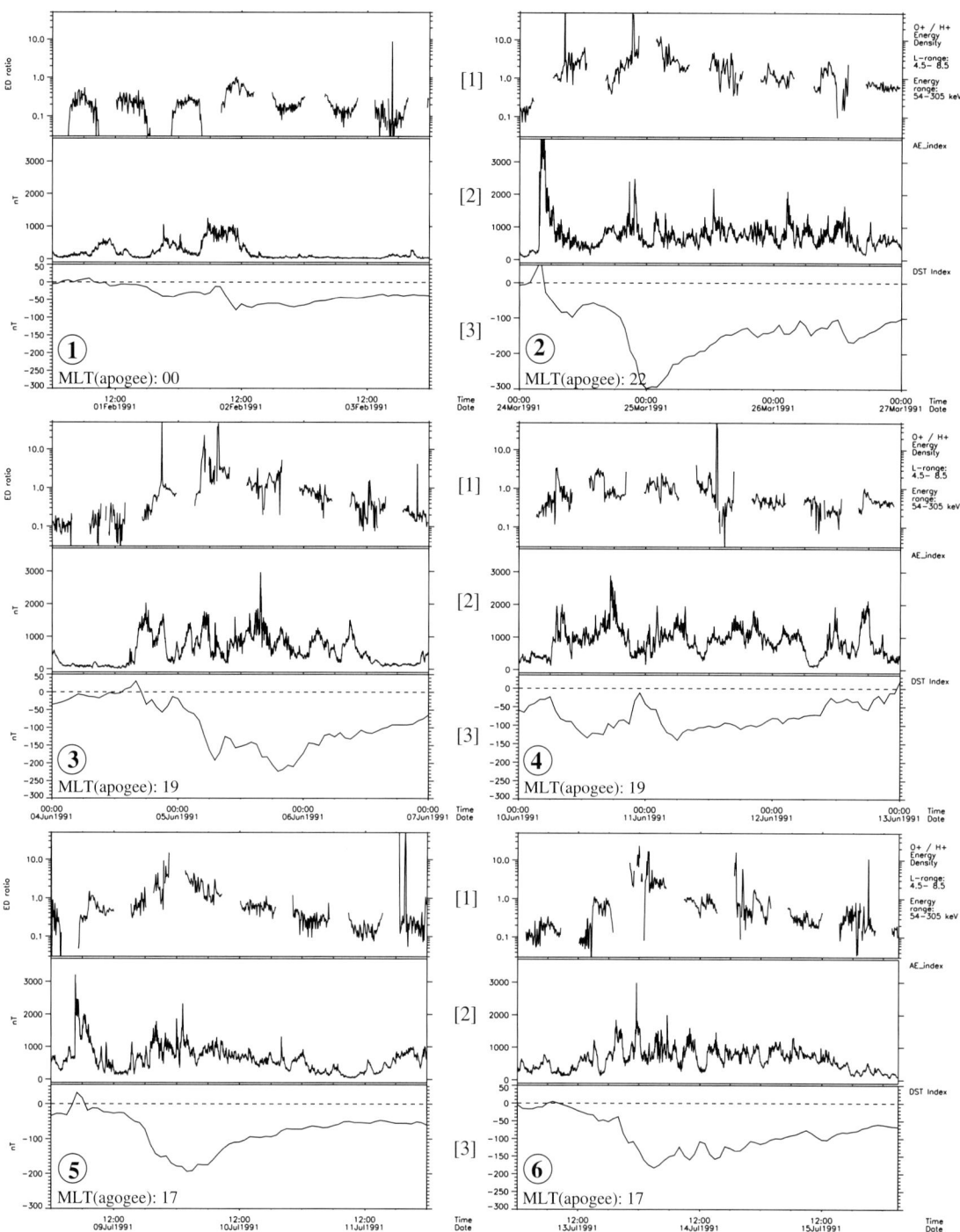

Figure 2. The O$^+$/H$^+$ energy density ratio (~ 54 to ~ 305 keV) during the six largest storms in the time frame from January to August 1991 during solar maximum of the solar cycle 22.

phase of the storm. The vertical solid lines mark the beginning of the ion injections for each observed substorm.

The ion and electron injections signatures are dispersionless, which indicates that the satellite is within the substorm injection region. Most of the injections line up with corresponding increases in the AE index, and those that do not could be due to substorms occurring away from CRRES, or CRRES being at lower L shells.

It is clear that with the beginning of the storm main phase substorms the oxygen energy density is increasing up to an order of magnitude. Likewise the O^+/H^+ energy density ratio increases with each substorm. The O^+/H^+ ratio shows its maximum near the minimum of the D_{st} value.

In contrast to the storm-time substorms investigated in Figure 3 the substorms shown in Figure 4 occurred during a quiet, non-storm period of very low geomagnetic activity ($D_{st} > -30$ nT) on January 24, 1991. The substorms occur around 16:51 UT and 17:01 UT (about 23.6 MLT) near $L = 6.2$. They can be identified in panel 4 by the intensification of the ion channels in the energy range of ~ 35 to 3200 keV (vertical dashed lines). Panels 2 and 3 display the energy density of singly charged oxygen and protons, respectively. The ratio of both is given in panel 1. The values for the ratio are averaged over 3 minutes, in order to smooth out spikes. The O^+/H^+ energy density ratio remained below 0.32.

Other investigations of non-storm time substorms on 7, 10, 12, and 14 February showed O^+/H^+ ratio between 0.15 and 0.65. This is much lower than the O^+/H^+ ratios of 10 and

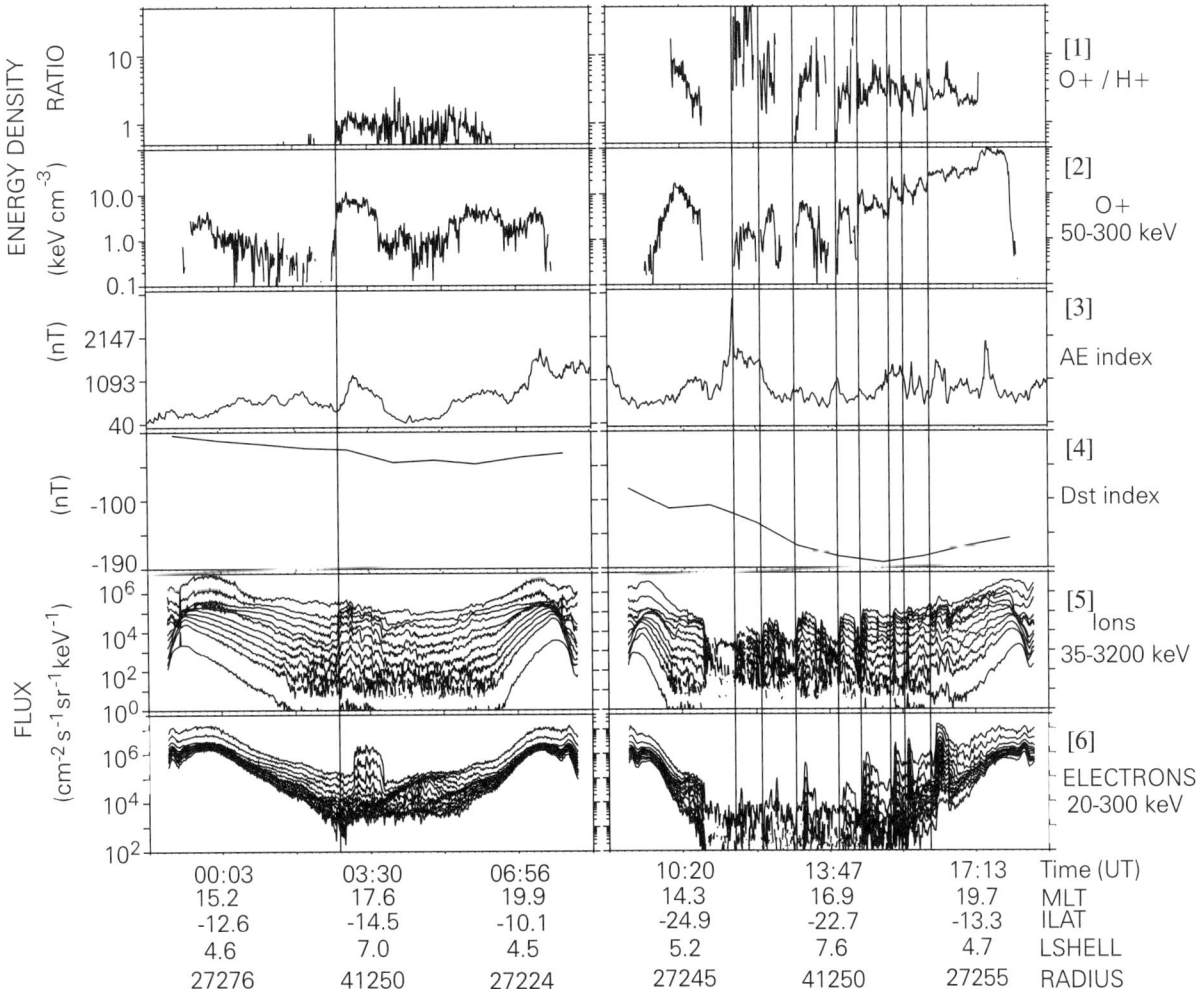

Figure 3. The O^+/H^+ energy density ratio of storm-time substorms during the main phase of the July 13, 1991 storm (marked ⑥ in Plate 1 and Figure 2). Data from two consecutive orbits (855 and 856).

higher that are observed for substorms occurring during the main phase of storms. A more detailed examination of the storm-time and non-storm time differences of the O^+/H^+ ratio can be found in *Daglis et al.* [1998] and *Korth et al.* [2001]

4. EFFECTIVENESS OF STORM VERSUS SUBSTORMS IN O^+ TRANSPORT

To further investigate the storm versus substorm contribution of O^+ we attempt to identify the cumulative effect of a series of substorms on O^+, and the cumulative effect of the storm. The main difficulty here is to develop a quantitative measure of *strength*—especially as applied to the effectiveness of transport. For both storms and substorms the access to the ring current is controlled by electric fields, large scale dawn-dusk fields in the case of storms, and more localized dipolarization fields for substorms.

How to measure these effects in the absence of global electric field measurements? CRRES only provides in situ electric field measurements which do not distinguish between large scale or localized electric fields; and more often CRRES is not in the region of substorm-induced electric fields, which are very localized in MLT, and do not often extend inside of the geosynchronous orbit [*Birn et al.*, 1998]. We use here the easily accessible D_{st} and AE indices as proxies for storm and substorm activity. While both indices have their limitations, and there is some mixing of phenomena they measure, they are widely used and understood in the space physics community.

4.1. Storm Transport

D_{st} broadly measures the strength of the ring current, which is dependent on the capture of fresh plasma sheet populations onto closed field lines [*Williams*, 1987]. The amount of capture depends on the large-scale convective electric field strength; D_{st} is thus proxy for the transport due to large-scale, storm-time dawn-dusk fields.

To obtain a measure of the effectiveness of this *storm-transport* not only the size of this electric field but also the duration is of relevance. To capture both we use a derived index, summing D_{st} for 12 hours before the minimum D_{st}, normalized by the time interval, yielding a value of $\Sigma D_{st}/h$ for each storm.

The summing period used here for D_{st} has been chosen to capture only the period of maximum large scale convective transport, which is limited to the period of onset to D_{st} minimum. This period is characterized by the largest in situ electric fields observed (see Panel 1 in Figure 2

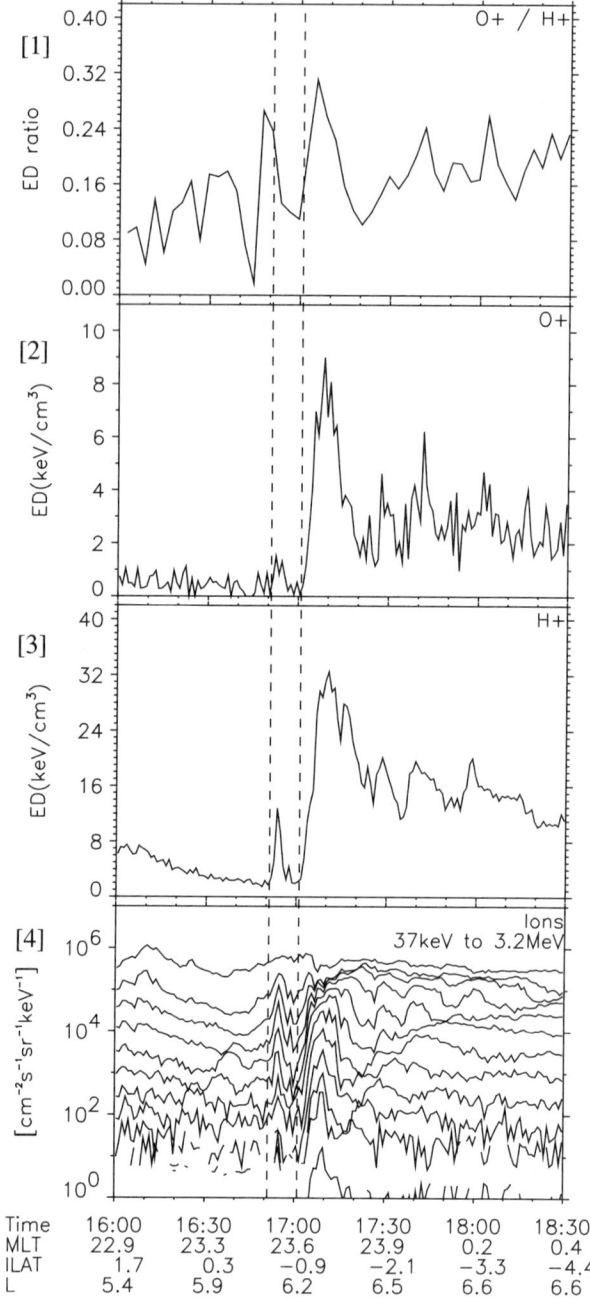

Figure 4. O^+/H^+, O^+, and H^+ observations during a non-storm-time substorm on January 24, 1991 (orbit 445).

and *Korth et al.* [2001]). While the length of this period varies from storm to storm, 12 hours is a good average compromise.

This value is plotted against the maximum value of the O^+/H^+ ratio observed during that storm in Figure 5.

Defining the *strength* of a storm in this way orders the data in an almost linear fashion, with the *stronger* storms leading to higher O^+/H^+ ratio. This is quite a remarkable correlation, given the range of D_{st} minima of the studied storms (–80 to –300 nT), and the variety in the temporal behaviour of the six events under discussion (see Section 3.2).

4.2. Substorm Transport

The AE index has been the traditional measure of substorm activity. Technically it measures the strength of the auroral electrojets, and thus the strength of the substorm current wedge [*Barfield et al.*, 1985]. This current wedge is a direct result of a localized reconfiguration of the magnetic field caused by substorm onset; it is this *dipolarization* [*Baker et al.*, 1981] of the field that causes the surge-like localized substorm induced convection electric fields which are responsible for the typical substorm particle injection features [*Birn et al.*, 1997, *Li et al.*, 1998] and associated transport of plasma sheet populations into the inner magnetosphere, where these populations can become trapped. AE is thus a proxy for these substorm convection electric fields, although the link maybe somewhat tenuous. AE is sensitive to any other ionospheric current systems near the auroral zone; during the main phase in particular it is difficult to detect individual substorms from the index. Nevertheless AE is the only global measure of substorm activity that is available, and is widely used as such.

Again, to obtain a measure of the effectiveness of this *substorm transport* not only the size of the substorm electric field but also the duration is of relevance. To capture both we use a derived index, summing AE for 12 hours before the minimum D_{st}, normalized by the time interval, yielding a value of $\Sigma AE/h$ for each storm. The averaging interval here is dictated by the choice made for D_{st}, since to compare the contributions of storms and substorms to the O^+/H^+ ratio we need to be considering the same interval for both. The 12 hour choice also avoids inclusion of AE enhancements associated with magnetopause compression, which are not substorm related (see event 2 and 5 of Figure 2). This value is plotted against the maximum value of the O^+/H^+ ratio observed during that storm in Figure 6.

Defining the *strength* of a substorm in this way orders the data along a double-valued curve. At first the observed O^+/H^+ ratio increases with increasing substorm activity. It is also clear by comparing to Figure 5, that substorm activity at first also increases with storm *strength*, exhibiting the known correlation between AE and D_{st} [*O'Brien et al.*, 2001]. Then for the strongest two storms (July 9th and March 24, 1991) the observed O^+/H^+ ratio continues to

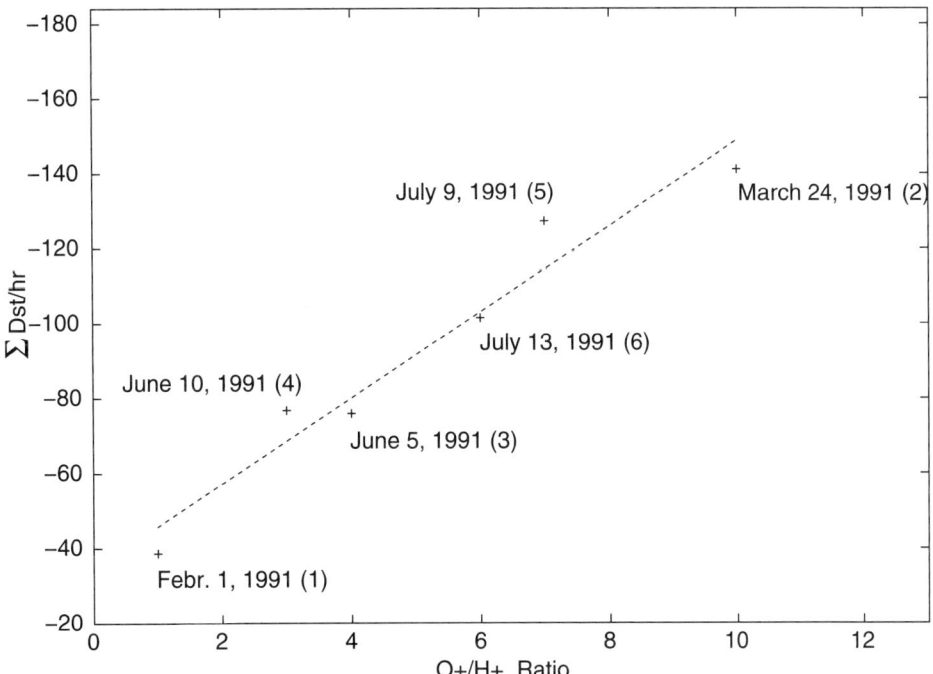

Figure 5. The sum of D_{st} per hour versus the O^+/H^+ energy density ratio for the six magnetic storms in Figure 2. The dashed line is a linear fit for the six events. D_{st} is used at 1 hour resolution.

increase while the amount of substorm activity falls off. This seems to indicate a threshold in storm *strength* for which substorm activity alone leads to additional O^+/H^+ ratio.

4.3. Interpretation

The first result, a good correlation between $\Sigma D_{st}/h$ and the maximum value of the O^+/H^+ ratio is not unexpected. An inspection of Panel 1 in Plate 1 gives some idea of the local dawn-dusk electric field measured at CRRES, and its correlation with the size of D_{st}. Unexpected is the apparent linearity of the relationship in Figure 5, given the ad-hoc nature of our derived index $\Sigma D_{st}/h$. Clearly absent is any saturation effect, at least from this limited sample of storms, indicating that the source population in the plasma sheet is not a limiting factor. No matter what the size of the convection, the plasma sheet is never *transported dry*. This indicates that the ultimate O^+ source, the high latitude ionosphere, can supply the near-Earth plasma sheet at timescales faster that any convection can remove.

The dependence of the maximum value of the O^+/H^+ ratio on substorms ($\Sigma AE/h$) is more complex. At first glance it seems that a given amount of substorm activity can lead to both a low and high value of the O^+/H^+ ratio, which is somewhat counter intuitive. However the data in Figure 6 do seem to lie along a well ordered curve, along which the strength of the storms is increasing. Figure 7 shows the relationship of storm strength to substorm strength. This shows clearly that our measure of substorms indicates less substorm activity for the larger D_{st} events of July, 9 and March, 24.

We offer three interpretations for this effect:

1. As storm strength increases the storm time large scale convective fields become large enough so that any additional transport from substorms becomes insignificant.
2. For the really large storms our measure of substorm activity is becoming less reliable.
3. As storm strength increases, there are simply less substorms. This would mean that the rate dissipation of energy input into the magnetosphere from the solar wind by storm activity alone is large and fast enough so that there is no need for additional dissipation by substorms. Assuming that our measure of substorms remains reliable, Figure 7 seems to support this interpretation.

Point 2 above in particular may offer some explanation of the roll-off effect or our measure of substorm activity for the

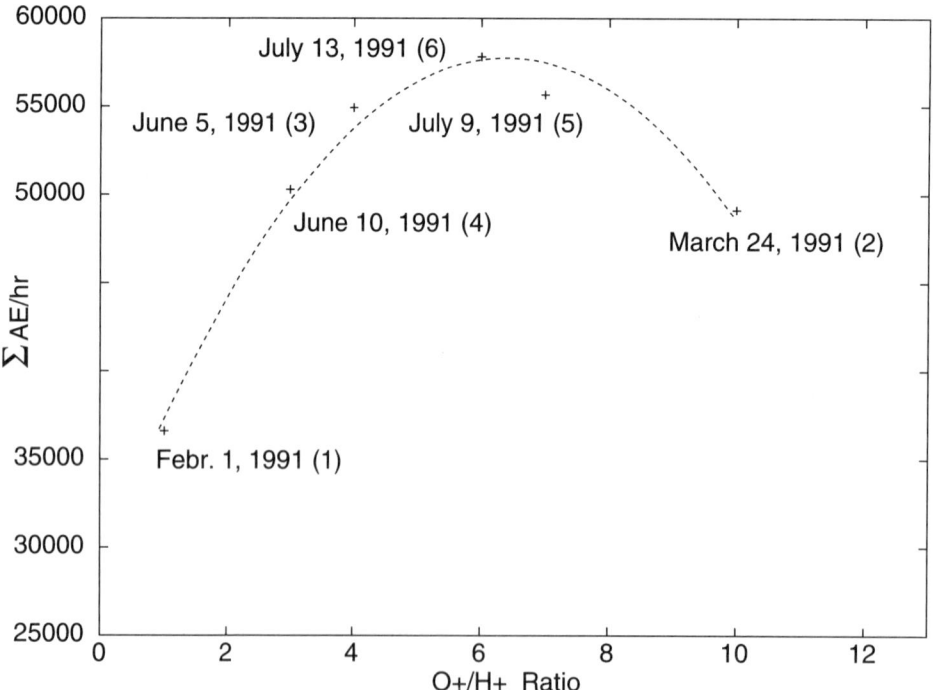

Figure 6. The sum of AE per hour versus the O^+/H^+ energy density ratio for the six magnetic storms in Figure 2. The dashed line is a second order polynomial fit for the six events. AE is used at 3 minute resolution.

large storms in Figures 6 and 7. Without the points from the March 24, 1991 point and to a lesser extend the July 9, 1991 storm, the curves could be interpreted as straight lines. For these *superstorms* the electrojets which are measured by AE move equatorward below the stations used to construct AE, thus making it a less reliable measure of substorm activity.

Both the results of the $\Sigma D_{st}/h$ and $\Sigma AE/h$ dependence are in some sense inconclusive. This is partly due to the small sample of storms used (6), and partly due to the apparent correlation between our two derived indices—in general, a larger storm leads to more substorm activity which leads to a higher value of the O^+/H^+ ratio.

5. DISCUSSION

While the evidence presented in this paper clearly shows that both storm and substorm activity can lead to an increase in the O^+/H^+ ratio in the ring current, it does not address the question of why this is the case, or why substorms occurring during storms are so much more efficient at transporting O^+ into the ring current region.

There are several factors that contribute to the increase of the O^+/H^+ ratio during storm-time substorms. First, we know that there is ionospheric outflow of H^+ and O^+ which is dependent on Kp [*Moore et al.*, 1999]. During a storm, there is a lot of auroral activity and hence a lot of outflow in the auroral zone, while the existence of large-scale storm induced electric fields facilitates good transport from the lobes towards the equatorial plasma sheet. This increases ionospheric ion access to the equatorial plasma sheet, but why would this lead to an increase of the O^+/H^+ ratio? All ion species are affected by the same auroral field-aligned electric field acceleration and $E \times B$ drift. Here one has to consider three factors:

1. The starting (or quiet, ambient) composition of the plasma sheet.
2. The ionospheric outflow *source* composition.
3. The species dependent transport characteristics of the outflow ions.

The effect of the first two factors is obvious. Unless the outflow and destination compositions are exactly the same, enhanced ionosphere → plasma sheet transport will change the plasma sheet composition. In a comprehensive study of DE-1 data *Yau et al.* [1988] showed that while both H^+ and O^+ outflow rates increase with magnetic activity as measured by D_{st} and AE, the O^+ rate outpaces H^+ for increased activity as measured by all of these indices. For D_{st} outflow of the 0.01–17 keV O^+ population

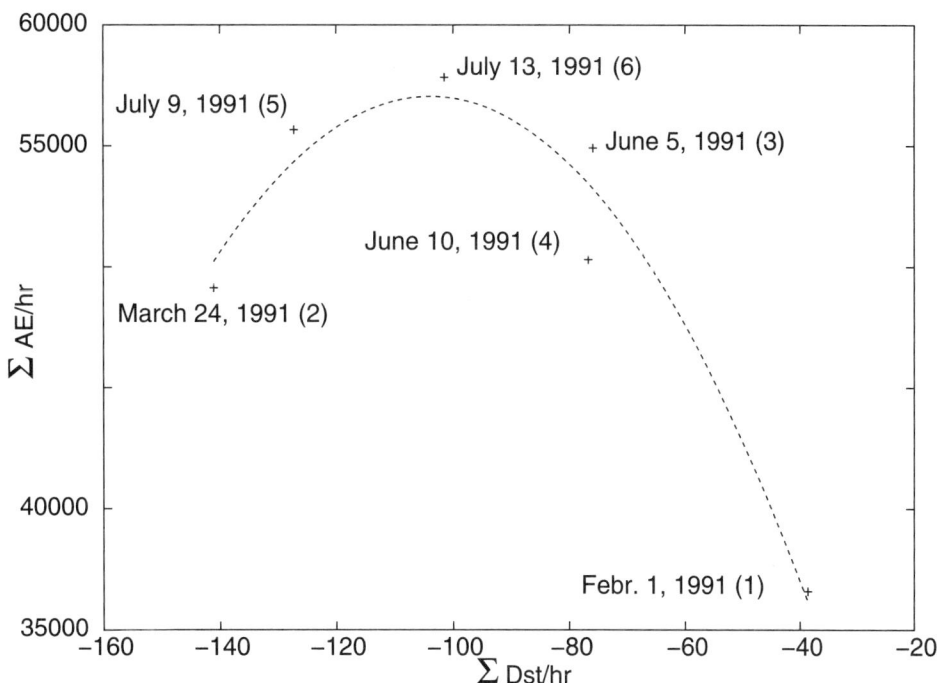

Figure 7. The sum of AE per hour versus the sum of D_{st} for the six magnetic storms in Figure 2. The dashed line is a second order polynomial fit for the six events.

is stronger than H$^+$ for D$_{st}$ < –40 nT (see Figure 8). For AE, O$^+$ outflow for the same population is not only stronger at most values of AE, but also increases at about twice the rate in log space than H$^+$ (see Figure 9). This shows that the O$^+$/H$^+$ ratio from the ionospheric outflow source is > 1 for anything from a moderate storm upwards, and > 1 for almost any substorm activity as measured by AE. If we assume that this ionospheric outflow is the primary controlling factor of plasma sheet composition (which is reasonable, as *normal* plasma sheet O$^+$/H$^+$ ratio is <1), then the results from *Yau et al.* [1988] alone indicate that substorms should be more effective at transporting in O$^+$ compared to storms, since they are more effective at inducing O$^+$ outflow. This result also explains why isolated substorms are less effective than storm-time substorms: Given the duration of sub-storm induced convection fields (~ 1–2 minutes) a single substorm cannot both cause higher ion outflow *and* inject these ions at the same time. During storms we have multiple substorms during the main phase, so that previous substorms can *seed* ionospheric ions for sub-sequent substorms to accelerate and transport inward. Storm-time substorms thus have a dual role: they produce enhanced ionospheric outflow *and* help transport and energize plasma sheet populations into the ring current.

The effect of the third factor is more direct. Taking a typical number for the energy of these ions in the auroral zone of around 1 keV, and converting this energy to velocity, then oxygen is four times slower than hydrogen (1 keV is 440 km/s for H$^+$ and 110 km/s for O$^+$). A strong E-field of about 2 mV/m and $E \times B$ drift bring these ions into the equatorial plane: O$^+$ reaches the equator at a distance of ~ 10R$_E$, while H$^+$ reaches the equator at a distance of ~ 40R$_E$. At the equator substorm processes accelerate the ions to ring current energies and transport these ions into the ring current. The substorm induced electric fields occur inward of any near-Earth neutral line, and are strongest in the equatorial region near 10R$_E$ [*Birn et al.*, 1998]. The O$^+$ contribution from the ionosphere is thus much more likely to be affected by substorms than the H$^+$ outflow: H$^+$ reaches the equator *tailward* of the neutral

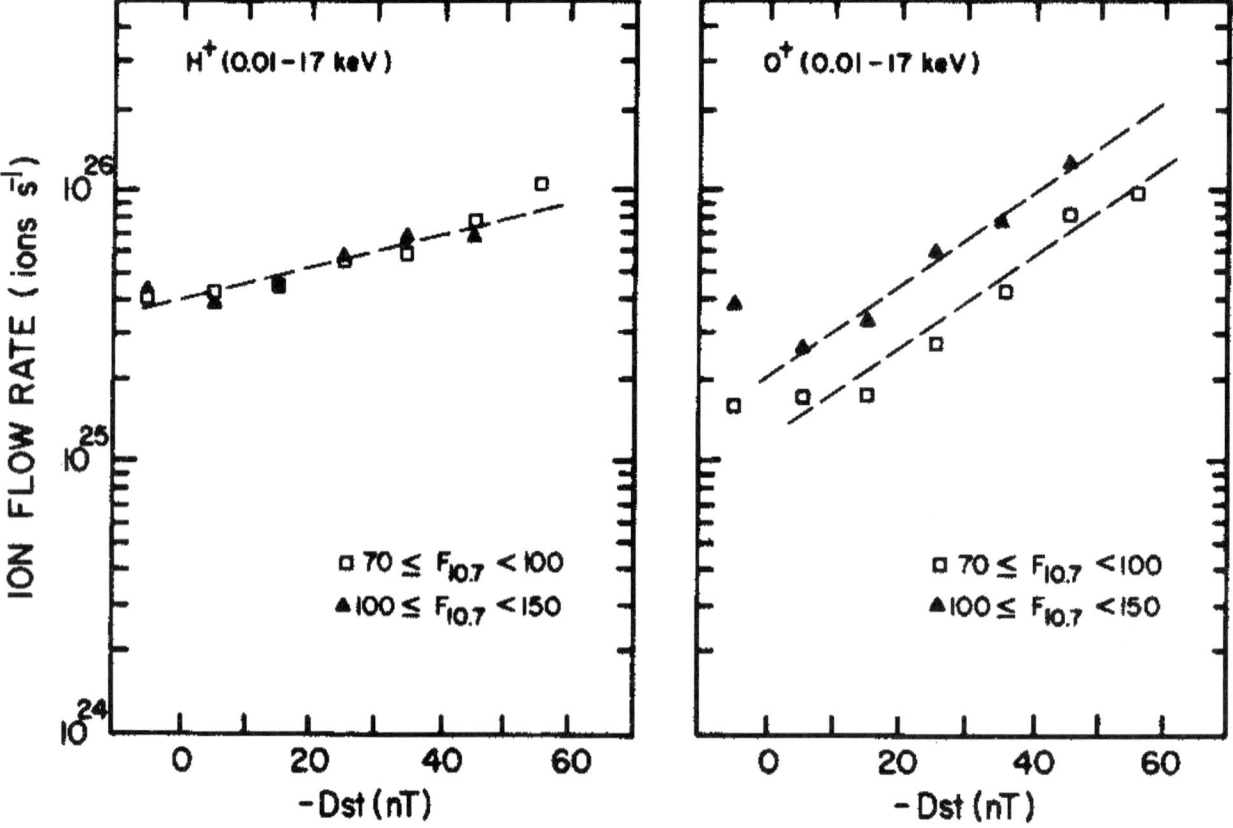

Figure 8. Ion outflow rates for H$^+$ and O$^+$ as a function of D$_{st}$ for two ranges of F$_{10.7}$; from *Yau et al.* [1988].

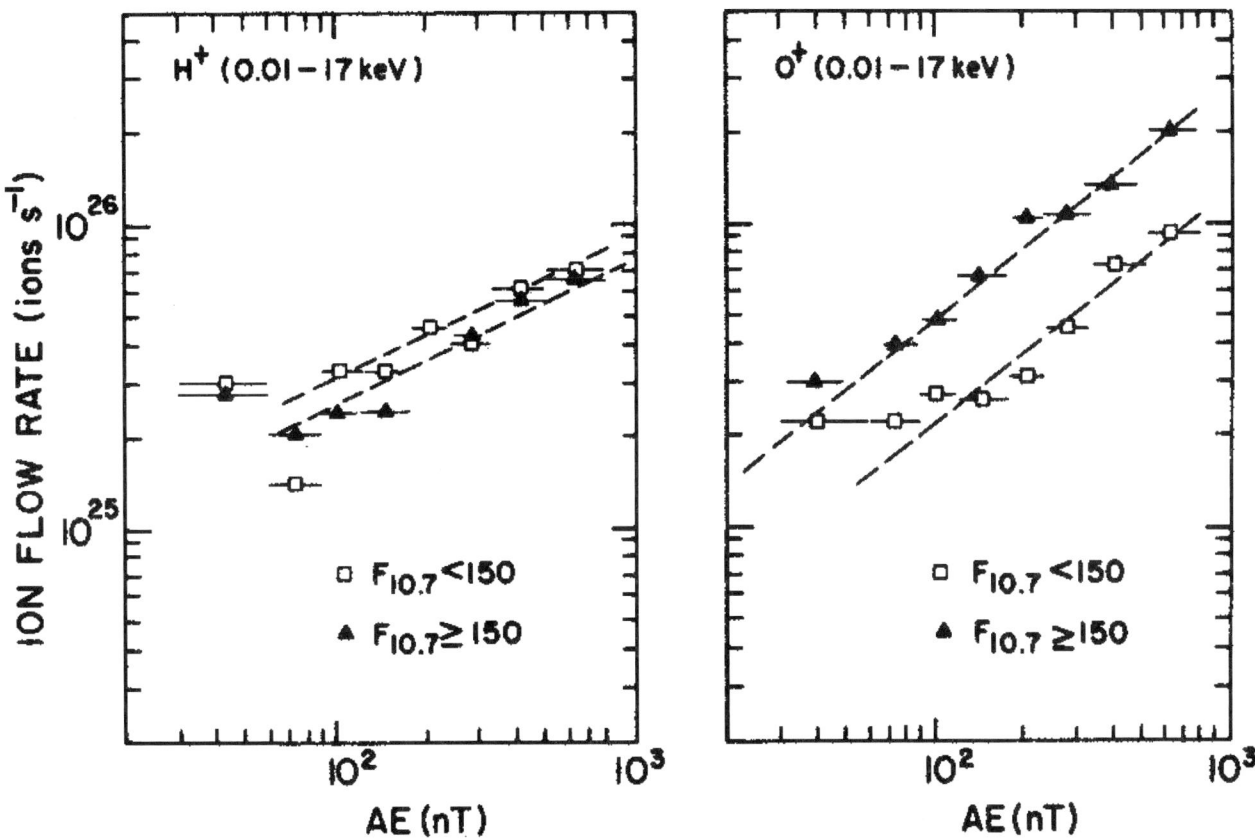

Figure 9. Ion outflow rates for H+ and O+ as a function of AE for two ranges of $F_{10.7}$; from *You et al.* [1988].

line and these ions are likely to be ejected tailward during a substorm. From this point of view the outflow *source* composition is of no consequence, since only the O+ part takes part in substorm convection; this will in all cases lead to an increase of the observed O+/H+ ratio.

However, the auroral zone is not the only source for H+ whereas for O+ it is. The main source for plasma sheet protons is the solar wind, which obviously is *not* modulated by terrestrial substorm activity.

6. SUMMARY

In this study we build on previous work [*Korth et al.*, 2001] which showed the enhanced effectiveness of substorms in transporting O+ into the ring current region during storms. We investigated here the relative role of storm versus substorm transport in producing the observed storm-time increases in the O+/H+ ratio by examining in detail 6 magnetic storms observed by CRRES, and discussed in detail the possible reasons for the observed O+/H+ ratio increases.

While the individual storm histories as measured by D_{st} are quite different, the most consistent signature throughout these events is that of the O+/H+ ratio. For all events the O+/H+ ratio increases throughout the main phase, reaching its maximum around the time of D_{st} minimum. The maximum ratio can be as high as 10 depending on the strength of the storm. The steepening of D_{st} just before reaching storm minimum is consistently associated with a burst in substorm (AE) activity.

We used two derived indices ($\Sigma D_{st}/h$ and $\Sigma AE/h$) to obtain a measure of the strength of storms and substorms. Even though these measures are not perfect (being somewhat correlated themselves, in that generally there is more substorm activity during larger storms), some useful conclusions can be drawn:

1. There is an almost linear relationship between storm strength and size of O+/H+ ratio increase.
2. There are no apparent saturation effects up to superstorm strength (min D_{st} –300 nT, March 1991). Bigger storm → more O+ transport into ring current. So there

is no apparent source-limitation of O^+ in the near-earth plasma sheet: ionospheric outflow can *always keep up with demand*.

3. Increased substorm activity around the main phase of storms leads to O^+/H^+ ratios up to some threshold in storm strength ($\Sigma D_{st}/h \sim -100$nT/h). For larger storms substorm activity decreases, while the O^+/H^+ ratio still increases. There are several possible interpretations for this:

- For the largest storms large scale convective fields become big enough so that any additional transport from substorms becomes insignificant.
- For large storms our measure of substorm activity is becoming unreliable.
- As storm strength increases, there are simply less substorms

In order to better determine the contribution of substorms versus storms to the O^+/H^+ ratio we would need to expand our study to include more storms beyond those seen by CRRES. A statistically more conclusive test would be an examination of storms which are all of a similar size, but which have different amounts of substorm activity associated with them. A limiting factor for such a study is the availability of equatorial composition data with sufficient time resolution in the inner magnetosphere. We intend to use the extensive POLAR CAMMICE composition measurements for such an expanded study.

Here we have explained much of the observed O^+/H^+ ratio increase during storms and the effective transport by storm-time substorms by drawing on earlier ionospheric outflow composition results, which showed the O^+ ion outflow outpaces H^+ for virtually all values of AE and all but the lowest values of D_{st}. But the strongest argument comes from the transport characteristics of these ions from the auroral ionosphere to the plasma sheet, which places O^+ ions much closer to the Earth than H^+—which for substorms would reliably place the ionospheric O^+ earthward of the near-Earth neutral line. In this picture only the ionospheric O^+ partakes in substorm energization and transport. The added effectiveness of storm-time substorms versus quiet time substorms in increasing the O^+/H^+ ratio in the ring current is explained by the increased frequency of substorms during the main phase, allowing subsequent substorms to transport earthward ionospheric O^+ transported out from the ionosphere by previous substorms.

Acknowledgments. The authors would like to thank John Wygant for making the CRRES electric probe data available to us and Michelle Thompsen for helpful comments and discussions.

REFERENCES

Baker, D. N., E. W. Hones Jr., P. R. Higbie, R. D. Velian, and P. Stauning, Global properties of the magnetosphere during a substorm growth phase: A case study, *J. Geophys. Res., 86*, 8941-8956, 1981.

Barfield, J. N., C. S. Lin, and R. L. McPherron, Observations of magnetic field perturbations at GOES 2 and GOES 3 during the March 22, 1979, substorms: CDAW 6 analysis, *J. Geophys. Res., 90*, 1289-1296, 1985.

Birn, J., M. F. Thomsen, J. E. Borovsky, G. D. Reeves, D. J. McComas, and R. D. Belian, Characteristic plasma properties during dipersionless substorm injections at geosynchronous orbit, *J. Geophys. Res., 102*, 2309-2324, 1997.

Birn, J., M. F. Thomsen, J. E. Borovsky, G. D. Reeves, D. J. McComas, and R. D. Belian, Substorm electron injections: Geosynchronous observations and test particle simulations, *J. Geophys. Res., 103*, 9235-9248, 1998.

Daglis, I. A., E. T. Sarris, and B. Wilken, AMPTE CCE CHEM observations of the energetic ion population at geosynchronous altitudes, *Ann. Geophys., 11* 685-696, 1993.

Daglis, I. A., S. Livi, E. T. Sarris, and B. Wilken, Energy density of ionospheric and solar wind origin ions in the near-Earth magnetotail during substorms, *J. Geophys. Res., 99*, 5691-5703, 1994.

Daglis, I. A., The role of magnetosphere-ionosphere coupling in magnetic storm dynamics, in *Magnetic Storms, Geophys. Monog.*, edited by B. T. Tsurutani et al., vol. 98, pp. 107-116, AGU, Washington, D. C., 1997.

Daglis, I. A., Y. Kamide, G. Katsotakis, C. Mouikis, B. Wilken, E. T. Sarris, and R. Nakamura, Ion composition in the inner magnetosphere: Its importance and its potential role as discriminator between storm-time and non-storm time substorms, in *Substorms-4*, edited by S. Kokubun and Y. Kamide, pp. 767-772, Terra/Kluwer Publications, Tokyo, 1998.

Daglis, I. A., G. Kasotakis, E. T. Sarris, Y. Kamide, S. Livi, and B. Wilken, Variations of the ion composition during an intense magnetic storm and their consequences, *Phys. Chem. Earth, 24*, 229-232, 1999.

Grande, M., C. H. Perry, A. Hall, Y. Kamide, R. Nakamura, J. Fennell, and B. Wilken, Superposed epoch analysis of magnetospheric composition and D_{st} during storm-time and quiet-time substorms, in *Substorms-4*, edited by S. Kokubun and Y. Kamide, pp. 773-778, Terra/Kluwer Publications, Tokyo, 1998.

Kamide, Y., N. Yokoyama, W. Gonzalez, B. T. Tsurutani, I. A. Daglis, A. Brekke, and S. Masuda, Two-step development of geomagnetic storms, *J. Geophys. Res., 103,* 6917-6921, 1998.

Kaye, S. M., R. G. Johnson, R. D. Sharp, and E. G. Shelley, Observations of transient H^+ and O^+ bursts in the equatorial magnetosphere, *J. Geophys. Res., 86*, 1335-1344, 1981.

Korth, A., G. Kremser, B. Wilken, W. Güttler, S. L. Ullaland, and R. Koga, Electron and proton wide-angle spectrometer (EPAS) on the CRRES spacecraft, *J. Spacecr. Rockets, 29*, 609-614, 1992.

Korth, A., R. H. W. Friedel, C. Mouikis, and J. F. Fennell, Storm/substorm signatures in the outer belt, in *Substorms-4*, edited by S. Kokubun and Y. Kamide, pp. 773-778, Terra/Kluwer Publications, Tokyo, 1998.

Korth, A., R. H. W. Friedel, C. G. Mouikis, J. F. Fennell, J. R. Wygant, and H. Korth, Comprehensive particle and field observations of magnetic storms at different local times from the CRRES spacecraft, *J. Geophys. Res., 105*, 18, 729-18, 740, 2000.

Korth, A., R. H. W. Friedel, F. Frutos-Alfaro, C. G. Mouikis, and Q. Zong, Ion composition of substorms during storm-time and non-storm-time periods, *J. Terr. Atmos. Phys.*, 2001, accepted.

Li, X., D. N. Baker, M. Temerin, G. D. Reeves, and R. D. Belian, Simulation of dispersionless injections and drift echoes of energetic electrons associated with substorms, *Geophys. Res. Lett, 25*, 3763-3766, 1998.

Lundin, R., and B. Hultqvist, Ionospheric plasma escape by high-altitude electric-fields: Magnetic-moment pumping, *J. Geophys. Res., 94*, 6665-6680, 1989.

McPherron, R. L., The role of substorms in the generation of magnetic storms, in *Magnetic Storms, Geophysical Monograph 98*, edited by V. T. Tsurutani, W. Gonzalez, Y. Kamide, and J. Arballo, pp. 131-147, AGU, Washington, D. C., 1997.

Moore, T. E., W. K. Peterson, C. T. Russell, M. O. Chandler, M. R. Collier, H. L. Collin, P. D. Craven, R. Fitzenreiter, B. L. Giles, and C. J. Pollock, Ionospheric mass ejection in response to a CME, *Geophys. Res. Let., 26*, 2339-2342, 1999.

O'Brien, T. P., R. L. McPherron, D. Sornette, G. D. Reeves, R. Friedel, and H. J. Singer, Which magnetic storms produce relativistic electrons at geosynchronous orbit?, *J. Geophys. Res., 106*, 15, 533-15,544, 2001.

Wilken, B., W. Weiss, D. Hall, M. Grande, F. Soeraas, and J. F. Fennell, Magnetospheric ion composition spectrometer onboard the CRRES spacecraft, *J. Spacecr. Rockets, 29*, 585-591, 1992.

Williams, D. J., The Earth's ring current: Present situation and future thrusts, *Phys. Scr., T18,* 140-151, 1987.

Wygant, J. R., P. R. Harvey, D. Pankow, F. S. Mozer, N. Maynard, H. Singer, M. Smiddy, W. Sullivan, and P. Anderson, CRRES electric field/Langmuir probe instrument, *J. Spacecr. Rockets, 29*, 601-604, 1992.

Yau, A. W., and B. A. Whalen, Auroral ion composition during large magnetic storms, *Canadian Journal Of Physic, 70*, 500-509, 1992.

Yau, A. W., and W. K. Peterson, and E. G. Shelley, Quantitative parametrization of energetic ionospheric ion out-flow, in *Modeling Magnetospheric Plasma*, edited by T. Moore and J. W. (Jr.), vol. 44 of Geophys. Monogr. Ser., pp. 211-217, AGU, Washington D. C., 1988.

A. Korth, Max-Planck-Institut für Aeronomie, Katlenburg-Lindau, Germany. (e-mail: korth@linmpi.mpg.de)

R. H. W. Friedel, Los Alamos National Laboratory, Los Alamos, New Mexico. (e-mail: friedel@lanl.gov)

M. G. Henderson, Los Alamos National Laboratory, Los Alamos, New Mexico.

F. Frutos-Alfaro, Physics Department of the University of Costa Rica, San José, Costa Rica. (e-mail: frutos@linmpi.mpg.de)

C. G. Mouikis Space Science Center, University of New Hampshire, Durham, New Hampshire. (email: cmouikis@atlas.sr.unh.edu)

What is the Effect of Substorms on the Ring Current Ion Population During a Geomagnetic Storm?

M. Grande, C. H. Perry, and A. Hall

Rutherford Appleton Lab, Chilton, Didcot, Oxfordshire, UK

J. Fennell

Aerospace Corporation USA

R. Nakamura and Y. Kamide

STEL, Nagoya University, Japan

We consider the properties of substorms during periods of high and low Dst, and their effect on Dst. Data from the MICS instrument on CRRES are used to examine composition changes of ring current ions in the energy range 72-400 keV/e. These we relate to the Dst index, generally taken to characterise the storm. A superposed epoch analysis using high time resolution Dst data shows that on a timescale of 30 minutes, Dst is largely unchanged by substorm onset. It also shows that there is some increase in the number density of ionospheric oxygen relative to solar wind origin He++. There is no systematic difference in the energisation between different species. The changes in energy density are primarily the result of changes in number density. We show that while the underlying ring current composition is a strong function of Dst, the energisation and increase in species number density at substorm onset is independent of Dst. Examination of long term survey plots of MICS data shows that during storms the bulk ring current composition changes. We see the emergence of a concentration of ionospheric material at low L-shells, and its rise to higher L-shells during the recovery. By contrast, solar wind material remains mainly at higher L-shells. The altitude of the peak in ionospheric material is a strong function of Dst. These major changes are always at lower L-shells than substorm particle injections. Overall, we conclude that there is no qualitative difference between stormtime substorms and those from less disturbed periods. During solar maximum when these measurements were made, substorms are not directly responsible for the injection of oxygen ions into the ring current. Although substorms may be important in initiating ionospheric upflow, substorm injected particles themselves play no part in modifying ring current composition during storms.

1. INTRODUCTION

It has been conventional to suggest that the stormtime ring current is built up as a result of a series of substorms. However, it was clear that the process would have to be more than simple addition [*Kamide*,1992], and recently an alternative explanation has emerged, suggesting that the stormtime ring current ions are the result of enhanced convection (See for example the review by *Sharma et al* 2002). The aim of this paper is to consider to what extent substorms are an essential feature of storms, whether they are directly responsible for feeding the ring current or perhaps only indirectly enable its growth. These issues have been discussed by a number of authors, for example see the review by *Kamide et al.,* 1998. Associated with this discussion is the question of whether stormtime and quiet time substorms have different characteristics. *Baumjohann et al.,* 1995, suggested that substorms during storm time are different from others, and it is these storm time events which have the classical (*Hones,* 1976] Near Earth Neutral Line (NENL) morphology. In less disturbed periods substorms would be initiated by current disruption near geostationary altitudes (Lui et al, 1991). Perry et al., 1996, in a statistical study found that there is a tendency for substorms to occur further to the west during stormtime than otherwise.

One possible reason for any difference might be composition. It is well known [e.g. *Hamilton* 1986; *Daglis,* 1997] that the ring current is substantially enriched in singly charged oxygen during large storms, the charge state indicating an ionospheric origin. The ring current during stormtime also contains high charge state ions. *Grande et al.,* 1997b have shown that ions with the frozen in charge signatures of distinct solar wind streams are observed to enter the magnetosphere on timescales as rapid as 30 minutes. They used the methods of *Chen et al.,* 1993, to examine the conditions needed for these ions to reach the location of the CRRES spacecraft on these fast timescales. It was concluded that this required the cross tail potential to be modeled, not as the steady average value inferred for a given Kp, but by a random time series of spikes, having the same average. This is the behavior that would be expected of a series of substorms, and suggested that indeed substorms may be necessary to produce the conditions required for the observed behavior of the inner magnetosphere during storms.

The question therefore arises whether substorms are necessary to produce the observed excess of oxygen seen during great storms. *Daglis et al.,* 1996, have examined evidence for a prompt ionospheric response to substorm activity. *Gazey et al.,* 1996 argued for a delayed response. Although the O^+ ion concentration during storms is greater, increased ionospheric upflow due to enhanced convection or the presence of an enhanced ring current [*Sheldon et al,* 1998], rather than the increased number of individual substorms may be responsible. *Baker et al.,* 1985, considered that a high concentration of O^+ could affect the triggering efficiency of substorms. *Daglis,* 1997, has argued that an essential cause of intense storms is the feeding of the inner magnetosphere with ionospheric O^+ due to substorm activity. In addition, there may be differences between the maximum and minimum of the solar cycle. Previous authors (e.g. *Young et al* 1982) show that there is far less O^+ present at solar minimum, except in the largest storms.

2. THE DATA SET

The CRRES spacecraft was launched into a geostationary transfer orbit on 25th July 1990. The 13-month mission thus took place near solar maximum, and a number of major Dst excursions were observed during this time. The eccentric nature of the orbit and the inclination of 18.1 degrees meant that CRRES sampled a range of L-shells up to L = 8 and magnetic latitudes within 30 degrees of the magnetic equator. The Magnetic Local Time (MLT) of apogee ranged from about 08:30 after launch to about 15:30 by the end of the mission. This spatial coverage of the magnetosphere has proven extremely useful, enabling the effect of position of observation on substorm particle injection signatures to be investigated. It also enabled us to survey of the whole inner magnetospheric ion population throughout the mission, with a data set large enough to permit meaningful statistical conclusions.

The details of the MICS sensor have been discussed elsewhere [*Wilken et al,* 1991]. MICS measured ions in the energy range 1–425 keV/e. Three parameters, ion total energy (E), Time-of-flight (TOF) and energy per charge (E/Q) were measured for each ion and were combined on board to extract the energy, charge (Q) and mass (M) of the ion. We concentrate on a period from 13 January to 12 September 1991 (days 13 to 255) during which the instrument remained in a constant operating regime. The data have been corrected for small instrumental drifts, and reduced to number densities (ND) and energy densities (ED) for each of the major species.

3. ANALYSIS OF DISPERSIONLESS SUBSTORM INJECTIONS

In order to characterise trends in the size, type and composition of substorm onset signatures, a database of events was created. The database contains a total of 110 events identified during the period of investigation between 13 Jan 1991 and 10 July 1991. Energy spectrograms were searched by eye for suitable dispersionless injections. Plate 1 shows an example of one substorm event from 14 Feb 1991 [*Grande et al* 1992]. It is a good example of the large changes in composition seen in individual events. The event is centered on midnight, and consists of two particle injections separated by about 10 minutes. The second has a higher O+ concentration. The derived number densities, which form the basis of the analysis in this paper, are shown in the center panel.

These "dispersionless" injections show little evidence of gradient-B drift, and are therefore assumed to be observed near the injection/energisation region. Events were examined at high time resolution to ensure that count rate enhancements on time scales of less than a minute were observed over the entire range of energies. We made no attempt to systematically compare events identified in this way with other methods of substorm identification, for example AE or energetic electrons in geostationary orbit. However it has been our experience from case studies that this is a valid method of onset recognition. Since the purpose of the present study is to identify the changes in the ring current produced by substorm associated energetic particle injections, this is an appropriate methodology. Some such events were identified prior to 13 Jan 1991, but problems with the instrumental gain during this period made unbiased selection more difficult and these events were not included in the current sample. After 10 July 1991, when the spacecraft was further from the midnight sector, only dispersed injections were seen.

The 30 second resolution spin-averaged raw count rate data were integrated over energies 72–400 keV/e, then smoothed over 90 seconds to reduce noise, and plotted for each ion species for each event. The half height was used to derive an onset time for each injection. The location of each injection is then specified in terms of its magnetic local time, magnetic latitude and L-Shell. Events in our sample cover MLT from 15:00 to 01:30, magnetic latitude from –28.5 to +23.0° and L-Shells from 4.7 to 8.2. Number density and energy density were evaluated, and hence we derived the peak flux value during the injection. The average reference pre-onset flux level, hereafter referred to as "background", was inferred for each species from the 20 minute period prior to onset.

We have defined substorm "size" as the difference between the peak and background, and calculated it for both He^{++} and O^+ ions separately. Scattering and loss processes should have minimal effect on the injected particle population if the data are taken in the injection region, and therefore for dispersionless injections the change in the ND or ED during an event is a good indicator of the actual size of the injection.

We classified events in a number of different ways. Some injections occur as a single enhancement against an otherwise quiet background level while others, such as the event in Plate 1, display a sequence of two or more enhancements, typically separated by 10–15 minutes. In our sample 53 events were classified as singles and 57 as multiples. For the purposes of the work presented here properties attributed to a multiple event are those of the first injection in the series. To distinguish between substorms during active and quiet solar conditions we have used the hourly Dst to classify events as stormtime (Dst <= –30 nT) or non-storm time (Dst > –30 nT). There were 59 stormtime events, and 51 non-storm time.

The number of stormtime events may be under-sampled. Individual dispersionless injections are not often observed during the most disturbed periods. This may be the result of event confusion, or may indeed represent a different phenomenology for energetic particle events during the most active periods of large storms. Frequent sharp boundary crossings, flux dropouts and recoveries are seen in the data, which are of similar appearance in some respects to dispersionless injections; they are removed from the data set.

Most injections are seen to contain a mix of both solar wind type material, of which we take He^{++} to be indicative, and ionospheric material, identified in this study by O^+ content. This mix is highly variable. By dividing the size of the injection seen in the alpha particles by the size seen in the O^+, a ratio representing the relative contributions of the solar wind and ionosphere to the injected particle population was calculated. This enabled the 110 injections to be further classified for use in the superposed epoch and statistical analysis described below. In our sample 61 events were categorised as He^{++} rich, 24 as mixed and 25 as O^+ rich. In Figure 1, we show the distribution in L shell and local time for injections split between storm and quiet time. It is seen that the distributions are similar, with the stormtime events biased further towards dusk [*Perry et al* 1996]. Also shown is the distribution corrected for sampling effects. There is no major difference in the L-shell distribution and no suggestion of a low L-shell population in either storm or quiet time.

We attempted to use the dataset classifications to find predictors for the composition changes in individual sub-

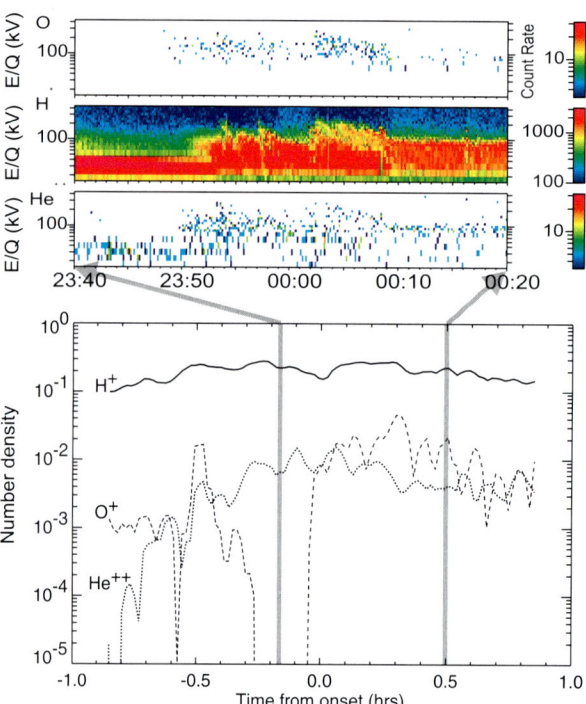

Plate 1. The top 3 panels show an example of a typical MICS observation of a dispersionless injection (14 Feb 1991) for O^+, total ions, and He^{++}. The lower panel shows the number densities of H^+, He^{++} and O^+ for the same event.

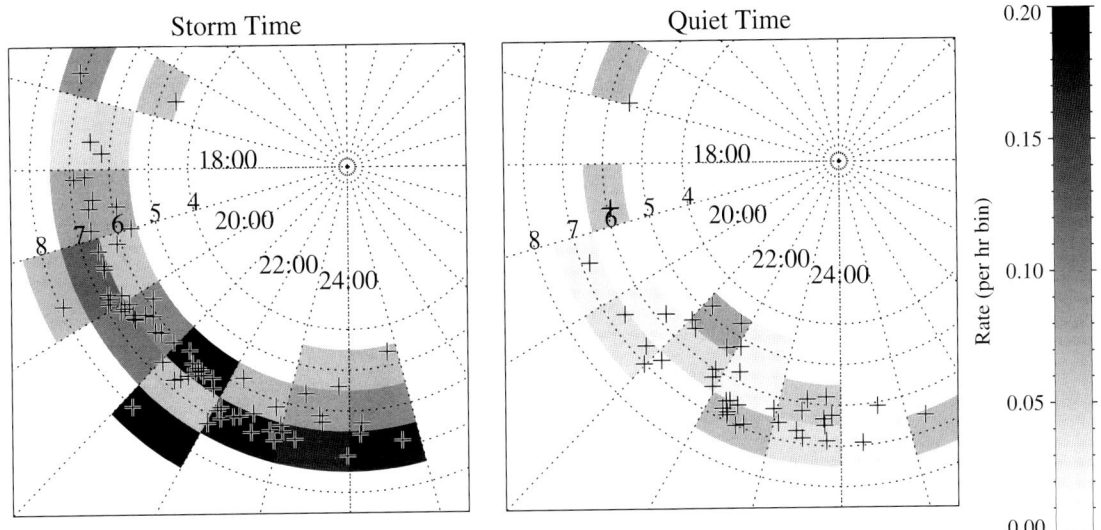

Figure 1. Location of substorm onset as a function of MLT and L-shell. Individual substorms are shown as points, and the average rate corrected for spacecraft occupancy are shown on the color scale.

storms. There is a correlation (R=0.55) between total (O^+ plus He^{++}) substorm injection size and Dst, indicating that larger substorms happen in stormtime. While for He^{++} size the trend is not strong (R= 0.30), there is a stronger correlation between O^+ and Dst (R=0.58. If we consider the change in total O^+ to He^{++} ratio for individual injections, the correlation with Dst is weak (R=0.28). We found a slightly stronger (R=0.30) correlation with the background O^+ to He^{++} ratio before onset. These observations may simply reflect that the background (pre-substorm) ring current is oxygen rich in stormtime. In no other case is there a strong correlation between composition and other characteristics such as position. Also, in no other case do we see any clear separation between storm and quiet time behavior. The standard deviations are large, and the differences in the storm and quiet-time averages small. It is possible to find numerous examples which go against the overall trend, emphasizing that case studies are not appropriate as a means of understanding the average composition changes in the inner magnetosphere during storms and substorms. For example there are many multiple injections which occur in non stormtime where high and increasing O^+ concentrations are observed, for example the double injection substorm in Plate 1. But on average, there is no tendency for the number of multiple (as against single) injections to increase with Dst, and no overall tendency for the oxygen concentration to increase in sequences of multiple injections as predicted by Gazey et al 1996. While prior energetic particle composition is the best predictor of substorm injection composition that we have identified, the spread in individual composition changes is much larger than the trend.

3.1 Superposed Epoch Dst Study

In an important previous study, Iyemori and Rao 1996 showed that the long term effect of substorms on the Dst index was to produce a slight reduction in the excursion, on a timescale of hours. They distinguished between substorms in the main phase and the recovery phase of a storm, finding that Dst decayed slowly after substorm onset during recovery, and grew slightly less rapidly after onset during main phase.

We have repeated the study, for substorm periods during the CRRES era, using high quality Dst- like data, synthesised from 6 (occasionally only 5) stations, at a 5 minute time resolution. We have carried out a superposed epoch analysis of this Dst index, over the 110 onsets of dispersionless substorm injections we observed with CRRES. We have then subdivided the data set in order to investigate the behavior of the different ion species and different magnetospheric activity regimes as defined by the Dst index. The results are shown in Figure 2. Events were aligned using the onset time, derived from the procedure described above. For each time step the mean was determined from the distribution of all events at that time. Due to the inclusion of "multiple" type injection events in both subsets, we examine only the behavior of the first onset of a sequence.

The first result is that on the timescale of order one hour or less there is no major change in the average behavior of

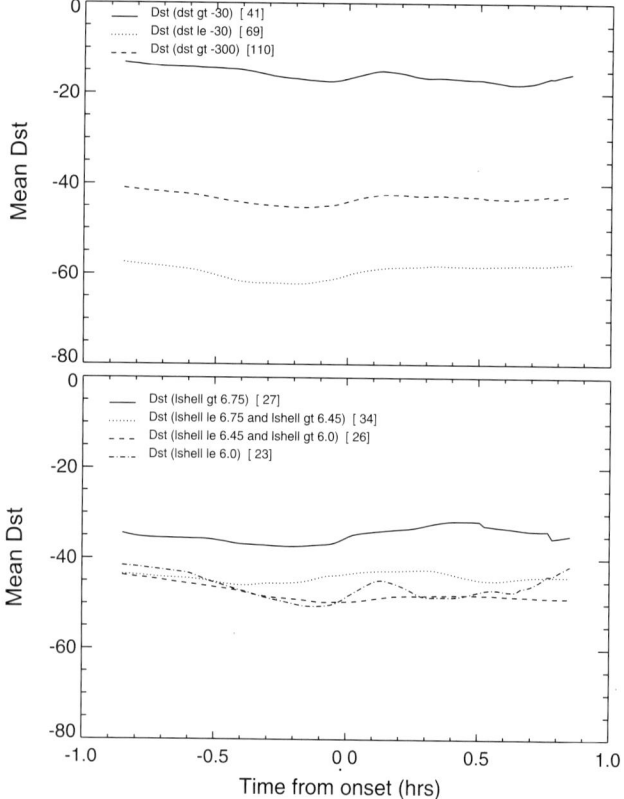

Figure 2. Superposed epoch analysis of Dst for 110 substorms. Top panel is subdivided into all (Dst>–300), quiet (Dst> –30) and storm, (Dst< –30) times. Lower panel shows the results of subdivision by L-shell.

the Dst index itself, when ordered by substorm onset. On average, there is no prompt change in Dst associated with substorm onset. This is in agreement with the analysis of *Iyemori and Rao* 1996. We conclude therefore, that there is no prompt connection between the injection of energetic particles at substorm onset and changes in the energetic particle population of the ring current, which would give rise to a change in Dst. If following *Gonzalez et al.* 1994 we subdivide the dataset, again using Dst < –30 nT to characterise storm time and Dst > –30 nT to characterise non-storm time, we find stormtime substorms have a slight tendency for an increase in the Dst excursion in the 30 minutes prior to onset. The explanation may be that during stormtime, there is a higher level of convection, which we would contend is the mechanism for populating the inner ring current, and that this increased level is likely to produce a substorm. We also see in the lower panel of Figure 2 that events at lower L-shell are associated with a higher Dst. The mean Dst five minutes before onset of the 27 events observed at L>6.75 is –35nT, of the 26 events for which 6.75>L>6.45 is –43 nT, of the 34 events for which 6.45>L>6.0 is –49 nT, and of the 27 events for which L<6.0 is –51nT. This indicates a tendency for stormtime substorm injections to occur deeper in the magnetosphere. However, as is seen in Figure 1 even after correcting for sampling effects, we found that substorm injections at L-shells below 5 were very rare. The lack of correlation of Dst change with substorm onset time would preclude a simple identification with substorms of the spikes in cross tail field used by *Grande et al* 1996 to model ion access to the ring current.

3.2 Superposed Epoch Analysis of Ion Species Abundances

Using the same set of dispersionless injection events, we then carried out a superposed epoch analysis of the number and energy density of ion species, aligned using onset time. Quantities were derived using particles of energies greater than 72 keV/e. The 110 substorm data set was subdivided in order to investigate the behavior of the different ion species and different magnetospheric activity regimes as defined by the Dst index. As before, we use Dst <= –30 nT to characterise storm time and Dst > –30 nT to characterise non-storm time.

Firstly, in order to derive the average energy per particle for four different species, H^+ He^+, He^{++} and O^+ we divide the energy density by the number density. Figure 3 shows the result. In each case, the solid line is for stormtime and the dotted line is for quiet time. We see that all species show very similar fractional increase in stormtime as in quiet time. The increases in average particle energy are on the order of 10% for all species. The increase in O^+ is the most peaked at onset, but settles back on a timescale of 10 minutes to a 10% increase. In general He^+ is only enhanced [*Grande et al* 1997a] at an extremely low level in substorm injections. There is no preferential energisation for different species on the 45 minute timescale. The average energy of He^{++} shows a larger difference between stormtime and non-stormtime than the other two. Whatever mechanism results in this relatively greater stormtime mean energy of He^{++} is clearly not substorm related. It appears that in studying population changes, ND and ED are therefore largely equivalent. Moreover, when considering charge exchange phenomena, ND conservation is maintained, and hence in general in what follows we will refer only to ND, rather than ED, although it is ED which contributes directly to the quantity Dst.

Figure 4 shows the ratio of the mean number density of O^+ and He^{++}, broken into subsets according to Dst. From top to bottom they are Dst less than –60, –60 to –30 and

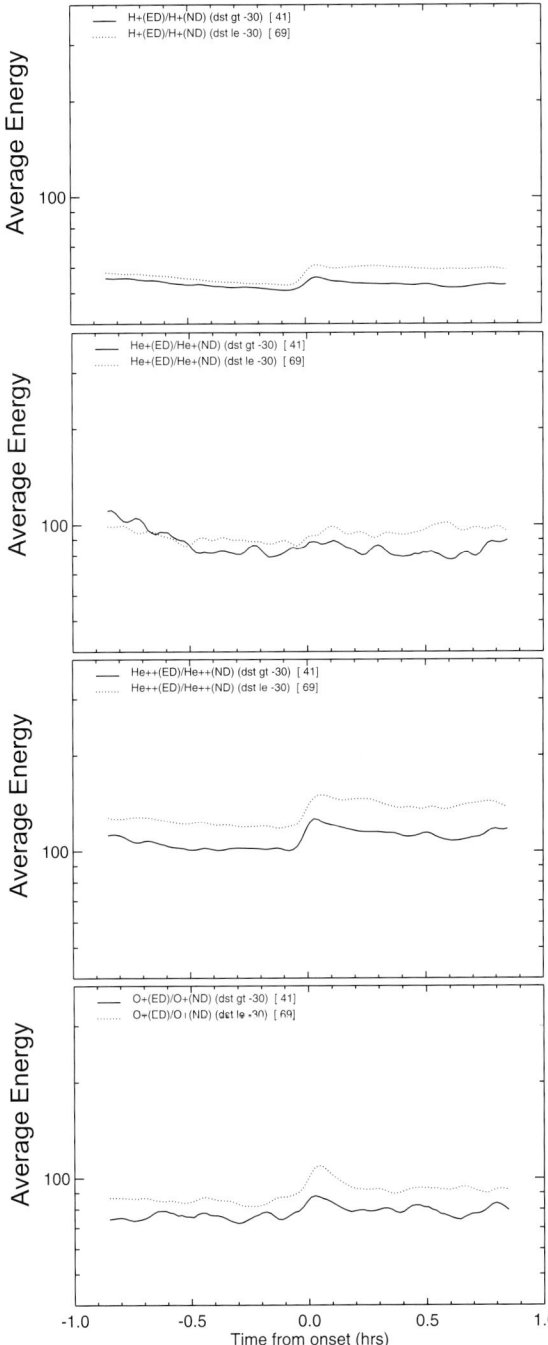

Figure 3. Superposed epoch analysis of 110 substorms. We show average energy per particle in KeV for different species, obtained by dividing the energy density by the number density. Top panel is H+ panel 2 He+, panel 3 He++ and bottom O+. In each case, the solid line is for stormtime and the dotted line is for quiet time. All species show the approximately the same fractional increase in stormtime as in quiet time. These increases in average particle energy are on the order of 10% for H+ and O+ and 20% for He++.

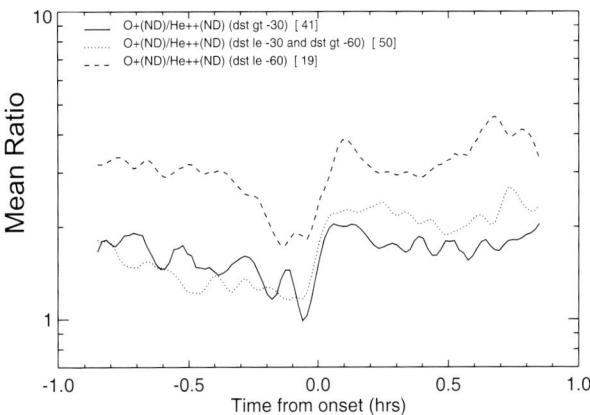

Figure 4. Ratio of the mean number density of singly charged oxygen and alpha particles, broken into subsets according to Dst. From top to bottom they are Dst less than –60, –60 to –30 and greater than –30. We see that the percentage changes are similar for all three cases, indicating that there is no change in composition evolution for different levels of Dst.

greater than –30. We see that the percentage changes are similar for all three cases, indicating that there is no change in composition evolution for different levels of Dst. In all cases there is a decrease in the relative abundance of oxygen before onset, perhaps due to its larger gyroradius in a thinning plasma sheet, followed by a recovery. The situation is summarised in Figure 5, where the change in number density at onset for O^+ (solid lines) and He^{++} (dashed lines) is plotted versus Dst. The densities before (lower curves) and at the peak of onset (upper curves) are shown. Very much the same trend is seen for both. Thus for each species the lower line shows the mean variation of the ring current with Dst. It is seen that the number density of both species increases with the magnitude of Dst, which is unsurprising since at these L-shells, as elsewhere, the numbers of ambient particles are higher during larger storms. We emphasize that these measurements relate to the position at which the substorm occurred, since we have selected the substorms as being dispersionless. It is the case that during storms the O^+ increases relative to He^{++} in the inner magnetosphere as a whole, but that increase is most pronounced at lower L-shells, as discussed in the next section.

The difference between the lower and upper of each pair of lines in Figure 5 represents the increase in the number of particles of that species injected at substorm onset. It is seen firstly that there is no major difference in the fractional increase of each species at injection. Moreover, there is no great difference in the behavior at different Dst levels, with perhaps a tendency for the substorm injection to have less relative effect on the ring current in stormtime than in quiet

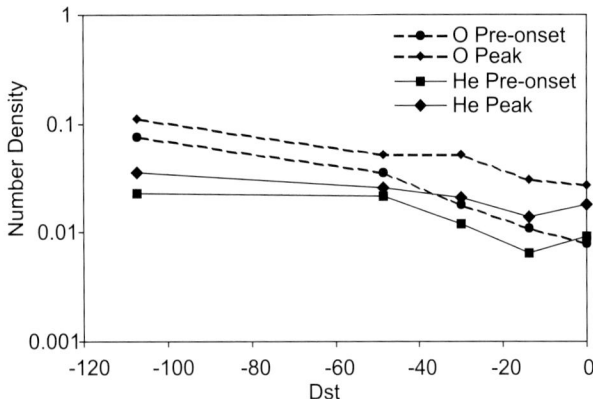

Figure 5. Number density before and after substorm onset plotted against Dst. Dotted lines represent O^+, solid lines He^{++}, For each pair of lines, the lower represents number density prior to onset, and the upper immediately after.

time. It should be noted that in order to remain insensitive to instrumental efficiency variations with energy, we are not including particles of energies below 72 KeV/e in our integration. Hence we could be missing population differences at lower energies. However, it has been pointed out by many authors (eg *Hamilton* 1986) that the bulk of the stormtime energy density is above this threshold. Overall, we see that the major change for different Dst levels is in the number of pre-existing particles. For each species, the substorm changes the abundance by the same fraction. Hence the substorm does not change the relative abundances of species. The substorm itself is not responsible for any large systematic change in the ratio of O^+ to He^{++}.

4. RING CURRENT COMPOSITION

We now turn to the overall composition of the ring current, and how this relates to Dst, and to the position of individual substorms. We have taken all data from the period between CRRES orbits 417 and 1067, corresponding to days 13 to 255 of 1991, when apogee was between 16:00 and 02:00 MLT, which is where the majority of substorms were observed, During this period the instrument was in a consistent gain configuration.

In Plate 2, each orbit contributes two traverses, corresponding to inbound and outbound passes, which are averaged, and ordered according to time and L-shell. Rows 1–3 of the left hand panels show the long term number density evolution of the ring current species He^+, He^{++}, and O^+. H^+ is not shown since there is an instrument mode change at around L=4.5 which makes it difficult to match up H^+ fluxes at lower and higher L-shells.

Comparison with Dst, plotted in row 4, shows that new ring current material is supplied by storms. In the right hand panels of Plate 2, we show an individual storm period in July 1991, in order to better illustrate the changes in population which take place. During a storm, the peak number density of the ionospheric material, represented by O^+, shown in row 3 of Plate 2, is initially at a low (3–4) L-shell; the larger the Dst the lower the L. The location of this peak flux rises in L as the storm decays, indicating that the ions decay more rapidly at low L. We suggest that this is because of charge exchange in the denser geocorona there.

During stormtime, O^+ (row 3) and He^+ (row 1) behave in a similar way, suggesting a common ionospheric source. However at much quieter times the He^+ is more persistent, particularly at high L shell, suggesting either a less efficient loss or a quiet time source, presumably charge exchange from He^{++}.

By contrast, the peak in the He^{++} number density (row 2), which should be characteristic of solar wind material, persists at L-shells greater than 5 while below L= 5 there is a rapid loss. Note that the enhancement at L=3–4 seen after three large storms, is ascribed contamination by penetrating relativistic particles. Detailed comparison of the He^+ (row 1) and He^{++} (row 2) plots shows that at later times, as the storm decays, He^{++} loss at high L(>5) is accompanied by He^+ increase (for example following day 206). This is compatible with the suggestion that the solar origin He^{++} is decaying in this region via charge exchange, and is thus responsible for the majority of the post storm He^+ population. However, it is only in the quietest periods with Dst above –20 that we observe evidence of this process, first described by Fritz and Spelvijk 1978.

The behavior of O^+ is very similar to He^+ with the exception that there is no evidence of a source in quiet times. We infer that during storm time both are supplied from a common ionospheric source.

From the ratio of He^+/He^{++}, and O^+/He^{++} (Plate 3) we find that the ratio of both show a peak in L-shell, and that this peak appears strongly correlated with Dst (shown as an overlaid white line), with low L corresponding to a large negative Dst excursion.

In order to quantify this striking correlation with Dst, for each orbit we have derived the mean L-shell for a given species, using the number density. In Figure 6 we plot this mean L-shell against Dst, for He^+, He^{++} and O^+. It is seen that there is a strong correlation (R=.613 and .485) with Dst for He^+ and O^+, the species generally considered to be ionospheric. There is a rather insignificant (R=0.201) correlation for He^{++}. The regression lines are overplotted. The suggestion is that Dst systematically alters the height profile

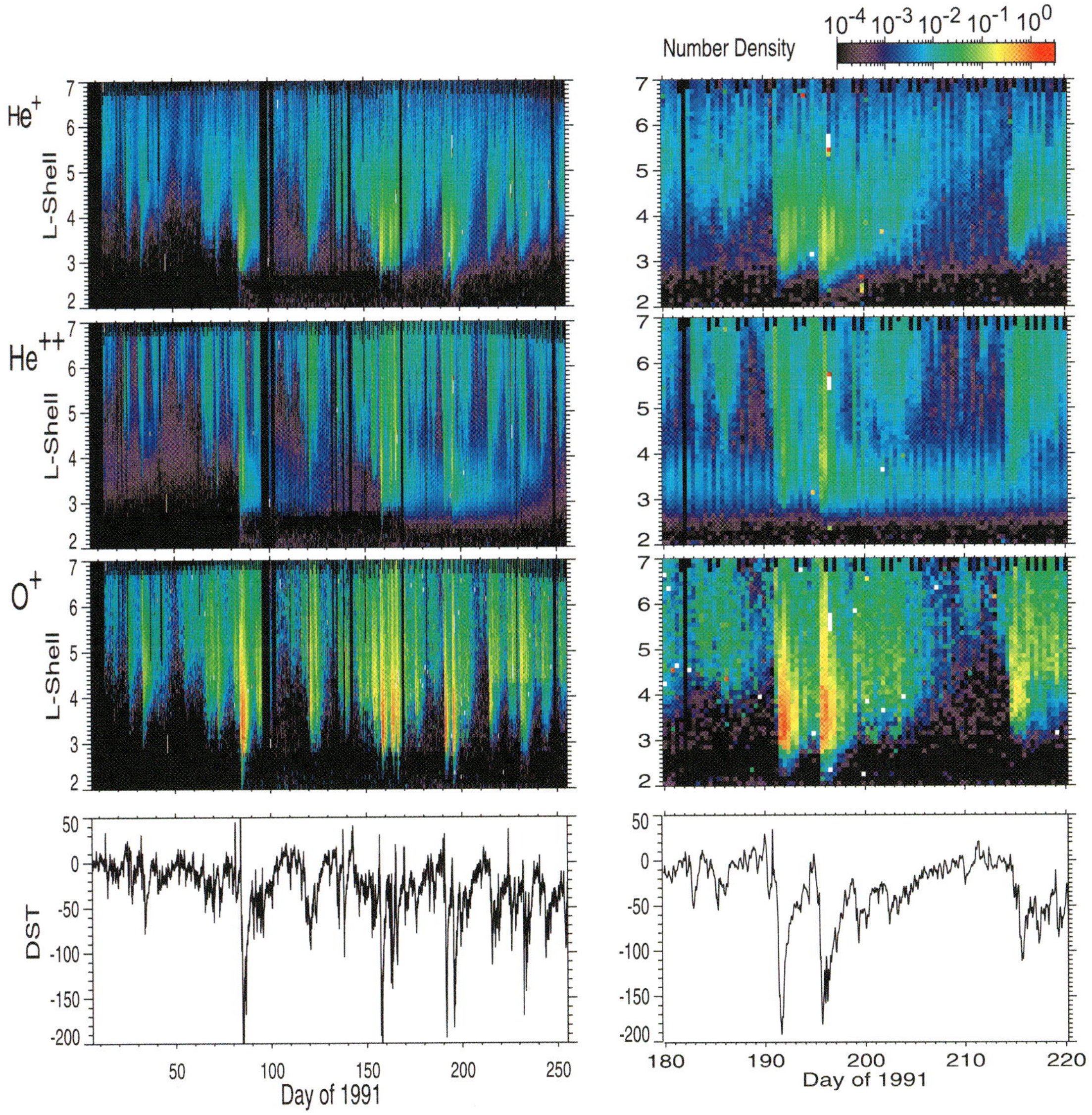

Plate 2. Magnetospheric energetic particle species ordered by time and L-shell. The left hand panels show a survey of particle species covering the period days 13 to 255 of 1991. The right hand panels show in more detail a storm time period, days 180 to 220 of 1991. Row 1 shows He$^+$ number density, row 2 He^{++} and row 3 shows O$^+$. The lowest sub-panel shows Dst for the same intervals.

84 EFFECT OF SUBSTORMS ON THE RING CURRENT ION POPULATION

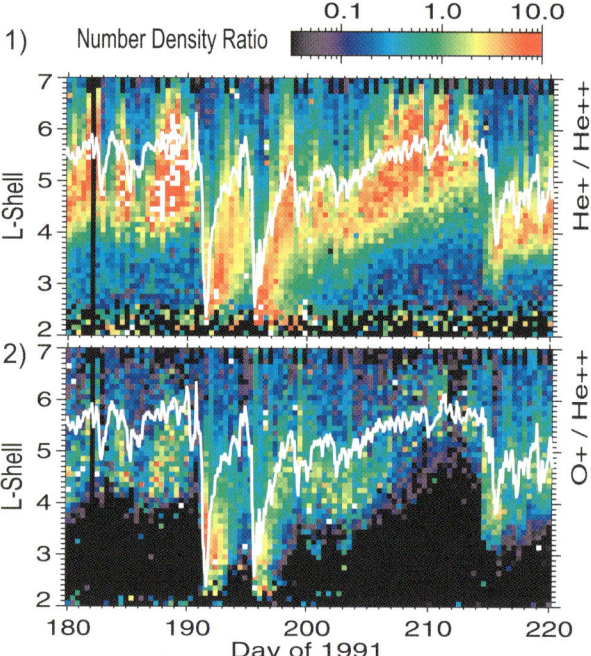

Plate 3. Ratios of magnetospheric energetic particle species ordered by time and L-shell. The period covered is days 180 to 220 of 1991. Panel 1 shows He^+/He^{++}, and panel 2 O^+/He^{++}. The value of Dst is overlayed as a white line.

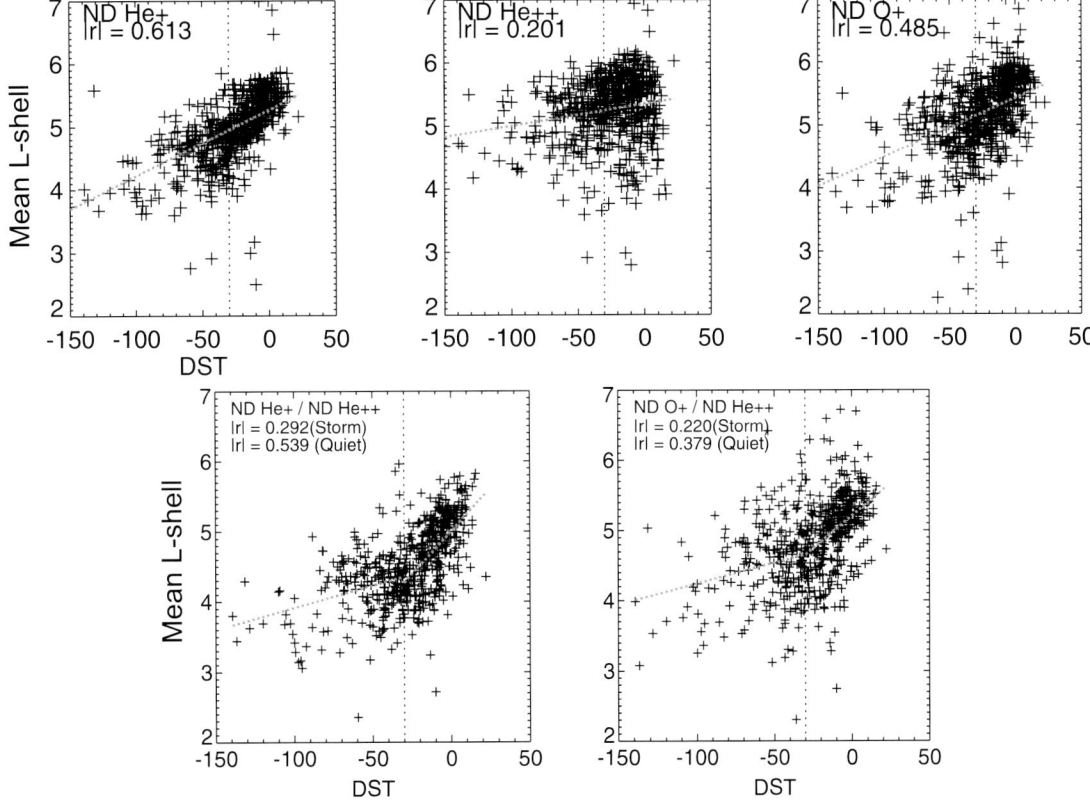

Figure 6. Derived mean value of L-shell against Dst, for He$^+$, He^{++} and O$^+$. It is seen that there is a strong correlation (R=.613 and .485) with Dst for He$^+$ and O$^+$, the species generally considered to be ionospheric. There is a rather insignificant (R=0.201) correlation for He^{++}. The regression lines are overplotted. When we plot the ratios He$^+$/He^{++} and O$^+$/He^{++}, we see a break between storm and non-storm time. There is a good correlation (R=0.539) between He$^+$/He^{++} ratio and Dst in non-storm time.

of the mix of ionospheric and solar wind origin plasma in the inner magnetosphere. We have also divided the plot into two halves, for Dst greater and less than –30, and calculated the individual regression lines. There is no strong difference in the trend for each half for individual species. There is however a break in the trend for the ratios of species, particularly the ratio of He$^+$ to He^{++}, which we take as supporting evidence for the occurrence of the charge exchange process in quiet times.

4.1 Reordered plots

In Plate 4 we have reordered the data from Plate 2, plotting it against Dst (instead of date). In a sense therefore, this gives an impression of the mean recovery of a stormtime ring current as Dst increases from left to right. Each vertical cut shows the variation with L of the average number density for that species, for a given Dst value. Horizontal cuts show how, for a given species, the number density changes with Dst at a given L-shell. The plotted quantity is always species number density. Thus where charge exchange is occurring, losses for one charge state should balance increases by another.

Individual species are shown on the left. Several trends are particularly noticeable. For He$^+$, we see that the peak composition varies smoothly from around L=3.4 at large Dst to L= 5.3 at small Dst. A very similar behavior is seen for O$^+$. In both cases, the densities become lower at L-shells above 5.5. By contrast, He^{++} number density is similar for all L-shells between 3 and 5.5 when Dst is larger than –100nT. In quieter periods with Dst > –100, He^{++} number density remains high above L~5, and seems to drop of quicker the ionospheric species at Lower L-shells.

On the right of Plate 4, we show composition ratios, again reordered by Dst. The L-shell variation of the peak in the He$^+$/He^{++} ratio with Dst is particularly striking, and appears

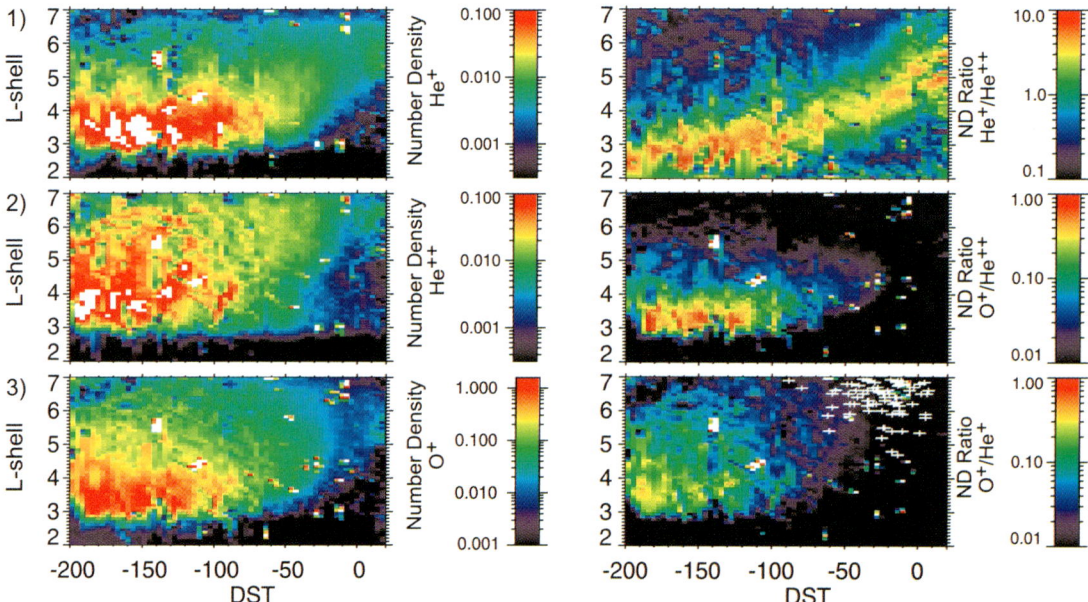

Plate 4. Ion data for the whole period days 13 to 255 of 1991, Reordered by Dst. See the text for details. Left Hand Panel 1 shows He^+, panel 2 He^{++} and panel 3 shows O^+. Right hand panels show ratios. Data is reordered by Dst. Panel 1 shows He^+/He^{++}, panel 2 O^+/He^{++} and panel 3 shows He^+/O^+. The positions of our sample 110 substorms is shown in the final (bottom right) panel as overplotted white crosses.

almost linear. We tentatively suggest that this is the result of a mixing between two separate He$^+$ source processes. These are ionospheric upflow, followed by inward convection, which predominates during stormtime, and charge exchange from He^{++} which is characteristic of very quiet times. O$^+$ behaves in a similar way to He$^+$, and indeed the plot of O$^+$ to He$^+$ ratio is constant at all locations, except for small Dst, where the He$^+$ fraction increases, again suggesting a separate source. In general He$^+$ [*Grande et al* 1997a] is only enhanced at an extremely low level in substorm injections, whereas O$^+$ is often and on average substantially enhanced. It therefore follows that the process which produces the similar distributions of these two ionospheric components cannot be direct substorm energisation.

In the final panel we show the location of the 110 dispersionless substorms. We note that they are not coincident with the location of the stormtime ring current populations, as would be expected if they were the same phenomenon. The substorms themselves generally occur at much higher L-shells than the L-shell of the largest Dst related composition changes. Thus the substorm injected ions cannot be contributing instantaneously to this changed ring current population. It of course is not excluded that the substorm injected ions may be part of the seed population that are subsequently transported to low L-shells during the magnetic storm onset to main phase. Fennell et al 1998 have observed convection features with Polar which show that some ions of all species get to very low L-shells even during relatively quiet times following substorms. It is very rare to see an injection below L=5, even after correcting our statistics for occupancy, whereas the main contribution to the ring current is at L =3 to 4.

5. DISCUSSION AND CONCLUSION

These observations of dispersionless injections at different locations and under a range of conditions lead to a number of conclusions on the storm-substorm relationship, which we summarise briefly. Dispersionless injections show a preponderance towards dusk, which is significantly increased during storm time. This may be a selection effect, linked to the duskward gradient curvature drift of energetic ions. However, since we have attempted to select only dispersionless injections, the lack of injections earlier than local midnight is surprising. The expected dispersion at L=6.6 between 80 and 250 keV/q ions is 2.4 minutes per hour of local time. Thus our criterion of dispersion less than 1 minute should restrict the effect to less than half an hour in local time, suggesting that what we are seeing is a tailward source, transported in to lower L-shells.

Storms show a peak in the O$^+$/He^{++} ratio with L-shell that is well correlated to Dst. There is no coincidence between the generally high L-shell at which substorms were observed, and the lower average value of L-shell where the ring current composition changes are greatest. We have taken care to ensure that this conclusion is not an observational effect.

The ring current as evidenced by a high time resolution Dst does not show any systematic response to substorm onset. An exception is a small increased excursion during the growth phase of large substorms, possibly related to conditions of enhanced convection with which they coincide.

Thus we conclude that substorms do not contribute to the generation of a stormtime ring current by directly injecting particles into the ring current. The indirect effect whereby substorms produce spikes in the convection potential that are the motor of ring current buildup is also precluded. We feel that the long term (10 hour) effects identified by *Iyemori and Rao* 1996 are not directly produced by individual substorms, but rather reflect the increased probability of substorms during stormtime.

From the superposed epoch analysis, we identified a small increase in the number density of ionospheric oxygen relative to solar wind origin He^{++} during a substorm. Since there is no systematic difference in the energisation between the different species, the changes in energy density are the result of changes in number density.

Large storms are responsible for producing a considerable upflow of ionospheric oxygen [*Kamide et al* 1997] and in a long lived storm this material will be recycled through the magnetosphere, and end up in the ring current. An oxygen atom has the mass of sixteen protons, and therefore the effect of oxygen on the energy density of the ring current is likewise magnified. Thus oxygen can contribute a large fraction to the measured Dst. But the oxygen ion itself is not behaving in a different way from other ions, and has no special influence on the physical processes involved. Stormtime substorms process the available ambient material, but do not greatly change its concentration over the substorm timescale of an hour. The properties of oxygen rich substorms examined in this and earlier [*Grande et al* 1997a] studies are not different from other substorms.

Possibly, the substorms inject a lower energy seed population, which MICS would not observe, that is energized by the cross L-shell transport during the storm process to energies >70 keV but at L-shell values well earthward of where the substorm onset signature is observed. There have been many discussions in the literature of ionospheric upflow and its association with substorm activity, see for example *Gazey et al* 1996 and references therein.

Singly charged helium behaves in a very similar way to O^+ during stormtimes, as evidenced by its constant ratio at all L-shells. Yet it is always a minor component of substorms injected population. In quiet times a separate charge exchange mechanism may be operating. Prompt entry of solar wind material does occur [Grande et al 1996] on the timescale of the superposed epoch analysis we have carried out here. But neither the ring current composition nor the Dst is affected on this timescale by substorms.

It is clear that the differences in average injection composition between storm and quiet time are small, and the variation between individual events is. So, while the O^+ concentration in the ring current increases during storms, the reason for this cannot however be attributed directly to the ions injected during a substorm.

CRRES flew at solar maximum. Observations during solar minimum periods, from Viking and shortly after the launch of the Polar satellite, show very little oxygen, and also relatively weak storms. Thus at solar minimum steady state charge exchange mechanisms may be more important.

Overall, we conclude that there is no great difference, as far as the energetic particle signatures are concerned, between stormtime substorms and those from less disturbed periods. Although substorms may be important in initiating ionospheric upflow, substorm injected particles do not contribute, on a timescale of an hour or less, to the ring current. Oxygen does not play a causative role in storms, except in as much as its mass means that it is a major contributor to Dst.

Acknowledgements. A. Hall was jointly supported under a CASE studentship by PPARC and CCLRC. The original CRRES work at RAL was supported in part by grant AFOSR-85-0237 from the US Air Force. The work of J.F. Fennell was supported by the USAF under contract F04701-891-C-0089.

REFERENCES

Chen, M W, Schulz M, L R Lyons, and D. J. Gorney, Stormtime transport of ring current and radiation belt ions, *J Geophys. Res.* 98 3835-3849, 1993

Baker D. N., T. A. Fritz, W. Lennartsson, B. Wilken, H. W. Kroehl, and J. Birn, The role of heavy ions in the localisation of substorm disturbances on March 22, 1979: CDAW 6, *J. Geophys. Res.*, *90*, 1273-1281, 1985.

Baumjohann W, Y. Kamide and R. Nakamura . Substorms, storms and the near earth tail. *J Geomag. Geoelectr.* 1995

Daglis, I. A. and Axford, W. I., Livi S. Wilken B, Grande M and Soraas F, Auroral Ionospheric Ion Feeding of the Inner Plasma Sheet during Substorms. *J. Geomag, Geoelec.* 48 729-740 1996

Daglis, I. A. and Axford, W. I., Fast ionospheric response to enhanced activity in geospace: Ion feeding of the inner magnetotail, *J. Geophys. Res.*, *101*, 5047-5065, 1996.

Daglis, I. A., The role of magnetosphere-ionosphere coupling in magnetic storm dynamics, in *Magnetic Storms*, Geophys. Monogr. Ser., vol. 98, edited by B. T. Tsurutani, W. D. Gonzalez, Y. Kamide, and J. K. Arballo, pp. 107-116, American Geophysical Union, Washington, DC, 1997.

Fennell, J.F, M. W. Chen, J. L. Roeder, W. K. Peterson, K. J. Trattner, R. Friedel, S. Livi, M. Grande, C. Perry, T. A. Fritz, and R. Sheldon: Multiple Discrete-Energy Ion Features in the Inner Magnetosphere: Polar Observations. *Physics of Space Plasmas*, No. 15, p 395-400, MIT Center for Theoretical Geo/Cosmo Plasma Physics, Cambridge, MA, 1998.

Spjeldvik W N , TA Fritz Theory for charge states of energetic oxygen ions in the Earth's radiation belts *J. Geophys. Res* vol.83, no.A4; 1 April 1978; p.1583-94. 1978

Gazey N G J, M. Lockwood M. Grande, C.H. Perry, P.N. Smith, S. Coles, A.M. Aylward, R.J. Bunting, H. Opgenoorth and B. Wilken. EISCAT/CRRES observations: Nightside Ionospheric ion outflow and oxygen-rich substorm injections. *Annales Geophysica* 14, 1032-1043, 1996.

Gonzalez, W.D., J.A. Joselyn, Y. Kamide, H.W. Kroehl, G.Rostoker, B.T. Tsurutani, V.M.Vasyliunas. What is a Geomagnetic Storm? *J. Geophys. Res.*, 99, 5771, 1994

Grande,M., C.,H.,Perry, D.,Hall, B.,Wilken, S.,Livi, F. Soraas, J.,F.,Fennell, Composition Signatures of Substorm Injections, in Proceedings of the International Conference on Substorms (ICS-1), Kiruna, Sweden, ESA SP-335), pp.,485-490, 1992.

Grande M., C.H.Perry, A.M.Hall, J.F.Fennell and B.Wilken Survey of Ring Current Composition During Magnetic Storms. *Adv. Space Res.* 20, 3,321-326 1997a.

Grande M., C.H.Perry, J.B.Blake, M.W Chen, J.F.Fennell and B. Wilken. Observations of Iron Silicon and other Heavy Ions in the Geostationary Altitude Region during the March 1991 Geomagnetic Storm. *J. Geophys. Res* Vol l01 a11 P24 707 1997b.

Hones E. W. The Magnetotail: its generation and dissipation, in *Physics of Solar-Planetary Environments*, v II, ed. D. J. Williams, p558, 1976.

Hamilton D.C; Gloeckler G; Ipavich F.M. Studemann-W; Wilken-B; Kremser-G, Ring current development during the great geomagnetic storm of February 1986 *J. Geophys. Res* vol.93, no.A12; p.14343-55. 1988;

Iyemori T; Rao D.R.K. Decay of the Dst field of geomagnetic disturbance after substorm onset and its implication to storm-substorm relation. *Annales-Geophysicae.* vol.14, no.6; June 1996; p.608-18. 1996.

Kamide, Y., Is substorm occurrence a necessary condition for a magnetic storm?, *J. Geomagn. Geoelectr.*, 44, 109, 1992.

Kamide, Y., W. Baumjohann, I. A. Daglis, W. D. Gonzalez, M. Grande, J. A. Joselyn, R. L. McPherron, J. L. Phillips, G. D. Reeves, G. Rostoker, A.S. Sharma, H. J. Singer, B.T. Tsurutani, and V.M. Vasyliunas, Current understanding of magnetic storms: Storm/substorm relationships, *J. Geophys. Res -Space Physics*, 1998, Vol.103, No.A8, pp.17705-17728

Lui, A. T. Y., R. E. Lopez, B.J. Anderson, K. Takahashi, L.J. Zanetti, R.W. McEntire, T. A. Potemera, D. M. Klumpar, E. M. Greene, R. Strangeway, Current disruptions in the near-earth neutral sheet region, *J Geophy. Res.*, 96 1991.

Perry C. H., M. Grande, A.M. Hall, B. Wilken. Statistical Survey of Dispersionless Substorm Injections Observed by the CRRES MICS Ion Spectrometer. Proc ICS-3 ESA SP389 1996

Sharma A. S., D. N. Baker, M. Grande, Y. Kamide, G. S. Lakhina, R. M. McPherron, G. D. Reeves, G. Rostoker, R. Vondrak, L. Zelenyi Storm-Substorm Relationship: Current Understanding and Outlook. This volume.

Sheldon- R.B; H.E Spence; J.F. Fennell, Observation of the 40 keV field-aligned ion beams *Geophysical-Research-Letters*. vol.25, no.10; p.1617-20. 1998

Wilken B., W. Weiß, D. Hall, M. Grande, F. Søraas, and J. F. Fennell, Magnetospheric Ion Composition Spectrometer onboard the CRRES spacecraft, *J. of Spacecraft and Rockets*, *29*, 585-591, 1992.

Young D.T., H. Balsiger and J. Geiss, Correlations of Magnetospheric Ion Composition with Geomagnetic and Solar Activity. *J. Geophys. Res* vol 87 A11, p9077 1982.

M. Grande, C. H. Perry, A. Hall, Rutherford Appleton Lab, Chilton, Didcot Oxon OX11 0QX, UK.

J.F. Fennell, Aerospace Corporation, PO Box 92957, Los Angeles, CA 90009-2957 USA.

Y Kamide, R Nakamura, Solar Terrestrial Environment Laboratory, Nagoya University, Toyokawa, Japan.

IMAGE, POLAR, and Geosynchronous Observations of Substorm and Ring Current Ion Injection

G. D. Reeves[1], M. G. Henderson[1], R. M. Skoug[1], M. F. Thomsen[1], J. E. Borovsky[1], H. O. Funsten[1],
P. C:son Brandt[2], D. J. Mitchell[2], J.-M. Jahn[3], C. J. Pollock[3], D. J. McComas[3], and S. B. Mende[4]

The geomagnetic storm of October 4–6, 2000 provides an exceptionally good opportunity to examine the role of substorms during the storm because of the relatively moderate solar wind driving and because of the excellent set of satellite observations. We show that the entire day of October 4 was characterized a sequence of substorms and a gradual build-up of the storm-time ring current. We examined one of those substorms in some detail showing that it had the expected signatures of a magnetospheric substorm. ENA observations and in situ measurements show that the substorms clearly do provide a mechanism for transporting energetic ions from the magnetotail to the inner magnetosphere. Substorm injections do not, however, provide the only mechanism. As other studies have suggested, the evidence from this storm shows that the presence of a quasi-steady convection electric field plays an additional important role. This is seen through the comparison of Dst which decreases slowly and smoothly over nearly 10 hours with geosynchronous particle injections which occur impulsively at roughly 2-hour intervals producing a "sawtooth" injection profile. Further evidence comes from ENA observations that show both impulsive injection during substorms and continued intensification and eastward expansion in response to convection electric fields. We also investigate the transport and symmetry of the ring current particles. We find that, even more than 10-hours into the storm when Dst had decreased below -100 nT and at least five clear substorm injections had occurred, the ring current remained highly asymmetric with essentially no ENA emissions from the dawn-to-noon sector. We also find that Dst, SYM-H, and ASY-H all respond approximately equally in spite of the fact that there is apparently no symmetric component to the ring current until later in the storm.

[1]Los Alamos National Laboratory, Los Alamos, New Mexico
[2]Johns Hopkins APL, Laurel Maryland
[3]Southwest Research Institute, San Antonio, Texas
[4]University of California, Berkeley California

Disturbances in Geospace: The Storm-Substorm Relationship
Geophysical Monograph 142
Copyright 2003 by the American Geophysical Union
10.1029/142GM09

1. INTRODUCTION

A central question in the storm-substorm relationship is whether the storm-time ring current is built up solely through a series of substorm injections, whether other processes are necessary, or even whether substorms themselves are necessary. It is now possible to investigate these questions using a powerful combination of ground-based observations, satellite measurements of the in situ ion populations, global auroral imagers, and global energetic neutral atom images.

The newest component in this "toolbox" are the Energetic Neutral Atom (ENA) images. ENAs are produced when a cold exospheric neutral atom gives up its electron to a trapped energetic ion producing a free energetic neutral atom which can be collected remotely and processed to reveal the global, time-dependent distribution of ions in the inner magnetosphere. Thus, we can supplement the more traditional means of investigating the storm-substorm relationship with ENA images which show the timing and location of substorm injections and their relationship to the build-up and trapping of the ring current. (Throughout this paper we use the term "ring current" to describe the magnetospheric current carriers regardless of whether a closed ring of current has been formed)

Despite it's relatively recent introduction, ENA imaging has already been applied quite successfully to the study of storms and substorms. The first application of ENA observations to geomagnetic storms was reported by Roelof [1987] who used ISEE-1 energetic particle observations to produce ENA images of the storm-time ring current. Later, ENA fluxes from the inner magnetosphere were shown to correspond closely to the ground magnetic perturbations represented by the Dst index [e.g. *Roelof et al.*, 1985; *Jorgensen et al.*, 1997, 2001; *C:son Brandt et al.*, this volume]. *Henderson et al.* [1997] used the Imaging Proton Spectrometer (IPS) on POLAR to show the first time-dependent images of the substorm injection and further investigated substorm injections using inversion techniques to determine the magnetospheric ion distributions [*Henderson et al.*, 1999, 2000]. Jorgensen et al. [2000] showed statistically that substorm typically produce "bursts" of ENA emissions. The first instrument specifically designed to measure ENAs was flown on the low-altitude Swedish microsatellite, ASTRID [e.g. *Barabash et al.*, 1997; *C:son Brandt et al.*, 2001 a, b] and the recently-launched IMAGE satellite now provides ENA images from a suite of detectors [e.g. *Burch et al.*, 2001] which are providing unprecedented spatial and temporal resolution of the injection and transport of inner magnetospheric ions during storms and substorms [e.g. *C:son Brandt et al.*, 2002].

Two recent studies have used ENA observations and supporting data to look explicitly at the storm-substorm relationship. Reeves and Henderson [2001] presented a study which compared 7 isolated substorm injections with 7 storm-time injections using POLAR ENA observations and in situ geosynchronous fluxes. One conclusion of that study was that, while main phase substorms can be difficult to identify, essentially all storms began with a clear substorm and a clear substorm injection. Further they found that the storm-time injections were essentially identical to isolated substorm injections. They were neither larger (in flux or local time) or more intense (e.g. in spectral hardness). What distinguished storm-times from isolated events was (a) continued injection activity for a period of hours following the initial injection, (b) a spreading of the local time extent of ion injection toward dawn—opposite to the direction of ion drift, and (c) an immediate response in Dst for the storm-time injections compared to no measurable response for the isolated events.

More recently *Lui et al.* [2001] used Geotail ENA observations of the ring current and SuperDARN radar observations of polar cap convection to study the storm-substorm relationship. They showed that the ring current intensified even at times when there were no substorm injections occurring but that there was simultaneous increase in polar cap convection. Both studies concluded that it was the presence of large-scale, externally-imposed, "convection" electric field superimposed on localized, inductive electric fields which differentiated storm-time particle injections from typical substorm injections—a hypothesis we further investigate in this paper.

A more complete review of recent ENA-based studies of storms and substorms is presented in the introductory paper to this volume [*Sharma et al.*, this volume]. Additionally *C:son Brandt et al*, [this volume] provide a complementary analysis of the October 2000 storm which uses inversion techniques to calculate the magnetospheric particle distributions and compare the time-dependent magnetospheric energy content with the Dst index.

2. THE OCTOBER 4–6, 2000 STORM

A particularly good storm for investigating the storm-substorm relationship occurred on October 4–6, 2000 (days 278–280). The event began on October 4 (day 278) with a southward turning of the IMF (Plate 1, panel a). B_Z remained moderate at around –10 nT until 04:30 UT on October 5. The moderate driving produced a very gradual build-up of the ring current with Dst decreasing steadily until it reached -142 nT at 19 UT on October 4 (Plate 1, panel b). During that time the solar wind velocity remained below 500 km/s and the density hovered around typical values of 10 particles/cm^3 (data not shown).

For this time period ACE was upstream of the Earth located at X=225, Y=30 R_E while Wind was leading the Earth in its orbit at X=32, Y=-213 R_E. Despite the large perpendicular separations to the Earth, the two spacecraft observed nearly identical solar wind conditions. Plate 1 shows this for B_Z but it was also true for the other parameters which are not shown. Therefore we can be confident that the solar wind conditions measured were very likely those experienced by the magnetosphere.

In many ways the IMF conditions and the Dst response were characteristic of those which produce so-called Steady

Magnetospheric Convection (SMC) events. However, the magnetospheric response was not at all steady as we see from the AE index (panel c) and geosynchronous ion injection data (panel d). Instead, the rather moderate energy input into the magnetosphere allowed substorms to occur at well-separated, roughly two-hour intervals. This gradual development makes it relatively easy to identify individual substorms and to investigate their relationship to the overall dynamics of the storm. (We will present more evidence that this is indeed substorm activity later.)

Throughout the event it is notable how closely the Dst signatures track the solar wind driving. After about 19 UT a gradual weakening of B_Z was accompanied by a weakening of the ring current. A subsequent decrease in B_Z to below 20 nT around 05 UT on October 5 resulted in another decrease in Dst. When the IMF turned northward for a period of about 4 hours, AE became quiet and Dst became less negative (but only recovered to a value around –100 nT). When the IMF turned moderately southward again, Dst reached its minimum of –187 nT around 13 UT. As the IMF B_Z gradually weakened to around zero the recovery of Dst changed slope and AE and injection activity ceased.

We also note the very "direct" response of the ground magnetic perturbations measured by the SYM-H and ASY-H indices which are also plotted in panel b of Plate 1. The SYM and ASY indices are produced from 6 stations and have 1-min time resolution. SYM-H is essentially a higher-resolution version of Dst, is calculated in the same way, and is typically interpreted as the symmetric component of the ring current. It is therefore unremarkable that SYM-H follows the Dst curve and magnitude so faithfully. Here we plot the negative value of ASY-H which is typically interpreted as the asymmetric component of the ring current. -ASY-H is plotted on the same scale as SYM-H and Dst and we see that it also tracks Dst and the solar wind input quite faithfully. While there are times that the magnitude of SYM-H and ASY-H differ by up to a factor of two, it is clear that the two indices are not measuring independent quantities.

We will return to theses points as we examine the main phase of the storm on October 4 in more detail. First we will examine the second substorm injection of the main phase to establish its characteristics and its pedigree as a "substorm" and then consider the effect of the sequence of substorms that occur during the main phase.

3. THE 0930 UT SUBSTORM INJECTION

In order to examine the role of substorms within the storm we first establish that the activity on October 4 has the characteristic signatures of substorms. Those signatures include auroral brightening and poleward expansion of auroral activity, injection of energetic particles into the inner magnetosphere, stretching and dipolarization of the near-Earth magnetic field, and disruption/diversion of the cross-tail current in the substorm current wedge.

The first injection event of the main phase of this storm took place at 0623 UT approximately 37 min after the southward turning was measured at Wind, which is approximately the time the solar wind would arrive at the magnetopause. The FUV WIC camera on IMAGE showed that, while some auroral activity had started at earlier times, the first auroral brightening and poleward expansion took place at 0611 UT. We will not present data for this substorm/injection event here but rather will concentrate on the second substorm/injection which has nearly identical characteristics but might be considered more representative simply because it is not the first of the sequence.

3.1 Geosynchronous Energetic Particle Injections

Plate 2 shows the geosynchronous measurements and the ground-based indices for the period from 07 to 13 UT on October 4, 2000. The top panel shows energetic electrons (50–315 keV) from spacecraft 1989-046 which was slightly pre-midnight. The next panel shows the protons (75–400 keV) from the same spacecraft. A clear, dispersionless injection occurred at 0938 UT and was simultaneous in both electrons and ions indicating that the satellite, at 2238 LT, was in the heart of the substorm injection region [*Reeves et al.*, 1991; *Birn et al.*, 1997].

Prior to the injection we see a gradual decrease of the energetic electron and ion fluxes from their typical quiet-time values to values about 100 times lower. This is characteristic of a "growth phase dropout" produced by the stretching of the field into a more tail-like configuration [*Baker et al.*, 1978, 1981; *Reeves et al.*, 1993]. However, it is somewhat longer (2 hours) than a typical substorm growth phase. A similar dropout of the geosynchronous energetic particle fluxes begins shortly after the 0938 UT injection producing a "sawtooth" profile.

3.2 Magnetic Field Stretching and Dipolarization

The next panel shows the magnetic field inclination angle from the GOES satellites with 0° representing a completely Earthward-directed field and 90° representing a dipole-like field. Between 0700 and 0938 UT, GOES-10, which was very close to midnight, saw the field stretch from a dipole-like inclination of ≈80° to a very tail-like inclination of ≈10°. Then, between 0934 and 0954 UT the field at GOES-10 recovered to ≈75°. This is the signature of a very clear and strong stretching and dipolarization. GOES-8, which

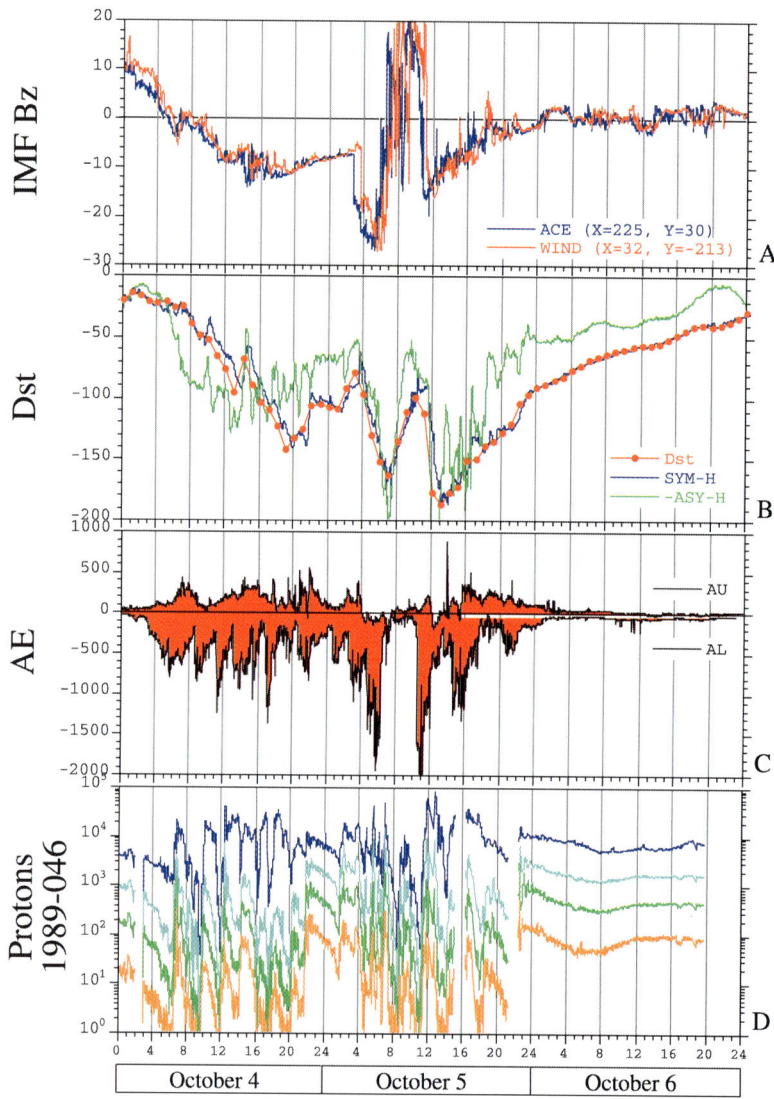

Plate 1. Overview of the October 4–6, 2000 geomagnetic storm. (A) the IMF B_Z component (courtesy of the ACE and WIND magnetometer teams). (B) The geomagnetic response measured by the preliminary 1-hour Dst index and by the 1-min. SYM-H and ASY-H components (courtesy of the University of Kyoto). (C) The preliminary AE index as specified by AU and AL (courtesy of the University of Kyoto). (D) The energetic proton flux (75–400 keV) measured by one of the LANL geosynchronous spacecraft.

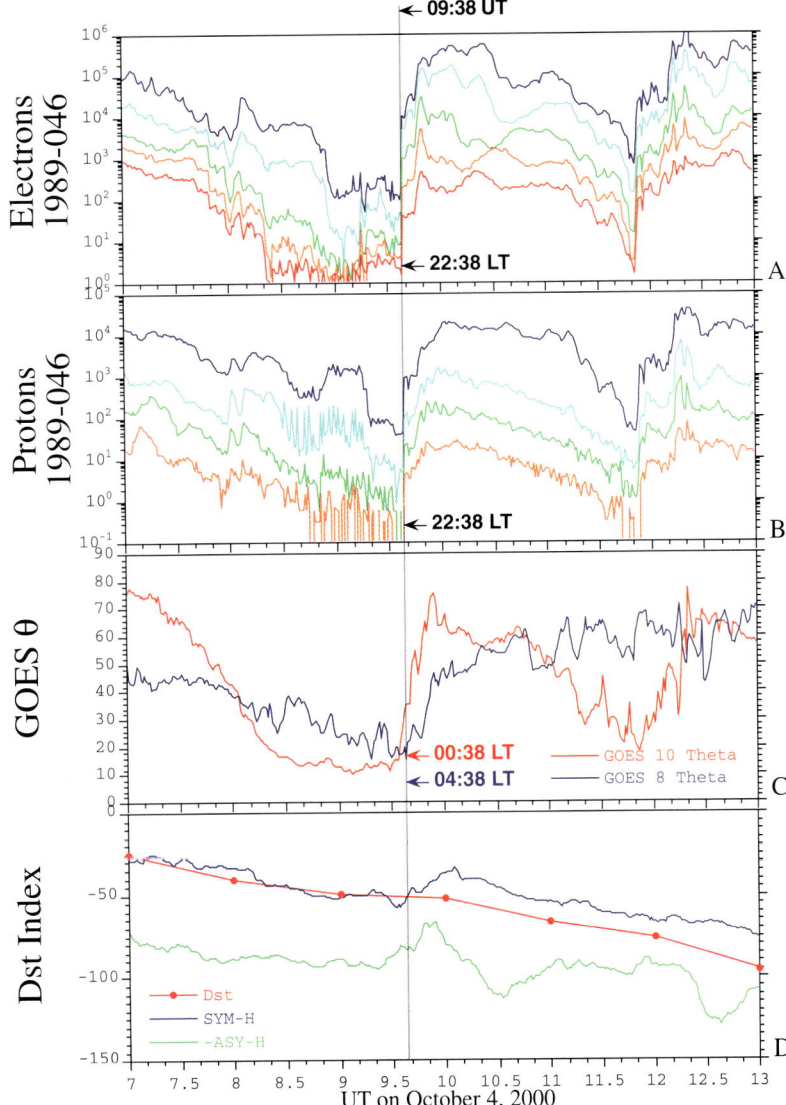

Plate 2. Characteristics of the 0938 UT substorm injections. Geosynchronous spacecraft 1989-046 was at 22:38 LT when it measured a dispersionless electron (A) and proton (B) injection. Both GOES spacecraft measured a nearly simultaneous dipolarization of the magnetic field (C) and the diversion of the cross-tail current into the ionosphere produced a mid-latitude H bay which also shows up in the SYM-H and ASY-H indices (D)

was located 4 hours further east observed qualitatively similar but less intense stretching and dipolarization. We also note that the nightside field almost immediately began to stretch again in response to continued dayside reconnection so, the nightside magnetic field inclination also exhibits a "sawtooth" profile similar to the energetic particle fluxes. The similarity is expected since the energetic particle "dropout" is produced by the stretching of the field.

3.3 Substorm Current Wedge

The bottom panel of Plate 2 shows the Dst, SYM-H, and –ASY-H indices for this time period. We note again how similar in magnitude and rate of change all three measures of the ground magnetic signatures of the ring current are. Also notable is the change of SYM-H and –ASY-H around the time of substorm onset. This signature is the well-known "mid-latitude H bay" which has long been used as a substorm signature. *Turner et al* [2000] further investigated this phenomena and also concluded the roughly 25 nT change in the ground magnetic signature is caused by the disruption and diversion of the cross-tail current into the ionosphere in the substorm current wedge.

3.4 Auroral Brightening and Expansion

Plate 3 shows auroral and Energetic Neutral Atom (ENA) images of the substorm collected by the IMAGE spacecraft [*Burch et al., 2001*]. The Far Ultra-Violet Wideband Imaging Camera (FUV-WIC) auroral imager provides images every two minutes but only representative images are shown here. The orientation is such that the sun is to the right and dusk is to the top of each image. The first image from 0705 UT shows the already expanded auroral oval produced by the earlier substorm. The next two images show the auroral activity dimming and retreating equatorward. The next auroral brightening takes place in the pre-midnight sector at the equatorward edge of the oval at 0930 UT. The auroral brightening expands rapidly poleward within the region of dimmer emissions left over from previous activity. The brightening also expands rapidly in local time until it extends from approximately dusk to well past midnight. By about 0950 a broad oval with regions of localized activity extends around the whole night side. This type if of activity continues until the next substorm in the sequence around 1200 UT.

3.5 ENA Images of the Injection

Plate 3 also gives a global view of the ion injection process seen through ENA images from the Medium Energy Neutral Atom (MENA) and High Energy Neutral Atom (HENA) instruments on IMAGE. Three energy ranges are shown, 5–12, 16–27, and 39–50 keV. In all cases the images show integrated ENA fluxes collected over four minutes and one image is shown each ten minutes beginning at 0930 UT. A "prestorm" image from 0600 UT is also shown to indicate the quiet-time fluxes and spatial distributions. The color scale is linear and shows the flux of ENA emissions from the line of sight. ENA fluxes are a convolution of the trapped magnetospheric ion distributions and the exospheric neutral density. (*C:Son Brandt et al.* [this volume] show inversions which determine trapped ion fluxes for this event.)

The MENA images show representative, dipole field lines at L=4 and 8 and the HENA images show dipole field lines at L=4, 8, and 12. Local noon is near the top of the images and is indicated in red for the MENA data and with the label 12 for the HENA data. Slightly different projections are used for the two data sets.

In all three energy ranges the 0930 UT images show effects of activity which began at 0623 UT with the first substorm of the main phase. The 0940 UT images show a clear injection of energetic ions in the pre-midnight region of the inner magnetosphere. To within the 4-minute resolution of these images, this is simultaneous with the injection observed at geosynchronous orbit at 0938 UT and not with the auroral onset which began approximately eight minutes earlier. The ENA emissions intensify between 0940 and 1010 UT which corresponds closely to the increase in ion fluxes measured in situ by the LANL instruments. Subsequently the ENA emissions fade as the particles drift and disperse but the fluxes remain elevated above the 0930 UT levels (See also *C:son Brandt et al.* [this volume]).

The region of intense ENA emissions also spreads in local time, extending both west, in the direction of ion drift, and east, opposite to the direction of ion drift. The eastward expansion of the ion injection region during storms was first reported by *Reeves and Henderson* [2001] and is a clear sign of continued Earthward particle transport even after the impulsive injection is over. Some evidence of westward gradient-curvature drift is also visible but during this time scattered sunlight and the viewing geometry make it difficult to unambiguously determine the extent of drift of particles to the dayside magnetosphere. We will examine that issue in more detail in the next section.

The observation of the substorm injection in the ENA emissions makes it clear that the "sawtooth" signatures seen in the geosynchronous observations are true injections and not simply an adiabatic response to the changing local magnetic field. Combined with the auroral images, the GOES magnetic field observations, and the ground magnetic perturbations this provides a fairly complete characterization of a storm-time substorm.

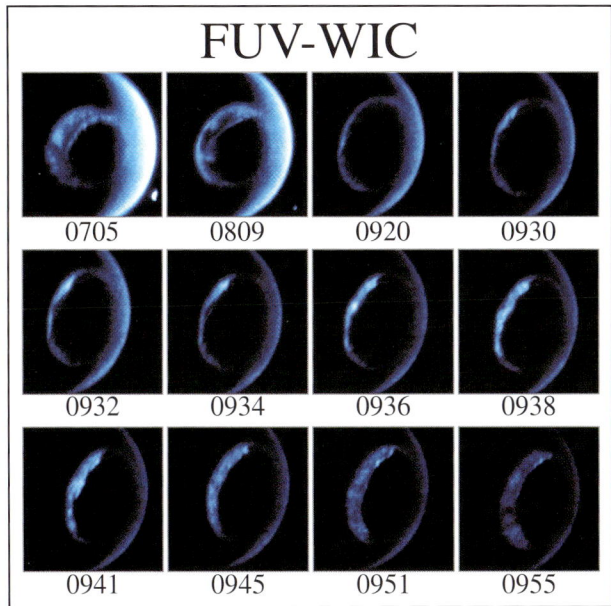
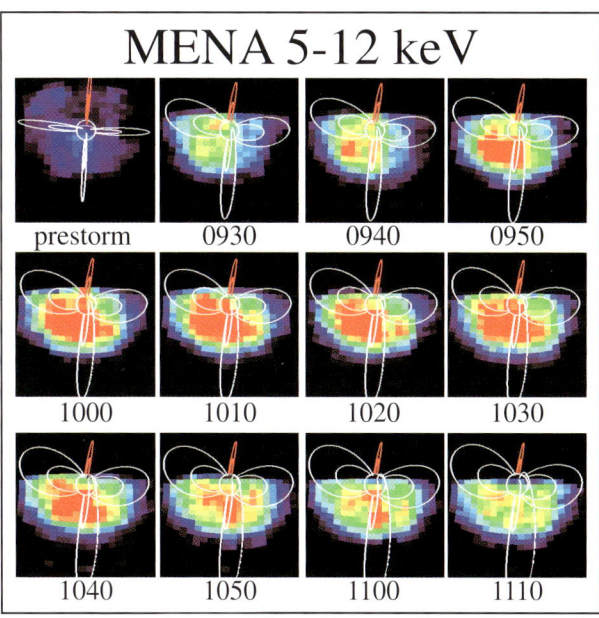
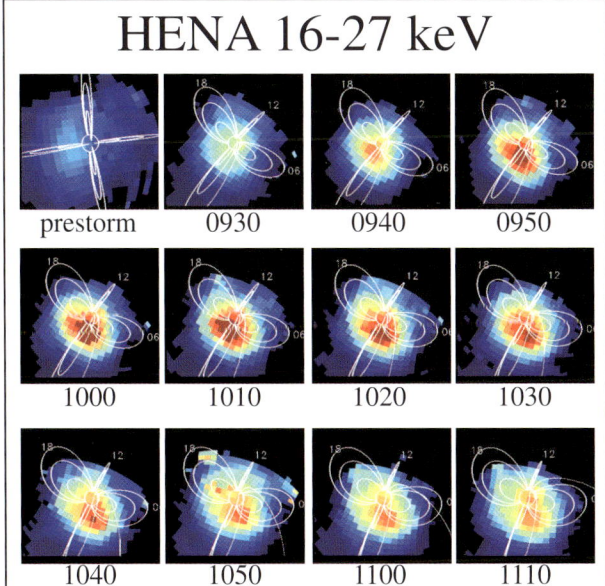
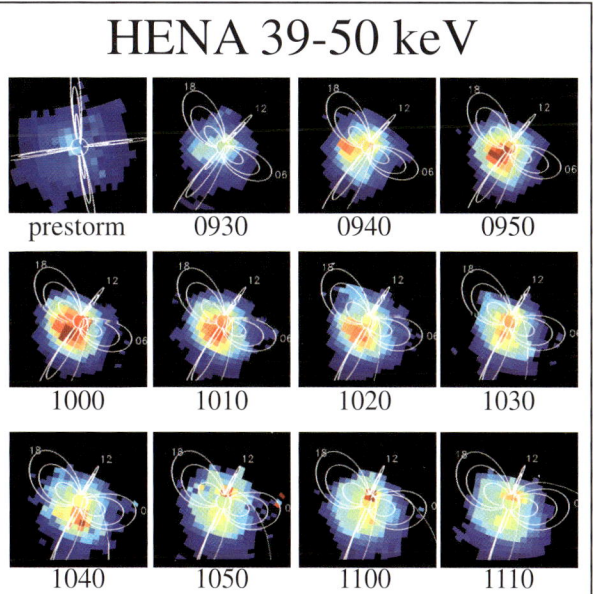

Plate 3. Auroral and ENA images from the IMAGE spacecraft. The auroral onset measured by the FUV instrument took place at 0930 UT and was followed by a rapid poleward and azimuthal expansion of activity. The substorm injection occurred at 0938 UT and is seen in the 0940 UT ENA images captured by both the MENA and HENA instruments over a wide range of energies.

There were eight such substorms during the storm main phase on October 4. Not all spacecraft were well-positioned to make good observations for all eight substorms but, whenever they were, the same signatures we have shown for this substorm were observed in the others and therefore may be typical of storm-time substorms at least under conditions of moderate driving. We note that similar physical processes probably occur under stronger solar wind driving but the rate of activations makes it difficult or impossible to separate the growth, onset, and recovery phases which characterize isolated substorms.

4. DEVELOPMENT OF THE RING CURRENT

Having examined the injection of ions by substorm processes, let us now look at the overall development of the ring current during the storm main phase on October 4, 2000. Plate 4 shows the geosynchronous energetic ion fluxes, the IMF B_Z, the Dst, SYM-H and –ASY-H indices, and the ENA images from IMAGE and POLAR for the time period from 06 to 24 UT. The ENA images are somewhat different from those in Plate 3. Here we show fluxes for 16-60 keV integrated for 10 minutes and displayed using a logarithmic color scale for the ENA fluxes. One 10-minute image is shown each two hours. Between the 12 and 16 UT images, IMAGE was in the radiation belts and unable to make ENA observations. However, during that time POLAR was well-positioned to make ENA observations (E>37.5 keV) from a similar vantage point. (POLAR made ENA observations at other times which were simultaneous with the IMAGE observations and the two sets of measurements were quite consistent.)

All images are taken from the northern hemisphere and local noon is located near the top of each image. Pixels with known contamination from scattered sunlight have been blacked out but some photon contamination remains in pixels on the dayside at L>8 and those fluxes should be ignored.

At 0600 UT we again see the fairly symmetric, low intensity ENA fluxes prior to the storm which are characteristic of quiet times. Over the next several hours, as Dst decreases, the ENA fluxes intensify. These images provide only a qualitative picture of the energetic ion energy density but *C:son Brandt et al*, [this volume] show a quantitative analysis which confirms the impression that the energy density in the inner magnetosphere corresponds well with the changes in Dst.

Notice that the geosynchronous ion fluxes (even at their peaks) do not show a long-term build up. However, as we have seen it is the injections at substorm onset that deliver material into the inner magnetosphere. This is understandable if we consider that geosynchronous orbit lies near the transition from dipole to tail field lines. Therefore the fluxes measured there show the particles that are passing through that region as they are transported from the plasmasheet to the inner magnetosphere.

The ENA emissions also allow us to infer the local time distribution of the ring current ions. The images at 0800, 1430 (POLAR), 1600, and 1800 provide particularly good viewing geometry. At other times, when IMAGE is viewing from the tail toward the dayside, it is not possible to distinguish between ENA fluxes from the dayside equatorial region and ENA fluxes from the high-latitude 'horns' of the nightside field lines without the use of modeling.

However, even given the caveats discussed above it is clear that the ring current development in this storm is highly asymmetric. Up to at least 1800 UT there is little or no evidence of ENA emissions from the dawn-to-noon quadrant of the magnetosphere. Hence there is no truly symmetric component to the ring current in spite of the fact that Dst has been decreasing steadily to less than –120 nT over a period exceeding 10 hours—much longer than ring current ion drift periods. It is likely that the majority of ions injected by substorms on the night side are lost to the magnetopause on the dayside as has been shown through modeling by *Liemohn et al*, [2001].

The 1600 UT image is particularly striking. ENA fluxes are strongest on the nightside where the injections occur. A "tail" of gradient-curvature drifting ions extends around dusk to the dayside but ends abruptly near Earth-Sun line at noon. Fluxes in the dawn-to-noon sector are extremely low, even as seen on this logarithmic scale. Yet, at 1600 UT, Dst≈104 nT and the magnitudes of SYM-H and ASY-H were nearly equal at approximately 87 nT. It is clear that all three geomagnetic indices are responding with nearly equal intensity to a highly asymmetric ring current distribution.

5. CONCLUSIONS

In this paper we have examined the geomagnetic storm of October 4–6, 2000. The storm provides an exceptionally good opportunity to examine the role of substorms during the storm because of the relatively moderate solar wind driving and because of the excellent set of satellite observations.

The period on October 4th was characterized by a sequence of substorms. We examined one of those substorms in some detail showing that it did, in fact, have the expected signatures of a magnetospheric substorm. The auroral brightening for this substorm was first observed at 0930 UT and the auroral activity subsequently expanded rapidly poleward and in local time. Prior to onset, the nightside magnetic field at geosynchronous orbit was highly stretched with an inclination angle around 10°. At about the same time as the auroral onset a rapid dipolarization of the magnetic field was observed. The diversion of the cross tail current in the substorm current wedge was also observed on the ground as a positive magnetic H bay which also shows up in the 1-minute resolution SYM-H and ASY-H indices.

The injection of energetic particles took place at 0938 UT shortly after the auroral onset and magnetic field dipolarization. The injection was seen in situ at geosynchronous altitudes and also remotely as a sudden brightening of ENA emissions from the pre-midnight inner magnetosphere.

What then is the role of these substorms in the large-scale spatial and temporal development of the storm-time ring current? It seems clear that one role is the rapid transport of large fluxes of energetic particles from the near-Earth tail into the inner magnetosphere through the effect of inductive electric fields associated with the dipolarization of the magnetic field. This is the traditional view of the role that substorms play in building up the ring current.

However, as we see from Plate 4, Dst had decreased to nearly –50 nT before the 0938 UT injection—after only one substorm injection had occurred—yet isolated substorm injections produce no change in Dst. These results support the conclusions of *Reeves and Henderson* [2001] who found the same result in a superposed epoch analysis of storm-time and isolated substorm injections.

Clearly other factors contribute to the build-up of the storm-time ring current. The study of Reeves and Henderson suggested that it was the continued injection of ions after onset that distinguished storm-time substorm injections. The ENA observations for October 4 provide further evidence for the role of continued injection. Plate 3 shows that the nightside ENA emissions remain strong and even intensify for almost an hour after onset—even while the magnetic field at GOES is re-stretching. Furthermore the region of nightside injection expands dawnward, opposite to the direction of ion drift, which can only be explained by continued injection.

Reeves and Henderson [2001] suggested that it was the action of the large-scale, quasi-steady "convection" electric field during storm times which was responsible for this continuing injection activity. *Lui et al.* [2001] reached a similar conclusion based on Geotail ENA and cross-polar cap potential measurements. The observations presented here provide even stronger evidence that it is the combined action of bursty inductive and quasi-steady convective electric fields which produce the storm-time ring current.

Are substorms therefore *essential* to storm development or are they *coincidental*? These observations cannot definitively answer that question but they do provide fodder for speculation. *Erickson and Wolf* [1980] showed conclusively that it is not possible to adiabatically convect magnetic flux from the distant plasmasheet to the near earth magnetosphere. If the magnetotail starts out with a substantial extent in the anti-sunward direction then substorms seem to be essential to rid the magnetotail of excess plasma and return magnetic flux to the dayside magnetosphere. It seems highly probable that all storms begin with a substorm and we know of no published evidence showing a storm that did not begin with a substorm.

However, if the neutral line remains not too distant from the Earth and magnetotail does not recover to an extended configuration it may be possible to avoid the Erickson and Wolf "pressure catastrophe". In that case it may be possible for transport of particles and flux to continue without impulsive "unloading" via substorms. This could, then, produce continuation of the storm without subsequent substorms – which has been referred to as steady magnetospheric convection (SMC) events. In fact the storm of October 2000 has the moderate solar wind driving conditions and slow, steady decrease of Dst which are often seen in other events which have been classified as SMCs.

It is sometimes assumed that storms are characterized by the development of a symmetric component to the ring current while substorm injections and SMCs do not trap particles and therefore produce only asymmetric ring current contributions. This is clearly not a useful distinction. For example, it is well-known that isolated substorm injections can produce "drift echoes" in which the injected particles drift through 360° to be observed again at the same spacecraft which is only possible if the particles are "trapped". On the other hand this event produced a dip in Dst (and SYM-H) to below –100 nT over a 10-hour time period without any evidence that the ring current ions were able to drift past midnight to form a symmetric component.

Indeed, while it is possible to separate any azimuthal distribution into symmetric and asymmetric components it may not be terribly informative to do so. This is particularly true if ground magnetic perturbations are used to define the symmetry of the ring current. However, as these observations begin to demonstrate, global observations of the ENA emissions from the ions which actually carry the "ring" current can remove some of the ambiguity which has been an ongoing source of debate. Much as global auroral images have become indispensable to the study of substorms, global ENA images are gradually becoming indispensable to the study of storms. The combination of both along with ground-based and in-situ observations holds great promise for better understanding of the storm-substorm relationship.

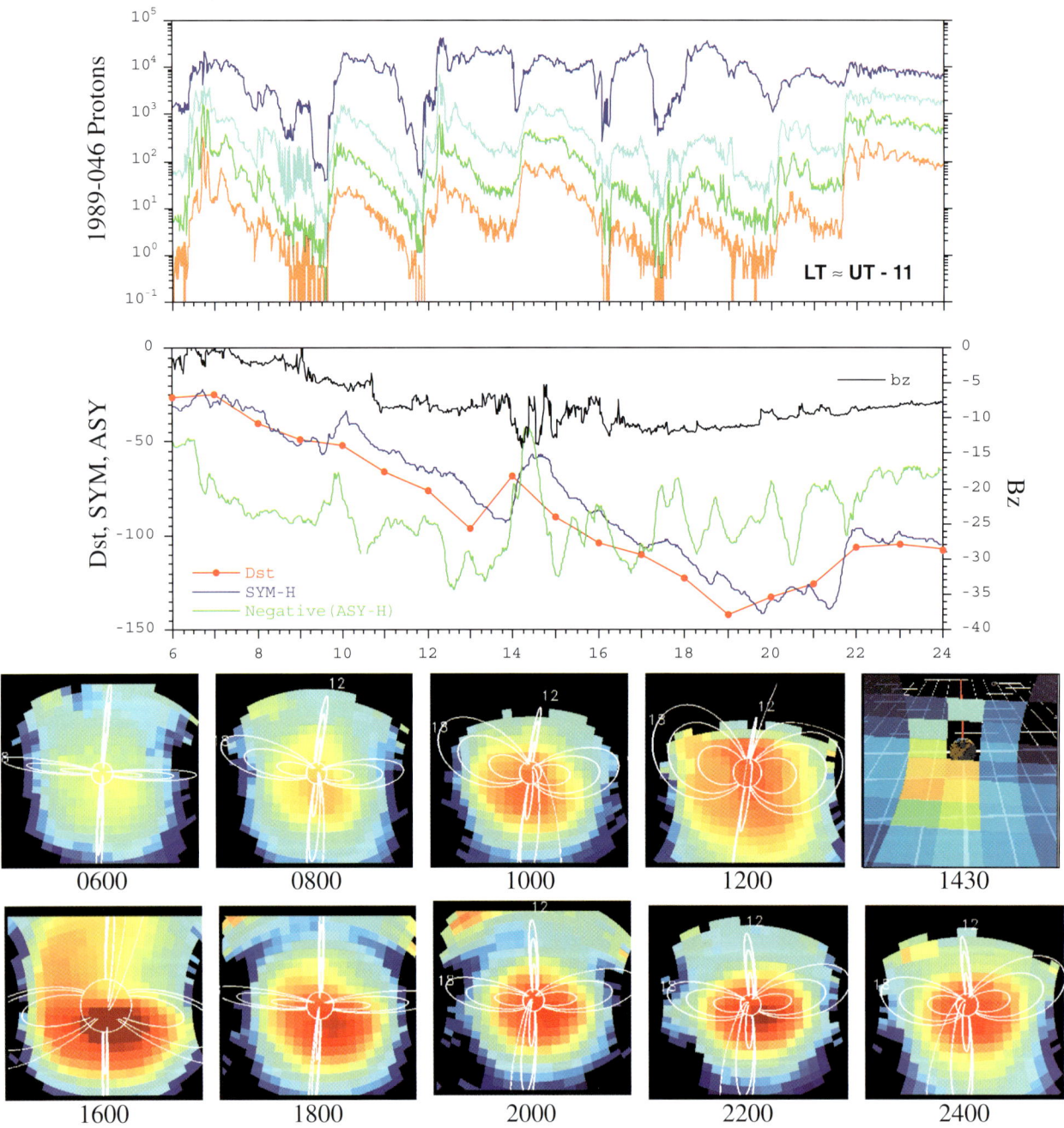

Plate 4. The development of the ring current relative to the substorm injections from 6-24 UT on October 4. This figure shows energetic proton data, IMF B_Z (from ACE), geomagnetic indices, and ENA images from IMAGE and POLAR. The temporal development and large asymmetry of the ring current are apparent.

Acknowledgments. We would like to thank the many individuals who have provide valuable data and physical insight including the ACE and WIND magnetometer teams, the University of Kyoto, NOAA Space Environment Center and especially John Stienberg, Bern Blake, Terry Onsager, and Howard Singer.

REFERENCES

Baker, D. N., E. W. Hones, Jr., P. R. Higbie, R. D. Belian, and P. Stauning, Global properties of the magnetosphere during a substorm growth phase: A case study, *J. Geophys. Res., 86*, 8941-9856, 1981.

Baker, D. N., P. R. Higbie, E. W. Hones Jr., and R. D. Belian, High-resolution energetic particle measurements at 6.6 Re, 3, Low-energy electron anisotropies and short-term substorm predictions, *J. Geophys. Res., 83*, 4864, 1978.

Barabash, S., P. C:son Brandt, O. Norberg, R. Lundin, E. C. Roelof, C. J. Chase, and B. H. Mauk, Energetic neutral atom imaging by the Astrid microsatellite, *Adv. Space Res., 20*, 1055-1060, 1997.

Birn, J., M. F. Thomsen, J. E. Borovsky, G. D. Reeves, D. J. McComas, and R. D. Belian, Characteristic plasma properties during dispersionless substorm injection at geosynchronous orbit, *J. Geophys. Res., 102*, 2309-2324, 1997.

Burch, J. L., J. L. Burch, S. B. Mende, D. G. Mitchell, T. E. Moore, C. J. Pollock, B. W. Reinisch, B. R. Sandel, S. A. Fuselier, D. L. Gallagher, J. L. Green, J. D. Perez, and P. H. Reiff, Views of Earth's Magnetosphere with the IMAGE Satellite, *Science, 291*, 629, 2001.

C:son Brandt, P., D. G. Mitchell, R. Demajistre, E. C. Roelof, S. Ohtani, J.-M. Janh, C. Polluck, and G. D. Reeves, Storm-substorm relationship during the 4 October, 2000 storm. IMAGE global imaging results, *this volume*.

C:son Brandt, P., R. Demajistre, E. C. Roelof, D. G. Mitchell, and S. Mende, IMAGE/HENA: Global ENA imaging of the plasmasheet and ring current during substorms, *J. Geophys. Res., submitted*, 2002.

C:son Brandt, P., S. Barabash, E. C. Roelof, and C. J. Chase, Energetic neutral atom imaging at low altitude from the Swedish microsatellite Astrid: Observations at low (<=10 keV) energies, *J. Geophys. Res., 106*, 24,663, 2001a.

C:son Brandt, P., S. Barabash. E. C. Roelof, and C. J. Chase, Energetic neutral atom imaging at low altitudes from the Swedish microsatellite Astrid: Extraction of the equatorial ion distribution, *J. Geophys. Res., 106*, 25,731, 2001b.

Erickson, G. M., and R. A. Wolf, Is steady convection possible in the Earth's magnetotail?, *Geophys. Res. Lett., 7*, 897-900, 1980.

Henderson, M. G., G. D. Reeves, A. M. Jorgensen, H. E. Spence, L. A. Frank, J. B. Sigworth, J. F. Fennell, J. L. Roeder, J. B. Blake, K. Yumoto, K. Shiokawa, and S. Bourdarie, POLAR CEPPAD/IPS energetic neutral atom (ENA) images of a substorm injection, *Adv. Space Res., 25*, 2407-2416, 2000.

Henderson, M. G., G. D. Reeves, K. R. Moore, H. E. Spence, A. M. Jorgensen, J. F. Fennell, J.B. Blake, and E. C. Roelof, Energetic neutral atom imaging with the Polar CEPPAD/IPS instrument: Initial forward modeling results, *Phys. Chem. Earth, 24*, 203, 1999.

Henderson, M. G., G. D. Reeves, H. E. Spence, R. B. Sheldon, A. M. Jorgensen, J. B. Blake, and J. F. Fennell, First energetic neutral atom images from Polar CEPPAD/IPS, *Geophys. Res. Lett., 24*, 1167, 1997.

Jorgensen, A. M., M. G. Henderson, E. C. Roelof, G. D. Reeves, and H. E. Spence, The charge-exchange contribution to the decay of the ring current measured by Energetic Neutral Atoms (ENAs), *J. Geophys. Res., 106*, 1931-1938, 2001.

Jorgensen, A. M., L. Kepko, M. G. Henderson, H. E. Spence, G. D. Reeves, J. B. Sigwarth, and L. A. Frank, The association of energetic neutral atom (ENA) bursts and magnetospheric substorms, *J. Geophys. Res., 105*, 18,753, 2000.

Jorgensen, A. M., H. E. Spence, M. G. Henderson, G. D. Reeves, M. Sugiura, and T. Kamei, Global energetic neutral atom (ENA) measurements and their association with the Dst index, *Geophys. Res. Lett., 24*, 3173-3176, 1997.

Liemohn, M. W., J. U. Kozyra, M. F. Thomsen, J. L. Roeder, G. Lu, J. E. Borovsky, and T. E. Cayton, The dominant role of the asymmetric ring current in producing the stormtime Dst*, *J. Geophys. Res., 106*, 10,883-10,904, 2001.

Lui, A. T. Y., R. W. McEntire, and K. B. Baker, A New Insight on the Cause of Magnetic Storms, Geophys. Res. Lett., 28, 3413, 2001.

Reeves, G. D., and M. G. Henderson, The storm-substorm relationship: Ion injections in geosynchronous measurements and composite energetic neutral atom images, *J. Geophys. Res., 106*, 5833-5844, 2001.

Reeves, G. D., T. A. Fritz, R. D. Belian, R. W. McEntire, D. J. Williams, E. C. Roelof, M. G. Kivelson, and B. Wilken, Structured plasma sheet thinning observed by Galileo and 1984-129, *J. Geophys. Res , 98*, 21,323, 1993.

Reeves, G. D., R. D. Belian, and T. A. Fritz, Numerical tracing of energetic particle drifts in a model magnetosphere, J. Geophys. Res., 96, 13,997, 1991.

Roelof, E. C., Energetic neutral atom image of a storm-time ring current, *Geophys. Res. Lett., 14*, 652, 1987.

Roelof, E. C., D. G. Mitchell, and D. J. Williams, Energetic neutral Atoms (E~50 keV) from the ring current: IMP 7/8 and ISEE 1, *J. Geophys. Res., 90*, 10991, 1985.

Sharma, A. S., and others, Introduction paper to "AGU Monograph on the Storm-Substorm Relationship" (still need the correct title), this volume.

Turner, N. E., D. N. Baker, T. I. Pulkkinen, and R. L. McPherron, Evaluation of the tail current contribution to Dst, *J. Geophys. Res., 105*, 5431-5439, 2000.

G. D. Reeves, Los Alamos National Laboratory, NIS-1 MS D-466, Los Alamos, NM 87545. (reeves@lanl.gov)

Storm-Substorm Relationships During the 4 October, 2000 Storm. IMAGE Global ENA Imaging Results

Pontus C:son Brandt[1], Donald G. Mitchell[1], Shin Ohtani[1], Robert Demajistre[1], Edmond C. Roelof[1], Jörg-Micha Jahn,[2] Craig Pollock[2], and Geoff Reeves[3]

Global ion distributions in the 1–200 keV energy range from the main phase of the geomagnetic storm on 4 October 2000 are presented and analyzed. Proton distributions have been obtained by inverting energetic neutral atom (ENA) images from the high energy neutral atom (HENA) instrument on board the IMAGE satellite using a constrained linear inversion technique. The storm is characterized by a 24-hour long main phase where the IMF B_z steadily decreases followed by a 2 day recovery. Several substorms occured during the main phase as can be seen from in-situ measurements from geosynchronous satellites (LANL, GOES). Substorm injections during the early main phase, when the dawn to dusk electric field was weak, ocurred on closed trajectories. A strong asymmetric ring current developed as the IMF B_z decreased gradually to about –10 nT. A substorm ocurred at about 17:30 UT which injected plasma onto open trajectories with no clear change in the morphology of the partial ring current. As the IMF B_z increased towards zero, substorms were observed to inject ions onto closed trajectories. The peak of the ring current moved from L=5 to L=3 during the entire main phase. A preliminary inspection of ~80–160 keV oxygen ENA fluxes reveal a one order of magnitude increase during the entire main phase, implying that O^+ contributed significantly to this storm. Rapid decrease followed by decay (~1 h) was superposed on the gradual increase of the oxygen ENA. Each one of these "bursts" are associated with a substorm onset. No burst-like features were present in the hydrogen data. In order to quantify the variations in the ring current energy content, the equivalent magnetic disturbance D_{ENA} is calculated for the L≤6 proton distributions using the Dessler-Parker-Sckopke relation. Our calculated D_{ENA} suggests that substorm proton injections did not increase the ring current energy content over the main phase. Together with the fact that the proton ring current was mostly partial, this shows that the dominant ring current energy increase must have been due to increased convection. However, the long-term increase in oxygen ENA fluxes suggests that O^+ may have been continuously extracted from the ionosphere throughout the main phase and subsequently energized at each substorm dipolarization to give rise to the oxygen ENA bursts. We also discuss implications of strong electric fields in the inner region L<4.

Disturbances in Geospace: The Storm-Substorm Relationship
Geophysical Monograph 142
Copyright 2003 by the American Geophysical Union
10.1029/142GM10

[1]The Johns Hopkins University Applied Physics Laboratory, Laurel, Maryland
[2]Southwest Research Institute, San Antonio, Texas
[3]Los Alamos National Laboratories, Los Alamos, New Mexico

1. INTRODUCTION

Historically it was believed that a geomagnetic storm was the effect of many substorms [*Akasofu*, 1968]. It was later recognized by *Gonzalez and Tsurutani* [1987] that the requirement for a geomagnetic storm to occurr was an IMF $B_z \leq -10$ nT for at least 3 h. More recent studies have shown that convection is the dominant driver in geomagnetic storms and that the main-phase ring current is mostly partial where ions drift on open trajectories out through the dayside magnetopause [*Liemohn et al.*, 2001]. However, there still remains a question of how much the substorm injections during a storm main phase contribute to the overall storm time energy content of the ring current.

The problem we investigate in this paper can be summarized as follows. The storm intensity has been characterized by the D_{st}, and more recently, by the SYM-H and ASY-H indices. These are indices directly calculated from the magnetic disturbance measured at the equatorial surface. The magnetic disturbance can be related to the total energy content of the ions that flow around the Earth, via the Dessler-Parker-Sckopke (DPS) relation [*Dessler and Parker*, 1959; *Sckopke*, 1966]. Now the problem is how much of the growth of the energy content during a geomagnetic storm can be attributed to substorms and how much can be attributed to an increase in the overall convection strength. Once plasma is injected during a substorm onto open trajectories it will not contribute further to the overall growth of the energy. This is because the injected particles will be lost through the magnetopause. So, the only way accumulated substorm injections can contribute to the overall growth is if the injections are onto closed drift trajectories.

On the other hand, the energy content of plasma being transported by the $\mathbf{E} \times \mathbf{B}$ (in other words, the partial ring current) can increase if the cross-tail (convectional) electric field increases. The reason for this is that a stronger cross tail electric field makes the tranport of particles be dominated by the $\mathbf{E} \times \mathbf{B}$ drift closer to Earth and so open trajectories are allowed closer to Earth. Another way of saying this is that the Alfven boundary (boundary between open and closed trajectories for particles with given magnetic moment) shrinks for higher cross tail electric field. This will in turn lead to adiabatic energization since the plasma is now transported into a region with higher magnetic field strength, and thus the energy content of the partial ring current will increase.

Since the IMF B_z decreases during a main phase of a storm, the convectional electric field increases and the location of the innermost open trajectory grows closer to Earth. Therefore, plasma cannot be trapped during the main phase of the storm. However, depending on how fast the convectional electric field turns off (the start of storm recovery phase) more or less plasma will be trapped. Thus, if the convectional electric field increases monotonically, any trapped population during a storm main phase could only come from substorm injections. We will examine the global ion distribution during such periods and investigate how much of the growth of the total energy is due to substorms injections or increased convection.

In this paper we will present ENA images from the main phase of the 4 October 2000 storm. The ENA images have been inverted to obtain an equatorial proton distribution, using a constrained linear inversion technique. In order to quantify how much the substorm injections contribute to the overall storm growth we compute the equivalent magnetic disturbance D_{ENA} for the global proton distributions during the main phase. By looking at the local time distribution of the obtained global proton distribution at a given energy we can infer if those protons are on closed or open trajectories. The contribution of those protons to the D_{ENA} relative to other energies will tell us how much it contributes to the over all growth of the total energy of the main-phase ring current. In addition to this we will discuss how the oxygen ENA flux increased over the main phase. We also discuss briefly the spectral features of the main phase indicating that there was a deep potential minimum on $L \leq 3$ implying a significant electric fields. We compare our results to a model derived electric field by *Ridley and Liemohn* [2002]. The purpose of this study is to investigate what restrictions can be put on different energy sources to the ring current energization by analyzing the global ion distributions we have obtained.

2. INVERSION TECHNIQUE

We use a constrained linear inversion technique that closely follows the method described by *Twomey* [1977] and also similar to *Perez et al.* [2001]. Previous studies [*Henderson et al.*, 1997; *C:son Brandt et al.*, 2001] have used a forward modeling technique based on a parameterized model of the ion distribution developed by *Roelof and Skinner* [2000]. This inversion technique was described in *C:son Brandt et al.* [2002a] and we only outline its main components here. The idea is to expand the line of sight (LOS) integral that describes the production of ENAs into sums of linear quadrature and then equating them with the observed image and in that way determine the quadrature coefficients. In the present formulation the pitch angle distribution (PAD) is described by one isotropic component and one linear component representing the field aligned and perpendicular shape of the PAD. In this paper we focus only on the isotropic component. The ion distributions were clamped down to zero at L=2 and L=16. We use here the day-midnight asymmetric exosphere based on the DE-1 measurements reported by *Rairden et al.* [1986] and also used by [*C:son Brandt et al.*, 2002a]. The absolute fluxes obtained by our algorithm appear to be somewhat overestimated. The calculated D_{ENA} calculated below should therefore not be taken as absolute, but rather as a relative indicator on the total energy content of the ring current. Fluxes should be correct in a relative sense.

3. CALCULATING MAGNETIC DISTURBANCE

We use the retrieved ion distributions to calculate the magnetic disturbance at the equator on the surface of the Earth. We do this by using the Dessler-Parker-Sckopke (DPS) relation [*Dessler and Parker*, 1959; *Sckopke*, 1966], which states that the horisontal magnetic perturbation ΔB at the equator can be written

$$\Delta B = -\frac{2}{3} \frac{E_{tot}}{E_m} B_0, \quad (1)$$

where B_0 is the nominal dipole magnetic field intensity at the surface. The magnetic energy E_m contained in the dipole magnetic field and can be written

$$E_m = \frac{4\pi}{3\mu_0} B_0^2 R_E^3. \quad (2)$$

The total energy of the particles E_{tot} is expressed as the volume integral over the energy density. Note that the magnetic disturbance at Earth as expressed in Equation (1) is independent of L. We can obtain the energy density ϵ by

$$\varepsilon = 4\pi \int p^2 f(p) E dp, \quad (3)$$

where E is the energy of the ions and m is the ion mass, where we have assumed protons. $p = mv$ is the momentum of the ions. The distribution function f (in momentum space) can be related to the ion flux (differential in energy) through

$$j(E) = p^2 f(p). \quad (4)$$

Transforming integral (3) to sums for an isotropic pitch angle distribution in ϕ and L space we can write Equation (1) to a first approximation

$$D_{ENA} \approx -\frac{64}{35} \frac{\mu_0 m^2}{B_0} \sum_{ijk} J_{ijk} \frac{E_i^2 L_j^2}{(2mE_i)^{3/2}} \Delta L_j \Delta \phi_k, \quad (5)$$

where μ_0 is the magnetic permeability in vacuum, J_{ijk} is the proton flux (cm^{-2} sr^{-1} s^{-1}) at energy E_i in a finite interval ΔE_i, L-shell L_j and local time angle ϕ_k. The bin size in L and ϕ is denoted ΔL and $\Delta \phi$. The approximation comes from the fact that we have neglected the L-dependence in our evaluation of the flux tube volumes. The error is about 20% at L≤3, but rapidly decreases as L increases. We will use this formula to calculate the magnetic disturbance from the proton distributions. Since the DPS relation assumes a pure dipole field we will only calculate it for ion distributions on L≤6.

4. GLOBAL STORM OBSERVATIONS

The 4 October storm main phase was characterized by a long and gradual decrease of the IMF B_z. It ended by some rapid fluctuations in the IMF and then a gradual recovery. About a half a dozen substorms ocurred during 4 October that showed up clearly in the Los Alamos National Laboratory (LANL) geosynchronous proton data as well as in the geomagnetic field components observed by the GOES satellite. We will only show ENA data from two of those substorms here. *C:son Brandt et al.* [2002a] have examined ENA and in-situ data from some of these substorm in more detail.

Plate 1 shows the SYMH (black line) and (negative) ASYH (green line) for the main phase we are studying in this paper. The red line is the electric field E_y set up by the solar wind ($v_x B_z$). Note that the E_y increases up until approximately 18:00 UT and then starts slowly decreasing. This indicates that the Earthward **E** × **B** drift feeds more and more plasma into the nightside magnetosphere and that the Alfven layer (for given magnetic moment it is the boundary between open and closed drift trajectories) continously shrinks up until 18:00 UT.

Also plotted in Plate 1 is the equivalent magnetic disturbance D_{ENA} from the proton distributions inverted from the ENA images as described above. Squares represent the individual energies as indicated and stars represent the total sum. The values of the total D_{ENA} have been scaled to fit on the same scale as SYMH and more specifically to coincide with the SYMH at 04:21 UT. We stress that our estimated D_{ENA} should not be taken as absolute but as a measure of how the energy content of ions inside L=6 varies. We discuss the implications in the Discussion section.

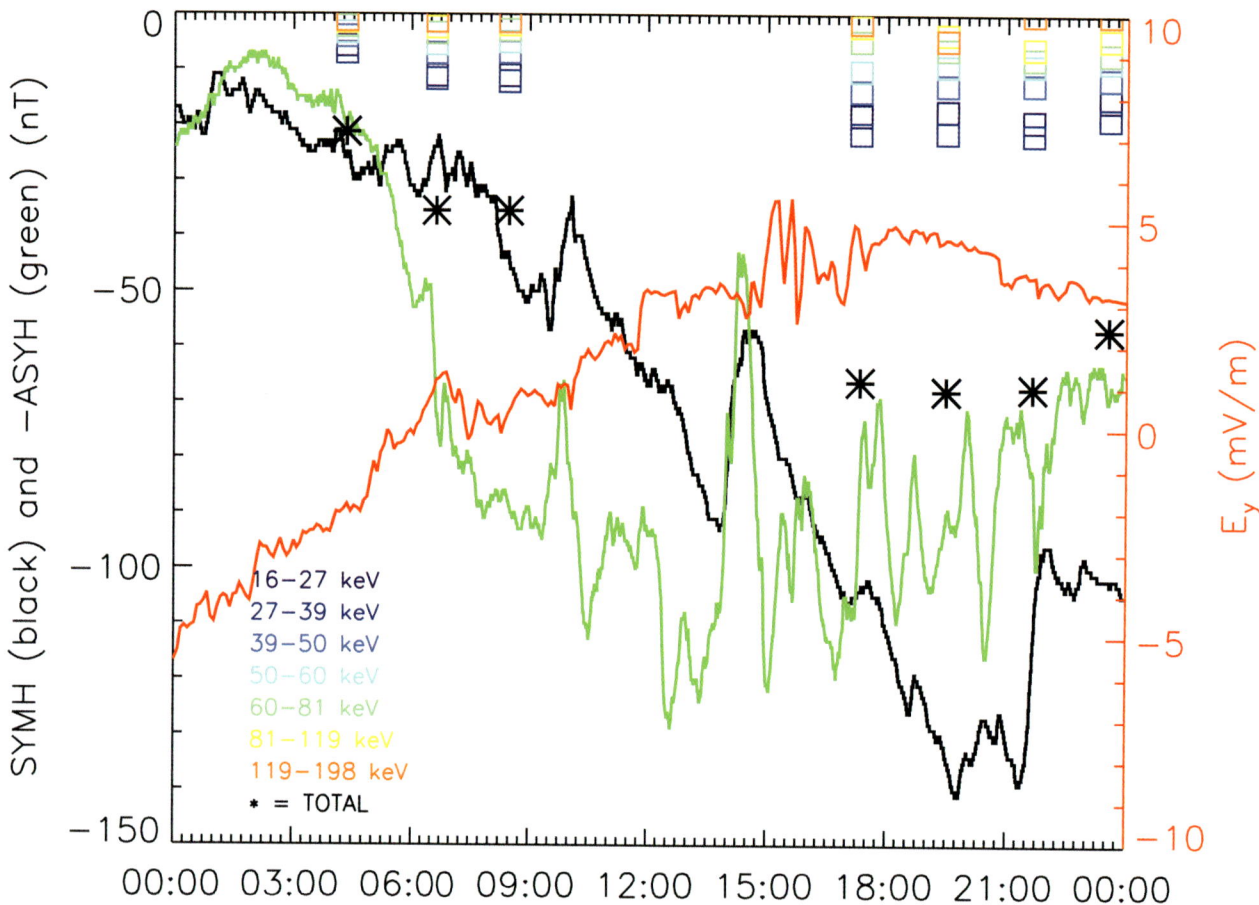

Plate 1. The SYM-H (black line) and the negative of the ASY-H index (green line) and the IMF E_y during the early main phase. Squares represent the differential energy contribution to the ring current as indicated by the color coding and stars represents the total energy contribution (see section 3).

All ENA images in this paper are presented in an azimuthal, equidistant projection of the sky hemisphere. The dipole field lines of L-shells 4, 8 and 12 are shown for reference and the MLTs are indicated by red numbers. The coordinates in rectangular (and spherical coordinates radius, latitude and longitude) solar magnetic (SM) coordinates in units of R_E of the IMAGE spacecraft for the images in Plates 2 and 4 are the following: For 06:40 UT {−1.7, 0.6, 7.8} (8.0, 76.8°, 160.0°); 08:30 UT {−3.6, 0.9, 7.2} (8.1, 62.7°, 166.1°); 17:21 UT {−0.4, −0.1, 5.5} (5.5, 86.0°, −161.9°); and 19:30 UT {−2.5, −0.9, 6.9} (7.4, 68.7°, −160.7°). In some images a narrow band of emissions runs horizontally across the upper portions. This is the solar contamination from residual sunlight hitting the detector plates. The LOSs to this contamination usually intersects L-shells much higher than where it can effect the magnetospheric ion distribution. All HENA images in this paper are obtained with a 10 min integration time and all MENA images with 30 min integration.

During the early main phase (when the E_y was still weak) the substorms appeared to inject plasma onto closed trajectories at 06:10 and 09:24 UT (not shown). Later in the main phase as E_y had increased to about 5 mV/m the ENA images indicated substorm injections onto open drift trajectories such as for the 12:10, 14:00 UT substorms (POLAR/IPS measurements, not shown) and the 17:30 UT substorm. Although the spacecraft was located on low latitudes on the nightside as the E_y stated to decrease slowly, ENA images of the injections at 20:00 and 21:30 UT (not shown) appeared to build up a more symmetric component of the ring current gradually.

4.1 Early Main Phase

Plate 2 shows the observations at 06:40 and 08:30 UT on 4 October in the 27–39 and 60–81 keV energy range. Plate 3 shows the corresponding ion distributions obtained by the inversion method described above. We see that the injection elevated the nightside ion fluxes at L=4, but fluxes remained low on the dayside. Later in Plates 3c and 3d, the nightside fluxes have decreased while the ring current appears to have become slightly more enhanced and symmetric on L=4. According to the calculated total D_{ENA} the ring current energy content of the protons during this time did not increase as can be seen in Plate 1. This implies that the O$^+$ may have contributed significantly to the SYMH and ASYH. We will discuss below in the Discussion section the O$^+$ abundance for this period.

There were two substorms at 06:10 and 09:22 UT, which both were preceded by elevated plasma sheet fluxes beyond 8 R_E. Their behavior have been reported by C:son Brandt et al. [2002a]. In Plate 3a plasma sheet fluxes are low, but not that in Plate 3c plasma sheet fluxes have increased which is consistent with the overall convection continuously feeding the plasma sheet with fresh plasma.

4.2. Late Main Phase

Plate 4 shows the observed ENA images in the 27–39 and 60–81 keV range for 17:21 UT and 19:30 UT. Plate 5 shows the equatorial proton distributions inverted from the ENA images in Plate 4. According to auroral FUV images obtained by the FUV camera on board IMAGE, a substorm onset occurred at approximately 17:20 UT. At 17:30 UT an ion injection was observed around midnight at geosynchronous altitudes. We can see that there is not much change in ion flux from 17:21 UT to 19:30 UT in either energy range.

If one considers that the curvature-gradient drift period of 70 keV protons is approximately 3 h, it is reasonable to expect that the dayside ions present at 19:30 UT are the ions from the injection at 17:30 UT. For all energies in Plate 5 there are ion fluxes extending past dusk to noon and weak signatures of ion fluxes extending out to L=8 around noon. This implies that the ions at these energies curvature-gradient drifted around to noon where they were lost through the dayside magnetopause and picked up by the magnetosheath flow. The magnetopause during this time was around 11 R_E on the dusk flank and was estimated to be inside L=8 at the subsolar point using the fits by Roelof and Sibeck [1993]. It is clear that there are almost no ion fluxes in the pre noon sector for 60–81 keV which implies that the drift trajectories were open at this energy. We also note that the MLT region spanned by the partial ring current leaves only a narrow sector in MLT with significantly lower ion fluxes. Thus a finite number of in-situ measurements at geosynchronous orbit could run a high risk of missing this minimum, making it look like the ring current was still closed during this time. It is also interesting to note that the 27–39 keV proton distribution at 19:30 in Plate 5c display higher intensities at L>8 on the nightside than at 17:21 UT. The reason for this may be related to the fact that a dipolarization occured around the 17:30 substorm which transported the plasmasheet ions to the inner magnetosphere. At 19:30 UT, however, the **E** × **B** drift from further down the tail has had time to transport fresh plasma in to the plasmasheet. This pattern can also be seen in Plates 3a and 3c.

19:30 UT is also the time of the deepest minimum of SYMH in Plate 1, but this does not seem to be reflected in the proton distributions and D_{ENA}. Again, the cause for this

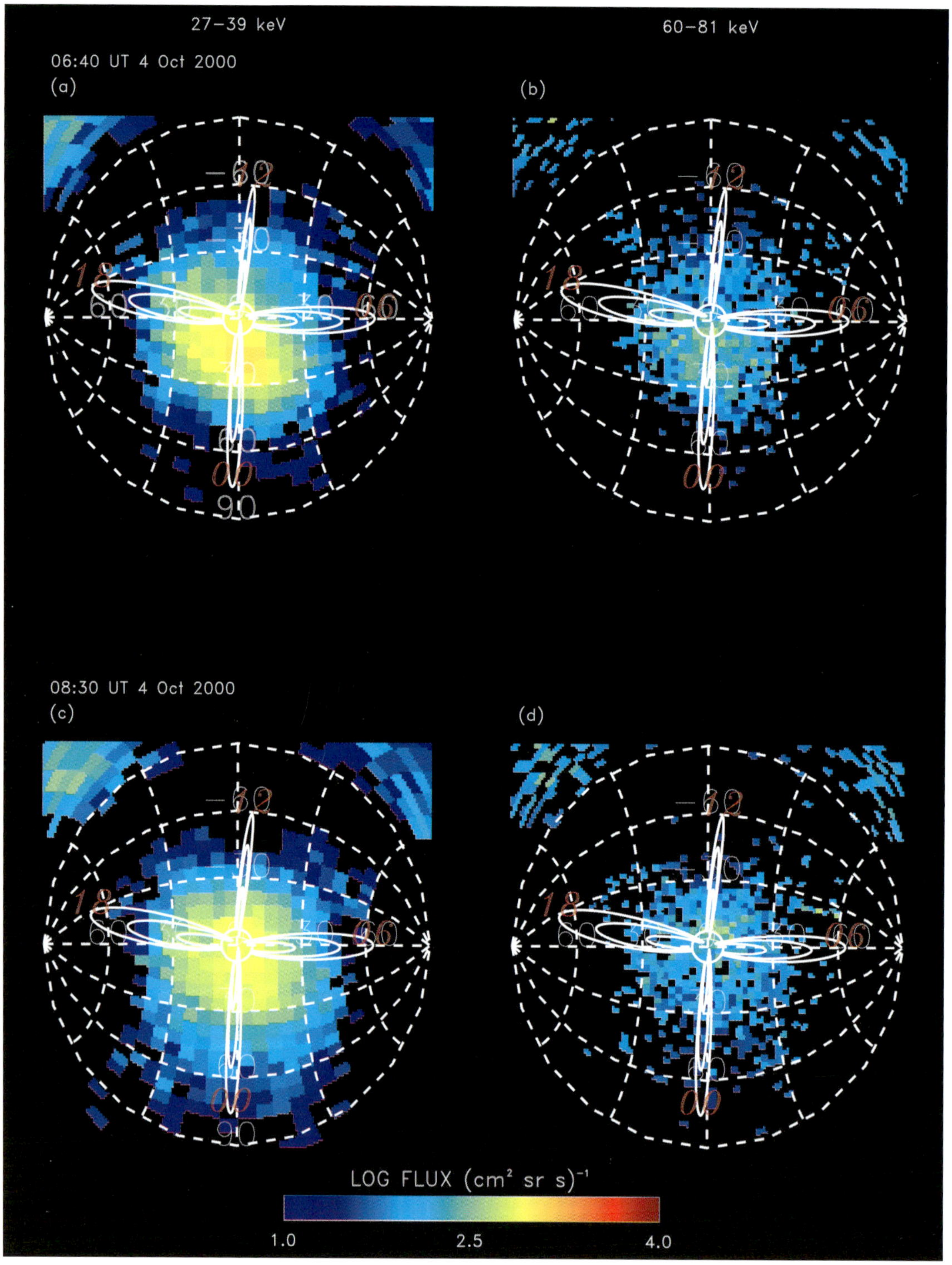

Plate 2. The first substorm injection during the early main phase occurred at 06:10 UT. This plate shows the hydrogen ENA images 30 min and 140 min after the injection. The ENA image at (a) 06:40 UT at 27–39 keV, (b) 06:40 UT at 60–81 keV (c) 08:30 UT at 27–39 keV, and (d) 08:30 UT at 60–81 keV. Note that the ENA flux decreases although the ASYMH increases.

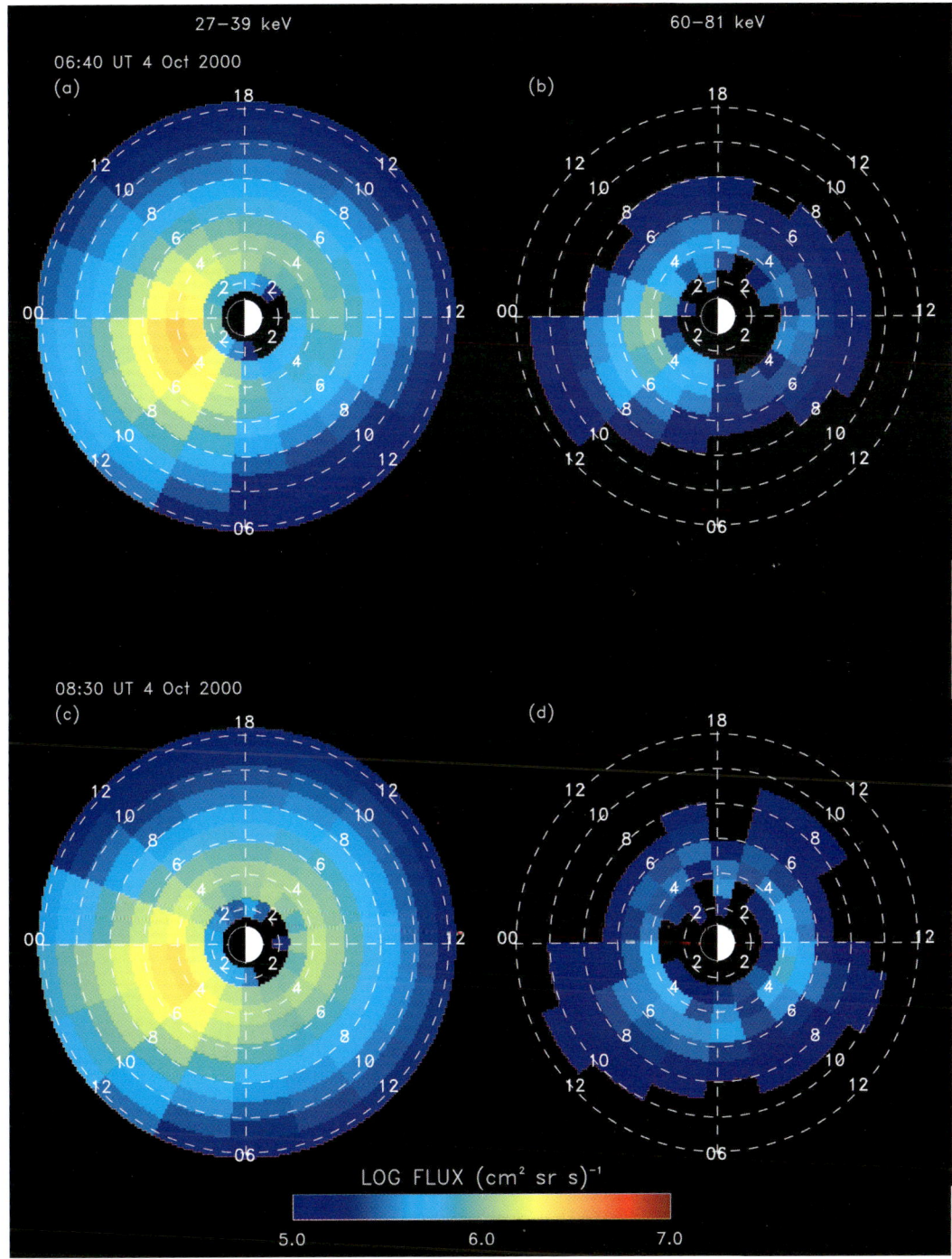

Plate 3. The proton fluxes obtained from the inversions of Plate 2 plotted in the equatorial plane. Only the isotropic pitch angle component is shown. (a) 06:40 UT and 27–39 keV. The injection is clearly visible on the nightside. A minimum appears in the pre noon sector. (b) 06:40 UT and 60–81 keV. Intensities lower but pattern almost unchanged. (c) 08:30 UT and 27–39 keV. High intensities on the nightside are still visible. Ion distribution appears more isotropic. (d) 08:30 UT and 60–81 keV. Weak maximum appears on nightside and around noon.

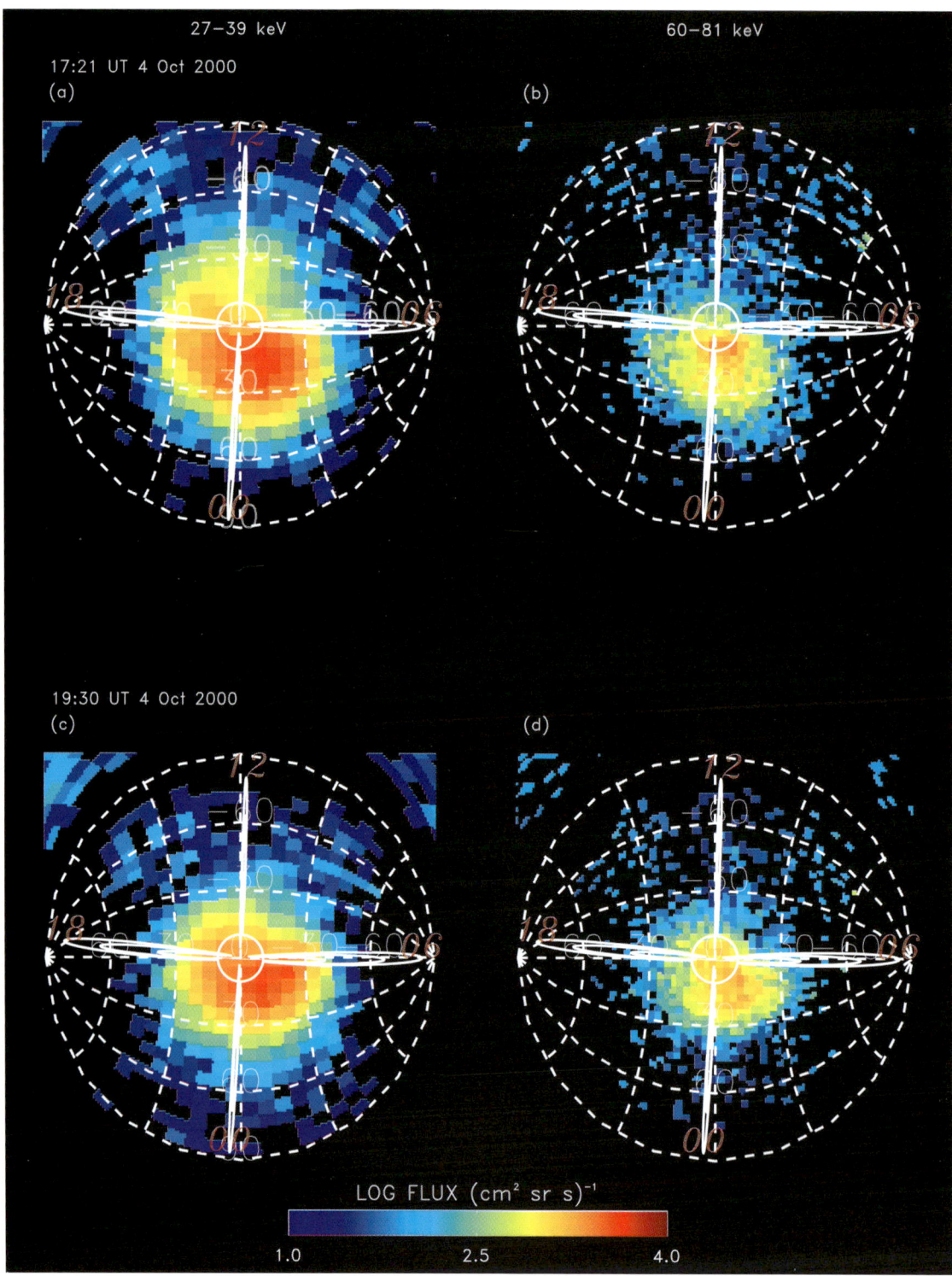

Plate 4. Observed hydrogen ENA images from (a) 17:21 UT and 27–39 keV, (b) 17:21 UT and 60–81 keV, (c) 19:30 UT and 27–39 keV, and (d) 19:30 UT and 60–81 keV.

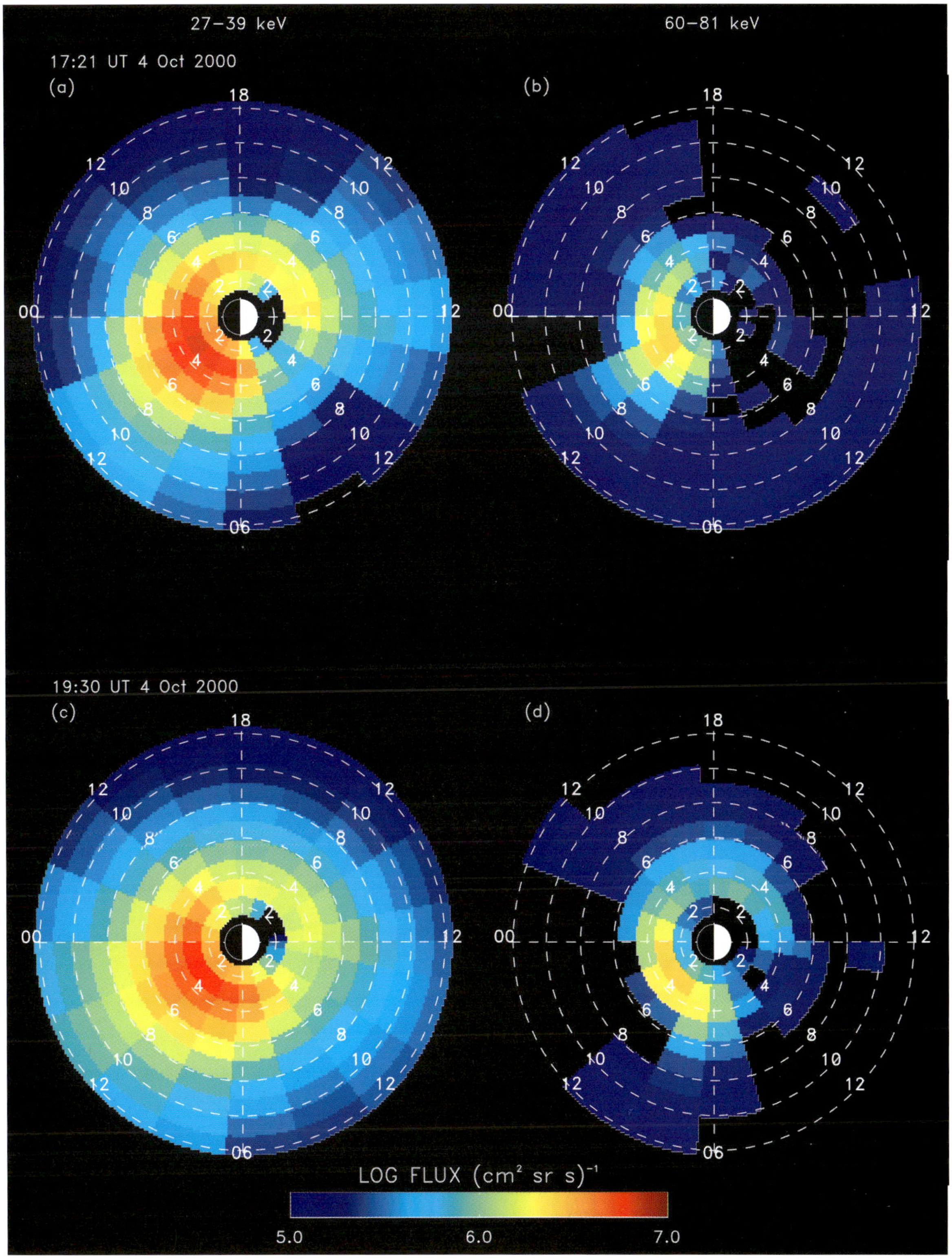

Plate 5. Equatorial ion distributions inverted from the ENA images in Plate 4.

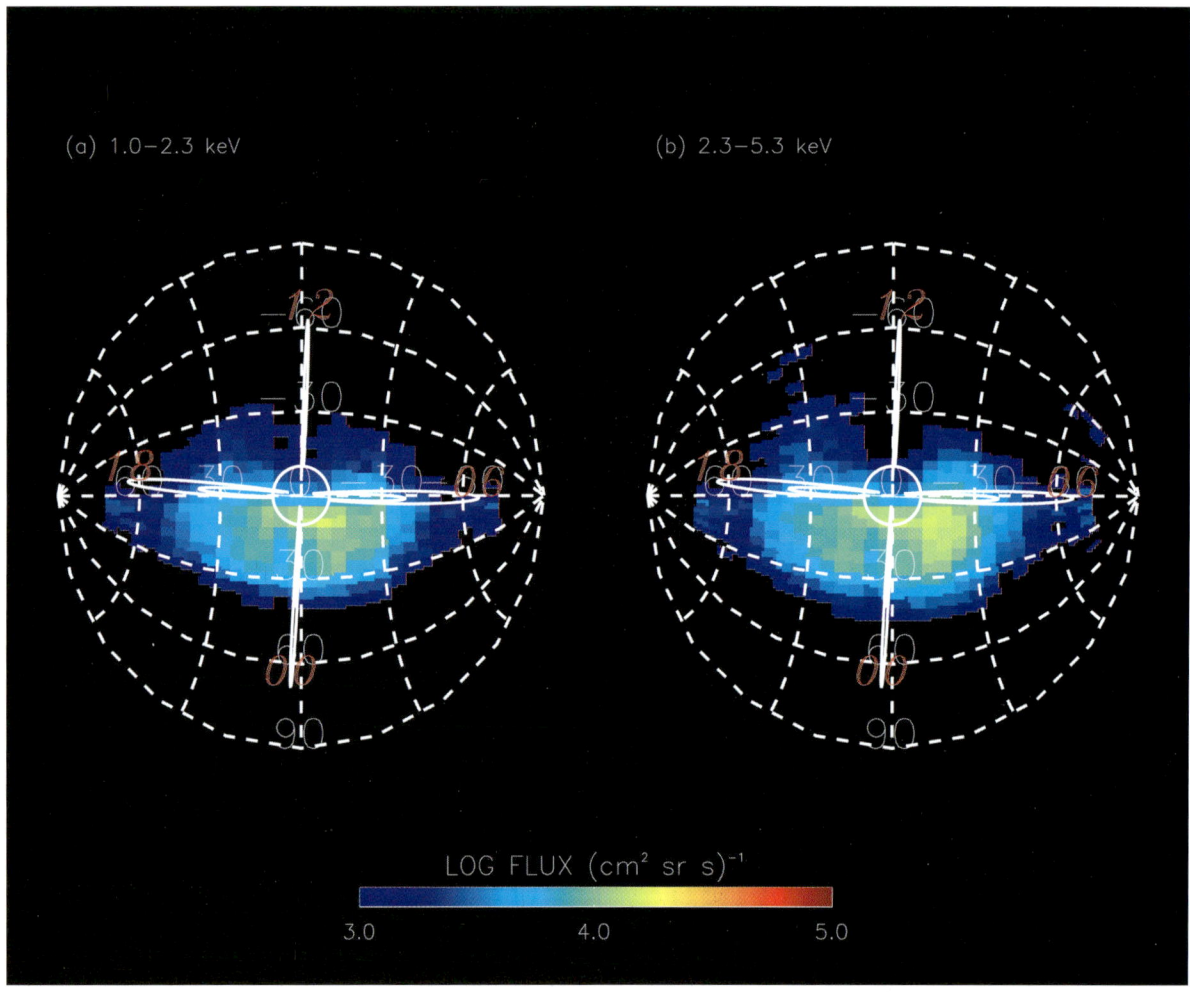

Plate 6. MENA images obtained in the (a) 1.0–2.3 keV range and (b) 2.3–5.3 keV range. The morphology is quite similar to that obtained by HENA in the higher energy range. See Plates 4a and 4b, but note the different colorbar.

is most likely the contribution from O^+ or possibly a tail current contribution.

Plate 6 shows the ENA images from the MENA imager in the 1.0–5.3 keV energy range. We immediately note the similarity between these and the ones observed by HENA at 27–39 keV (Plate 4a). We have not inverted the MENA images, so the comparison has to be qualitative. The observations at MENA energies are essential since the ions are dominated by the electric drifts at these energies. From Plate 6 we see that there is perhaps a weak maximum in the post-midnight sector, and that the ENA fluxes decrease rapidly once beyond dusk. In Plates 4a and 4b we see that the ENA fluxes continue beyond dusk. This is reasonable since at higher energies the curvature-gradient drift should be more pronounced. However, the change of the morphology over the entire energy range is not drastic, and we will discuss below how this can imply significant electric field magnitudes deep in the inner magnetosphere.

5. DISCUSSION

In order to answer the original question about how much substorms contribute to the growth of the geomagnetic storm, consider Plates 3 and 5. The proton distributions are clearly asymmetric in MLT. This means that the substorms did not induce electric fields sufficiently strong to inject protons onto stably trapped orbits. Also, ENA images (not presented here) from POLAR/IPS at 11:30–14:30 UT show the development of another two substorms, supporting the above conclusion. The first substorm ocurred at approximately 12:10 UT and the second around 14:00 UT. Neither of these injections showed any indications of being on closed trajectories. This implies that convection responsible for increasing the energy of proton ring current during this main phase.

It is also clear from Plate 1 that the substorm injections at 17:30, 20:00, and 21:30 UT did not increase the proton ring current energy. We will discuss the O^+ content next. If we consider our calculated D_{ENA} in Plate 1 we can see that it does not decrease as much as the SYMH. At the time of these observations the mass resolving capability of HENA had not been optimized. However, we know that the <10 keV/nucleon hydrogen channel is dominated by ~80–160 keV oxygen, which has the same velocity as <10 keV hydrogen. The reason for this is that the energy of a <10 keV/nucleon hydrogen atom is too low to penetrate the front foil of HENA.

Figures 1 and 2 show the image integrated ENA flux in the lowest energy per nucleon channel over the entire main phase. We see that the total oxygen ENA flux increases about one order of magnitude over the entire main phase. This shows that the ring current energy increased significantly in the O^+ population. Superposed on this long term increase there are intensifications at every substorm onset that decay away during an hour. This is consistent with the idea that the substorm dipolarizations energize O^+ in the plasmasheet and trap a fraction of O^+, which is reflected as the gradual increase of the oxygen ENA flux over the main phase. The decay of the oxygen ENA "bursts" can be explained as the part of the O^+ that is drifting out from the magnetosphere on open trajectories. Consequently, the more substorms, the more O^+ is added and the higher the energy of the ring current becomes. If we take the oxygen ENA fluxes at face value, the ring current energy should decrease shortly after the 17:30 UT injection until 19:30 UT when a second substorm injection appears to occur. The second injection or "burst" decays away until around 21:30 UT when a third burst appears that reaches its maximum at around 22:00 UT and then gradually decays away.

The bursts of oxygen ENAs and the lack of the same signature in the hydrogen ENAs are worth discussing. Recent modeling results by *Delcourt* [2002] show that the induced dipolarization electric field occurs in timescales of the gyroperiod of ~1 keV O^+, which leads to a significant diabatic energization of the O^+ up to ~100 keV. This energization mechanism is not efficient for ~10 keV protons. A reasonable scenario for the occurence of the oxygen bursts would therefore be one where O^+ is, more or less, continuously extracted from the ionosphere to the plasmasheet during southward IMF. The O^+ would reach about <1 keV in the plasmasheet, and at every dipolarization it would be energized up to ~100 keV and transported to the inner magnetosphere where it would show up as bright bursts of oxygen ENA. This is a topic that is currently under deeper investigation and more results will soon appear in *Space Science Reviews*.

Comparing the oxygen ENA flux with the SYMH there is an increase at around 17:30 UT, then the SYMH decreases (ring current energy increase) until about about 19:30 when SYMH minimum is reached around which a new SYMH increase is seen. Then, again, shortly after 21:00 UT there is a new SYMH increase. In summary, it seems almost as if the oxygen ENA bursts track the SYMH, which is inconsistent if the SYMH was only measuring the energy content of the ring current. We think the explanation for the inconsistency is that at substorm onset the magnetic signature of the disruption of the tail current is strong enough to cause an increase of the SYMH as was discussed by *Ohtani et al.* [2001]. It is not our intention to fully explore that here, but rather outline as a possibility.

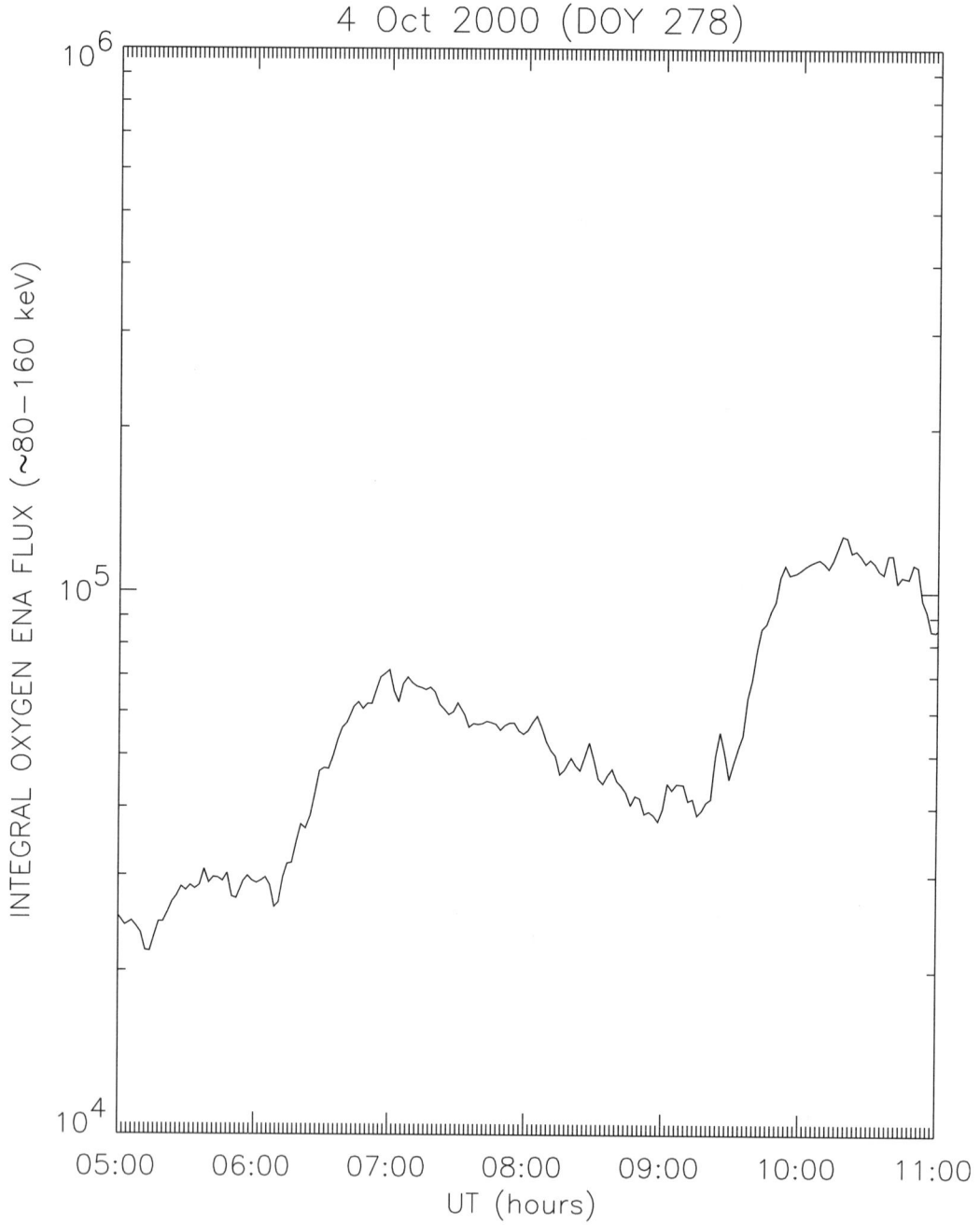

Figure 1. The integrated oxygen ENA flux in the ~80–160 keV range over the first half of the main phase.

At this stage we can only say that we would expect a significant long term contribution of the O$^+$ ions because of the one order of magnitude increase in oxygen ENA flux. So, in this sense substorms appear to play an important role in the long term increase of the O$^+$ energy content in the magnetosphere. ENA image inversion at these energies is difficult to interpret in detail due to a larger angular scattering (up to 20°) in the front foil.

A natural question arising is why the injections at 12:10, 14:00 (POLAR/IPS ENA images) and 17:30 UT (Plate 5) did not become trapped. At first glance the substorm induced electric field should decrease the size of the

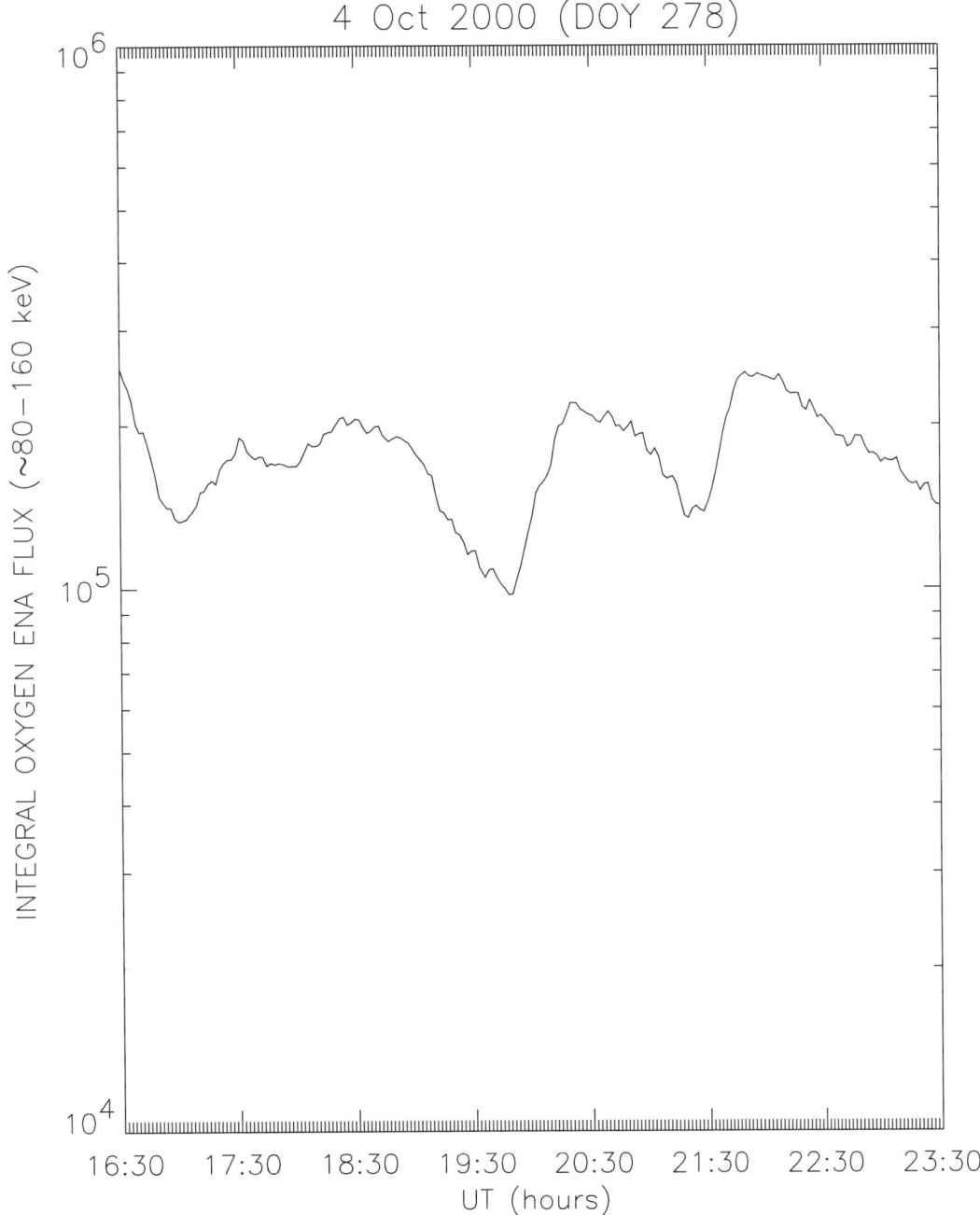

Figure 2. The integrated oxygen ENA flux in the ~80–160 keV range over the second half of the main phase.

Alfven boundary, and that same field would transport particles to its boundary. As the induced electric field decreased, the Alfven boundary then would be expected to increase rapidly and leave the particles inside it, thus placing them on closed drift paths. This simple scenario is valid only if the shielding of the electric field was constant in time. In fact under shielding is expected to occur during substorms. If the shielding decreases during the substorm and the lower values of the shielding are maintained (even after the injection is complete), the Alfven layer will stay at the smaller size, allowing ions to still drift on open trajectories.

Another interesting scenario can be realized if one considers the fact that there exist significant electric fields up to 6 mV/m between L=2 and L=4 [*Wygant et al.*, 1998]. The electric drift velocity in such high electric fields at L=4 would be comparable to the gradient-curvature drift of ≤100 keV ions. This means that ≤100 keV ions injected onto L<4 would experience electric as well as magnetic drifts and may not become trapped. It is therefore relevant to briefly discuss the implications of the electric field pattern that can be inferred from our observations. Consider the pattern of the ion distribution during the later main phase in Plate 5 and Plate 6. The intense ion fluxes around midnight and L=3 are persistent features for all energies 1–200 keV. According to the POLAR/IPS observation, it appears this feature had been reasonably stable over the 14:00–22:00 UT period.

In the 1.0–5.3 keV range the motion of the ions is dominated by electric drifts. Therefore important implications can be made by studying the ENA images in this energy range (Plate 6). In the superposition of a pure dawn to dusk electric field plus a corotation electric field, the electric drifts would carry the low energy ions straight through on the dawn side. Inspecting Plate 6, it appears that the electric field is configured such that ions will be deflected both at dawn and dusk.

From inspection of the higher energies in Plates 5a and 5b, it is evident that ions have succeeded in drifting to the dayside, but they have decreased in flux considerably. This implies that the high energy ions either (1) are being deviated from their curvature-gradient drift trajectories and are lost through the magnetopause in the post-noon sector, (2) decrease their energy so that their differential flux decreases at a given energy, or a combination of these two.

In conclusion we can say that there must have been a strong enough field to first bring the ions in to L=3 and that the electric field there changed dramatically to have no, or a very small, positive y-component that would otherwise allow low energy ions to drift across the dawn to dusk meridian, or perhaps the electric field is twisted as a function of radial distance. The implications for strong electric fields in the inner magnetosphere has been discussed and reported by *C:son Brandt et al.* [2002b].

We would like to draw attention to a study by *Ridley and Liemohn* [2002] where they estimated the inner magnetospheric electric field that is set up by the asymmetric ring current. Using the kinetic ring current model by *Liemohn et al.* [2001] they calculated the electric field pattern in the ionosphere due to the region 2 currents of the asymmetric ring current. The ionospheric potential was then mapped back out to the equatorial plane assuming the magnetic field lines to be infinitely conducting. They found strong electric fields inside L=3 with strong eastward and outward radial components in the post-midnight sector and equally strong westward and outward radial components in the pre-midnight sector (see their Plate 2). Superposed on this was also the over all dawn to dusk field and corotational field. This would stretch their patterns slightly more towards dusk. Such strong electric field on low L-shells have been reported by *Wygant et al.* [1998]. This type of self-consistent modeling was pioneered by [*Wolf*, 1983] in their Rice Convection Model (RCM) who also noticed the same type of westward skewing that we have observed. *Fok et al.* [2001] has succesfully coupled a kinetic ring current model with the RCM. About three different storm observations has been compared to runs of this model with a very good agreement.

Low energy ions drifting in such an electric field would therefore drift eastward and then inward until they come closer to the Earth near dawn where the outward pointing electric field would transport them westward past local midnight. As the low energy ions reach dusk the eastward electric field component would cause the ions to drift outward and be lost through the duskside/afternoon magnetopause. This general pattern is consistent with our observations.

At high energies ions will start to curvature-gradient drift strongly once they are convected into L=3. As they drift around towards dusk our observations show that they decrease their intensity drastically. However, there are still weak ion fluxes close to noon, as can be seen in Plates 5a and 5b. At these energies the dominating transport is most likely curvature-gradient drift, but our observations imply that there is indeed a significant electric field such that even the high energy ions may be lost through the dusk to afternoon magnetopause. Without detailed kinetic modeling it is difficult to answer how much of the flux decrease is due to transport to weaker magnetic field strength, and how much can be attributed to loss through the magnetopause.

Although the electric field pattern obtained by *Ridley and Liemohn* [2002] qualitatively agrees with our observations, it does not answer how the ions were transported inward to L=3 in the first place. The dawn to dusk electric field during 4 October was ≤5 mV/m which is not extreme by any means, but may have been sufficient to transport the ions all the way in to L≤3 where the electric field driven by the asymmetric ring current would take over.

6. SUMMARY AND CONCLUSIONS

We have presented the global equatorial proton distributions in the 16–198 keV energy range for the main phase of the 4 October 2000. The proton distributions were obtained by applying a constrained linear inversion technique [*Twomey*, 1977] to ENA images observed by IMAGE/HENA. The IMF B_z decreased steadily from early on the 4 October and slowly started to increase again around 19:00 UT. During the entire 4 October about a half a dozen substorms occurred with regular intervals of which two were described in this paper. In the early main phase, when the dawn to dusk electric field was still weak, the substorm appeared to inject ions onto closed trajectories. When E_y increased in strength, substorms injected particles onto open trajectories. As E_y started to slowly decrease at around 19:00 UT, substorm injections appeared to become trapped.

Up to the time of minimum SYMH we found that no significant symmetric component of the proton ring current had developed, which implies that substorms did not build up a durably trapped proton population and therefore did not contribute directly to the long term energy increase of the main-phase ring current. Also, from the time of E_y=0 to the time of maximum E_y the peak ring current moved in from L=5 to L=3. This is consistent with the increase in E_y which decreases the size of the Alfven boundary. In this respect the increase in solar wind driven convection is the long term contributor to increased energy content of the (partial) proton ring current.

In order to estimate the contribution to the energy content of the ring current from the obtained proton distributions we calculated the equivalent magnetic disturbance (D_{ENA}) at the Earth's surface at the equator using the DPS relation [*Dessler and Parker*, 1959; *Sckopke*, 1966] for the L≤6 proton distributions. The D_{ENA} was calculated for the 16–198 keV energy range in seven energy bins as indicated in Plate 1. We found that our proton distributions did not fully account for the SYMH depression. We expect that the O^+ will contribute significantly to the SYMH. Some contribution from enhanced tail currents may also have to be considered.

A preliminary inspection of the ~80–160 keV oxygen ENA fluxes revealed a one order of magnitude gradual increase over the entire main phase, which we expect would have a significant contribution to the ring current energy build up. Superimposed on the gradual increase, oxygen ENA fluxes displayed ~1 h increases and decreases starting at every substorm onset. This is consistent with low-energy ionospheric O^+ in the plasma sheet being energized up to ~100 keV during each substorm dipolarization [*Delcourt*, 2002]. The observed gradual increase of oxygen ENA during the main phase is consistent with the idea that O^+ is being trapped. In this respect, substorms contributed to the long term ring current energization.

The fact that the substorm around 17:30 UT injected protons onto open trajectories (with very little change to the asymmetric ring current pattern), raises the question of whether strong electric fields were present at L<4. We discussed studies by *Wygant et al.* [1998] and *Ridley and Liemohn* [2002] and found that our observations were consistent with their conclusions and observations of 5–10 mV/m electric fields in the inner region L<4.

Acknowledgments. Thanks to M. W. Liemohn, University of Michigan and A. T. Y. Lui at the Johns Hopkins University Applied Physics Laboratory, Maryland for stimulating discussions. Thanks also to Dominique Delcourt, Centre National de la Recherche Scientique, France, for modeling results of the response of O^+ to substorms.

REFERENCES

Akasofu, S. I., *Polar and magnetospheric substorms*, D. Reidel, Norwell, Mass., 1968.

C:son Brandt, P., S. Barabash, E. C. Roelof, and C. J. Chase, ENA imaging at low altitudes from the Swedish microsatellite Astrid: Extraction of the equatorial ion distribution, *J. Geophys. Res.*, *106*, 25, 731-25, 744, 2001.

C:son Brandt, P., R. Demajistre, E. C. Roelof, D. G. Mitchell, and S. Mende, IMAGE/HENA: Global ENA imaging of the plasmasheet and ring current during substorms, *J. Geophys. Res.*, 107, A12, 1454, doi 10.1029/2002JA009307, 2002a.

C:son Brandt, P., S. Ohtani, D. G. Mitchell, M. C. Fok, E. C. Roelof, and R. Demajistre, Global ena observations of the storm mainphase ring current: Implications for skewed electric fields in the inner magnetosphere, *Geophys. Res. Lett.*, 29, 20, 1954, doi 10.1029/2002GL015160, 2002b.

Delcourt, D. C., Particle acceleration by inductive electric fields in the inner magnetosphere, *J. of Atmosph. and Solar-Terr. Phys.*, *64*, 551-559, 2002.

Dessler, A. J., and E. N. Parker, Hydromagnetic theory of geomagnetic storms, *J. Geophys. Res.*, 64, 2239-2252, 1959.

Fok, M. C., R. A. Wolf, R. W. Spiro, and T. E. Moore, Comprehensive computational model of Earth's ring current, *J. Geophys. Res.*, *106*, 8417-8424, 2001.

Gonzalez, W. D., and B. T. Tsurutani, Criteria of interplanetary parameters causing intense magnetic storms (D_{st} <–100 nt), *Planetary and Space Sci.*, *35*, 1101, 1987.

Henderson, M. G., G. D. Reeves, H. E. Spence, R. B. Sheldon, A. M. Jorgensen, J. B. Blake, and J. F. Fennell, First energetic neutral atom images from polar, *Geophys. Res. Lett.*, 24, 1167-11700, 1997.

Liemohn, M. W., J. U. Kozyra, M. F. Thomsen, J. L. Roeder, G. Lu, J. E. Borovsky, and T. E. Cayton, The dominant role of the asymmetric ring current in producing the stormtime *Dst*, *J. Geophys. Res.*, *106*, 10,883-10,904, 2001.

Ohtani, S., M. Nose, G. Rostoker, H. Singer, A. T. Y. Lui, and M. Nakamura, Storm-substorm relationship: Contribution of the tail current to dst, *J. Geophys. Res.*, *106*, 21,199-21,209, 2001.

Perez, J. D., G. Kozlowski, P. C. Brandt, D. G. MItchell, J. M. Jahn, C. J. Pollock, and X. Zhang, Initial ion equatorial pitch angle distributions from energetic netural atom images obtained by IMAGE, *Geophys. Res. Lett.*, *28*, 1155-1158, 2001.

Rairden, R. L., L. A. Frank, and J. D. Craven, Geocoronal imaging with dynamics explorer, *J. Geophys. Res.*, *91*, 13,613-13,630, 1986.

Ridley, A. J., and M. W. Liemohn, A model-derived stormtime asymmetric ring current driven electric field description, *J. Geophys. Res.*, 107, A8, 1151, doi 10.1029/2001JA000051, 2002.

Roelof, E. C., and D. G. Sibeck, Magnetopause shape as a bivariate function of interplanetary magnetic field B_x and solar wind dynamic pressure, *J. Geophys. Res.*, *98*, 21,421-21,450, 1993.

Roelof, E. C., and A. J. Skinner, Extraction of ion distributions from magnetospheric ENA and EUV images, *Space Sci. Rev.*, *91*, 437-459, 2000.

Sckopke, N., A general relation between the energy of trapped particles and the disturbance field near the Earth, *J. Geophys. Res.*, *71*, 3125-3130, 1966.

Twomey, S., *Introduction to the mathematics in remote sensing and indirect measurements*, Developments in geomathematics 3, 1st ed., Elsevier scientific publishing company, 1977.

Wolf, R., *Solar Terrestrial Physics*, pp. 303-368, D. Reidel, Hingham, MA, 1983.

Wygant, J., D. Rowland, H. J. Singer, M. Temerin, F. Mozer, and M. K. Hudson, Experimental evidence on the role of the large spatial scale electric field in creating the ring current, *J. Geophys. Res.*, *103*, 29,527-29,544, 1998.

The Role of Substorms in Storm-time Particle Acceleration

Ioannis A. Daglis

Institute for Space Applications and Remote Sensing, National Observatory of Athens, Greece

Yohsuke Kamide

Solar-Terrestrial Environment Laboratory, Nagoya University, Japan

The terrestrial magnetosphere has the capability to rapidly accelerate charged particles up to very high energies over relatively short times and distances. Acceleration of charged particles is an essential ingredient of both magnetospheric substorms and space storms. In the case of space storms, the ultimate result is a bulk flow of electric charge through the inner magnetosphere, commonly known as the ring current. Syun-Ichi Akasofu and Sydney Chapman, two of the early pioneers in space physics, postulated that the bulk acceleration of particles during storms is rather the additive result of partial acceleration during consecutive substorms. This paradigm has been heavily disputed during recent years. The new case is that substorm acceleration may be sufficient to produce individual high-energy particles that create auroras and possibly harm spacecraft, but it cannot produce the massive acceleration that constitutes a storm. This paper is a critical review of the long-standing issue of the storm-substorm relationship, or—in other words—the capability or necessity of substorms in facilitating or driving the build-up of the storm-time ring current. We mainly address the physical effect itself, i.c. the bulk acceleration of particles, and not the diagnostic of the process, i.e. the *Dst* index, which is rather often the case. Within the framework of particle acceleration, substorms retain their storm-importance due to the potential of substorm-induced impulsive electric fields in obtaining the massive ion acceleration needed for the storm-time ring current buildup.

1. INTRODUCTION

Following a rather long time period of consensus on the storm-substorm relationship as defined by S. Chapman and S.-I. Akasofu, a lively dispute on this issue has emerged in recent years [see recent review by *Kamide et al.*, 1998]. The role/function/efficiency of substorms in building up the ring current in the magnetosphere, which is the major element of space storms, has been questioned in a number of recent studies. It is noteworthy that the substorm dispute relates to two coupled, yet distinct issues, which are often confused: the effects of substorms on the ring current and the effects of substorms on *Dst* variations.

The classical "substorm-added-value" paradigm conceived by S. Chapman is the notion that a magnetic storm is the result of the superposition of individual substorms. It is assumed in this scenario that each substorm involves a partial ring current, and that before this partial ring current has died away, a second substorm produces a second partial ring current, and so forth. If substorms occur frequently enough, particles of the partial ring current accumulate in the trapping region, forming a complete ring and causing a significant decrease in the intensity of low-latitude geomagnetic field over the entire local-time range. In other words, an intense ring current can be created only when intense particle injection occurs successively. The essence of this scenario is that the interplanetary magnetic field (IMF) and the solar wind are important in generating substorms, which in turn create the storm-time ring current, although Chapman did not of course mention anything about the IMF.

In contrast, a number of studies have concluded that storms are not a consequence of substorm occurrence, but the result of the solar wind driven magnetospheric convection (see section 3). Enhanced magnetospheric convection is caused by magnetic reconnection at the dayside magnetopause, which happens to also increase the occurrence probability of substorms. Accordingly, relevant studies [e.g., *Kamide*, 1992; *Iyemori and Rao*, 1996; *McPherron*, 1997] suggested that substorm occurrence during storms is incidental and does not have any causal relation to storm development [also see relevant comment by *Siscoe*, 1997].

However, such studies have not considered any cumulative effect of substorms on the ring current growth [e.g., *Daglis*, 2001]. In a recent innovative approach, *Metallinou et al.* [2002] deviated from the standard line of work, which has been the use of time series of the respective storm and substorm indices. Instead, *Metallinou et al.* used the integrals of the squares of the indices, on the basis of the assumption that substorms, which are episodic phenomena, have a cumulative effect on the ring current buildup. To explore the existence of this effect, *Metallinou et al.* investigated the correlation between time series representing the energy associated with substorms and storms. Accordingly, they calculated the integrals of the squares of the *SYM-H* and *AL* indices ($SYM-H_i^2$, AL_i^2), for the time interval from the start of the storm to the end of its recovery phase. Figure 1 shows the relation of the two time series for a group of 25 space storms. The best fit yields a correlation coefficient of r = 0.789, indicating that energy dissipation for the ring current development and the (substorm) auroral activity are closely related.

The present paper addresses details of the storm-substorm relationship and, particularly, the effects of substorms on the bulk particle acceleration that leads to the buildup of the ring current. We also review a number of studies concerned with the diagnostic rather than with the phenomenon itself, i.e. with the *Dst* index.

2. ANTECEDENTS OF THE STORM-SUBSTORM DISPUTE

Akasofu and Chapman [1961] and *Chapman* [1962] were the first to use the term substorm, in order to describe the rapid, repeatable magnetic variations observed in polar regions during space storms (we use the term "space storm" instead of the classical "magnetic storm" throughout this paper). The association of these variations with the main phase of the space storm suggested that they were a discrete phenomenon and that several in a sequence create the storm time ring current. Therefore, the original conception of the relationship between storms and substorms was that of the whole and the part: only storm-time substorms were named "sub-storms".

In Chapman's own words: "A magnetic storm consists of sporadic and intermittent polar disturbances, lifetime being usually one or two hours. These I call polar substorms" [*Chapman*, 1962]. Chapman noted that although polar substorms occur most often during magnetic storms, they appear also during rather quiet periods when no significant storm is in progress. These were called "magnetic bays" by Chapman. In other words, he distinguished between storm-time substorms (which he designated as "substorms") and non-storm substorms (which he called "bays").

Accordingly, *Akasofu* [1968] postulated that storms are the result of a superposition of successive "sub-storms". The name "sub-storm" was presumably inspired by the early work of Kristian Birkeland, who used the term "elementary polar magnetic storms" to describe the same phenomena [*Birkeland*, 1913]. In the meanwhile, however, we have come to know that Sydney Chapman's "sub-storm" and "magnetic bay" is essentially the same phenomenon, namely the magnetospheric substorm. It remains to answer if storm-time substorms have any special characteristics, which make them more efficient in building up a ring current, that is, in creating a space storm.

The storm-substorm connection suggested by Chapman and Akasofu received further support by the studies of *Davis and Partharathy* [1967] and *Davis* [1969], who calculated the energy injection rate for the ring current and compared its time profile to the *AE* index, which they found to be remarkably similar. *Davis and Partharathy* concluded that the energy responsible for the growth of the ring current and that for the polar substorm are injected simultaneously.

Kamide [1979] was the first to challenge the "classical" storm-substorm relationship paradigm and to suggest that the frequent occurrence of substorms alone may not be sufficient to generate space storms. *Kamide* [1979] argued that there might be an "efficiency" of ring current growth asso-

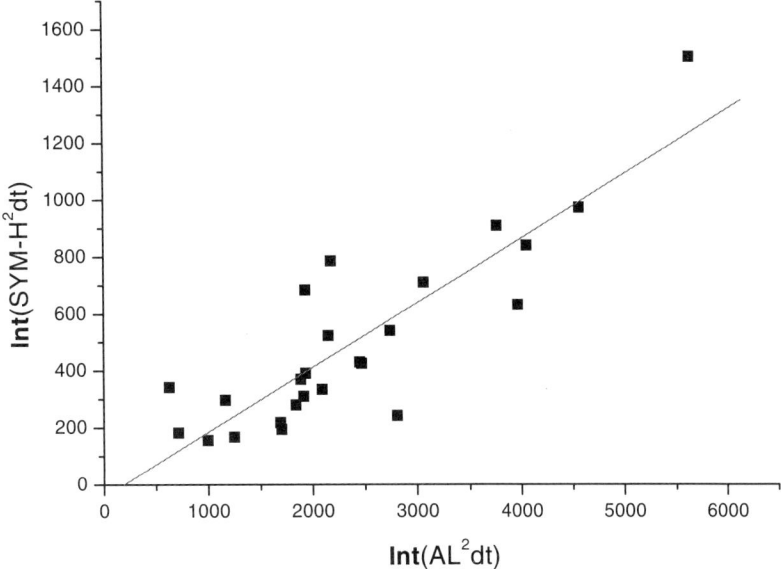

Figure 1. The relation of the integrals of the squares of *SYM-H* and *AL* indices over the time interval from the start of a storm to the end of its recovery phase; each point represents one storm. The best fit, with a correlation coefficient of r = 0.789, is also plotted in the diagram [*Metallinou et al.*, 2002].

ciated with each substorm. That is, the development of the ring current (storm) and the development of the auroral electrojets (substorm) are indeed closely related, but the partition of energy injected into the ring current and into the high-latitude ionosphere is not always in constant proportion. Actually, a first hint to this direction was the study of *Kamide and Fukushima* [1971] which showed a considerable difference in the ratio of the ring current energy injection rate (based on *Dst*) to *AE*, when calculated for the main phase and the recovery phase of a storm. This fact indicated that the energy injection rate into the ring current is not simply proportional to the substorm activity throughout the space storm. *Kamide and Fukushima* [1971] suggested that the ring current grows more easily during the main phase than during any other times for the same degree of substorm intensity. The efficiency of ring current development suggested by *Kamide and Fukushima* [1971] is indeed what we now know as southward IMF.

The arguments of *Kamide* [1979] were later confirmed in part by studies on high intensity long-duration continuous *AE* activity (HILDCAAs), which showed that ongoing substorm activity does not necessarily lead to space storms [*Tsurutani and Gonzalez*, 1987; *Tsurutani et al.*, 1990].

A work that became rather important for the storm-substorm dispute was published by *Burton et al.* [1975]. Burton et al. presented an algorithm for predicting the storm *Dst* signature only from information on the velocity and density of the solar wind and the southward component of the IMF. Assuming the injection rate to be linearly proportional to the dawn-to-dusk component of the interplanetary electric field, the algorithm resulting from the energy budget equation reproduced the observed *Dst* rather successfully. This success later led to the suggestion that storms are a direct consequence of the solar wind electric field, without the need for intermediate substorm action [e.g., *McPherron*, 1997]. According to this view, substorms occur coincidentally during the main phase of storms and the development of the ring current is not a result of the accumulation of substorm-driven partial ring currents.

3. HOT ISSUES IN THE STORM-SUBSTORM DISPUTE

During the last years, the storm-substorm dispute has focused on a couple of hot, "partial" issues:

a. the correlation between (substorm) auroral electrojet activity and space storms in comparison to the correlation between the large-scale magnetospheric convection and space storms.
b. the influence of substorm activity on the ring current build-up, or—to be more precise—the influence of substorm expansion on the *Dst* variations.

Interestingly enough, most efforts seem to concentrate on the diagnostics rather than on the phenomena themselves.

The two issues above are often considered equivalent to each other and relevant to the Chapman-Akasofu paradigm "substorms-create-storms". "Technically", issue a. has involved the global storm disturbance index Dst; the rectified solar wind electric field (vB_s), representing the rate of dayside reconnection and nightside convection (i.e. external driving); and the westward auroral electrojet index AL, representing substorm expansion level (i.e. internal driving). *McPherron* [1997], who is often cited in this context, found that the best solar wind-magnetosphere coupling function (very nearly the rectified solar wind electric field) predicted 76% of the Dst variance during 1979, while AL predicted 71% of the Dst variance. He moreover showed that the combined solar wind dynamic pressure and electric field account for over 85% of the variance of Dst. Accordingly he concluded that the ring current growth is not the result of substorm expansion, but rather the result of direct solar wind driving through large-scale magnetospheric convection.

Issue b. also has involved the AL and Dst indices, in an effort to investigate the storm response at substorm expansion onset. *Iyemori and Rao* [1996] identified a total of 28 space storms "containing" substorm expansion onsets identifiable through mid-latitude geomagnetic disturbance indices. As an indicator of the storm intensity they used the $SYM-H$ index, which is essentially a high-time resolution (1-min) Dst index. *Iyemori and Rao* found out that substorm expansion onsets were accompanied by negative SYM-H responses in some cases and by positive responses in other cases. Assuming that substorm expansions are the causes of ring current development and that the $SYM-H$ index precisely indicates negative Dst response, Iyemori and Rao expected a negative response of the $SYM-H$ index after each substorm onset, which was not the case. Moreover, *Iyemori and Rao* made a superposed epoch analysis of the 28 storms, which showed that the average $SYM-H$ index did not show any negative development after substorm onset. In contrary, the slope of $SYM-H$ indicated a change to positive direction rather than the expected negative. The authors concluded that the ring current development is not the result of the frequent occurrence of substorms, but the result of enhanced convection caused by large southward IMF.

Along the same line, *McPherron* [1997] noted that visual inspections of AL and Dst time series during storms generally show that substorms occur at times when there is no ring current development or when Dst is recovering. For a particular storm in April 1979, *McPherron* emphasized that Dst began to decrease concurrently with the start of a substorm growth phase, almost an hour before the first substorm expansion onset. As a consequence, *McPherron* concluded that the inductive electric field of the substorm expansion phase is not likely to be the primary source of ring current particles or energy; it could however be responsible for scattering the ring current particles from open to closed drift paths. The role suggested by *McPherron* for substorms can be perceived as a breakwater for the flow of particles convecting from the magnetotail, drifting along the duskside magnetosphere and escaping from the dayside magnetopause. We will further elaborate on this idea in the next section.

Regarding the apparent lack of Dst response to substorm expansion [noted by *Iyemori and Rao*, 1996, and *McPherron*, 1997], *Daglis et al.* [2000] examined the intense storm of June 5, 1991 and reached different conclusions. *Daglis et al.* examined the second main phase of this particular storm with respect to substorm occurrence, Dst variations and compositional variations. They showed that the Dst change rate increased after substorm occurrence. The relative abundance of O^+ ions in the ring current also increased after substorm occurrence. Distinct, large compositional changes have been shown to accompany the main phase of intense storms. The importance of this finding and its implications for the storm-substorm relationship is the main topic of this paper and is elaborated in the next section.

At this point we need to recall two important facts that influence the apparent storm-substorm relationship, but are often neglected:

i. Dst is not a pure ring current index

The Dst variation is influenced by several factors—not just the intensity of the symmetric ring current. There is a variable, yet significant influence from the other magnetospheric current systems: the magnetopause current, cross-tail current, induced Earth currents, and the substorm current wedge [e.g., *Alexeev et al.*, 1996]. According to a recent evaluation by *Turner et al.* [2000] the tail current alone, on the basis of Tsyganenko's magnetospheric model, contributes about 25% of the measured Dst variation. Therefore statistical studies like the ones by *Iyemori and Rao* [1996] and *McPherron* [1997] have inherent limitations and may be misleading.

ii. Substorm expansion has mixed effects on Dst variation

Substorm expansion has a double effect: injection of energetic particles to the inner magnetosphere, with a negative effect on Dst, and disruption of the cross-tail current in the near-Earth plasma sheet, with a positive effect on Dst. The field-aligned current system associated with this tail current disruption produces positive and negative ΔH effects on the nightside and dayside, respectively. The net effect may well be positive in Dst, despite the fact that the injected particles do contribute to the ring current buildup [*Ohtani et al.*, 2001]. Careful modeling can help us in evaluating the actual contribution of substorm particle injections to the *Dst* variations.

4. SUBSTORMS AND STORM-TIME PARTICLE ACCELERATION

Both space storms and magnetospheric substorms are characterized by the efficient acceleration of charged particles and their injection into the inner magnetosphere. There is however a significant difference in the extent of acceleration: non-storm substorms remain far behind storms regarding quantity and inward penetration of accelerated particles. This difference makes up for the storm-time ring current, a belt of energized particles drifting around the Earth and producing global geomagnetic disturbances on the ground [*Daglis*, 2001].

Some recent studies have questioned not only the traditional perception of substorms as "individual, indispensable bricks of the storm-wall", i.e. the necessity of substorm occurrence for storm development. They have moreover questioned the very ability of substorms to efficiently accelerate large number of charged particles to ring current energies and to thus influence the storm build-up [e.g., *Korth et al.*, 2002]. The common argument of this "school of thought" is that ion energization and earthward penetration is much larger during storms than during substorms. However this is not a real argument—it is simply the defining difference between storms and substorms. Hence it does not prove anything regarding substorm "modular" functionality during storms. Actually the anti-substorm polemic is based on the (usually untold) *a priori* assumption that storm-time substorms do not differ from non-storm substorms, hence the "inability" of non-storm substorms to produce storms condemns all substorms to "storm-impotence". However, there are no sound research results that could justify this assumption and therefore it is still too premature to dismiss substorms as particle accelerating processes.

Actually the dispute must refer to the relative importance of the large-scale convection electric field and the substorm-associated impulsive electric fields in the energization and transport of ions into the ring current. Short-lived impulsive electric fields reported in the literature are induced by magnetic field reconfigurations at substorm onset: i.e., "dipolarizations" from stretched tail-like to dipole-like configuration [e.g., *Moore et al.*, 1981]. *Wygant et al.* [1998] showed that during the large March 1991 storm, the large-scale electric field repeatedly penetrated earthward, maximizing between $L=2$ and $L=4$ with magnitudes of 6 mV/m. Such magnitudes are 60 times larger than quiet-time values. Furthermore, *Wygant et al.* also noted that strong impulsive electric fields with amplitudes of up to 20 mV/m (i.e., more than three times the largest convection electric field) were observed during magnetic field dipolarizations in the inner magnetosphere, i.e. during substorm expansions or intensifications. Consequently, substorm-induced electric fields do "compete" with the convection electric field in particle acceleration during storm development: they are episodic, but on the other side they are much stronger.

The problem has also been addressed in comprehensive computer simulations. *Chen et al.* [1994] used spike-like enhancements of the convection electric field to attempt to simulate the effect of individual substorms, while *Fok et al.* [1996], who used an inductive localized electric field, tied to successive cycles of stretching and dipolarization of the Tsyganenko model magnetic field. The results of both efforts suggested that the substorm contribution was subtle, and possibly negative to the development of a ring current.

More recently however, *Fok et al.* [1999] made some substantial modifications to their model in order to make it more realistic. First, the range of the ring current model was extended out to 12 R_E, in the nightside magnetosphere, setting the boundary condition well outside of geosynchronous orbit, at the outer limits of the region of validity of the adiabatic bounce-averaged ring current code. This provides a plasma input that is realistically influenced by substorm-dipolarization electric fields in the inner plasma sheet. Second, a 3D test particle code was used to construct the ion velocity distribution by backtracking particles from a velocity space grid to source regions assumed to have constant properties independent of the storm/substorm process. *Fok et al.* found that the substorm-associated induced electric fields significantly enhance the ring current by redistributing plasma pressure earthward. Accordingly, it was concluded that global convection and substorm dipolarizations do cooperate to inject plasma energy more deeply into the magnetosphere than either would individually.

Analysis of single-particle dynamics in simulations of magnetic field dipolarizations reveals prominent short-lived accelerations of plasma sheet ions during the expansion phase of substorms [*Delcourt*, 2002]. *Delcourt* has shown that at low latitudes, these ions are centrifugally accelerated toward the mid-plane and locally experience parallel energization up to several tens of keV (Figure 2). Furthermore, under the effect of the electric field induced by relaxation of the magnetic field lines, ions with gyroperiods comparable to the field variation time scale can experience dramatic

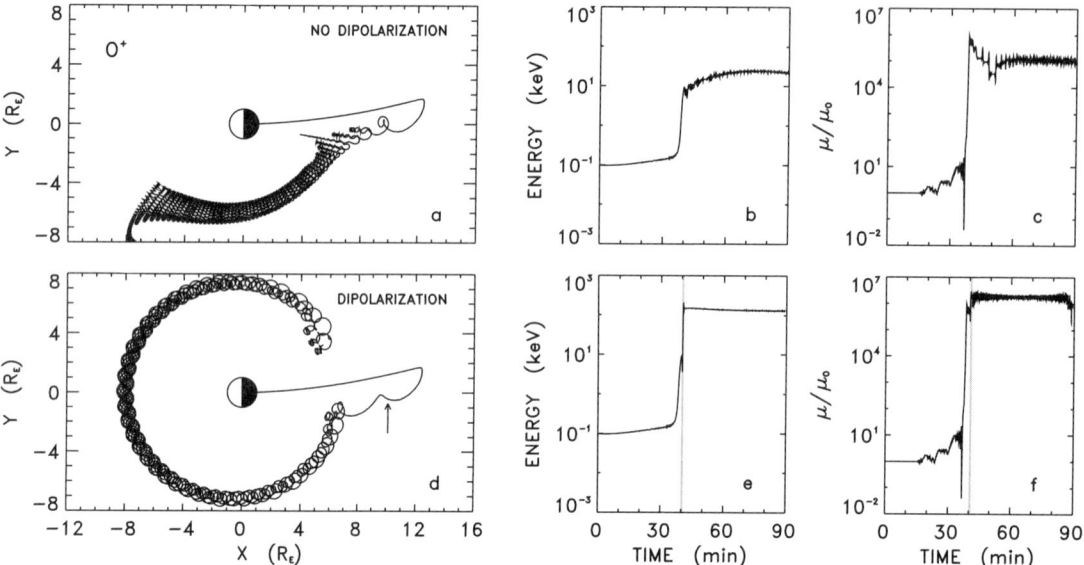

Figure 2. Trajectories of test O$^+$ launched from the nightside auroral zone (0000 MLT) with an initial energy of 100 eV and considering either steady state (top panels) or 1-min dipolarization (bottom panels) during transport. Left panels show the trajectory projection in the X-Y plane, whereas center and right panels show the kinetic energy and magnetic moment (normalized to the initial value) versus time, respectively. In the bottom panels, dipolarization occurs after 40-min time-of-flight (shaded area in panels (e) and (f)). The arrow in (d) indicates the O$^+$ position at the dipolarization onset. Reprinted from *Delcourt* [2002] with permission from Elsevier Science.

nonadiabatic heating. For instance, when considering a relatively smooth 1-min magnetic transition, low-energy O$^+$ originating from the terrestrial ionosphere are found to be accelerated up to a few hundreds of keV during earthward injection. These newly accelerated ions, which subsequently drift rapidly around Earth, evidently can provide a significant—or even major—part of the ring current. By performing such intense energization, inductive electric fields accordingly are of considerable importance for the storm-time dynamics of the inner magnetosphere.

Delcourt's results are consistent with repeated observations that acceleration mechanisms for ions in the near-Earth plasma sheet are mass dependent. Such studies have shown that the energy density of O$^+$ increases much more drastically than the energy density of H$^+$ or He^{++} during substorms, and even more so during the main phase of space storms (Figure 3).

Massive outflow and preferential acceleration of ionospheric O$^+$ ions is outstanding during intense storms, when the oxygen to proton energy density ratio can reach values of up to 400% [*Daglis*, 1997a,b; *Daglis et al.*, 1999]. This is also a persistent feature of substorms, certified by relevant studies with measurements from the AMPTE and CRRES missions [e.g., *Daglis et al.*, 1994, 1996]. The close correlation of O$^+$ energy density enhancements with enhancements of the auroral electrojet intensity led *Daglis and Axford* [1996] to suggest a fast (that is, on a timescale of a substorm phase) response of the ionosphere to increased magnetospheric activity. Such a fast response makes the ionosphere a more active part of the dynamic geospace.

The increased O$^+$ abundance associated with substorm expansion has an important implication for the storm-substorm relationship. Observations and modeling studies have indicated a feedback between O$^+$ injections and substorm breakups (and associated induced electric fields) moving progressively duskward and earthward. *Baker et al.* [1982] had suggested an O$^+$ influence on the localization of substorm onset due to the potential role of O$^+$ in ion tearing instability growth. Actually, a substorm case study showed that the suggested O$^+$ influence was consistent with an earthward and duskward displacement of substorm onset associated with loading of the inner magnetosphere with O$^+$ [*Baker et al.*, 1985]. Later, a model by *Rothwell et al.* [1988] predicted that a higher concentration of O$^+$ in the nightside magnetosphere would permit substorm onset at lower L-values. Lower-energy (<30 keV) O$^+$ ions that are injected directly to the inner ($L < 6$) midnight-duskside magnetosphere [e.g., *Strangeway and Johnson*, 1983], will be energized very efficiently by substorm induced electric fields [e.g., *Delcourt*, 2002]. Hence, the combination of, or, more-

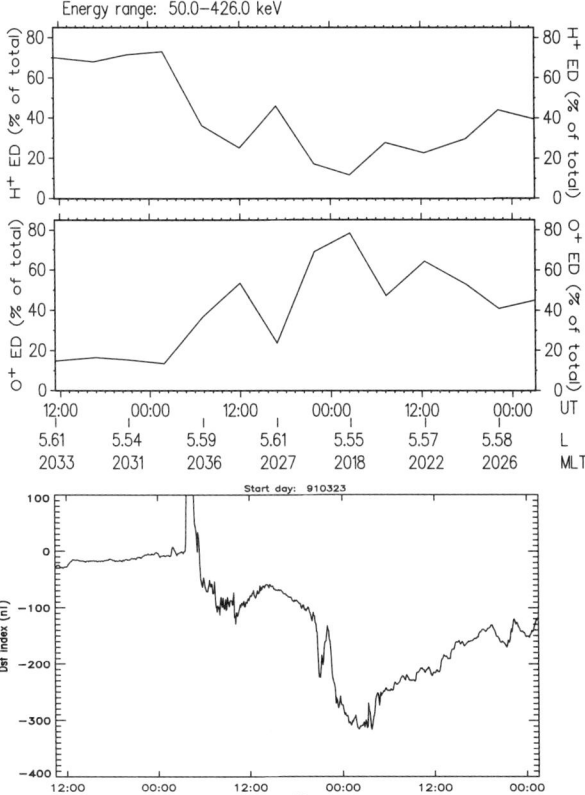

Figure 3. O^+ ions dominate the ring current during the main phase of intense storms. In the case of the March 1991 storm, O^+ ions alone contributed nearly 80% of the ring current energy density at storm maximum. This figure shows the time profile of the H^+ (top panel) and O^+ (middle panel) contribution to the total energy density in the energy range 50-426 keV and the L-range 5-6. The bottom panel shows the time profile of a 5-min resolution Dst index. It is remarkable that the Dst minima are concurrent with O^+ maxima, indicating that the ring current intensifications are due to the acceleration and transport of new ionospheric ions into the inner magnetosphere.

over, the feedback between substorm-breakups and O^+ injections successively proceeding to the duskside and to lower L shells, will substantially contribute to a rapid enhancement of the ring current. This feature is also consistent with the penetration of intense electric fields to low altitudes during intense storms [*Wygant et al.*, 1998].

Interestingly, a relatively old storm study by *Konradi et al.* [1976] had shown that the substorm injection boundary was displaced earthward with each successive substorm during the storm.

Combining model predictions with observations, and considering the fact that O^+ abundance increases with storm size [*Daglis*, 1997a], one comes up with a scenario of a feedback between enhanced (in quantity and spatial extension) O^+-feeding of the plasma sheet and/or the inner magnetosphere and series of intense substorms occurring at progressively lower L-shells. Such a combination of successive substorms and continuous O^+ supply would facilitate successive inward penetration of substorm ion injections, consistent with the *Rothwell et al.* [1988] model and with the observations reported by *Konradi et al.* [1976] and *Daglis* [1997a]. The result of successive inward penetration of substorm injections would be the transport of increasingly more energetic ions into the inner magnetosphere, resulting in the intensification of the storm-time ring current. This scenario can explain why some substorms seem to influence the storm-time ring current growth, while others don't: substorms resulting in weak inward penetration of injections will not contribute much to the ring current growth. An experimental verification would be possible through detailed global imaging of storms.

5. DISCUSSION

The storm-substorm dispute has resided within a well-defined frame, dealing with a number of widely recognized sub-issues and tools. The main argument against substorm involvement in storm development, has been the result of studies on the relationship between Dst and AL time series, combined with the assumption that AL is the most appropriate "substorm index". Moreover, there is the usually untold assumption that all $|AL|$ enhancements are substorms, although this is not the case [*Kamide*, 2001].

However, as pointed out by a number of authors [*Akasofu et al.*, 1983; *Allen and Kroehl*, 1975; *Kamide and Akasofu*, 1983; *Feldstein et al.*, 1997], the AL index has an inherent deficiency: it is not reliable during intense space storms. Since the auroral electrojets expand equatorwards during intense storms, the AE ground magnetogram stations will not provide the actual value of the electrojet intensity, since they will miss a significant part of the electrojets, which will be larger the greater the storm is. In other words, the greater the storm, the worse the underestimate of the auroral electrojet (substorm) intensity will be. Accordingly, prudence in drawing conclusions from studies based on such simplifying tools is imperative.

One line of thought has been to compare storm-time with non-storm-time substorms, in order to reveal possible differences or similarities [e.g., *Daglis et al.*, 1998]. It is obvious that in such studies the "strength" of the substorm expansion phase, which is commonly represented by the AL index, is of prime importance. It is meaningful to compare storm-time and non-storm substorms and reach conclusions about their differences, only if their size, their intensity, is the same, or at least similar. Or else, we face the possibility that their apparent differences are just due to their different size, and not due to their different "nature". It is rather doubtful that the issue of differences between storm-time and non-storm substorms can be fully addressed by statistical studies. In order to assess the issue of these differences, it is necessary to certify that the examined substorm events do indeed belong to the two different categories. Simple statistical studies that include all available measurements simply mix every single substorm observation, and therefore fail to use appropriate, classified substorms. Unfortunately, this is one of the main problems in such an effort. There are very few clearly defined observations of same- or similar-size substorms that belong to the two different categories. Intense storm-time substorms are relatively rare themselves (because intense storms are rare).

Daglis et al. [1998] performed a "first-order investigation" of a storm-time substorm and a non-storm substorm with approximately the same size (as measured by the *AL* index). Their results show that there are indeed differences between the two kinds of substorms, especially concerning compositional characteristics. Although compositional changes do occur in both cases, the storm-time substorm features more intense and temporally longer-lived changes than the non-storm substorm. O^+ provides 15% of the energy density in the non-storm substorm, and 80% of it in the storm-time substorm. This difference has been interpreted as the result of a cumulative effect of a series of successive substorms [*Daglis*, 2001; *Metallinou et al.*, 2002].

Storm-time mass loading by extraordinarily high fluxes of ionospheric O^+ was shown for the outer ring current of one intense storm by *Hamilton et al.* [1988] and for the total ring current of all intense storms of 1991 by *Daglis* [1997a] and *Daglis et al.* [1999]. The implications of mass loading have been discussed in many different studies, and what is probably of main concern to storm-substorm relationship is the feedback between the O^+ abundance and the location of substorm breakups/intensifications. Both observational and modeling studies show that in the presence of high O^+ fluxes, substorm breakups move progressively duskward and earthward [*Baker et al.*, 1985; *Rothwell et al.*, 1988]. The physical reason behind this is the ability of O^+ ions in driving plasma instabilities critical for substorm breakup [e.g., *Baker et al.*, 1985; *Rothwell et al.*, 1994; *Lakhina and Tsurutani*, 1997]. Near-Earth substorm onsets are associated with magnetic field dipolarization and impulsive induced electric fields that reach very high values during intense space storms [e.g., *Wygant et al.*, 1998]. Accordingly, the increasing abundance of O^+ ions during storms can lead to progressively earthward substorm breakups, permitting the transport and acceleration of ions to ever lower altitudes, leading to a stronger ring current and establishing a feedback between O^+ extraction/acceleration and earthward displacement of substorm breakups. Frequent substorm breakups will further have a cumulative effect on particle acceleration and ring current buildup [*Daglis*, 2001; *Metallinou et al.*, 2002].

Intrigued by the apparent lack of storm-substorm relationship shown by studies such as *Iyemori and Rao* [1996] or *McPherron* [1997], *Sun and Akasofu* [2000] suggested that it is more appropriate to examine the relationship between the corrected ring current intensity *Dst** and the upward field-aligned current density, instead of the standard *Dst* and *AE* indices. *Sun and Akasofu* proposed the new approach in order to accommodate the dominance of ionospheric ions in the ring current during intense storms [*Daglis*, 1997a; *Daglis et al.*, 1999]. Using the Method of Natural Orthogonal Components [*Sun et al.*, 1998], *Sun and Akasofu* showed that the directly driven component (DD) of the upward field-aligned currents is poorly correlated to the corrected *Dst* index (correlation coefficient of 0.33), while the unloading component (UL) correlates much stronger (correlation coefficient of 0.81). This indicates that the upward field-aligned currents during substorms play an important role in the formation of the ring current. *Sun and Akasofu* further concluded that the poor correlation between DD and *Dst* indicates that the formation of the ring current is not the result of enhanced convection. The strong correlation between UL and *Dst*, on the other side, is consistent with the observational evidence that the magnetosphere-ionosphere coupling plays an important role in the ring current growth.

Finally, an issue that has been poorly addressed is the influence of substorm recovery on ring current buildup. Traditionally, only substorm expansion has been thought of contributing to the storm development. Studies of DMSP (Defense Meteorological Satellite Program) measurements [*Shiokawa and Yumoto*, 1993] showed that high-latitude field-aligned potential drops, which are efficient agents of ionospheric ion acceleration, exhibit strong enhancements during the substorm recovery phase. Such enhancements will increase the rate of ionospheric ion feeding of the inner plasma sheet during substorm recovery, especially in a series of successive substorms. The obvious result is a net growth of the ring current far from substorm expansion. The criticism against substorm build-up of the ring current has focused on substorm expansion, leaving the case open as to whether other substorm phases influence the ring current growth. A powerful complementary tool to thoroughly investigate substorm influence on ring current growth is global neutral atom imaging of the inner magnetosphere.

Initial results are encouraging. Recently, *Lui et al.* [2001] used IMAGE data to show that the ring current intensified during a substorm with a reduction in convection. Further work is needed in this direction.

6. SUMMARY

The paper addressed the storm-substorm relationship with its main emphasis on the particle acceleration that is necessary for the buildup of the storm-time ring current.

The foundations of the storm-substorm dispute lie in correlation studies between auroral electroject indices and the *Dst* index. Where early studies showed a causal relationship between substorms and storms, more refined investigations that followed showed a direct relationship between solar wind conditions and storm development. These investigations concluded that storms are the result of large-scale magnetospheric convection, with no effect from substorm occurrence.

It is time however, to investigate the physical processes themselves instead of their diagnostics. The role of substorm induced impulsive electric fields may be critical for obtaining the massive ion acceleration needed for the storm-time ring current buildup. A small number of studies have investigated this aspect, showing that indeed substorm particle acceleration is very efficient, and that large-scale convection and substorm dipolarizations do cooperate to inject plasma energy more deeply into the magnetosphere than either would individually.

Furthermore, the fact that intense storms tend to be dominated by ionospheric O^+ ions has to be accomodated by any storm theory, and simple convection doesn't do the job.

Acknowledgments. This paper grew out of discussions initiated in 1998, when the first author (IAD) was a visiting associate professor of STEL. The paper is dedicated to the late colleague and friend Yuri Galperin, an excellent scientist, with whom the first author (IAD) had the pleasure to co-chair an Advanced Study Institute on Space Storms and Space Weather Hazards in June 2000. Yuri was also staying at STEL, discussing magnetospheric questions including ring current development during geomagnetic storms with the second author (YK), until two weeks before his sudden death.

REFERENCES

Akasofu, S. -I., *Polar and magnetospheric substorms*, D. Reidel, Dordrecht, 1968.

Akasofu, S.-I., and S. Chapman, The ring current, geomagnetic disturbance, and the Van Allen radiation belts, *J. Geophys. Res.*, *66*, 1321-1350, 1961.

Akasofu, S.-I., B. H. Ahn, Y. Kamide, and J. H. Allen, A note on the accuracy of the auroral electrojet indices, *J. Geophys. Res.*, *88*, 5769-5772, 1983.

Alexeev, I. I., E. S. Belenkaya, V. V. Kalegaev, Y. I. Feldstein, and A. Grafe, Magnetic storms and magnetotail currents, *J. Geophys. Res.*, *101*, 7737-7747, 1996.

Allen, J. H., and H. W. Kroehl, Spatial and temporal distributions of magnetic effects of auroral electrojets as derived from *AE* indices, *J. Geophys. Res.*, *80*, 3667-3677, 1975.

Baker, D. N., E. W. Hones Jr., D. T. Young, and J. Birn, The possible role of ionospheric oxygen in the initiation and development of plasma sheet instabilities, *Geophys. Res. Lett.*, *9*, 1337-1340, 1982.

Baker, D. N., T. A. Fritz, W. Lennartsson, B. Wilken, H. W. Kroehl, and J. Birn, The role of heavy ions in the localization of substorm disturbances on March 22, 1979: CDAW 6, *J. Geophys. Res.*, *90*, 1273-1281, 1985.

Birkeland, K., The Norwegian Aurora Borealis Expedition, H. Aschehoug and Co., Christiania, sect. 2, 1913.

Burton, R. K., R. L., McPherron, and C. T. Russell, An empirical relationship between interplanetary conditions and *Dst*, *J. Geophys. Res.*, *80*, 4204-4214, 1975.

Chapman, S., Earth storms: Retrospect and prospect, *J. Phys. Soc. Japan*, *17*, Suppl. A-I, 6-16, 1962.

Chen, M. W., L. Lyons, and M. Schultz, Simulations of phase space distributions of storm time proton ring current, *J. Geophys. Res.*, *99*, 5745-5759, 1994.

Daglis, I. A., The role of magnetosphere-ionosphere coupling in magnetic storm dynamics, in *Magnetic Storms*, Geophys. Monogr. Ser., vol. 98, edited by B. T. Tsurutani, W. D. Gonzalez, Y. Kamide, and J. K. Arballo, pp. 107-116, AGU, Washington, DC, 1997a.

Daglis, I. A., Terrestrial agents in the realm of space storms: Missions study oxygen ions, *Eos Trans. AGU*, *24*, 245, 1997b.

Daglis, I. A., The storm-time ring current, *Space Sci. Rev.*, *98*, 343-363, 2001.

Daglis, I. A., and W. I. Axford, Fast ionospheric response to enhanced activity in geospace: Ion feeding of the inner magnetotail, *J. Geophys. Res.*, *101*, 5047-5065, 1996.

Daglis, I. A., S. Livi, E. T. Sarris, and B. Wilken, Energy density of ionospheric and solar wind origin ions in the near-Earth magnetotail during substorms, *J. Geophys. Res.*, *99*, 5691-5703, 1994.

Daglis, I. A., W. I. Axford, S. Livi, B. Wilken, M. Grande, and F. Søraas, Auroral ionospheric ion feeding of the inner plasma sheet during substorms, *J. Geomagn. Geoelectr.*, *48*, 729-739, 1996.

Daglis, I. A., Y. Kamide, G. Kasotakis, C. Mouikis, B. Wilken, E. T. Sarris, and R. Nakamura, Ion composition in the inner magnetosphere: Its importance and its potential role as a discriminator between storm-time substorms and non-storm substorms, in *Fourth International Conference on Substorms (ICS-4)*, edited by S. Kokubun and Y. Kamide, Terra/Kluwer Publications, Tokyo, pp. 767-772, 1998.

Daglis, I. A., G. Kasotakis, E. T. Sarris, Y. Kamide, S. Livi, and B. Wilken, Variations of the ion composition during a large magnetic storm and their consequences, *Phys. Chem. Earth*, *24*, 229-232, 1999.

Daglis, I. A., Y. Kamide, C. Mouikis, G. D. Reeves, E. T. Sarris, K. Shiokawa, and B. Wilken, "Fine structure" of the storm-substorm relationship, *Adv. Space Res.*, *25* (12), 2369-2372, 2000.

Davis, T. N., Temporal behavior of energy injection into the geomagnetic ring current, *J. Geophys. Res.*, *74*, 6266, 1969.

Davis, T. N., and R. Parthasarathy, The relationship between polar magnetic activity DP and growth of the geomagnetic ring current, *J. Geophys. Res.*, *72*, 5825-5836, 1967.

Delcourt, D. C., Particle acceleration by inductive electric fields in the inner magnetosphere, *J. Atmos. Sol. Terr. Phys.*, *64*, 551–559, 2002.

Feldstein, Y. I., A. E. Levitin, S. A. Golyshev, L. A. Dremukhina, U. B. Vestchezerova, T. E. Valchuk, and A. Grafe, Ring current and auroral electrojets in connection with interplanetary medium parameters during magnetic storm, *Ann. Geophys.*, *12*, 602-611, 1994.

Feldstein, Y. I., A. Grafe, L. I. Gromova, and V. A. Popov, Auroral electrojets during geomagnetic storms, *J. Geophys. Res.*, *102*, 14,223-14,235, 1997.

Fok, M.-C., T. E. Moore, and M. E. Greenspan, Ring current development during storm main phase, *J. Geophys. Res.*, *101*, 15,311-15,322, 1996.

Fok, M.-C., T. E. Moore, and D. C. Delcourt, Modeling of inner plasma sheet and ring current during substorms, *J. Geophys. Res.*, *104*, 14,557-14,569, 1999.

Hultqvist, B., On the acceleration of positive ions by high-latitude, large amplitude electric field fluctuations, *J. Geophys. Res.*, *101*, 27111-27121, 1996.

Iyemori, T., and D. R. K. Rao, Decay of the *Dst* field of geomagnetic disturbance after substorm onset and its implication to storm-substorm relation, *Ann. Geophys.*, *14*, 608-618, 1996.

Kamide, Y., Relationship between substorms and storms, in *Dynamics of the magnetosphere*, edited by S.-I. Akasofu, pp. 425-443, D. Reidel, Boston, Mass., 1979.

Kamide, Y., Is substorm occurrence a necessary condition for a magnetic storm?, *J. Geomagn. Geoelectr.*, *44*, 109, 1992.

Kamide, Y., Interplanetary and magnetospheric electric fields during geomagnetic storms: What is more important, steady state fields or fluctuating fields?, *J. Atmos. Sol. Terr. Phys.*, *63*, 413-420, 2001.

Kamide, Y., and S.-I. Akasofu, Notes on the auroral electrojet indices, *Rev. Geophys. Space Phys.*, *21*, 1647-1656, 1983.

Kamide, Y., and N. Fukushima, Analysis of magnetic storms with DR-indices for equatorial ring current field, *Rep. Ionos. Space Res. Japan*, *25*, 125-162, 1971.

Kamide, Y., et al., Current understanding of magnetic storms: Storm/substorm relationships, *J. Geophys. Res.*, *103*, 17,705-17,728, 1998.

Konradi, A., C. L. Semar, and T. A Fritz, Injection boundary dynamics during a geomagnetic storm, *J. Geophys. Res.*, *81*, 3851-3865, 1976.

Korth, A., R. H. W. Friedel, F. Frutos-Alfaro, C. G. Mouikis, and Q. Zong, Ion composition of substorms during storm-time and non-storm-time periods, *J. Atmos. Sol. Terr. Phys.*, *64*, 561-566, 2002.

Lui, A. T. Y., R. W. McEntire, K. B. Baker, A new insight on the cause of magnetic storms, *Geophys. Res. Lett.*, *28*, 3413-3416, 2001.

McPherron, R. L., The role of substorms in the generation of magnetic storms, in *Magnetic Storms*, Geophys. Monogr. Ser., vol. 98, edited by B. T. Tsurutani, W. D. Gonzalez, Y. Kamide, and J. K. Arballo, pp. 131-147, AGU, Washington, DC, 1997.

Metallinou, F.-A., I. A. Daglis, and J.-H. Seiradakis, Study of the relation between storms and substorms in the terrestrial magnetosphere, in *Proceedings of the Fifth Astronomical Conference of The Hellenic Astronomical Society*, edited by D. Hatzidimitriou, in press, 2002.

Moore, T. E., R. L. Arnoldy, J. Feynman, and D. A. Hardy, Propagating substorm injection fronts, *J. Geophys. Res.*, *86*, 6713-6726, 1981.

Moore, T. E., M. O. Chandler, Mr. R. Collier, H. L. Collin, R. Fitzenreiter, B. L. Giles, W. K. Peterson, C. J. Pollock, and C. T. Russell, Ionospheric mass ejection in response to a coronal mass ejection, *Geophys. Res. Lett.*, *26*, 2339-2342, 1999.

Ohtani, S., M. Nosé, G. Rostoker, H. Singer, A. T. Y. Lui, and M. Nakamura, Storm-substorm relationship: Contribution of the tail current to *Dst*, *J. Geophys. Res.*, *106*, 21199-21209, 2001.

Rothwell, P. L., L. P. Block, M. B. Silevitch, and C.-G. Fälthammar, A new model for substorm onsets: The pre-breakup and triggering regimes, *Geophys. Res. Lett.*, *15*, 1279-1282, 1988.

Shiokawa, K., and K. Yumoto, Global characteristics of particle precipitation and field-aligned electron acceleration during isolated substorms, *J. Geophys. Res.*, *98*, 1359-1375, 1993.

Siscoe, G., Big storms make little storms, *Nature*, *390*, 448, 1997.

Strangeway, R. J., and R. G. Johnson, Mass composition of substorm-related energetic ion dispersion events, *J. Geophys. Res.*, *88*, 2057-2064, 1983.

Sun, W., and S.-I. Akasofu, On the formation of the storm-time ring current belt, *J. Geophys. Res.*, *105*, 5411-5418, 2000.

Sun, W., W.-Y. Xu, and S.-I. Akasofu, Mathematical separation of directly driven and unloading components in the ionospheric equivalent currents during substorms, *J. Geophys. Res.*, *103*, 11695, 1998.

Tsurutani, B. T., and W. D. Gonzalez, The cause of high intensity long-duration continuous *AE* activity (HILDCAAs): Interplanetary Alfvén wave trains, *Planet. Space Sci.*, *35*, 405, 1987.

Tsurutani, B. T., and W. D. Gonzalez, The interplanetary causes of magnetic storms: A review, in *Magnetic Storms*, Geophys. Monogr. Ser., vol. 98, edited by B. T. Tsurutani, W. D. Gonzalez, Y. Kamide, and J. K. Arballo, pp. 77-89, American Geophysical Union, Washington, DC, 1997.

Tsurutani, B. T., T. Gould, B. E. Goldstein, W. D. Gonzalez, and M. Sugiura, Alfvén waves and auroral (substorm) activity, *J. Geophys. Res.*, *95*, 2241, 1990.

Turner, N. E., D. N. Baker, T. I. Pulkkinen, and R. L. McPherron, Evaluation of the tail current contribution to *Dst*, *J. Geophys. Res.*, *105*, 5431-5439, 2000.

Wygant, J., D. Rowland, H. J. Singer, M. Temerin, F. Mozer, and M. K. Hudson, Experimental evidence on the role of the large spatial electric field in creating the ring current, *J. Geophys. Res.*, *103*, 29,527-29,544, 1998.

I. A. Daglis, Institute for Space Applications and Remote Sensing, National Observatory of Athens, Metaxa & Vas. Pavlou Str., Penteli, 152 36 Athens, Greece.

Y. Kamide, Solar-Terrestrial Environment Laboratory, Nagoya University, Toyokawa 442-8507, Japan

Role of Plasma Instabilities Driven by Oxygen Ions During Magnetic Storms and Substorms

G.S. Lakhina and S.V. Singh

Indian Institute of Geomagnetism, Dr. Nanabhai Moos Marg, Colaba, Mumbai-400005, India

Ionospheric-origin O^+ ions consititute an important and some times dominant part of the ring current and the near-Earth plasma sheet region during geomagnetic storms and substorms. Low-frequency instabilities excited by the energetic oxygen ions in the near-Earth plasma sheet and ring current region during the geomagnetic storms/substorms are investigated. It is shown that the presence of ionospheric-origin oxygen ion beams with anisotropic pressure can excite helicon mode instability in the near-Earth plasma sheet region provided their Alfvénic Mach numbers lie in a certain range. The helicon modes are easily excited under the conditions when the usual long wavelengths fire-hose modes are stable. On the other hand, low-frequency quasi-electrostatic loss-cone type instabilities can be driven by the energetic oxygen ions in the storm time ring current region. These instabilities may scatter ring current particles and contribute to the ring-current decay.

INTRODUCTION

Observations in the various regions of the Earth's magnetosphere have shown the presence of protons and other energetic heavier ion distributions. Several observations suggest that ionospheric-origin O^+ ions consititute an important part of the inner magnetosphere and the near-Earth plasma sheet region [*Peterson et al.*, 1981; *Sharp et al.*, 1981; *Shelley et al.*, 1982; *Lockwood et al.*, 1985; *Delcourt et al.*, 1989; *Hultqvist*, 1991; *Daglis et al.*, 1991, 1993, 1994; *Cladis and Francis*, 1992; *Lennartsson*, 1994; *Wilken et al.*, 1995] during substorms/storms. Recently, it has been shown that the auroral ionospheric ion feeding of the inner plasma sheet during substorms can be fast (i.e., ~ characteristic substorm timescales), and that the ionosphere could actively influence the substorm energization processes by responding to the increased solar wind-magnetosphere coupling [*Daglis and Axford*, 1996]. The ionospheric O^+ ion flux in the near-Earth plasma sheet ($X \sim -6R_E$ to $-15 R_E$) is observed to increase dramatically during the growth phase of substorms [*Daglis et al.*, 1990, 1991; *Kistler et al.*, 1992], and hence it could influence the dynamical evolution of the plasma sheet. However, the dominant mechanisms for the injection of ionospheric O^+ into the magnetosphere, and their subsequent acceleration to ring current energies during a storm are yet to be fully understood [*Daglis et al.*, 1999; *Hultqvist et al.*, 1999].

Lakhina and Tsurutani [1997, 1998] have shown that sufficiently strong fluxes of O^+ ions of ionospheric origin can excite an electromagnetic helicon mode instability in the near-Earth plasma sheet region. This instability can facilitate the excitation of tearing instability leading to substorm onset. During storms, plasma sheet oxygen ions can be accelerated by the helicon mode waves and injected earthwards and become part of the ring current.

It has been proposed that enhanced densities of ionospheric O^+ ions in some localized region in the plasma sheet

---

Disturbances in Geospace: The Storm-Substorm Relationship
Geophysical Monograph 142
Copyright 2003 by the American Geophysical Union
10.1029/142GM12

may also lead to the excitation of the ion tearing instability [*Schindler*, 1974; *Baker et al.*, 1982], velocity shear instabilities [*Cladis and Francis*, 1992], or firehose type instabilities [*Verheest and Lakhina*, 1991; *Yoon et al.*, 1993; *Lakhina*, 1995, 1996], which could influence the substorm processes.

Recent observations by AMPTE/CCE have shown that energetic O^+ ions of ionospheric origin can become important constituents in the inner magnetosphere with the development of magnetic storms [*Krimigis et al.*, 1985; *Hamilton et al.*, 1988; *Roeder et al.*, 1996, *Daglis et al.*, 1993, *Daglis* 1997a,b]. During the geomagnetic quiet periods, the fractional concentration of the energetic oxygen ions in terms of energy density is about 10% [*Gloeckler and Hamilton*, 1987]. However, during the main phase of storms, the energy contents of the geomagnetically trapped particles increases and the fractional concentration of the oxygen ions is also increased. *Daglis* [1997a,b] has reported that the energy density of O^+ ions exceeds that of H^+ ions during intense storms with Dst of less than –200 nT, while the energy density of O^+ ions contributes only to 30% during a moderate magnetic storm. It is important to know how the thermal oxygen ions are extracted from the ionosphere and then energized to ring current energies of ~ a few keV to hundreds of keV. It has been suggested that the O^+ injection processes may involved either perpendicular heating or field-aligned acceleration with potential drops between the ionosphere and the magnetosphere [*Peterson et al.*, 1993].

An important issue related to the magnetic storm is the ring current decay. The ring current particles can be removed by various collisional processes as well as by scattering of charged particles by the waves. The loss mechanisms of ring current ions involving charge exchange with neutral atoms and Coulomb collisions with thermal plasma have been studied by several workers [*Fok et al.*, 1991, 1993, 1995; *Smith and Bewtra*, 1978; *Chen et al.*, 1997; *Ebihara and Ejiri*, 1999]. All ring current ions are subject to charge-exchange decay. However the decay rate depends on the ion mass and energy. It is found that high-energy O^+ are lost much faster than H^+, and field-aligned pitch-angle distributions experience larger losses than pancake like distributions [*Daglis et al.*, 1993]. The losses of heavier ions due to Coulomb collisions can become significant below energies of ~ 10 keV [*Fok et al.*, 1991; *Noel and Prolss*, 1997]. Large scale models incorporating the effects of charge exchange and Coulomb scattering tend to overestimate the flux of protons above tens of keV, and also yield a pitch-angle distribution which is too flat for energies of the order of 100 keV and higher [*Fok et al.*, 1995]. Pitch-angle scattering by the plasma waves has been suggested as a possible mechanism to account for these discrepancies [*Fok et al.*, 1996].

However, little work has been done on the role of wave-particle interaction on the ring current decay. Intense plasma waves could provide an efficient mechanism for energy transfer between different ion species and may prove important for selectively heating and accelerating thermal heavy ions [*Gendrin and Roux*, 1980; *Thorne and Horne*, 1994]. In fact, assessing the integrated effect on storm time ring current losses due to the scattering of ions by waves is one of the unsolved problem concerning the magnetic storm. The fact that the occurrence of particular plasma wave modes is usually limited in time or confined to localized regions, casts doubts on the ability of the wave scattering processes to significantly affect the energy balance of the ring current globally. The energetic protons and heavier ions can excite the low frequency waves such as electromagnetic ion cyclotron (EMIC) waves [*Cornwall et al.*, 1970, 1977; *Horne and Thorne*, 1993; *Kozyra et al.*, 1997]. Interaction between the ring current ions and the EMIC waves is considered as an important ring current loss process [*Cornwall et al.*, 1970, 1977; *Kozyra et al.*, 1997]. The time scale for scattering of ions into the loss cone during resonant interactions with EMIC waves can be rapid [*Lyons and Thorne*, 1972]. This could explain the rapid ring current energy loss on the time scales of 0.5–1.0 hrs during the main phases of intense-to-great geomagnetic storms [*Gonzalez et al.*, 1989; *Prigancova and Feldstein*, 1992].

In addition to the EMIC waves, quasi-electrostatic instabilities driven by loss-cone distributions of the ring current ions can also occur during magnetic storms [*Coroniti et al.*, 1972; *Bernstein et al.*, 1974; *Lakhina*, 1976, *Bhatia and Lakhina*, 1980a]. Recently, *Singh et al.* [2003] have studied quasi-electrostatic instabilities driven by anisotropic loss-cone oxygen ions. These instabilities can also scatter ring current ions rather efficiently leading to the ring current decay. These instabilities have not attracted much attention as compared to the EMIC waves, and there exist no global simulation model to ascertain the impact of these instabilities on the ring current energy budget.

HELICON MODES DRIVEN BY OXYGEN IONS IN THE NEAR-EARTH PLASMA SHEET REGION

In an electron-proton plasma, the dispersion relation for the right-hand polarized low-frequency modes, i.e., $\omega \ll \Omega_p$ (here ω and Ω_p represent the wave and the proton cyclotron frequencies, respectively), propagating parallel to the magnetic field, $\mathbf{B}_0$, gives the MHD Alfvén modes. In this case the proton Hall current completely cancels the electron Hall current, and the wave is maintained by the proton polarization current [*Papadopoulos et al.*, 1994]. However, in the presence of oxygen ions, the ion (both proton and oxygen) Hall currents cannot completely cancel the electron Hall current unless $\omega \ll \Omega_o$ (Ω_o being the oxygen ion cyclotron frequency). Therefore for the case when O$^+$ ions are weakly magnetized or unmagnetized, they carry negligible Hall current, and the resultant ion Hall current is not sufficient to neutralize the electron Hall current. This situation could give rise to helicon waves [*Papadopoulos et al.*, 1994; *Zhou et al.*, 1996; *Lakhina and Tsurutani*, 1997, 1998; *Lakhina*, 2001]. The helicon waves could lead to the fast current and flux penetration across the plasma sheet [*Papadopoulos et al.*, 1994], thus affecting substorm dynamics. *Lakhina and Tsurutani* [1997] showed that the presence of cold O$^+$ ion beams can excite helicon mode instability in the near-Earth plasma sheet region. *Lakhina* [2001] has shown that the helicon instability persists even when the O$^+$ ion beams are treated as hot.

Following *Lakhina* [2001], a dispersion relation for the electromagnetic modes propagating parallel to the magnetic field, $\mathbf{B}_0 = B_0 \mathbf{z}$ in a multispecies plasma can be written as,

$$\omega^2 = c^2 k^2 - \sum_s \omega_{ps}^2 \left[\frac{\omega - k U_s}{k \alpha_{\|s}} Z(\eta_s) + \left(\frac{\alpha_{\perp s}^2}{\alpha_{\|s}^2} - 1 \right) \{1 + \eta_s Z(\eta_s)\} \right], \quad (1)$$

where $\omega_{ps} = (4\pi q_s^2 N_s/m_s)^{1/2}$ and $\Omega_s = q_s B_0 /m_s c$ are the plasma and the gyrofrequency of the sth species, with $s = e, p$ and o for electrons, protons and the oxygen ions respectively, N_s, q_s and m_s are the density, charge and mass of the sth species, U_s is the drift velocity of the sth species, c is the speed of light, and $\alpha_{\|s} = (2 T_{\|s}/m_s)^{1/2}$, $\alpha_{\perp s} = (2 T_{\perp s}/m_s)^{1/2}$ are respectively the parallel and perpendicular thermal velocities with respect to $\mathbf{B}_0$, $Z(\eta_s)$ is the plasma dispersion function with the argument $\eta_s = (\omega - k U_s \pm \Omega_s)/k a_{\|s}$, and k is the wave number. The $\pm$ sign in η_s denotes the RH (+ sign) and the LH (– sign) modes. In writing (1), we have considered the drifting bi-Maxwellian distribution functions for the plasma species. In the equilibrium state, the charge neutrality is maintained by taking $N_e = N_p + N_o$.

In order to solve (1), we consider an unneutralized hot O$^+$ ion beam passing through the near-Earth plasma sheet region (corresponding to a finite field-aligned current in the system and $U_p = 0$, and $U_e = 0$). The background plasma is treated as cold, i.e., $\eta_e \gg 1$ and $\eta_p \gg 1$, and the ionospheric O$^+$ ion beam as hot, i.e., $\eta_o < 1$. Further, considering $(\omega - k U_s)^2 \ll \Omega_s^2$ for $s = e$ (electrons) and p (protons), and $\omega^2 \ll c^2 k^2$, we obtain the following dispersion relation for the helicon modes

$$k^2 V_{Ap}^2 \left(1 - \frac{A_e}{2} - \frac{A_p}{2} \right) - R\Omega_o^2 \left[-\frac{A_o}{\beta_{\|o}} + \frac{2\Omega_o(\Omega_o \mp k U_o)}{k^2 \alpha_{\|o}^2} \right]$$

$$\mp R\Omega_o \omega \left(1 + \frac{2\Omega_o^2}{k^2 \alpha_{\|o}^2} \right) - \omega^2 \quad (2)$$

$$- i\sqrt{\pi} R\Omega_o^2 \exp(-\eta_o^2) \left[\frac{\beta_{\perp o}}{\beta_{\|o}} \frac{(\omega - k U_o)}{k \alpha_{\|o}} \mp \frac{A_o}{\beta_{\|o}} \frac{\Omega_o}{k \alpha_{\|o}} \right] = 0.$$

Here, $A_s = (\beta_{\|s} - \beta_{\perp s})$ denotes the pressure anisotropy of the plasma species and $\beta_{\|s}$ and $\beta_{\perp s}$ are respectively the parallel and the perpendicular plasma beta for the sth species, and $V_{Ap} = B_0/(4\pi \rho_p)^{1/2}$ is the Alfvén speed with respect to the proton mass density $\rho_p = N_p m_p$, and $R = (N_o m_o/N_p m_p)$ is the relative oxygen ion mass density with respect to protons.

Assuming $\omega^2 \ll R\Omega_o \omega$, and writing $\omega \ll \omega_r + i\gamma$, where $\gamma^2 \ll \omega_r^2$, then from (2), we obtain the following expressions for the real frequency, ω_r, and growth rate, γ,

$$\omega_r = \pm \frac{\left\{ \frac{N_e}{N_o} \left(1 - \frac{A_e}{2} - \frac{A_p}{2} \right) \frac{k^2 c^2}{\omega_{pe}^2} \Omega_e + \Omega_o \left[\frac{A_o}{\beta_{\|o}} - \frac{2\Omega_o(\Omega_o \mp k U_o)}{k^2 \alpha_{\|o}^2} \right] \right\}}{\left(1 + \frac{2\Omega_o^2}{k^2 \alpha_{\|o}^2} \right)}, \quad (3)$$

$$\gamma \approx \frac{\sqrt{\pi} \Omega_o \exp(-\eta_{or}^2)}{\left(1 + \frac{2\Omega_o^2}{k^2 \alpha_{\|o}^2} \right)} \left[\frac{\beta_{\perp o}}{\beta_{\|o}} \frac{(\omega_r - k U_o)}{k \alpha_{\|o}} \mp \frac{A_o}{\beta_{\|o}} \frac{\Omega_o}{k \alpha_{\|o}} \right], \quad (4)$$

where $\eta_{or} = (\omega_r - k U_o \pm \Omega_o)/k \alpha_{\|o}$. The $\pm$ sign in (3) and η_{or} denotes the RH (+ sign) and the LH (– sign) modes.

When the second term in the curly brackets is neglected, (3) becomes very similar in structure to the dispersion relation for the usual helicon mode,

$$\omega_H = \frac{k^2 c^2}{\omega_{pe}^2} \Omega_e, \quad (5)$$

in electron-proton plasma. Equation (5) shows that electron dynamics dominates the interaction between the electromagnetic waves and the multi-ion plasma.

Figure 1 shows the real frequency, ω_r, and the growth rate, γ versus the normalized wave number for the helicon modes for some typical parameters of the near-Earth plasma sheet. For parameters listed in the caption of Figure 1, it is found that the helicon mode instability occurred for the RH mode only. The assumption of treating the O$^+$ ion beam as hot breaks down at smaller values of normalized wavenum-

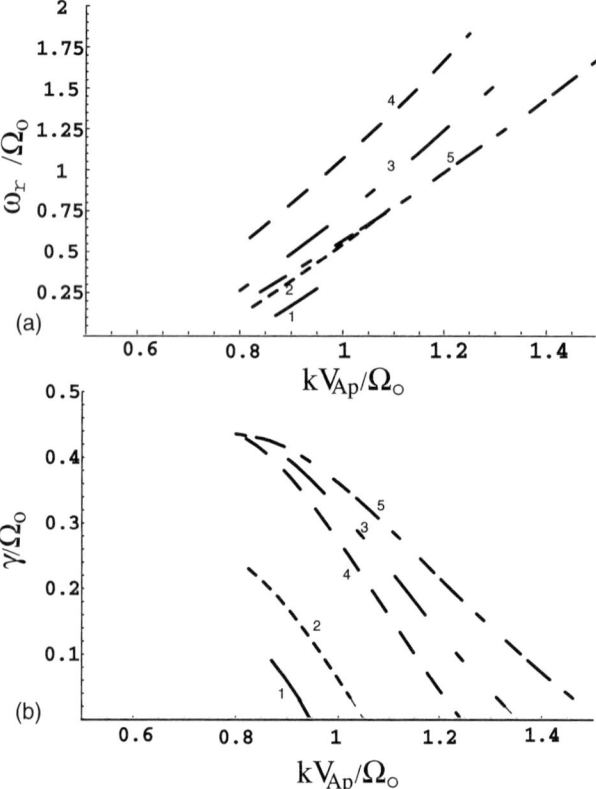

Figure 1. Variation of normalized real frequency ω_r/Ω_o (a), and growth rate γ/Ω_o (b) versus normalized wavenumber kV_{Ap}/Ω_o for the helicon mode instability driven by non-neutralized O^+ ions in the CPS region for $M = U_o/V_{Ap} = 0.25$, $R=1.0$, $A_e = A_p=0$, and $\beta_{\|o} = 3.5$. The curves 1, 2, and 3 are respectively for $A_o = (\beta_{\|o} - \beta_{\perp o}) = 0.1$, 0.5, and 2.0. The curves 4 and 5 are for $A_o = 2$, $M=0.25$, $R=1.0$, and $A_p=-0.5$ and 0.5, respectively. For the parameters considered here, the LH mode instability does not exist.

bers, kV_{Ap}/Ω_o where the peak in the growth rate occurs. Therefore in Figure 1, only those portion of the growth rate (and the corresponding real frequency) curves where all the assumption made in the analysis are satisfied are shown. It is clear from Figure 1 that ranges of real frequencies and growth rate increase when A_o is increased (cf. Curves 1, 2 and 3). Positive (negative) value of proton anisotropy has a destabilizing (stabilizing) effect (cf. Curves 4 and 5). Similar analysis for the case of neutralized O^+ ion beam (i.e., no field-aligned current) shows that in general, both the growth rate and real frequency are smaller as compared to the non-neutralized O^+ ion beam case. Further, the instability occurs for higher values of the normalized wave number in the case of neutralized O^+ ion beam as compared to the non-neutralized O^+ ion beam case. Therefore, the unsta-

ble modes for the neutralized beam case would have shorter wavelengths as compared to the latter case.

We consider the following parameters for the central plasma sheet (CPS) in the region $X \sim -10\, R_E$ to $-15\, R_E$: $A_0 = 0.1$ to 2.0 [*Daglis et al.*, 1991; *Lennartsson*, 1994; *Cladis and Francis*, 1992], $R = 1 - 10$ [*Wilken et al.*, 1995], and $|U_o - U_p| \approx 10 - 60$ km s^{-1} [*Peterson et al.*, 1981; *Orsini et al.*, 1985]. Then, for $B_0 = 10$ nT and $N_p = 0.5$ cm^{-3}, we get typical Mach numbers $M = U_o/V_{Ap} = 0.025–0.25$ in the CPS region. Then, the real frequencies, growth rates, and unstable wavelengths, $\lambda = 2\pi/k$, for the helicon mode instability are found to be $\omega_r = (1.0 - 19.0)$ mHz, $\gamma = (0.5 - 4.5)$ mHz, and $\lambda = (0.5 - 2)\, R_E$, respectively. The typical e-folding time of the instability is about 3 to 15 minutes at wavelengths of $\lambda \approx 0.5$ to $2\, R_E$, which is reasonably short. Therefore, these modes could attain saturation as the enhanced convection events may last for a few hours.

The large amplitude helicon mode may be responsible for some of the low-frequency RH polarized electromagnetic noise in the ULF – ELF frequency range observed in the CPS and plasma sheet boundary layer [*Russell*, 1972; *Tsurutani et al.*, 1985, 1987; *Bauer et al.*, 1995). The large scale fluctuating z and y components of B, associated with the helicon modes could twist the equilibrium magnetic field into flux ropes. Further, the large amplitude helicon modes could produce localised minima in B near the neutral axis that are the potential site for the excitation of the tearing mode instabilities. The tearing instabilities could lead to the onset of the expansion phase of substorms. Moreover, the low-frequency turbulence due to the helicon modes could scatter electrons trapped in the CPS region, thus further enhancing the growth of the tearing modes at these sites.

It has recently been noted that O^+ and H^+ ions in the dayside boundary layer/cusp are heated/accelerated essentially all of the time by intense broadband plasma waves. During southward interplanetary magnetic field events, these ions could be convected in the anti-sunward direction across the polar cap and into the plasma sheet [*Tsurutani et al.*, 1998]. There can also be a direct injection of O^+ ions from the auroral region as suggested by [*Daglis et al.*, 1994]. Such energetic oxygen ions could provide an ideal situation for the excitation of the helicon mode instability and substorm onset. These O^+ ions would be energized, along with the H^+ and other minor ions, during the substorm expansion phase. Consequential injection of this plasma inward into the Earth's night side magnetosphere by the storm electric fields could then lead to enhanced oxygen ion fluxes in the storm-time ring current. The enhancement of the O^+ ions in the ring current region could have a strong controlling effect on the excitation of EMIC waves, and thus on the ring current decay during storms [Thorne and Horne, 1997].

RING CURRENT INSTABILITIES DRIVEN BY ENERGETIC OXYGEN IONS

Electromagnetic Ion Cyclotron Waves

The AMPTE/CCE spacecraft frequently observed EMIC waves in the outer magnetosphere beyond $L = 7$ [Anderson et al., 1990, 1992]. The wave spectral properties and frequency of occurrence depends on local time and can be explained by the resonant instability of anisotropic ring current H^+ ions in the variable background thermal plasma [Cornwall, 1970; Horne and Thorne, 1993, 1994). The presence of oxygen ions during the storms can affect the excitation of EMIC and other low-frequency instabilities. Thorne and Horne [1994] showed that during the geomagnetic storms when the oxygen ion content is significantly enhanced, the absorption of EMIC waves become efficient and may lead to the acceleration of O^+ ions of ionospheric origin to ring current energies. Horne and Thorne [1994] studied the convective instabilities of the EMIC waves in the outer magnetosphere and discussed that the waves below the cyclotron frequency of the helium (He^+) ions are not reflected when the O^+ ion concentration is small. Thorne and Horne [1997] studied the modulation of EMIC instability due to the interaction with ring current O^+ ions during a geomagnetic storms. They demonstrated that during the modest storm, strong EMIC excitation can occur in the frequency band above the oxygen gyrofrequency due to cyclotron resonance with the anisotropic ring current H^+ ions. The energy of the excited wave is efficiently converted into the perpendicular heating of the O^+ ions thus making it more anisotropic during the main phase of the magnetic storm.

Jordanova et al. [1996] have modeled the global impact of ion cyclotron waves on the ring current by using the Ring Current-Atmosphere Interaction Model (RAM) which follows the evolution of three major ring current ion species (H^+, He^+, and O^+) considering adiabatic drift motion, Coulomb collisions, charge exchange, and pitch-angle scattering of protons in the field of EMIC waves. The model produced order- of -magnitude enhancements in the ion precipitation as a result of diffusion in the ion cyclotron waves within the unstable region. However, no significant impact of the wave losses was seen in the global energy balance though the waves reduced the anisotropy in the proton pitch angle distributions locally. Kozyra et al. [1997] implemented an improved scheme for estimating the global distribution and amplitude of EMIC activity using a warm plasma ray tracing program (HOTRAY) [Horne and Thorne, 1993] in the RAM model. The integrated energy loss from the ring current, due to the scattering of protons into the loss cone, caused an additional ~ 8 nT recovery in the Dst index and thus was found to be important for the global energy balance of the ring current.

Low-Frequency Quasi-Electrostatic Instabilities

Electrostatic noise in the frequency range of (20–500) Hz has been reported by Anderson and Gurnett [1973] just outside the plasmapause. Labelle et al. [1988] have observed electrostatic noise from (30–100) Hz in the ring current region. Olsen et al. [1987] observed a broadband electrostatic noise in frequency range of 1.8 Hz to 500 Hz and beyond peaking near the lower hybrid frequency (~ 100 Hz) in the ring current region. A quasi-electrostatic instability excited by loss-cone distribution of ring current protons [Coroniti et al., 1972, Bernstein et al., 1974, Lakhina, 1976, Bhatia and Lakhina, 1980a] can generate frequencies close to the lower hybrid frequency or below the ion plasma frequency. The instability is found to be strongest for $\beta \sim 1$, low-density region just outside the plasmapause and is nearly stable for auroral regions and high density plasmasphere for $\beta \gg 1$ [Coronoti et al., 1972].

We study the obliquely propagating low-frequency quasi-electrostatic waves generated by the anisotropic oxygen ions in the ring current region during the main phase of the storms. Ring current plasma is considered to be consisting of Maxwellian distributed electrons, and protons, and energetic oxygen ions having Dory-Guest-Harris (DGH) [Dory et al., 1965, Coroniti et al., 1972] type of distribution given by

$$f_o = \frac{N_o}{\pi^{3/2}} \frac{1}{J! \alpha_{\parallel o}^3} \left(\frac{v_\perp}{\alpha_{\parallel o}}\right)^{2J} exp\left[-\frac{v_\perp^2 + v_\parallel^2}{\alpha_{\parallel o}^2}\right], \quad (6)$$

where $J = \left(\frac{T_{\perp o}}{T_{\parallel o}} - 1\right)$ defines the temperature anisotropy, and $\alpha_{\parallel o} = (2T_{\parallel o}/m_o)^{1/2}$ is the parallel thermal velocity of the oxygen ions with respect to ambient magnetic field $\mathbf{B}_o = B_o \mathbf{z}$. Here, we treat electrons and protons as magnetized, and oxygen ions as unmagnetized, i.e., $\Omega_o < \omega < \Omega_p \ll \Omega_e$. The protons and oxygen ions are considered to be electrostatic, i.e., $\omega_{pp}^2/c^2 k^2 \ll 1$, and $\omega_{po}^2/c^2 k^2 \ll 1$ and electrons response is electromagnetic, i.e., $\omega_{pe}^2/c^2 k^2 \gg 1$. Under the above approximations, a linear dispersion relation for the low-frequency waves propagating obliquely to the magnetic field, $\mathbf{B}_o$ can be obtained by solving the linearized Vlasov equation and Maxwell's equations and is written as (*Davidson et al.*, 1977, *Bhatia and Lakhina*, 1977, 1980a, b)

$$1 + \frac{\omega_{pe}^2}{\Omega_e^2}\left[\frac{1-I_0(\lambda_e)e^{-\lambda_e}}{\lambda_e}\right] + \frac{\omega_{pe}^2}{\Omega_e^2}\frac{\omega_{pe}^2}{c^2k^2}\left[\frac{[\{I_0(\lambda_e)-I_1(\lambda_e)\}e^{-\lambda_e}]^2}{1+\beta_e\{I_0(\lambda_e)-I_1(\lambda_e)\}e^{-\lambda_e}}\right]$$
$$-\frac{k_\parallel^2}{k_\perp^2}\frac{\omega_{pe}^2}{\omega^2}\frac{I_0(\lambda_e)e^{-\lambda_e}}{1+\frac{\omega_{pe}^2}{c^2k^2}I_0(\lambda_e)e^{-\lambda_e}} - \sum_{n=1}^{\infty}\frac{4n^2\omega_{pp}^2\Omega_p^2}{k_\perp^2\alpha_p^2(\omega^2-n^2\Omega_p^2)}I_n(\lambda_p)e^{-\lambda_p} \quad (7)$$
$$-\frac{\omega_{po}^2}{k^2\alpha_{\parallel o}^2}\frac{(-1)^J}{J!}\mu^J\left[\frac{d^J}{d\mu^J}[Z'(\frac{\omega}{k}\sqrt{\mu})]\right] = 0,$$

where $Z'\left(\frac{\omega}{k}\sqrt{\mu}\right)$ is the derivative of the plasma dispersion function with respect to its argument, and $\lambda_{e,p} = k_\perp^2\alpha_{e,p}^2/2\Omega_{e,p}^2$, $\mu^{-1} = \alpha_{\parallel o}^2$ and I_n is the modified Bessel function of order n, and $\alpha_{e,p} = (2T_{e,p}/m_{e,p})^{1/2}$ is the thermal velocity of electrons and protons respectively. We expand the plasma dispersion function in the limit $\omega\sqrt{\mu}/k \le 1$ for the resonant oxygen ions, and consider $n = 1$ terms for the protons in the above dispersion relation (7). The latter assumption restricts the analysis to frequencies smaller than the proton cyclotron frequencies. Substituting for $\omega = \omega_r + i\gamma$ in (7) and separating real and imaginary parts, we obtain following expressions for real frequency (ω_r) and the growth rate (γ),

$$\omega_r^2 = \frac{B \pm \sqrt{B^2 - 4C}}{2}, \quad (8)$$

and,

$$\gamma = -\frac{\sqrt{\pi}}{P}\frac{\omega_r^2\omega_{po}^2}{\alpha_{\parallel o}^3 k^3}\left[\frac{(-1)^J}{J!}\mu^{J-1/2}\frac{d^J}{d\mu^J}\left\{\mu^{1/2}e^{-\mu\frac{\omega_r^2}{k^2}}\right\}\right], \quad (9)$$

where,

$$B = \left[\Omega_p^2 + \frac{\omega_{pe}^2}{P}\frac{k_\parallel^2}{k_\perp^2}\frac{I_0(\lambda_e)e^{-\lambda_e}}{1+\frac{\omega_{pe}^2}{c^2k^2}I_0(\lambda_e)e^{-\lambda_e}} + \frac{4}{P}\frac{\Omega_p^2\omega_{pp}^2}{k_\perp^2\alpha_p^2}I_1(\lambda_p)e^{-\lambda_p}\right], \quad (10)$$

$$C = \frac{1}{P}\frac{k_\parallel^2}{k^2}\omega_{pe}^2\Omega_p^2\frac{I_0(\lambda_e)e^{-\lambda_e}}{1+\frac{\omega_{pe}^2}{c^2k^2}I_0(\lambda_e)e^{-\lambda_e}}, \quad (11)$$

and,

$$P \simeq 1 + \frac{\omega_{pe}^2}{\Omega_e^2}\left[\frac{1-I_0(\lambda_e)e^{-\lambda_e}}{\lambda_e}\right] + \frac{\omega_{pe}^2}{\Omega_e^2}\frac{\omega_{pe}^2}{c^2k^2}\left[\frac{[\{I_0(\lambda_e)-I_1(\lambda_e)\}e^{-\lambda_e}]^2}{1+\beta_e\{I_0(\lambda_e)-I_1(\lambda_e)\}e^{-\lambda_e}}\right]$$
$$+ 2\frac{\omega_{po}^2}{k^2\alpha_{\parallel o}^2}\frac{(-1)^J}{J!}\mu^J\sigma \quad (12)$$

where $\sigma = 1$ for $J = 0$ and $\sigma = 0$ for $J \ge 1$. It may be noticed from the growth rate expression (9) that for $J = 0$, i.e., when perpendicular and parallel temperatures of oxygen ions are equal, the γ is negative and hence, waves are damped and the system is stable. It is expected as for $J = 0$, the plasma system is isotropic and there is no free energy source to excite the instability. For $J \ge 1$, the energetic oxygen ions have finite anisotropy which can drive the low frequency instability.

The growth rate expression for $J = 1$ is given by

$$\gamma = \frac{\sqrt{\pi}}{2P}\frac{\omega_r^2\omega_{po}^2}{\alpha_{\parallel o}^3 k^3}\left[1 - 2\frac{\omega_r^2}{k^2\alpha_{\parallel o}^2}\right]e^{-\frac{\omega_r^2}{k^2\alpha_{\parallel o}^2}}, \quad (13)$$

and for $J = 2$ growth rate is given by

$$\gamma = \frac{\sqrt{\pi}}{8P}\frac{\omega_r^2\omega_{po}^2}{\alpha_{\parallel o}^3 k^3}\left[1 + 4\frac{\omega_r^2}{k^2\alpha_{\parallel o}^2} - 4\frac{\omega_r^4}{k^4\alpha_{\parallel o}^4}\right]e^{-\frac{\omega_r^2}{k^2\alpha_{\parallel o}^2}}, \quad (14)$$

and similarly for $J = 3$ growth rate is given by

$$\gamma = \frac{\sqrt{\pi}}{16P}\frac{\omega_r^2\omega_{po}^2}{\alpha_{\parallel o}^3 k^3}\left[1 + 2\frac{\omega_r^2}{k^2\alpha_{\parallel o}^2} + 4\frac{\omega_r^4}{k^4\alpha_{\parallel o}^4} - \frac{8}{3}\frac{\omega_r^6}{k^6\alpha_{\parallel o}^6}\right]e^{-\frac{\omega_r^2}{k^2\alpha_{\parallel o}^2}}, \quad (15)$$

It can be seen from the growth rate expression (13) for $J = 1$, that for the growth of the waves a condition $\omega_r^2/k^2\alpha_{\parallel o}^2 < 1/2$ must be satisfied which is consistent with our assumption used in expanding the plasma dispersion function for the oxygen ions. The chosen parameters for numerical calculations for $J = 1$ case, are $k_\parallel/k = 0.05$, $T_{\parallel o}/T_e = 30$, $T_{\parallel o}/T_p = 1$, $\omega_{pe}/\Omega_e = 2$, $T_e = 1$ keV, $N_e = 10$ cm^{-3} and $\alpha_{\parallel o}/c \simeq 2 \times 10^{-3}$ which are representative of storm time ring current, and have been used for the results presented in the Figures 2–4.

Figure 2 shows the variation of normalized real frequency ω_r/Ω_o with $\lambda_o = k_\perp^2\alpha_{\parallel o}^2/2\Omega_o^2$ for fractional oxygen ion concentration, $N_o/N_e = 0.2$. The real frequency shows marginal increase with the increase of oxygen ion concentration. Figure 3 shows the variation of normalized growth rate γ/Ω_o with λ_o for various values of the fractional oxygen ion density N_o/N_e as shown on the curves. It may be noted that as the oxygen ion concentration increases, the growth rate increases. The range of unstable wavenumbers decreases with increase in oxygen ion concentration.

Figure 4 shows the comparison of the growth rates for $J = 1, 2$ and 3 values of the temperature anisotropies for the above mentioned parameters. The peak growth rate increas-

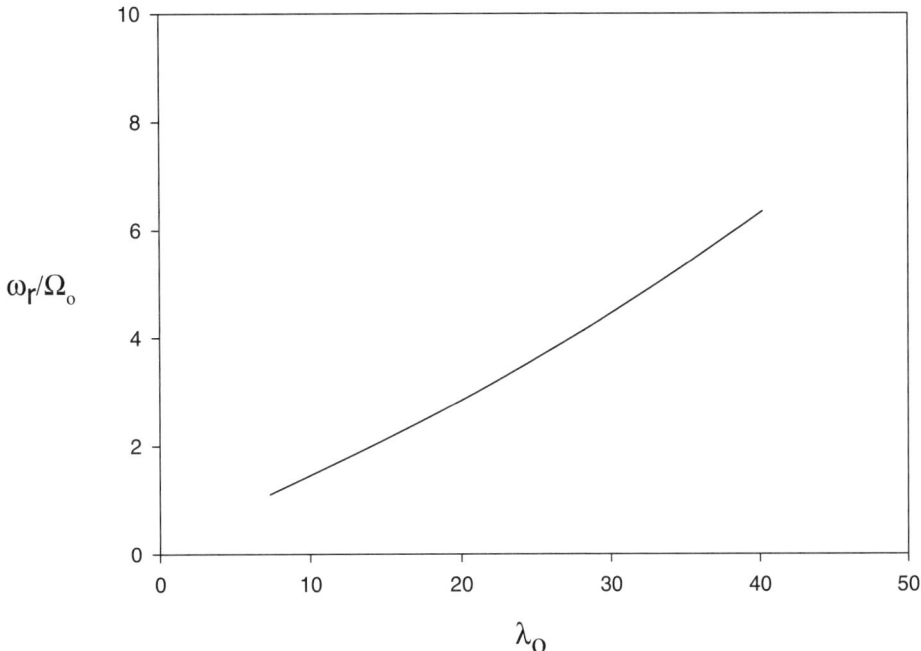

Figure 2. shows the normalized real frequency ω_r/Ω_o versus $\lambda_o = k_\perp^2 \alpha_{\|o}^2/2\Omega_o^2$ for $J = 1$ for fractional oxygen ion concentration $N_o/N_e = 0.2$. The typical parameters are, $k_\|/k = 0.05$, $T_{\|o}/T_e = 30$, $T_{\|o}/T_p = 1$, $\omega_{pe}/\Omega_e = 2$, and $\alpha_{\|o}/c \simeq 2 \times 10^{-3}$ for $J = 1$ case.

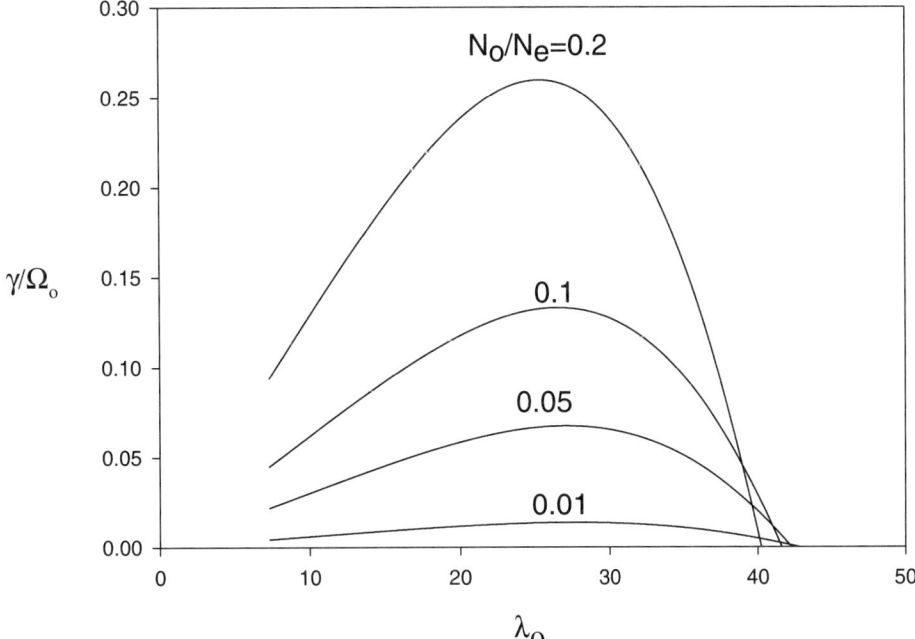

Figure 3 shows the variation of normalized growth rate γ/Ω_o with λ_o for various values of the fractional oxygen ion density N_o/N_e as shown on the curves. Other parameters are same as in Figure 2.

Figure 4 shows the comparison of the growth rates for $J = 1, 2$ and 3 for $N_o/N_e = 0.2$. Other parameters are same as for Figure 2.

es with the increase of the anisotropy. However, real frequency is not affected by the change in anisotropy. It is also found that the range of unstable wavenumbers increases with the increase in the anisotropy.

Thorne and Horne [1997] have provided the Maxwellian fits to the energetic H^+ and O^+ ion distributions obtained from the ring-current atmosphere interaction model (RAM) and have given the equatorial parameters. We have carried out the numerical calculations of the growth rate, real frequency etc. for the three different values of anisotropic index J for the storm time ring-current parameters. For $J = 1$, the frequency of the wave $\omega_r \sim (0.5 - 3)$ Hz and the corresponding to the growth rate $\gamma \sim (40 - 125)$ mHz as shown in Figure 4. The corresponding perpendicular wavelength ($\lambda_\perp$) of the excited waves is $\sim (140 - 350)$ km. For $J = 2$, the corresponding frequency and growth rate of the waves are in the range $\sim (0.5 - 5.5)$ Hz and $\sim (15 - 290)$ mHz respectively. The perpendicular wavelength of these waves is $\sim (109 - 350)$ km. For $J \approx 3$, the real frequency and growth rate are $\sim (0.5 - 5.5)$ Hz and $\sim (7 - 400)$ mHz respectively for $\lambda_\perp \sim (109 - 350)$ km. Thus, it can be concluded that for the higher values of anisotropy, the peak growth rate will be greater.

The saturation electric field amplitude of the waves can be estimated by comparing the linear wave growth rate with the trapping frequency of the oxygen ions [*Coroniti et al.*, 1972]. Thus, assuming that the modes are stabilized by the trapping of O^+ ions by the waves when their amplitude becomes large, the saturation electric field (E_s) for the low-frequency waves is given by

$$E_s = \frac{\pi m_o \omega_r^4 \omega_{po}^4}{e\alpha_{\parallel o}^4 k^7 P^2} \left[\frac{(-1)^J \mu^J}{J!} \frac{d^J}{d\mu^J} \left\{ \sqrt{\mu} \exp\left(-\frac{\omega_r^2}{k^2}\mu\right) \right\} \right]^2 . \quad (16)$$

To estimate the diffusion time for an oxygen ion to reach the loss cone, a diffusion coefficient $D = (\Delta v_\perp)^2/2\Delta t$ is constructed, where $\Delta v_\perp = eE_s\Delta t/m_o$ and Δt is the wave particle correlation time. If we assume $\Delta t \approx \gamma^{-1}$, then D is given by

$$D = \frac{\gamma^3}{2k^2} \quad (17)$$

The saturation electric fields for the $J = 1, 2,$ and 3 as calculated from (16) are $\sim (0.5 - 3)$ mV/m, $\sim (80\mu$V/m $- 11$ mV/m) and $\sim (0.2 \ \mu$V/m $- 20$ mV/m) respectively and the corresponding diffusion coefficients as calculated from (17) are of the order of $\sim (2.4 \times 10^{11} - 2 \times 10^{12})$ cm^2/s^3, $\sim (1.3 \times 10^{10} - 1.2 \times 10^{13})$ cm^2/s^3 and $\sim (1.25 \times 10^9 - 2.65 \times 10^{13})$ cm^2/s^3 respectively. Thus, saturation electric field and diffusion coefficients will increase with the increase in the oxygen ion anisotropy. The time during which an oxygen ion can diffuse into loss cone with a velocity $\alpha_{\|o}$ is ~ $(10^2 - 10^6)$ s.

CONCLUSION

The ionospheric origin oxygen ions dominate the ring current and near Earth plasma sheet region during the storms and substorms. The oxygen ion beams with anisotropic pressure can excite the helcon mode instability for the RH mode in the near Earth plasma sheet region. The e-folding time of the instability is about (3–15) minutes and these modes could attain saturation as the enhanced convection events last few hours. The low-frequency turbulence due to the helicon modes could scatter electrons trapped in the central plasma sheet region and enhance the growth of the tearing modes.

Presence of O^+ ions during the storms facilitates the absorption of EMIC waves and lead to the acceleration of O^+ ions of ionospheric origin to ring current energies. During the storms the EMIC waves can heat the O^+ ions and thus make them more anisotropic.

It is possible to excite the low-frequency quasi-electrostatic wave with anisotropic oxygen ions. The frequency of these waves is found to be between the oxygen ion and proton cyclotron frequencies. These waves have larger growth rate for nearly perpendicular propagation. Low-frequency waves are found to have a higher growth rate in the presence of higher anisotropy of oxygen ions but the corresponding real frequency is not affected by the anisotropy. These quasi-electrostatic low frequency waves considered here may explain the lowest frequency component (< 10 Hz) of the electrostatic waves observed in the ring current region beyond the plasmapause by *Olsen et al.* [1987]. Growth time of these waves is few tens of seconds whereas geomagnetic storm main phase lasts for several hours. The low-frequency waves may scatter the ring-current particles and lead to the ring-current decay. Thus, the interaction between the particles and the low-frequency waves studied here can provide a mechanism for the ring current decay that is complimentary to the charge exchange mechanism. However, the study has not taken into account the effect of cold plasma which become necessary when hot ring current penetrate the plasmapause during the main phase of the magnetic storm. We expect that the cold plasma population may reduce the growth rate of the low frequency waves in the ring current region as the fractional concentration of oxygen ions will be reduced. There is also a possibility that the presence of cold plasma species may allow the excitation of new modes. However, a detailed study is required to find out the effect of cold plasma on the growth rate and the range of the unstable wave numbers of the low frequency quasi-electrostatic wave modes. Further, our analysis treats the ring current system as a uniform plasma. This assumption is justified since the typical transverse wavelengths, $\lambda_\perp$ ~ 100–350 kms are much smaller than the inhomogeneity scale lengths in the ring current region.

REFERENCES

Anderson, B.J., R.E. Erlandson, and L.J. Zanetti, A statistical sudy of Pc 1-2 magnetic pulsations in the equatorial magnetosphere, 1, Equatorial occurrence distributions, *J. Geophys. Res., 97*, 3075, 1992a.

Anderson, B.J., R.E. Erlandson, and L.J. Zanetti, A statistical sudy of Pc 1-2 magnetic pulsations in the equatorial magnetosphere, 2, Wave properties, *J. Geophys. Res., 97*, 3089, 1992b.

Anderson, R.R. and D.A. Gurnett, Plasma wave observations near the plasmapause with S^3-A satellite, *J. Geophys. Res., 78*, 4756, 1973.

Anderson, B.J., K. Takahashi, R.E. Erlandson, and L.J. Zanetti, Pc 1-2 pulsations observed by AMPTE/CCE in the Earth's outer magnetosphere, *Geophys. Res. Lett., 17*, 1853, 1990.

Baker, D. N., Hones, Jr., E. W., Young, D. T., Birn, J., The possible role of ionospheric oxygen ion in the initiation and development of plasma sheet instabilities, *Geophys. Res. Lett., 9*, 1337, 1982.

Bauer, T. M., Baumjohann, W., Treumann, R. A., Sckopke, N., Lühr, H., Low-frequency waves in the near-Earth plasma sheet, *J. Geophys. Res., 100*, 9605, 1995.

Bernstein, W., B. Hultqvist, and H. Borg, Some implications of low altitude observations of isotropic precipitation of ring current protons beyond the plasmapause, *Planet. Space Sci., 22*, 767, 1974.

Bhatia, K.G. and G.S. Lakhina, Generation of hiss beyond plasmapause by magnetosonic waves, *Planet. Space Sci., 25*, 833, 1977.

Bhatia, K.G. and G.S. Lakhina, Drift loss cone instability in the ring current and plasma sheet, *Proc. Indian Acad. Sci. (Earth and Planet. Sci.), 89*, 99, 1980a.

Bhatia, K.G. and G.S. Lakhina, Instability near proton-cyclotron harmonics due to anti-loss cone proton distributions, *Astrophys. and Space Sci., 68*, 175, 1980b.

Chen, M. W., M. Schulz, and L. R. Lyons, Modeling of ring current formation and decay: A review in Magnetic Storms, *Geophys. Monogr. Ser., 98*, edited by B. T. Tsurutani, W. D. Gonzalez, Y. Kamide, and J. K. Arballo, pp. 173-186, American Geophysical Union, Washington, DC, 1997.

Cladis, J. B., and W. E. Francis, Distribution in magnetotail of O^+ ions from cusp/cleft ionosphere: a possible substorm trigger, *J. Geophys. Res., 97*, 123-130, 1992.

Cornwall, J.M., F.V. Coroniti, and R.M. Thorne, Turbulent loss of ring current protons, *J. Geophys. Res., 75*, 4699, 1970.

Cornwall, J. M., On the role of charge exchange in generating unstable waves in the ring current, *J. Geophys. Res., 82*, 1188, 1977.

Coroniti, F.V., R.W. Fredricks, and R. White, Instability of the ring current protons beyond the plasmapause during injection events, *J. Geophys. Res., 77*, 6243, 1972.

Daglis, I. A., E.T. Sarris, and G. Kremser, G., Indication for ionospheric participation in the substorm process from AMPTE/CCE observations, *Geophys. Res. Lett., 17*, 57-60, 1990.

Daglis, I. A., E.T. Sarris, and G. Kremser, G., Ionospheric contribution to the cross-tail current during the substorm growth phase, *J. Atmos. Terr. Phys., 53*, 1091-1098, 1991.

Daglis, I. A., E.T. Sarris, and B. Wilken, AMPTE/CCE CHEM observations of the ion population at geosynchronous altitudes, *Ann. Geophys., 11*, 685-696, 1993.

Daglis, I. A., S. Livi, E.T. Sarris, and B. Wilken, Energy density of ionospheric and solar wind origin ions in the near-Earth magnetotail during substorms, *J. Geophys. Res., 99*, 5691-5703, 1994.

Daglis, I. A., and W.I. Axford, Fast ionospheric response to enhanced activity in geospace: Ion feeding of the inner magnetotail, *J. Geophys. Res., 101*, 5047-5065, 1996.

Daglis, I. A., The role of magnetosphere-ionosphere coupling in magnetic storm dynamics, in Magnetic Storms, *Geophys. Monogr. Ser., 98*, edited by B. T. Tsurutani, W. D. Gonzalez, Y. Kamide, and J. K. Arballo, pp. 107-116, American Geophysical Union, Washington, DC, 1997a.

Daglis, I. A., Terrestrial agents in the realm of space storms: Missions study oxygen ions, *EOS Trans. AGU, 78* (24), 245, 1997b.

Daglis, I. A., R.M. Thorne, W. Baumjohann, W., and S. Orisini, The terrestrial ring current: Origin, formation, and decay, *Rev. Geophys., 37*, 407, 1999.

Davidson, R.C., N.T. Gladd, C.S. Wu, and J.D. Huba, Effects of finite plasma beta on the lower-hybrid-drift instability, *Phys. Fluids, 20*, 301, 1977.

Delcourt, D. C., C.R. Chappell, T.E. Moore, Jr. J.H. Waite, A three-dimensional numerical model of ionospheric plasma in the magnetosphere, *J. Geophys. Res., 94*, 11893-11920, 1989.

Dory, R.A., G.E. Guest, and E.G. Harris, Unstable electrostatic plasma waves propagating perpendicular to a magnetic field, *Phys. Rev. Lett., 14*, 131, 1965.

Ebihara, Y. and M. Ejiri, Quantitative ring current model: Overview and comparison with observations, *Adv. Polar Upper Atmos. Res., 13*, 1-36, 1999.

Fok, M-C., J.U. Kozyra, A.F. Nagy, and T.E. Cravens, Lifetime of ring current particles due to Coulomb collisions in the plasmasphere, *J. Geophys. Res., 96*, 7861, 1991.

Fok, M-C., J.U. Kozyra, A.F. Nagy, Ramussen, and G.V. Khazanov, Decay of equatorial ring current ions and associated aeronomical consequences, *J. Geophys. Res., 98*, 19381-1993.

Fok, M-C., T.E. Moore, J.U. Kozyra, G.C. Ho, and D.C. Hamilton, Three dimensional ring current decay model, *J. Geophys. Res., 100*, 9619, 1995.

Fok, M-C., T.E. Moore, and M.E. Greenspan, Ring current development during storm main phase, *J. Geophys. Res., 101*, 15311, 1996.

Gendrin, R. and A. Roux, Energization of helium ions by proton induced hydromagnetic waves, *J. Geophys. Res., 85*, 4577, 1980.

Gloeckler, G., and D.C. Hamilton, AMPTE ion composition results, *Phys. Scr., T18*, 73, 1987.

Gonzalez, W.D., B.T. Tsurutani, A.L.C. Gonzalez, E.J. Smith, F. Tang, and S.-I. Akasofu, Solar-wind magnetosphere coupling during intense magnetic storms, *J. Geophys. Res., 94*, 8835, 1989.

Hamilton, D. C., G. Glockler, F.M. Ipavich, W. Stüdemann, B. Wilken, and G. Kremser, G., Ring current development during the great geomagnetic storm of February 1986, *J. Geophys. Res., 93*, 14343, 1988.

Horne, R. B., and R. M. Thorne, On the preferred source location for the convective amplification of ion cyclotron waves, *J. Geophys. Res., 98*, 9233, 1993.

Horne, R. B., and R. M. Thorne, Convective instabilities of electromagnetic ion cyclotron waves in the outer magnetosphere, *J. Geophys. Res., 99*, 17259, 1994.

Hultqvist, B., Extraction of ionospheric plasma by magnetospheric processes, *J. Atmos. Terr. Phys., 53*, 3-15, 1991.

Hultqvist, B., M. Øieroset, G. Paschmann, and R.A. Treumann, (editors) Magnetospheric plasma sources and losses, *ISI Space Science Series, 6*, Kluwer Academic Publishers, Dordrecht, 1999.

Jordanova, V.K., L. M. Kistler, J.U. Kozyra, G.V. Khazanov, and A.F. Nagy, Collisional losses of ring current ions, *J. Geophys. Res., 101*, 111, 1996.

Kistler, L. M., E. Möbius, E., W. Baumjohann, G. Paschmann, and D.C. Hamilton, Pressure changes in the plama sheet during substorm injections, *Geophys. Res. Lett., 97*, 2973-2983, 1992.

Kozyra, J. U., V. K. Jordanova, R. B. Horne, and R. M. Thorne, Modeling of the contribution of electromagnetic ion cyclotron (EMIC) waves to stormtime ring current erosion, in Magnetic Storms, *Geophys. Monogr. Ser., 98*, edited by B. T. Tsurutani, W. D. Gonzalez, Y. Kamide, and J. K. Arballo, pp. 187-202, American Geophysical Union, Washington, DC, 1997.

Krimigis, S.M., G. Gloeckler, R.W. McEntire, A. Potemera, F.L.Scarf, and E.G. Shelly, Magnetic storm of September 4, 1984: A synthesis of ring current spectra and energy densities measured with AMPTE/CCE, *Geophys. Res. Lett., 12*, 329, 1985.

Labelle, J., R.A. Treumann, W. Baumjohann, G. Haerendel, N. Sckopke, and G. Paschman, G., The duskside plasmapause/ring-current interface: Convection and plasma wave observations, *J. Geophys. Res., 93*, 2573, 1988.

Lakhina, G.S., Gradients and thermal effects on ring current loss-cone instability, *Planet. Space Sci., 24*, 609, 1976.

Lakhina, G. S., Excitation of plasma sheet instabilities by ionospheric O+ ions, *Geophys. Res. Lett., 22*, 3453-3456, 1995.

Lakhina, G. S., Near-Earth low-frequency modes driven by ionospheric ions, in *Proc. 3rd International Conference on Substorms, ICS-3*, Versailles, May 12-17, pp. 441-445, 1996.

Lakhina, G.S., and B.T. Tsurutani, Helicon Modes driven by ionospheric O^+ ions in the plasma sheet region, *Geophys. Res. Lett., 15*, 1463-1466, 1997.

Lakhina, G.S., and B.T. Tsurutani, Role of Helicon Modes in Substorm processes, *SUBSTORMS-4*, Edited by S. Kokubun and Y. Kamide, Terra Scientific Publishing Company/Kluwer Academic Publishers, pp. 551-516, 1998.

Lakhina, G.S., Role of Helicon Modes in the injection of oxygen ions in the ring current, *J. Atmos. Solar-Terr. Phys.*, *63*, 481, 2001.

Lennartsson, O. W., Tail lobe ion composition at energies of 0.1 to 16 keV: Evidence of mass-dependent density gradients, *J. Geophys. Res.*, *99*, 2387-2401, 1994.

Lockwood, M., Jr. J. H. Waite, T.E. Moore, J.F.E. Johnson, and C.R. Chappell, A new source of suprathemal O^+ near the dayside polar cap boundary, *J. Geophys. Res.*, *90*, 4099-4116, 1985.

Lyons, R.L., and R.M. Thorne, Parasitic pitch angle diffusion of radiation belt particles by ion cyclotron waves, *J. Geophys. Res.*, *77*, 5608, 1972.

Noel, S., and G.W. Prolss, A Monte Carlo model of the ring current decay, *Adv. Space Res.*, *20* (3), 335-338, 1997.

Olsen, R.C., S.D. Shawhan, D.L. Gallagher, J.L. Green, C.R. Chappell, and R.R. Anderson, Plasma observations at the Earth's magnetic equator, *J. Geophys. Res.*, *92*, 2385, 1987.

Orsini, S., E. Amata, M. Candidi, H. Balsiger, M. Stokholm, C. Huang, W. Lennartsson, and P.A. Lindqvist, Cold streams of ionospheric oxygen in the plasma sheet during the CDAW 6 event of March 22, 1979, *J. Geophys. Res.*, *90*, 4091-4098, 1985.

Papadopoulos, K., H.B. Zhou, and A.S. Sharma, The role of helicons in magnetospheric and ionospheric physics, *Comments Plasma Phys. Controlled Fusion*, *15*, 321, 1994.

Peterson, W. K., R.D. Sharp, E.G. Shelley, R.G. Johnson, and H. Balsiger, Energetic ion composition in the plasma sheet, *J. Geophys. Res.*, *86*, 761-767, 1981.

Prigancova, A., and Y.I. Feldstein, Magnetospheric storm dynamics in terms of energy output rate, *Planet. Space Sci.*, *40*, 581, 1992.

Roeder, J.L., J.F. Fennell, M.W. Chen, M. Schulz, M. Grande, and S. Livi, CRRES observations of the composition of the ring current ion population, *Adv. Space, Res.*, *17*, 17, 1996.

Russell, C.T., Noise in the geomagnetic tail, *Planet. Space Sci.*, *20*, 1541-1553, 1972.

Schindler, K., A theory of the substorm mechanism, *J. Geophys. Res.*, *70*, 2803-2810, 1974.

Sharp, R. D., D.L. Carr, W.K. Peterson, and E.G. Shelley, Ion streams in the magnetotail, *J. Geophys. Res.*, *86*, 463-4648, 1981.

Shelley, E. G., W.K. Peterson, A.G. Ghielmetti, and J. Geiss, The polar ionosphere as a source of energetic magnetospheric plasma, *Geophys. Res. Lett.*, *9*, 941-944, 1982.

Singh, S.V., A.P. Kakad, R.V. Reddy, and G.S. Lakhina, Low-frequency instabilities due to energetic oxygen ions in the ring current region, *J. Plasma Phys.*, submitted, 2003.

Smith, P.H., and N.K. Bewtra, Charge exchange lifetimes for ring current ions, *Space Sci. Revs.*, *22*, 301, 1978.

Thorne, R. M., and R.B. Horne, Energy transfer between ring current H^+ and O^+ by electromagnetic ion cyclotron waves, *J. Geophys. Res.*, *99*, 17275, 1994.

Thorne, R. M., and R.B. Horne, Modulation of electromagnetic ion cyclotron Instability due to interaction with ring current O^+ during magnetic Storms, *J. Geophys. Res.*, *102*, 14, 155-14, 163, 1997.

Tsurutani, B. T., I.G. Richardson, R.M. Thorne, W. Butler, E.J. Smith, S.W.H. Cowley, S.P. Gary, S.-I. Akasofu, and R.D. Zwickl, Observations of the right-hand resonant ion beam instability in the distant plasma sheet boundary layer, *J. Geophys. Res.*, *90*, 12, 159-12, 172, 1985.

Tsurutani, B. T., M.E. Burton, E.J. Smith, and D.E. Jones, Statistical properties of magnetic field fluctuations in the distant plasma sheet, *Planet. Space Sci.*, *35*, 289-293, 1987.

Tsurutani, B. T., G.S. Lakhina, C.M. Ho, J.K. Arballo, C. Galvan, A. Boonsiriseth, J.S. Pickett, D.A. Gurnett, W.K. Peterson, and R.M. Thorne, Broadband palsma waves observed in the polar cap boundary layer: Polar, *J. Geophys. Res.*, *103*, 17, 351-17, 366, 1998.

Verheest, F., and G.S. Lakhina, Nonresonant low-frequency instabilities in multi—beam Plasmas: applications to cometary and plasma sheet boundary layers, *J. Geophys. Res.*, *96*, 7905-7810, 1991.

Wilken, B., Q.C. Zong, I.A. Daglis, T. Doke, S. Livi, K. Maezawa, Z.Y. Pu, S. Ullaland, and T. Yamamoto, Tailward flowing energetic oxygen ion bursts assiciated with multiple flux ropes in the distant magnetotail: GEOTAIL observations, *Geophys. Res. Lett.*, *22*, 3267-3270, 1995.

Yoon, P. H., C.S. Wu, and A.S. de Assis, Effect of finite ion gyroradius on the fire-hose instability in a high beta plasma, *Phys. Fluids*, *B 5*, 1971–1979, 1993.

Zhou, H. B., K. Papadopoulos, and A.S. Sharma, Electron magnetohydrodynamic response of a plasma to external current pulse, *Phys. Plasmas*, *3*, 1484, 1996.

G.S. Lakhina and S.V. Singh, Indian Institute of Geomagnetism, Dr. Nanabhai Moos Marg, Colaba, Mumbai—400005, India, e-mail: lakhina@iig.iigm.res.in; satyavir@iig.iigm.res.in

The Relationship of Storms and Substorms Determined from Mid-latitude Ground-based Magnetic Maps

C. Robert Clauer, Michael W. Liemohn, Janet U. Kozyra, Michelle L. Reno

University of Michigan, Ann Arbor, Michigan

Using data from a worldwide chain of mid- and low-latitude magnetic observatories, we construct maps of the magnetic disturbance field that display the temporal and spatial development of the storm time ring current, partial ring current and the substorm current wedge. While there is significant variability between storms, in general, the ring current produces an asymmetric depression in the northward magnetic field that becomes symmetric in the late recovery phase. The initial depression, attributed to a partial ring current, is largest on the afternoon and night side. Magnetic substorms generally produce an enhancement in the northward magnetic field within a limited longitudinal region corresponding to the location of the substorm current wedge. From the mid-latitude maps of the magnetic disturbance we can measure the magnitude, location, and extent of the substorm expansion phase disturbance. We find that large substorm expansions are generally associated with an enhancement of the dusk-side partial ring current, but they appear superposed upon an independently developing ring current. This is consistent with recent model simulations using the Michigan Ring current Atmosphere interaction Model (RAM). Growth of the ring current appears more related to the enhancement of the cross-magnetospheric electric field than to individual substorm expansions.

INTRODUCTION

A magnetic storm has been defined as an intensification of the ring current above some key threshold in response to enhanced and sustained solar wind energy input that results from a long interval of southward interplanetary magnetic field [cf. *Gonzales et al.*, 1994]. Such intervals of geoeffective southward IMF may be embedded in a variety of solar wind structures which are oftentimes preceded by high solar wind dynamic pressure at the leading edge. The arrival of the high dynamic pressure rapidly compresses the magnetosphere, producing an enhancement in the field measured by low latitude stations at the Earth's surface. This is called a sudden impulse—or SI. If the pressure enhancement is followed by southward IMF thus producing a magnetic storm, it is called a sudden commencement or SC. The period of elevated D_{st} following a SC is called the initial phase of the magnetic storm. Not all storms have sudden commencements, however. Storms without a SC are called "gradual onset" storms. Whether SC storms are significantly different than gradual onset storms has not yet been determined.

The main phase of the storm is characterized by a dramatic decrease in the horizontal component of the magnetic field measured by mid- and low-latitude stations. This world-wide field decrease has been attributed to a ring of

Disturbances in Geospace: The Storm-Substorm Relationship
Geophysical Monograph 142
Copyright 2003 by the American Geophysical Union
10.1029/142GM13

current around the Earth with the dominant current carriers being high energy ions (mainly H+, O+, and He+) in the energy range 10 keV/q to 300 keV/q that are convected into the inner magnetosphere during periods of enhanced crosstail electric field. Gradient and curvature drift produce differential motion between the ions and electrons.

The commonly used Dessler-Parker-Sckopke (DPS) formula [*Dessler and Parker*, 1959; *Sckopke*, 1966], is a linear relationship between ring current kinetic energy and the resulting magnetic field perturbation. The DPS relation is an analytical solution of the Biot-Savart law, given by:

$$\frac{\Delta B}{B_s} = \frac{-2K}{3U_M}$$

where ΔB is calculated at the center of the Earth, U_M is the total energy in the dipole magnetic field outside the Earth's surface and K is the total kinetic energy of the ring current. After inserting all the proper constants, the expression becomes:

$$\Delta B(nT) = -3.98 \cdot 10^{-30} K(keV)$$

This relationship assumes a steady state and that the magnetic field generated by the ring current is small compared to the background field. It also only accounts for the perturbation from the currents flowing across field lines in the magnetosphere. That is, the field-aligned and ionospheric closure currents are neglected. However, large local time asymmetries during the main phase and early recovery phase are now widely documented both in observations [*Roelof*, 1987; *Hamilton et al.*, 1988; *Grafe*, 1999; *Greenspan and Hamilton*, 2000] and in models [*Takahashi et al.*, 1990; *Kozyra et al.*, 1998; 2001; *Liemohn et al.*, 1999; 2001a,b].

The D_{st} index is defined to be the axially symmetric component of the horizontal disturbance field at any given UT measured by 4 midlatitude stations widely spaced in local time [*Sigura and Kamei*, 1991]. The D^*_{st} index results when the effects of the magnetopause currents, induced ground currents and the quiet time ring current are removed. D^*_{st} is therefore thought to be a measure of the disturbed ring current. Since the defining characteristic of magnetic storms is the development of the ring current, the D_{st} or D^*_{st} indices have been adopted as a measure of magnetic storm severity. Using the DPS relationship, the D^*_{st} has also taken on a physical interpretation as a measure of the kinetic energy in the ring current particle population.

According to *Burton et al.*, [1975], the D_{st} index can be corrected for the magnetopause currents and the effects of the quiet time magnetopause and ring current systems. The corrected index is written as:

$$D^*_{st} = D_{st} - b\sqrt{P_{dyn}} + c$$

The second term on the right side gives the magnetopause current correction where P_{dyn} is the dynamic pressure of the solar wind and b is empirically determined to be

$$b = 0.2nT / (eVcm^{-3})^{0.5}.$$

The c gives the effects of the quiet time ring current and magnetopause currents and is empirically determined to be

$$c = 20nT.$$

Dessler and Parker, [1959] note that the effects of the diamagnetic Earth require that induced shielding currents will be produced by any externally impressed field. For an impressed uniform field ΔB they show the change which would be measured at the Earth's surface due to shielding currents would be $3/2\Delta B \cos \Lambda$ where Λ is the magnetic latitude. Thus, an additional correction should be applied giving

$$D^*_{st} = \frac{D_{st} - b\sqrt{P_{dyn}} + c}{1.5 \cos \Lambda}.$$

D^*_{st} has become an important index because it potentially measures the behavior of the global ring current system during magnetic storms. However, its physical interpretation relies critically upon assumptions and knowledge regarding the current systems which combine to produce the low latitude axial disturbance field. In particular, it has been thought that a symmetric ring current makes the dominant contribution to D^*_{st} during magnetic storm conditions. Recent results, however, indicate that the partial ring current is not only the characteristic feature of the storm main phase, but actually exceeds the symmetric ring current contribution for the entire main phase and the early recovery phase for many storms [*Liemohn et al.*, 2001a].

To better capture both the temporal and spatial development of the ring current during storms, we have extended the analysis procedure developed by *Clauer and McPherron* [1974; 1980] to the investigation of several large geomagnetic storms which have been chosen by the Geospace Environment Modeling Workshop for community investigation. A description of the GEM storms is given at the GEM Website http://www-ssc.igpp.ucla.edu/gem/

Welcome.html. The analysis which we utilize here is based upon the use of data from a world wide longitudinal chain of mid-latitude magnetic stations.

Mid-latitude magnetic data have a long history of use in the study of storms and substorms. For example, *Kamide and Fukushima* [1971] used the H variation at mid-latitudes to calculate the parameters of simple line current models of the symmetric and partial ring current. *Clauer and McPherron* [1980] examined mid-latitude measurements to investigate the association of partial ring current development with isolated substorms, finding that the partial ring current may begin to develop prior to the onset of the substorm expansion, but a southward turing of the interplanetary magnetic field consistently precedes the onset of partial ring current development.

Figure 1 from *Clauer and McPherron* [1974] shows a simple schematic model of a field-aligned current wedge and the results of a Biot-Savart integration to obtain the magnetic effects of the current wedge at the surface of the Earth at 30° magnetic latitude. The bottom left panel of the figure shows the local time profiles of the disturbance field X (dashed line) and Y component (solid line).

In the work to be discussed here, we find that examinations of the full temporal and spatial disturbance field during magnetic storms raises concerns about all of the assumptions described above regarding the D_{st} index, ring current energy density, and the conventional view about the relationship between storms and substorms. In the conventional view, originally expressed when Chapman coined the term magnetic storm, the ring current results from the accumulation of many successive elementary disturbances called substorms [*Chapman and Bartels*, 1940]. Each substorm produces an enhanced westward electric field near the outer ring current boundary. This enhanced induction electric field brings particles in from the plasma sheet—a process called the substorm injection. The injected particles become trapped on closed azimuthal drift paths. Since the trapped ions and electrons drift in opposite directions due to their gradient and curvature drifts, a westward azimuthal ring current is produced. Another substorm effect is the apparent enhancement of the plasma sheet population with ionospheric particles which then also drift and contribute to the ring current.

A new view, however, asserts that ring current development results from a sustained enhancement of the convection electric field. This view has been discussed, in parts, by many authors including *Clauer and McPherron*, [1980]; *Liemohn et al.*, [1999; 2001a; 2001b] and references therein. In this view, most of the ring current magnetic perturbations are due to a partial ring current which closes, in part, through the ionosphere, and in part, through the magnetopause. The effect of the enhanced cross-magnetospheric electric field is to move the Alfvén layers inward with the consequence of further energizing plasma and also moving the ring current closer to the Earth. Only after the enhanced field is reduced do particles find themselves on closed azimuthal drift paths and the ring current becomes symmetric. This occurs during the decay phase. The 2-phase decay is explained now with the rapid initial decay due to plasma on open drift paths convecting out of the system and the slow decay due to charge exchange within the trapped population.

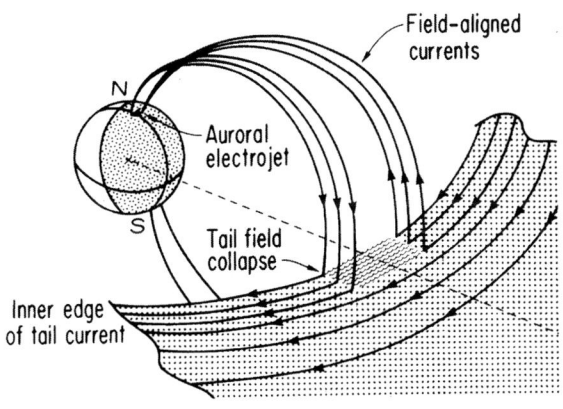

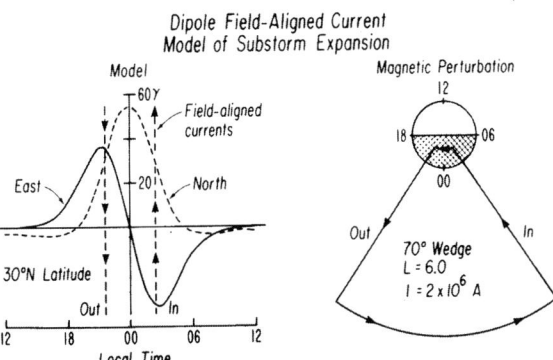

Figure 1. Schematic diagram of the substorm current wedge and resulting magnetic disturbance measured on the Earth's surface at 30° magnetic latitude. The lower left corner shows the local time profile of the disturbance field for the northward (dashed line) and eastward (solid line) components.

METHODOLOGY

Midlatitude magnetic variations have been used by many investigators as tools for the study of magnetospheric substorms and partial ring current formation. For example, the onset of the midlatitude positive H bay near midnight has been used as a signature of the beginning of a substorm expansion phase [*Iijima and Nagata*, 1972; *Hones et al.*, 1971; *Mende et al.*, 1972; *McPherron*, 1973; *Nishida*, 1973; *Clauer and McPherron*, 1974; 1980]. The sign of the east–west component associated with the midlatitude perturbation was used by *Meng and Akasofu* [1969] to infer whether field-aligned currents flow into or out of the ionosphere at the longitude of a given station. *McPherron*, [1973] shows that auroral zone magnetic activity may either coincide with or precede midlatitude activity. However, he finds that onset times based on midlatitude magnetic positive bays in H appear to order magnetospheric observations better than auroral onsets. Of relevance to this study, midlatitude negative H bays in the afternoon and evening sector have been used to make inferences about the partial ring current [*Kamide and Fukushima*, 1971; *Clauer and McPherron*, 1980].

The method of midlatitude magnetic field analysis utilized here follows that developed by *Clauer and McPherron*, [1974; 1980]. The availability of digital data from an extensive chain of midlatitude stations makes possible a computerized analysis of such measurements at high time resolution. Table 1 provides the names and locations in geographic and corrected geomagnetic (cgm) coordinates of about 25 stations for which high resolution digital data is available during the years 1991—present. Geomagnetic coordinates in Table 1 are calculated in accordance with *Gustafsson et al.* [1992]. Note that by assuming the magnetic equator to be a plane of symmetry, we may use

Table 1. Magnetic Observatories.

Station Name	Station Code	Geographic E. Longitude	Geographic Latitude	CGM Longitude	CGM Latitude
Karachi	KRC	67.1	25.0	139.2	19.0
Alibag	ABG	72.9	18.6	144.8	11.7
Learmonth	LRM	114.1	−22.2	185.2	−33.1
Gnangara	GNA	116.0	−31.8	186.2	−44.4
Beijing	BJI	116.2	40.0	188.3	33.9
Lunping	LNP	121.2	25.0	192.5	17.8
Kanoya	KNY	130.9	31.4	202.6	24.3
Kakioka	KAK	140.2	36.2	211.2	28.9
Mizusawa	MIZ	141.0	39.0	212.3	31.8
Chichijima	CBI	142.2	27.1	212.8	19.4
Memambetsu	MMB	144.2	43.9	215.2	36.8
Guam	GUA	144.9	13.6	215.3	5.6
Canberra	CAN	149.4	−35.3	226.5	−45.7
Eyrewell	EYR	172.4	−43.4	256.1	−50.2
Honolulu	HON	202.0	21.3	269.2	21.8
Pamatai	PPT	210.4	−17.6	284.7	−16.3
Fresno	FRN	240.3	37.1	303.2	43.2
Tucson	TUC	249.2	32.2	313.8	39.9
Boulder	BOU	254.8	40.1	318.7	49.2
Del Rio	DLR	259.1	29.5	325.9	39.0
Bay St. Louis	BSL	270.6	30.4	339.7	41.6
Fredericksburg	FRD	282.6	38.2	356.7	49.8
San Juan	SJG	293.8	18.1	9.3	29.1
Vassouras	VSS	316.4	−22.4	22.8	−16.2
San Fernando	SFS	353.8	36.5	70.3	28.4
Tamanrasset	TAM	5.5	22.8	78.3	5.7
Hermanus	HER	19.2	−34.4	81.8	−42.3
L'Aquila	AQU	13.3	42.4	87.5	36.2
Hartebeesthoek	HBK	27.7	−25.9	95.4	−34.6
Misallat	MLT	30.9	29.5	102.2	21.5
Tananarive	TAN	47.6	218.9	116.4	−29.1
Amsterdam Is.	AMS	77.6	−37.8	138.0	−49.0

stations in the southern hemisphere provided that we change the sign of the east component (Y). Not all stations have complete digital data available, so for any given event it is likely that a subset of these stations will be utilized. Figure 2 shows the location of these stations in a graphical format.

We utilize the magnetic data in dipole geomagnetic coordinates, with X north and Y east and Z vertical down. For ring current investigations, we wish to obtain the axial magnetic field perturbation. To obtain the axial perturbation field, we divide the X component at each station by cos Λ where Λ is the geomagnetic latitude of the station. This is done for both disturbed and quiet days. The quiet day is subtracted from the data at each station to remove the main field and diurnal variation. The difference between the disturbed and quiet X components (normalized by station latitude) then gives a profile of the disturbance field as a function of local time around the world. This profile can be fit using a variety of techniques. Profiles at successive times can then be plotted in a map as a function of universal time and local time to display the temporal and spatial variation of the midlatitude disturbance field due to substorms and storms. Similarly, the difference in the Y component can be investigated for the disturbance variations in the east–west direction. We do not apply a latitude normalization to the Y perturbations, because their source is primarily from field-aligned currents. Therefore the Y perturbation maps exhibit more structure, depending on the relative latitudes of the magnetometer stations to the field-aligned currents.

Figure 3 shows a stack plot of magnetograms for September 24–25, 1998 one of the storms selected for community investigation. We use these data here to illustrate the analysis. In this stack plot, the quiet day variation has been subtracted from each of the magnetometer traces and divided by cos Λ to obtain the axial component of the field.

The next step in the analysis is to create local time profiles. For each 1-minute universal time step, we fit a local time profile of the disturbance field. Figure 4 is an example of the profile obtained at 0410 UT on September 25, 1998 showing the large asymmetric depression of the field during the storm main phase.

The collection of profiles are contoured to produce the Universal Time—Local Time maps of the disturbance field and this is shown in the top panel of Plate 1. Note that these values have not been corrected for magnetopause currents or induced currents within the Earth. A local time average of these values is therefore comparable to D_{st} rather than D^*_{st}. The bottom panel shows the results of the same process

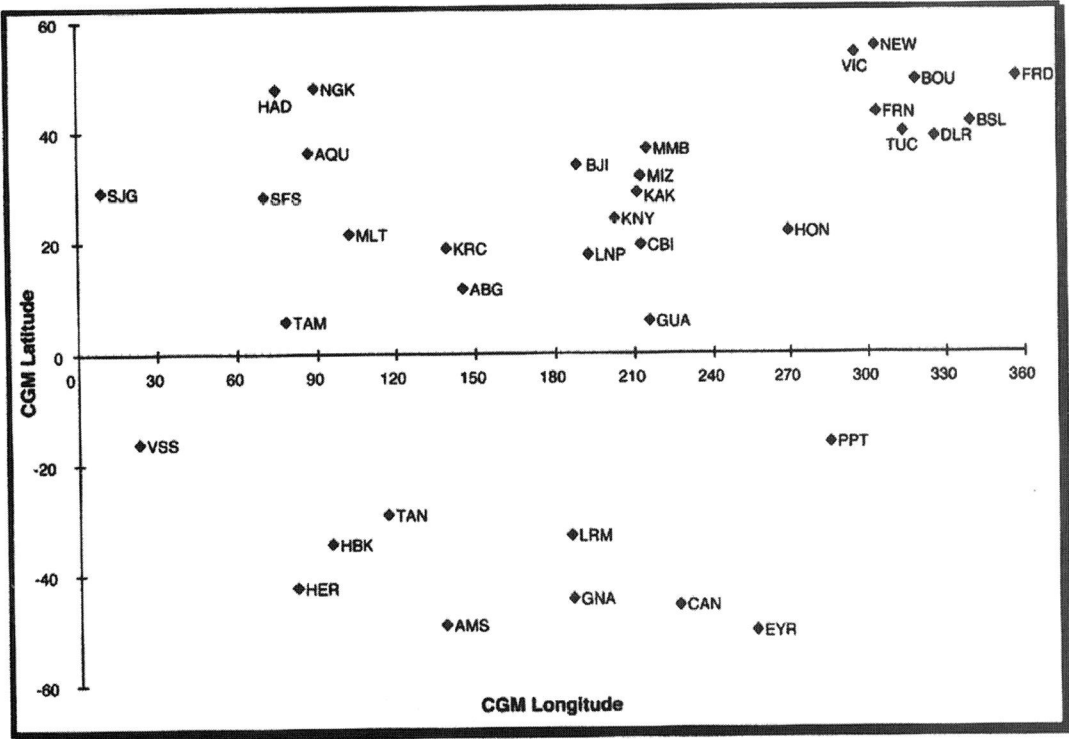

Figure 2. Locations of low- and midlatitude geomagnetic observatories in corrected geomagnetic coordinates.

148 TEMPORAL AND SPATIAL DEVELOPMENT OF THE RING CURRENT

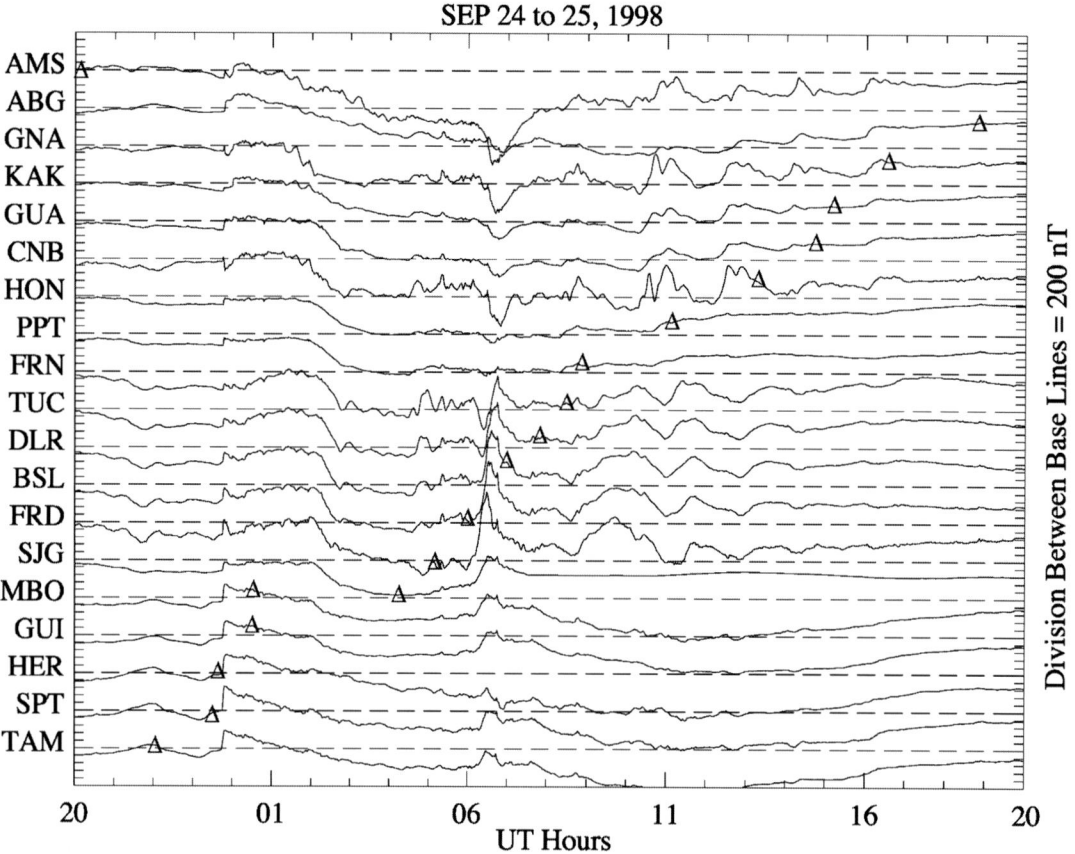

Figure 3. Stack plot of mid latitude magnetograms for the storm day September 24–25, 1998. The dashed base lines are spaced at 200 nT separations. Δ symbols on the traces mark local magnetic midnight at each station.

applied to the Eastward component of the disturbance field. These plots will be discussed in greater detail in the following sections.

DATA ANALYSIS

Figure 5 shows the interplanetary magnetic field (IMF) and solar wind plasma density, velocity and dynamic pressure for the September 24–25 storm shown in Plate 1 and Figure 3. The magnetospheric response is indicated in the bottom two traces which are the Kyoto quick-look provisional AE index and a 1-minute D_{st} index computed from the stations plotted in Figure 3, respectively. At 23:45 UT on September 24, 1998, an interplanetary shock and simultaneous northward turning of the IMF hit the magnetosphere. The time delay between the satellite and magnetosphere is about 15 minutes. The IMF subsequently turned southward at about 01:25 UT and the geomagnetic storm main phase developed thereafter.

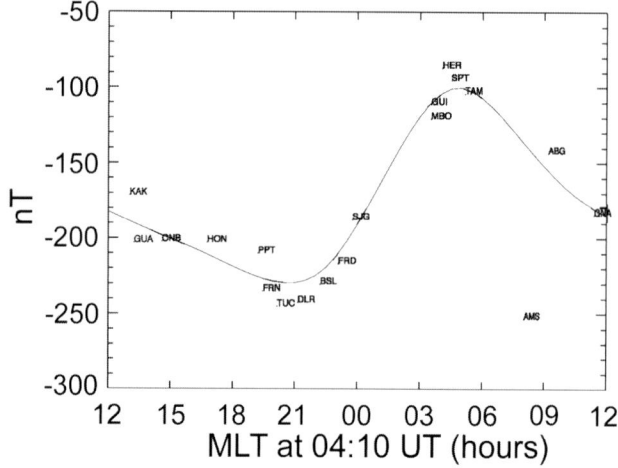

Figure 4. Local time profile of the low latitude disturbance field at 0410 UT on Sept. 25, 1998. Station name codes show the measured disturbance at each of the stations used in the analysis.

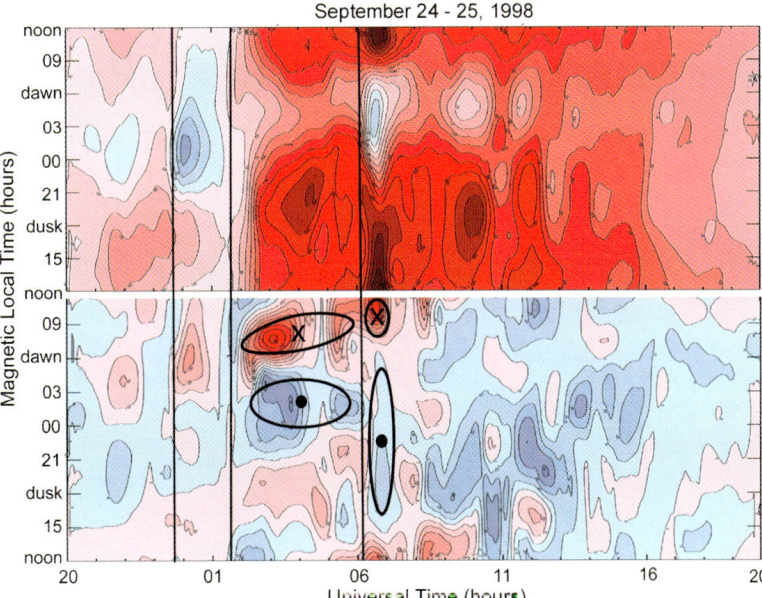

Plate 1. UT-LT map of the magnetic disturbance field on 20 UT September 24, 1998 to 20 UT September 25, 1998. The top panel shows the disturbance in the component direction parallel to the dipole axis. Contours are every 25 nT, with positive values shaded blue and negative values in a shaded red. The bottom panel shows the disturbance in the magnetic eastward component. Contours are every 10 nT with positive values shaded blue and negative values shaded red. The positive values are generally associated with out-ward field and negative values with inward field aligned current.

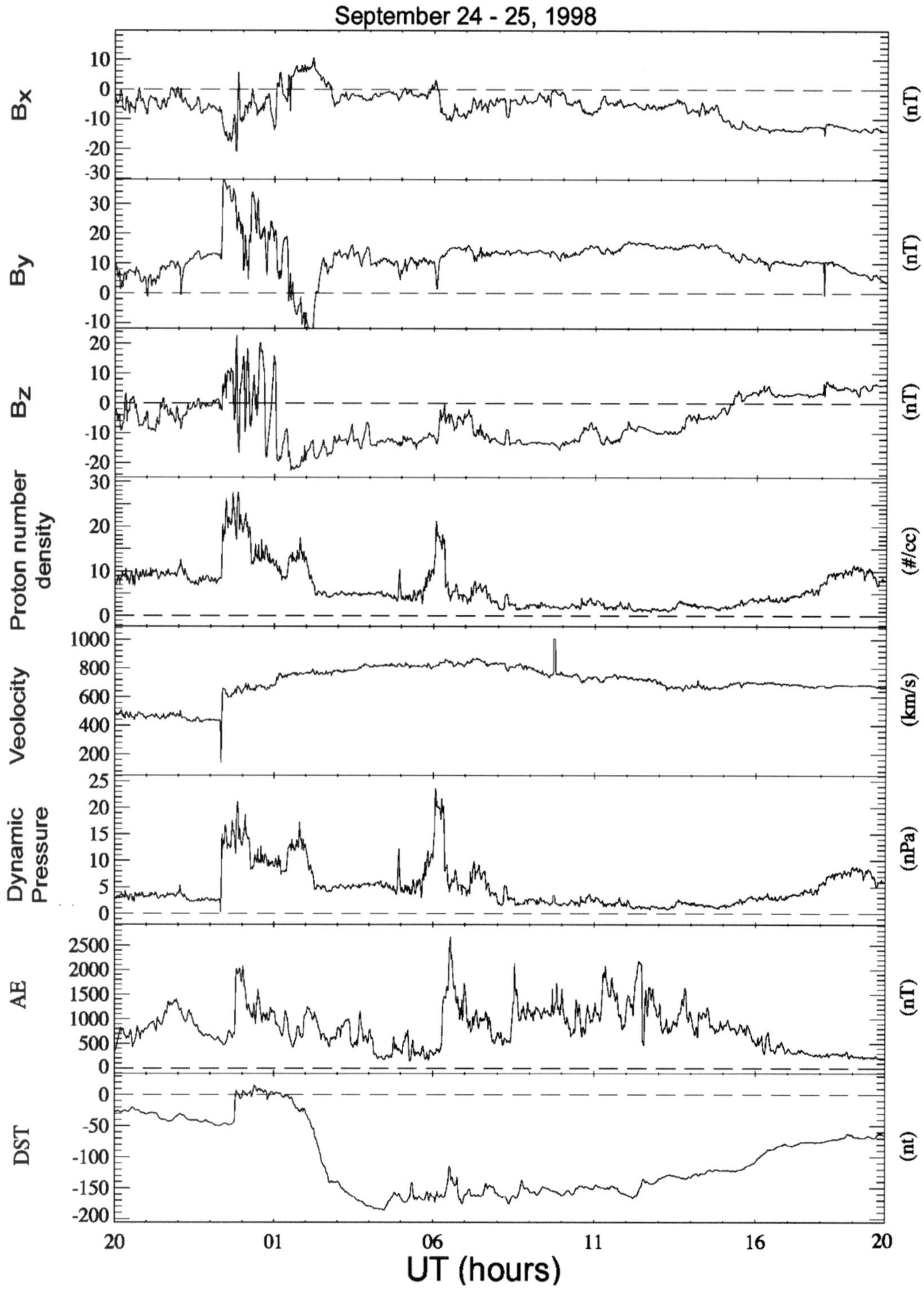

Figure 5. Interplanetary magnetic field and solar wind plasma data for September 24–25, 1998 measured by the IMP-8 satellite. From the top are the IMF B_x, B_y, and B_z components in GSM coordinates. The third trace is the solar wind proton number density followed by the velocity and dynamic pressure. The bottom two traces show the magnetospheric response as seen in the AE and Dst indices (here, Dst is the LT average from the upper map in Plate 1).

The mid-latitude ground response to the shock is an unusual signature in which the magnetic field is enhanced on the night side but shows little enhancement, or even a small depression, on the dayside. This is shown quite clearly in the stack plot shown in Figure 3 and in the UT-LT map shown in the top panel of Plate 1. The vertical line at 23:45 UT in Plate 1 marks the initiation of the shock response seen in the ground magnetic data. This magnetic response is quite different from that of most shock compressions of the magnetosphere which show a world wide sudden enhancement in the magnetic field. At present, this unusual response is thought to be the result of a transition current system which develops in response to the northward turning of the IMF. This is discussed further along with an analysis of this shock in *Clauer et al.*, [2001]. We will not discuss this further here since the primary focus of this report is on the development and decay of the storm main phase.

The magnetic field response following the southward turning of the IMF is a worldwide decrease of the field which is nonuniform in local time. The vertical bar at 01:40 UT on Plate 1 marks the beginning of the response measured in the ground magnetic data. The greatest depression is centered near 20 MLT and reaches a peak depression around 04:30 UT. This development in the magnetic field depression can be seen in the top panel of Plate 1. The nonuniform development of the magnetic field decrease during the initial part of a magnetic storm is generally attributed to the development of a partial ring current. In this case, we can see that this is the primary feature during the entire development of the main phase from 01:25–04:30 UT. The bottom panel of Plate 1 shows the magnetic disturbance measured in the Y or eastward component of the field. Perturbations in the eastward component at low latitudes are generally attributed to field aligned currents at slightly higher latitudes, with positive perturbations indicating an outward field aligned current and negative perturbations indicating an inward field aligned current. During the interval 01:40–06:00 UT we have marked on the bottom panel the indication of an inward field aligned current centered near 08 MLT and an outward field aligned current centered near 02 MLT. These would form the field aligned segments which connect the partial ring current to the ionosphere.

As shown by *Clauer and McPherron*, [1974], substorms can be identified from the UT-LT maps. They produce a positive X perturbation at the LT of the current wedge, flanked on either side by positive and negative Y perturbations (see Figure 1). It is clear that during the interval from 01:40–06:00 UT there is no midlatitude positive bay which would mark the signature of a magnetospheric substorm expansion. Thus, we can infer from the worldwide ground-based magnetic observations at mid-latitudes that the partial ring current develops in response to the southward turning of the IMF without the indication of any substantial substorm expansions. At 06:05 there is a sudden change in IMF B_z and solar wind dynamic pressure which can be associated with the formation of a midlatitude positive bay signature centered near 02 MLT. There is also a simultaneous and distinct signature in the auroral zone reflected in the AE index as well. This is the signature of a large substorm expansion, and associated with the expansion is an enhancement of the partial ring current. In this case the enhancement and decay of the partial ring current appears to occur on the same time scale as the substorm expansion and recovery. The substorm associated partial ring current enhancement and decay also seem to be superposed on the independently developing partial ring current system which was initiated at 01:25 UT.

After the 06:05 UT substorm, the asymmetric ring current signature in the X perturbation map persists until the IMF turns northward (at roughly 15:00 UT). Then there is a distinct morphological change to a more symmetric LT perturbation. This is the change from partial ring current dominated to symmetric ring current dominated perturbations.

MODEL SIMULATION RESULTS

We are able to utilize the results of the Michigan version of the Ring current—Atmosphere interaction Model (RAM) to compute magnetospheric and field-aligned current distributions in near-Earth space. From this knowledge, we can calculate the magnetic field produced by the various ionospheric and magnetospheric current systems. For instance, the energetic ion distribution in the inner magnetosphere can be integrated to obtain the bulk pressures, which in turn can be used to calculate the currents generated from these particles (that is, the symmetric and partial ring currents) [*Parker and Stewart*, 1967; *Akasofu and Chapman*, 1961; *Lui et al.*, 1987; *Takahashi et al.*, 1991; *Ebihara and Ejiri*, 1998; *Liemohn et al.*, 2001a]. A Biot-Savart law integration of these currents yields a magnetic field perturbation at some location of interest. A simplified analytical solution for the perturbation at the Earth's center was derived by *Dessler and Parker*, [1959], in order to relate the magnetic perturbation to the total ring current energy. *Sckopke*, [1966] [*Sckopke* (1966)] then showed that this result is valid for an arbitrary ring current distribution. However, as discussed earlier, the validity of the DPS relation has been called into question because of the assumptions used to obtain the analytical solution [c.f. *Carovillano and Siscoe*, 1973]. In particular, any magnetic perturbations due to field-aligned currents associated with the partial ring current are not accounted for in the DPS relationship.

Lately, *Takahashi et al.*, [1991] and then *Ebihara and Ejiri*, [1998, 2000] have used the Biot-Savart law integration technique on results from their respective particle drift models. They calculated the distribution of the energetic ions everywhere in near-Earth space, and then obtained the field perturbation from this result. These calculations are a breakthrough, showing that large-scale theoretical model results can be used to obtain the field perturbation for comparison with the primary features observed during magnetic storms. However, *Takahashi et al.*, [1991] obtained unrealistically large ionospheric closure current contributions to the perturbation, while *Ebihara and Ejiri*, [1998, 2000] only performed the calculation for $J_\perp$ and not the field-aligned and ionospheric components of the current system. In addition, neither compared directly with ground based perturbation observations. Therefore, there is still much work to be done in this field.

A similar computational technique has been developed for use with our kinetic ring current transport code, the Michigan version of RAM. Descriptions of RAM are given elsewhere [*Jordanova et al.*, 1994; 1996; 1997; *Liemohn et al.*, 1998; 1999; 2001b], and so only a brief description of the code and numerical technique are presented here. The program solves the time-dependent, gyration- and bounce-averaged kinetic equation for the phase-space distribution function $f(t, R, \varphi, E, \mu_0)$ of a chosen ring current species. The five independent variables are, in order, time, geocentric distance in the equatorial plane, magnetic local time, kinetic energy, and cosine of the equatorial pitch angle. The code includes collisionless drifts, energy loss and pitch angle scattering due to Coulomb collisions with the thermal plasma, charge exchange loss with th hydrogen geocorona, and precipitative loss to the upper atmosphere. Solution of the kinetic equation is accomplished by replacing the derivatives with second-order accurate, finite volume, numerical operators. The source term for the distribution function is the outer nightside simulation boundary, where observed particle fluxes from geosynchronous orbiting satellites (such as those maintained by Los Alamos National Laboratory) are applied as input functions. Because the MPA [*McComas et al.*, 1993] and SOPA [*Belian et al.*, 1992] instruments on the LANL satellites do not resolve ion mass, the composition of the incoming particles is determined from the statistical relationships derived by *Young et al.*, [1982] from previous geosynchronous orbit measurements.

The transport model also requires a specification of the magnetospheric electric and magnetic field. The magnetospheric convection electric field used in the work reported here is Volland-Stern field [*Volland*, 1973; *Stern*, 1995] which has a potential of

$$\Phi_{VS} = AR^2 \sin\phi.$$

A is the activity-dependent strength coefficient given by

$$A = \frac{0.0449}{\left(1. - 0.159K_p + 0.0093K_p^2\right)^3} \left[\frac{kV}{R_E^2}\right]$$

where a shielding factor of 2 has been assumed [*Maynard and Chen*, 1975]. This gives a purely dawn-dusk electric field at large R, but a weaker field near the Earth. Our calculations use a dipole magnetic field. The use of this Kp-dependent convection electric field does not directly demonstrate the connection between southward IMF and partial ring current formation. However, Kp responds to the increased activity generated by a southward IMF, and this field description has been shown to be an adequate analytical representation of electric field strength in the inner magnetosphere [e.g. *Korth et al.*, 1998].

We have developed a method of calculating the field perturbation from RAM simulation results at an arbitrary location of a "virtual magnetometer." It solves the Biot-Savart law relating the magnetic perturbation at a certain location to the current density within a volume of space,

$$\Delta B(r) = \frac{\mu_0}{4\pi} \int_V \frac{J(r') \times (r - r')}{|r - r'|^3} dr'$$

where r is the location of the virtual magnetometer and r' is the location of the volume element within the total volume V where the current density **J** exists everywhere inside of the $L = 6.625$ dipole shell. This equation is numerically integrated over the simulation domain for azimuthal, radial, and field-aligned currents in the magnetosphere as well as for the closure current in the ionosphere (assumed to be at 120 km altitude). **J**(r') is found using the technique of *Liemohn et al.*, [2001a] which projects the bounce averaged phase space densities from RAM along each field line to obtain the local plasma pressure gradients. Our technique closes the currents primarily in the north-south direction with the region 1 field-aligned currents (FACs), rather than through an east-west electrojet between the two regions of partial ring current FACs. This yields high-latitude influences on the mid-latitude stations contributing to the asymmetry of the perturbations, but with a small average influence (that is, close to zero). The technique also includes the effects of the high-latitude closure currents. Thus, we can compute the

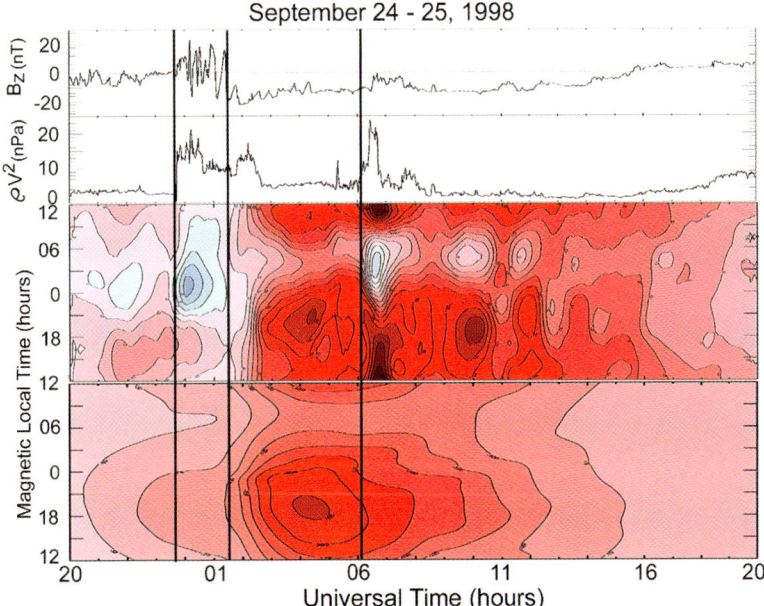

Plate 2. (top) Interplanetary magnetic field B_z GSM component and solar wind dynamic pressure measured by the IMP-8 satellite. (center) UT-LT map of the magnetic disturbance field at mid latitudes in the field component parallel to the dipole axis (the same as the top panel of Plate 1). (bottom) Simulated UT-LT map of the axial component of the low latitude magnetic disturbance field obtained from the Michigan RAM kinetic ring current model. The contour increment in both UT-LT maps is 25 nT.

mid-latitude magnetic perturbation field on the Earth's surface which is produced by the model for direct comparison with data presented in the form of the UT-LT map. To produce the map, we use 24 virtual stations located every hour in local time at 35° magnetic latitude.

Plate 2 shows a comparison of the RAM generated magnetic field disturbance with the measured disturbance during the September 24–25, 1998 storm. The top panel of the plate shows the upstream IMF B_z GSM component and dynamic pressure measurements. The middle panel shows the worldwide magnetic disturbance in the axial component measured at mid-latitude ground stations in the form of a UT-LT map. The bottom panel shows the simulated ground magnetic disturbance produced by the RAM in a similar display format.

There is considerable structure in the observed ground disturbance map which is not reflected in the simulated disturbance. This structure is the result of processes which are not accounted for in the RAM such as the magnetospheric reaction to dynamic pressure changes and to substorm expansions. What is shown fairly clearly is that the model response to the southward IMF marked by the vertical line at 01:25 UT is an intensification of the partial ring current which very closely matches the observations. The partial ring current was already developing prior to this due to the ramp up in Kp from a southward IMF before the shock arrival. The ring current development between 01 UT and 06 UT is entirely a dusk centered partial ring current which develops in response to enhanced convection (as inferred from the large southward IMF) during an interval without significant substorm expansions. This is the case both in the RAM simulation and in the observations.

In the RAM simulation, the partial ring current decays during the interval 06–16 UT while the observed partial ring current disturbance remains fairly strong until 12:30 UT and then the decay in the partial ring current is observed to progress slowly becoming symmetric sometime after 20 UT. Some of this discrepancy can be accounted for by the magnetopause currents and the induced currents in the Earth, which are present in the observed values but not in the simulation. Including these into the model results, for example, will deepen the perturbation everywhere (the influence of the induced currents) and will lessen the perturbation during times of high solar wind pressure (the influence of the magnetopause currents). In addition, the observations show geomagnetic activity after 06 UT which appear to help sustain the partial ring current. This activity has been investigated at geosynchronous orbit and has been referred to as sawtooth oscillations. The activity shows a stretching and relaxation of the geomagnetic field, but also seems distinctly different than substorm expansions in character. Thus, this is an unresolved issue which requires further investigation [*Borovsky*, private communication]. Since the RAM does not directly include substorm activity or sawtooth activity, these effects are missing in the simulation. The decay of the ring current observed in the simulation results from reduction in the source population measured at geostationary satellites, we believe. That these reductions in the source population observed at geostationary orbit are not reflected in the ring current strength measured by ground stations is an interesting issue which must be resolved in future studies.

Plate 3 shows results from the RAM at various times during the simulation. The top three rows shows the field-line averaged values of the plasma pressure, the azimuthal current density, and the radial current density, while the bottom row shows the ionospheric field-aligned current density. While this last row is calculated at 120 km altitude in the ionosphere, it is shown in the equatorial plane for similarity with the other values. Along the top of the plot we indicate the time during the storm for which these outputs are obtained. These plots show also that the pressure and azimuthal current were very asymmetric during the main phase and early recovery phase. Only in the late recovery phase do the azimuthal current and pressure become symmetric to form a symmetric ring current. Comparing Plates 2 and 3 shows that the modeled pressures and azimuthal currents are far more asymmetric than the magnetic perturbations. This is because a westward current in the near-Earth equatorial plane produces a negative perturbation all around the Earth at low and mid latitudes. the Biot-Savart law magnetic perturbation has a $1/R^2$ dependence on the distance between the current and the measurement point. An azimuthally localized westward current at 4.5 R_E geocentric distance (the peak of the partial ring current in this simulation) produces a negative perturbation that is 2.5 times stronger on the near-side of the planet than on the far side, but the perturbation is negative at both locations. The average perturbation from this current (that is, its contribution to D_{st}) would be 70% of the near-side perturbation. Even a completely asymmetric current system produces a D_{st} signature. In fact, the magnetopause and tail currents are also azimuthally asymmetric, but their contributions to the magnetic perturbation at Earth are often assumed to be symmetric because these currents are so far from the planet.

SUMMARY AND CONCLUSIONS

This new view, however, is not complete and several challenges remain to be solved. *Fok et al.*, [1996], for example,

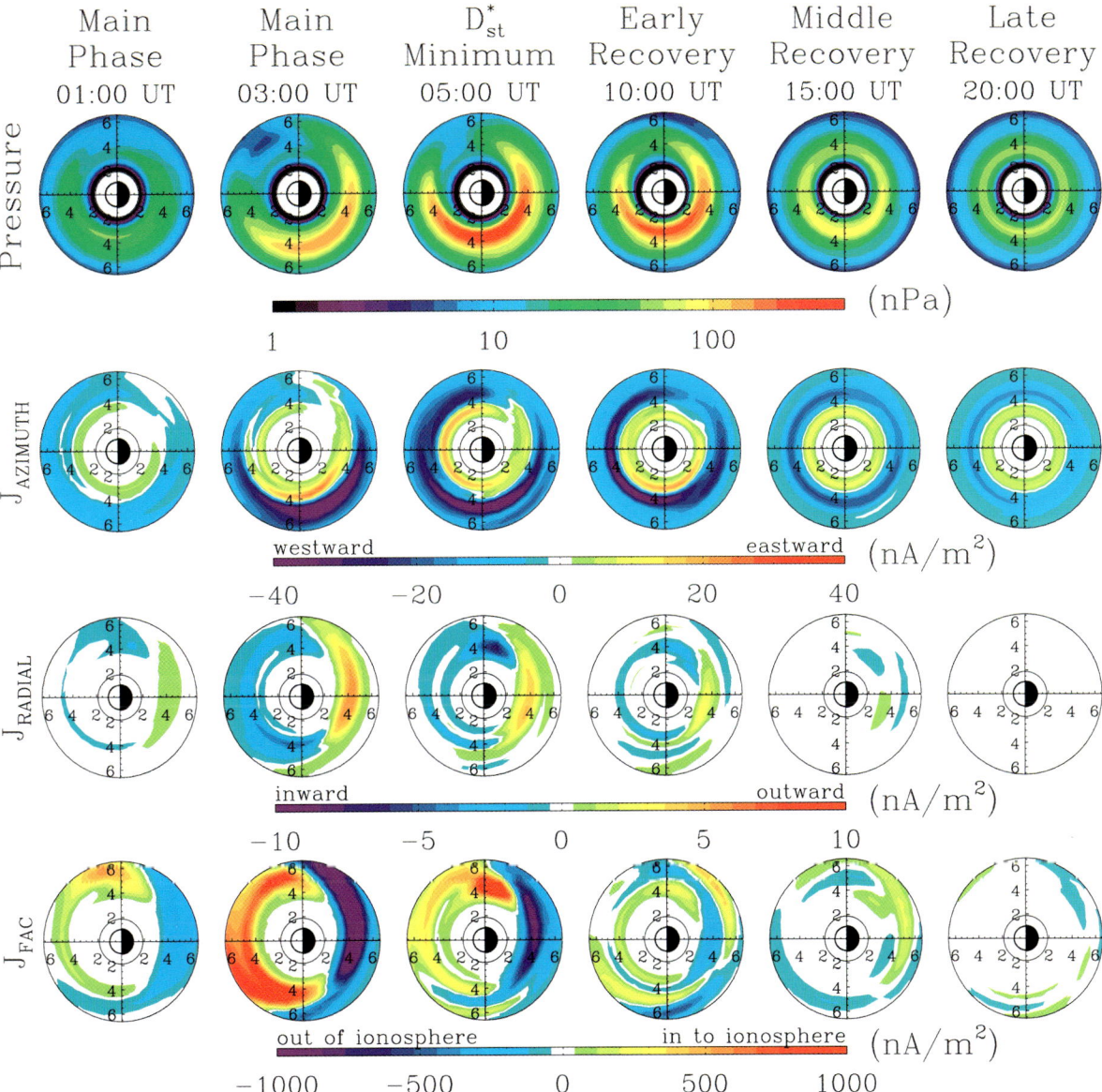

Plate 3. Dial plots of the simulation results for September 25, 1998. The 6 columns are at the times indicated across the top. The 4 rows are, from the top, field-line averaged ion pressure (H^+ and O^+), azimuthal current, radial current, and the ionospheric (120 km) value of the field-aligned current shown in the magnetospheric equatorial plane. Note that each row has its own colorscale, the first one logarithmic and the others are linear. In each plot, the view is looking down over the north pole, with noon to the left and distances given in R_E.

show through a numerical simulation that enhancement of the convection electric field alone can be effective in the buildup of the ring current for only about the first 3 hours. The continued presence of an enhanced cross-magnetospheric electric field does not further increase the ring current, since the strong convection removes as many particles from the magnetosphere through the dayside magnetopause as it adds from the nightside plasmasheet. Indeed, there are also observed periods of steady enhanced magnetospheric convection without substorm signatures which show little partial ring current development in contrast to similar IMF conditions which produce substorms together with substantial partial ring current development. It is clear that the plasma sheet density must be enhanced somehow in order for convection to drive a prolonged ring current buildup. Substorms are a clear choice for the mechanism that increases the density. However, the September 1998 storm is an example of a large ring current growth without prior or concurrent major substorm activity. In fact, the major substorm at 06:50 UT on September 25 might be responsible for depleting the plasma sheet density via a large plasmoid ejection.

A synthesis of new and old ideas may well provide the correct understanding for storm-time ring current development and decay. However, it will require the analysis of high quality global data sets combined with the use of global model simulations to achieve this understanding.

Acknowledgment. This work has been supported by the National Science Foundation through grants ATM9804941, ATM9980075, OPP9876473, ATM0090165 and NASA grants NAG5-10297 and NAG5-12176.

REFERENCES

Akasofu, S.-I., and S. Chapman, The ring current, geomagnetic disturbance and the Van Allen radiation belts, *J. Geophys. Res.,* 66, 1321, 1961.

Belian, R. D., G. R. Gisler, T. Cayton, and R. Christensen, High–Z energetic particles at geosynchronous orbit during the great solar proton event series of October 1989, *J. Geophys. Res., 97,* 16897, 1992.

Burton, R. K., R. L. McPherron, and C. T. Russell, An empirical relationship between interplanetary conditions and D_{st}, *J. Geophys. Res., 80,* 4204, 1975.

Carovillano, R. L., and G. L. Siscoe, Energy and momentum theorems in magnetospheric processes, *Rev. Geophys. and Space Phys., 11,* 289, 1973.

Chapman, S., and J. Bartels, *Geomagnetism,* Clarendon, Oxford, 1940.

Clauer, C. R., and R. L. McPherron, Mapping the local time – universal time development of magnetospheric substorms using mid-latitude magnetic observations, *J. Geophys. Res., 79,* 2811, 1974.

Clauer, C. R., and R. L. McPherron, The relative importance of the interplanetary electric field and magnetospheric substorms on partial ring current development, *J. Geophys. Res.,* 85 6747, 1980.

Clauer, C. R., I. I. Alexeev, E. S. Belenkaya, and J. B. Baker, Special features of the 24–27 September, 1998 storm during high solar wind dynamic pressure and northward IMF, *J. Geophys. Res., 106,* 25,695-25,711, 2001.

Dessler, A. J., and E. N. Parker, Hydromagnetic theory of geomagnetic storms, *J. Geophys. Res., 64,* 2239, 1959.

Ebihara, Y., and M. Ejiri, Modeling of solar wind control of the ring current buildup: A case study of the magnetic storms in April 1997, *Geophys. Res. Lett., 25,* 3751-3755, 1998.

Ebihara, Y., and M. Ejiri, Simulation study on fundamental properties of the storm-time ring current, *J. Geophys. Res., 105,* 15,843-15,859, 2000.

Fok, M.-C., T. E. Moore, and M. E. Greenspan, Ring current development during storm main phase, *J. Geophys. Res., 101,* 15311, 1996.

Gonzalez, W. D., J. A. Joselyn, Y. Kamide, H. W. Kroehl, G. Rostoker, B. T. Tsurutani, and V. M. Vasyliunas, What is a geomagnetic storm?, *J. Geophys. Res., 99,* 5771, 1994.

Grafe, A., Are our ideas about D_{st} correct?, *Annales Geophysicae,* 17, 1-10, 1999.

Greenspan, M. E., and D. C. Hamilton, A test of the Dessler-Parker-Sckopke relation during magnetic storms, *J. Geophys. Res., 105,* 5419, 2000.

Gustafsson, G., N. E. Papitashvili, and V. O. Papitashvili, A revised corrected geomagnetic corrdinate system for epoch 1985 and 1990, *J. Atmos. Terr. Physics, 54,* 1609, 1992.

Hamilton, D. C., G. Gloeckler, F. M. Ipavich, W. Studemann, B. Wilkey, and G. Kremser, Ring current development during the great geomagnetic storm of February 1986, *J. Geophys. Res., 93,* 14343-14355, 1988.

Hones, E. W., S. Singer, L. J. Lanzerotti, J. D. Pierson, and T. J. Rosenberg, Magnetospheric substorm of August 25–26, 1967, *J. Geophys. Res., 76,* 2977, 1971.

Iijima, T., and T. Nagata, Signature for substorm development of the growth phase and expansion phase, *Planet. Space Sci., 20,* 1095, 1972.

Jordanova, V. K., J. U. Kozyra, G. V. Khazanov, A. F. Nagy, C. E. Rasmussen, and M.-C. Fok, A bounce-averaged kinetic model of the ring current ion population, *Geophys. Res. Lett., 21,* 2785, 1994.

Jordanova, V. K., L. M. Kistler, J. U. Kozyra, G. V. Khazanov, and A. F. Nagy, Collisional losses of ring current ions, *J. Geophys. Res., 101,* 111, 1996.

Jordanova, V. K., J. U. Kozyra, A. F. Nagy, and G. V. Khazanov, Kinetic model of the ring current—atmosphere interactions, *J. Geophys. Res., 102,* 14279, 1997.

Kamide, Y., and N. Fukushima, Analysis of magnetic storms with DR-indices for equatorial ring current field, *Rep. Ionos. Space Res. Jpn., 25,* 125, 1971.

Korth, A., R. H. W. Friedel, C. Mouikis, and J. F. Fennell, Storm/Substorm signatures in the outer belt, in *Substorms-4,* edited by S. Kokubun, and Y. Kamide, p. 779, Terra Scientific Publishing Co/Kluwer Academic Publishers, 1998.

Kozyra, J. U., M. Fok, E. R. Sanchez, D. S. Evans, D. C. Hamilton, , and A. F. Nagy, The role of precipitation losses in producing the rapid early recovery phase of the great magnetic storm of February 1986, *J. Geophys. Res.*, p. 6801, 1998.

Kozyra, J. U., M. W. Liemohn, C. R. Clauer, A. J. Ridley, M. F. Thomsen, J. E. Borovsky, J. L. Roeder, V. K. Jordanova, and W. D. Gonzalez, Multistep D_{st} development and ring current composition changes during the 4–6 june 1991 magnetic storm, *J. Geophys. Res., 106*, in press, 2001.

Liemohn, M. W., G. V. Khazanov, and J. U. Kozyra, Banded electron structure formation in the inner magnetosphere, *Geophys. Res. Lett., 25*, 877, 1998.

Liemohn, M. W., J. U. Kozyra, V. K. Jordanova, G. V. Khazanov, M. F. Thomsen, and T. E. Cayton, Analysis of early phase ring current recovery mechanisms during geomagnetic storms, *Geophys. Res. Lett., 26*, 2845-2848, 1999.

Liemohn, M. W., J. U. Kozyra, C. R. Clauer, and A. J. Ridley, Computational analysis of the near-earth magnetospheric current system during two-phase decay storms, *J. Geophys. Res., 106*, in press, 2001a.

Liemohn, M. W., J. U. Kozyra, M. F. Thomsen, J. L. Roeder, G. Lu, J. E. Borovsky, and T. E. Cayton, Dominant role of the asymmetric ring current in producing the stormtime D^*_{st}, *J. Geophys. Res., 106*, 10883, 2001b.

Lui, A. T. Y., R. W. McEntire, and S. M. Krimigis, Evolution of the ring current during two geomagnetic storms, *J. Geophys. Res., 92*, 7459-7470, 1987.

Maynard, N. C., and A. J. Chen, Isolated cold plasma regions: Observations and their relation to possible production mechanisms, *J. Geophys. Res., 80*, 1009, 1975.

McComas, D. J., S. J. Bame, B. L. Barraclough, J. R. Donnart, R. C. Elphic, J. T. Gosling, M. B. Moldwin, K. R. Moore, and M. F. Thomsen, Magnetospheric plasma analyzer: Initial three-spacecraft observations from geosynchronous orbit, *J. Geophys. Res., 98*, 13453, 1993.

McPherron, R. L., Satellite studies of magnetospheric substorms on August 15, 1968, 1, State of the magnetosphere, *J. Geophys. Res., 78*, 3044, 1973.

Mende, S. B., R. D. Sharp, E. G. Shelley, G. Haerendel, and E. W. Hones, Coordinated observations of the magnetosphere: The development of a substorm, *J. Geophys. Res., 77*, 4682, 1972.

Meng, C.-I., and S.-I. Akasofu, A study of polar magnetic substorms, *J. Geophys. Res., 74*, 4035, 1969.

Nishida, A., Reply to comments by K. Kawasaki, *Astrophys. Space Sci., 24*, 455, 1973.

Parker, E. N., and H. A. Stewart, Nonlinear inflation of a magnetic dipole, *J. Geophys. Res., 72*, 5287, 1967.

Roelof, E. C., Energetic neutral atom image of storm-time ring current, *Geophys. Res. Lett., 14*, 652-655, 1987.

Sckopke, N., A general relation between the energy of trapped particles and the disturbance field near the Earth, *J. Geophys. Res., 71*, 3125, 1966.

Sigiura, M., and T. Kamei, Equatorial D_{st} index 1957–1986, in *IAGA Bulletin No. 40*, ISGI, Saint-Maur-des-/fosses, France, 1991.

Stern, D. P., The motion of a proton in the equatorial magnetosphere, *J. Geophys. Res., 80*, 595, 1995.

Takahashi, S., T. Iyemori, and M. Takeda, A simulation of the storm-time ring current, *Planet. Space Sci., 38*, 1133-1141, 1990.

Takahashi, S., M. Takeda, and Y. Yamada, A simulation of the storm-time partial ring current system and the dawn-dusk asymmetry of geomagnetic variation, *Planet. Space Sci., 39*, 821-832, 1991.

Volland, H., A semiempirical model of large-scale magnetospheric electric fields, *J. Geophys. Res., 78*, 171, 1973.

Young, D. T., H. Balsiger, and J. Geiss, Correlations of magnetospheric ion composition during solar minimum, *J. Geophys. Res., 87*, 9077, 1982.

C. R. Clauer, Space Physics Research Laboratory, University of Michigan, 2455 Hayward, Ann Arbor, MI 48109-2143. (e-mail: bob.clauer@umich.edu)

A Wavelet Analysis of Storm-substorm Relationships

W. B. Cade III, J. J. Sojka, and L. Zhu

Center for Atmospheric and Space Sciences, Utah State University, Logan, Utah

Y. Kamide

Solar-Terrestrial Environmental Laboratory, Nagoya University, Toyokawa, Japan

While Fourier analysis is good for spectrum analysis of signals with more or less constant frequency components, its usefulness is limited when signals with impulsive components are involved. Wavelet analysis, on the other hand, is perfectly suited to finding frequency components that may not be sinusoidal, such as with pulsed signals. From this perspective, then, wavelets have a promising potential when applied to geophysical systems, such as magnetic storm records, which are composed of impulsive inputs of a non-sinusoidal nature, such as intense substorms during the main phase of magnetic storms. First results of applying wavelet analysis to a set of ground-based magnetometer data are presented. Distinct low- and high-frequency components are found in the individual magnetograms and even in *SYMH*, with the high-frequency component possibly related to substorms. An effort to separate the two components seems to confirm the result that clear substorm signatures are present in the low- and mid-latitude measurements.

1. INTRODUCTION

The issue of storm-substorm relationships has been a topic of debate for some time and a variety of techniques have been used to try to separate the influences on surface magnetic measurements. Wavelet Analysis offers a relatively new technique to study geophysical systems driven by impulsive or one or two cycle wave trains. Such dynamics cannot be studied by the commonly used Fourier techniques. Wavelet Analysis has already been used to study such geophysical phenomena as ionospheric plasma turbulence [*Lagoutte et al.*, 1992], current disruption events [*Lui and Najmi*, 1997], and plasma shock profiles [*Gedalin et al.*, 1998].

With the ability to provide a time-dependent frequency breakdown of signals, wavelet analysis may help us gain more insight into the storm-substorm relationship The specific challenge is to use large scale wavelets to identify the slowly varying storm component, remove this from the signal, and then use small scale wavelets to extract short term dynamics which would be associated with substorms accompanied by asymmetric ring currents. In this first study the emphasis is placed on demonstrating the potential of wavelet analysis using one specific storm period.

2. DATA

The data we will analyze is for the geomagnetic storm that occurred over 24–27 Apr 79. The data consists of 5-minute surface magnetograms for 5 observatories (Table 1) as well as *SYMH* derived from these observations. Plots of these observations are shown in Figure 1. For certain wavelet application it is convenient to have a data set with

Table 1. Geomagnetic Station Locations

Station Name	Geographic Latitude	Geographic Longitude	LT Separation (hours)*
Hermanus	–34.42	19.23	3.5
Boulder	40.13	254.77	8.3
Honolulu	21.32	202.00	3.5
Kakioka	36.23	140.18	4.1
Tashkent	41.33	69.62	4.7

* The local time separation is computed relative to the nearest station to the east.

a number of data points equal to a power of two; therefore our series is from 0000 UT on 24 Apr 79 to 1315 UT on 27 Apr 79 (1024 data points). Because of missing data, the Honolulu magnetogram only covers until 1445 UT on 26 Apr 79, after which time we fill in the remaining period with constant values in order to produce a useful wavelet analysis. For the higher frequency results of our analysis near the storm peak, this padding of the data will have no effect. Our analysis was focused on the individual magnetograms with the rationale that these are probably closer to representing inputs of direct physical measurements more so than *SYMH*, which is an average of these measurements.

The selected geomagnetic storm, Figure 1, appears to be quite complicated, characterized by a series of large fluctuations associated with shorter term events, perhaps produced by individual substorms. In reducing these data with a wavelet analysis, reference will be made to the *AL* index. This will provide reference to a commonly used substorm index. It is evident that during the main phase of the geomagnetic storm, say from Time 25 to Time 37, intense substorms took place very frequently, or almost continuously.

3. RESULTS

CWT Coefficients

First, a Continuous Wavelet Transform (CWT) was performed on the available magnetograms using the Reverse Biorthogonal 2.8 wavelet. This wavelet, pictured in Figure 2, is similar to the Mexican Hat wavelet but is 'sharper', and was chosen in order to better correspond to the 'spiky' nature of the signals. These and all subsequent wavelet analyses were performed using MATLAB. The resulting CWT coefficients for pseudo-periods up to 20 hours are shown in Plate 1. What stands out is a series of higher frequency inputs

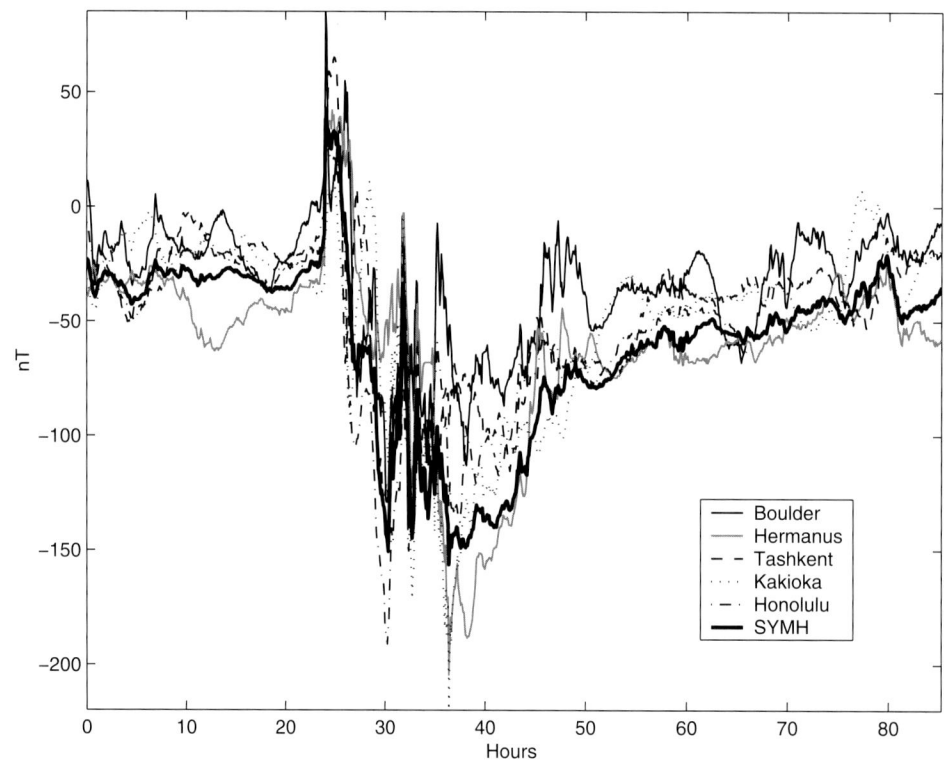

Figure 1. Magnetic Observatory Data for the time period 0000 UT/24 Apr 79 to 1315 UT/27 Apr 79.

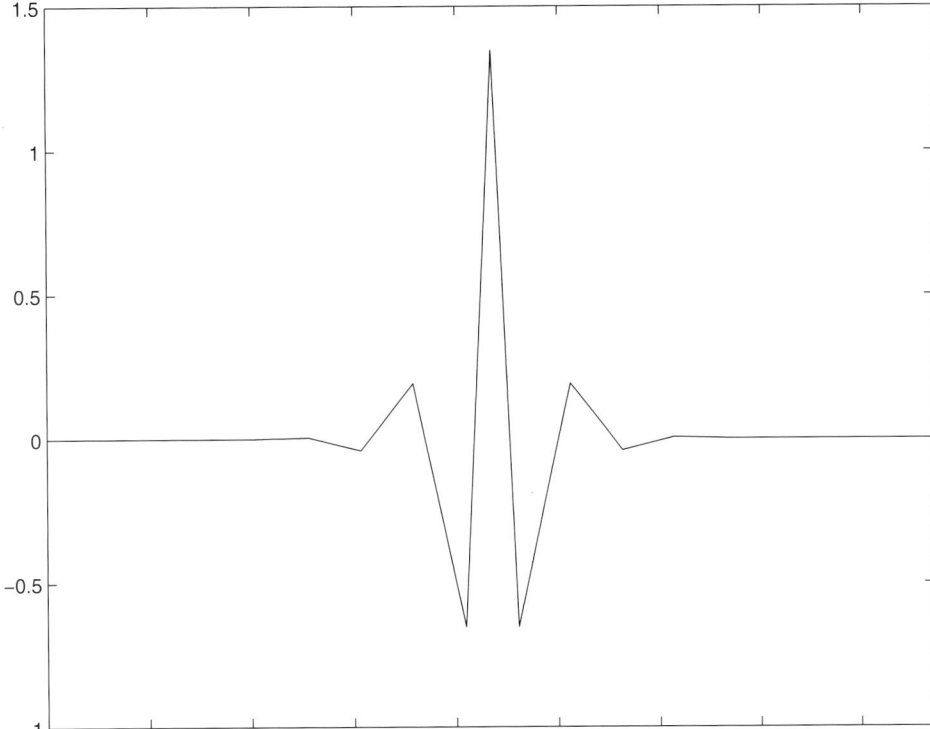

Figure 2. The Reverse Biorthogonal 2.8 Wavelet.

(preceding the lowest *H* excursion) that corresponds to about a period of about 3-hours. At Kakioka, for example, this train begins at 24 hours into the period and 11 intensity peaks are present throughout 41 hours. Each oscillation is represented by two intensity peaks since the absolute value of the wavelet coefficients are plotted in Plate 1. Power in this frequency range is evident in all magnetograms, with the exception of Hermanus (the only southern hemisphere station), and are confined to the region around the storm main phase. Note that this component even appears in *SYMH*. It would be reasonable to conclude that this component is due to substorm activity since the lifetime of substorms is typically two to three hours. Wavelet analysis has allowed us to clearly isolate this characteristic component in the magnetograms. Without additional southern hemisphere data, it is unclear if there is any reason to assume that this input is restricted to the northern hemisphere.

Separation of High and Low Frequencies

The Discrete Wavelet Transform (DWT) is especially suited to the filtering of signals by the suppression of selected scales in the reconstruction of the decomposed signal. One way this can be used is to remove the highest frequencies from a signal to recover only the low-frequency component. This was done for the magnetograms for this storm period and the results are shown in Figures 4 and 5.

Figure 3 represents the low-frequency approximations representing a de-noising using the Stationary Wavelet Transform version of the DWT. While difficult to assign a specific cutoff frequency (see Appendix A1), it is clear from Plate 1 that the higher frequency variations are below a characteristic time of 16 hours, and these have been removed. The resulting approximations were then corrected for an equivalent equatorial current by dividing the original *H* component by the cosine of the latitude. These approximations are quite smooth, but all show a character that is similar to the 'classic' *Dst* storm shape. This would seem to indicate that the lowest frequency part of the magnetograms are due to the slowly-varying symmetric ring current, as we would expect.

The higher frequency parts, obtained by subtracting the low-frequency approximations from the original signals, are shown in Figure 4 compared to the *AL* index that has been adjusted (*AL*/10) to fit on the same scale. Solar wind data are also included (because they are incomplete, wavelet analysis was not possible). Aside from some clearly solar wind pressure-associated increases at ~24 and ~31 hours,

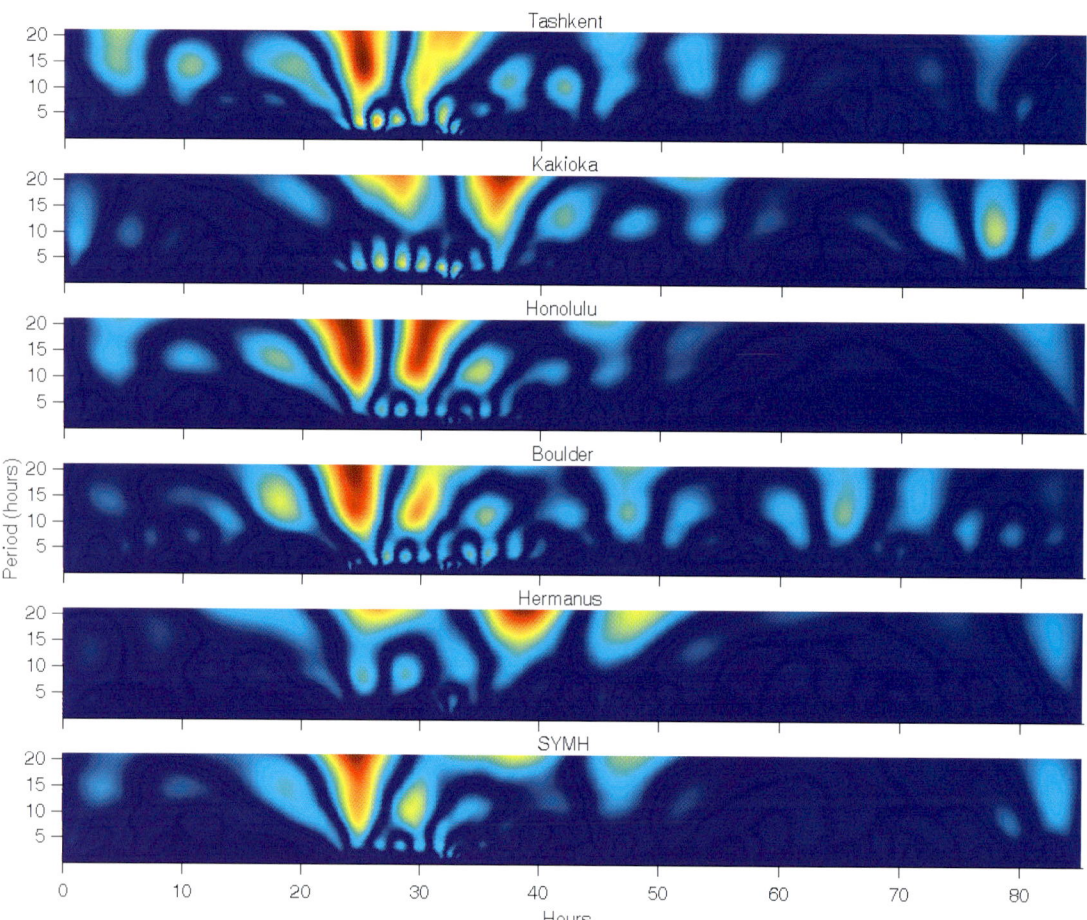

Plate 1. Absolute value of the Continuous Wavelet Transform Coefficients using the Reverse Biorthogonal 2.8 Wavelet.

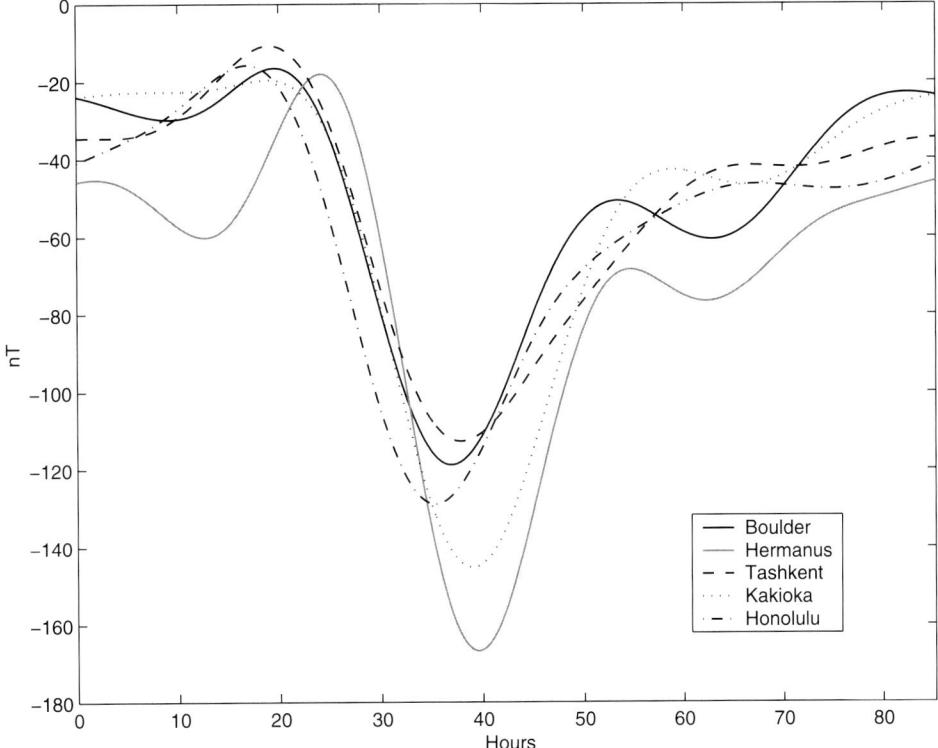

Figure 3. Reconstructed magnetograms with higher frequencies removed and then corrected for latitude.

this higher frequency component exhibits some similarities with the substorm-like current enhancements in *AL*, although some differences would be expected since *AL* is global while the residuals are local. For example, the *AL* minimums at 26 and 30 hours show coincident decreases at Tashkent, Kakioka, Boulder, and Honolulu. *AL* shows several smaller minimums between 39 and 49 hours that match coincident decreases at Kakioka. The sign of the *H* component perturbations at each observatory depends on whether the particular observatory was located inside or outside the substorm wedge current, which is connected to the partial ring current. For example, if observatories were inside the current wedge in the midnight sector, they must have observed intense positive *H* perturbations, while observatories outside the wedge would record weak negative perturbations [*Clauer and McPherron*, 1974; *McPherron*, 1997; *Kamide et al.*, 1998]. Aside from the solar wind-associated fluctuations, these perturbations are generally in the tens of nT range, which is what is normally expected of substorm signatures at low- and mid-latitudes.

While we cannot rule out that these residuals are directly influenced by the auroral currents themselves, it is more likely affected by the field-aligned currents which are connected to the partial ring current, especially since many see the measured low-latitude asymmetry as due to an imbalance of the field-aligned currents [*Harel et al.*, 1981a, b; *Crooker and Siscoe*, 1981; *Chen et al.*, 1982; *Sun et al.*, 1984; *Grafe et al.*, 1986; *Takahashi et al.*, 1991; *Nakabe et al.*, 1997]. Our result seems to corroborate the suggestion of *Feldstein* [1992], who ascribed the shorter period low-latitude asymmetry fluctuations to auroral electrojets and the connected field-aligned currents, noting that during strong storms the electrojets expand equatorward. *De Michelis et al.* [1997] attributed the ring current asymmetry to the effects of storm/substorm injections, ignoring the effect of field-aligned currents altogether. Regarding the cause of the main phase of geomagnetic storms, two different interpretations are possible [*Kamide et al.*, 1998]. One is more or less the definition of the storm-substorm relationship by *Chapman* [1962] in which individual substorms creating 'mini' ring currents contribute to the growth of the symmetric ring current during the main phase. On the other hand, the occurrence of substorms at the main phase is rather by a coincidence in another view. It is the dawn-dusk electric field in the solar wind that is in charge of the main phase, pumping plasma sheet plasmas toward the inner magnetos-

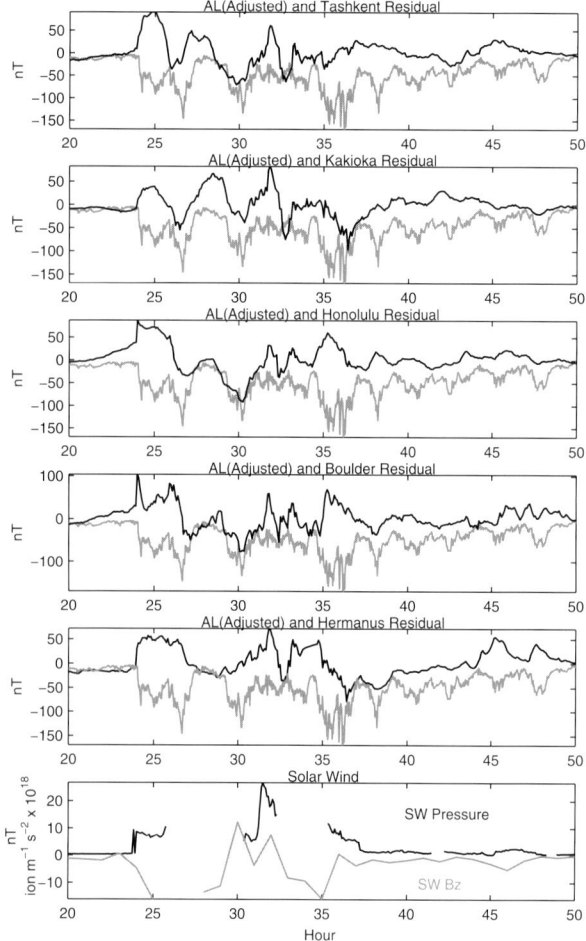

Figure 4. High-frequency residuals (dark line) at each observatory compared to adjusted *AL* (gray line). Solar wind data for the period is shown at the bottom.

phere, generating the ring current. Changes in the solar wind would generate individual substorms (unfortunately the solar wind data is not complete so little can be inferred from it). Any of these views could explain our similarity of the high-frequency residuals and *AL*. Because of the complex nature and interactions of the various current systems involved, it is impossible in this initial application to attribute these signatures to a single current system.

4. DISCUSSION

Fourier analysis has been used in the past on surface magnetic records to find *Sq* and induction effects [e.g., *Chapman and Price*, 1930; *Chapman and Bartels*, 1940], to study symmetric and asymmetric components in disturbed times [e.g., *Suguira and Chapman*, 1960; *Crooker and Siscoe*, 1971; *Crooker* 1972], and to find characteristic spectra in quiet and disturbed times [e.g., *Campbell*, 1976a, b, 1997]. However, geophysical measurements made during individual geomagnetic storms, such as in Figure 1, tend to be rather irregular in nature with a frequency spectrum that changes with time. As such, they are not well suited to the use of Fourier analysis, which decomposes signals into infinite sine and cosine components. Discrete Fourier analysis may offer more hope, but the signal is still assumed to contain a relatively constant frequency spectrum within the windows used to break up the signal. Further, the Discrete Fourier transform has inherent spectral and time resolution problems, especially with limited data sets. Wavelet analysis, on the other hand, was created partly to analyze signals of great complexity that contain a time-dependent spectrum. The Continuous Wavelet Transform is especially useful in obtaining a time-frequency visualization of how the signal's frequency components evolve in time; it can also be used to determine the relative energy content and power spectra of the signal. A further advantage is that, with an infinite number of wavelet choices, one can choose a wavelet that is best suited for the particular application. For example, the Mexican Hat Wavelet is used for single event characterization, the Morlet Wavelet is used to detect wave activity, and the First Derivative Gaussian Wavelet is used to detect gradients. For our purpose, we chose the Reverse Biorthogonal 2.8 wavelet. It is similar to the Mexican Hat wavelet and so is useful in localizing impulsive events, but has a shape that is more characteristic of the events we are measuring than the much smoother Mexican Hat wavelet.

One issue with wavelet analysis is that the concept of frequency is less clear that it is with Fourier analysis. Since wavelets are finite in time, the meaning of frequency is less obvious. This issue is addressed in the appendix, but we will mention that with wavelets, it makes more sense to talk of 'characteristic times' or 'pseudo-frequencies' than true frequencies.

5. CONCLUSION

The first objective of using a wavelet analysis to de-trend or separate the high and low frequency components of ground-based magnetometer responses to storm-substorm dynamics has been achieved. We are still at the stage of understanding the best way to use the new tool in this application. Based on this initial analysis, we can conclude the following:

1) Wavelet analysis shows promise as a tool to extract new information from storm measurements.

2) Wavelet decomposition of magnetometer data seems to show an expected underlying low-frequency current system upon which higher frequency perturbations are superimposed.

3) The higher frequency part seems to show some similarity with *AL* variations, indicating a possible connection between this component and substorm activity.

This study is continuing with the focus on i) improved understanding and hence usage of the wavelet technique; ii) application to many storms in order to develop experience with storm decomposition by this technique; and iii) a continued evaluation of what the higher frequency wavelets can reveal about the question of substorms, field-aligned currents, asymmetric ring currents, and the relative contributions of the various current systems during storms.

APPENDIX

A discussion of this subject has already been provided in numerous publications [*Lagoutte et al.*, 1992; *Graps*, 1995; *Lui and Najmi*, 1997; *Gedalin et al.*, 1998; *Torrence and Compo*, 1998], so here we will focus on the important issues related to this particular application.

Continuous Wavelet Transform (CWT).

The continuous wavelet transform is given by

$$CWT_f^\psi(\tau, s) = \frac{1}{\sqrt{|s|}} \int_{-\infty}^{\infty} f(t) \psi^* \left(\frac{t-\tau}{s} \right) dt \quad (1)$$

where $f(t)$ is our function to be analyzed and $\Psi_{s,\tau}(\tau)$ is known as the 'mother wavelet'—'mother' implying that it is a prototype for generating other comparative functions, and 'wavelet' meaning small wave. Unlike the infinite sines and cosines of Fourier analysis, the mother wavelet is finite in time. The parameters τ (translation) and s (scale) move the wavelet in time and make the wavelet larger or smaller in order to compare it to the function $f(t)$. The integral in effect decomposes the signal into a combination of scaled and translated versions of the Mother Wavelet. Figure A1 shows the Second Derivative Gaussian Wavelet (more popularly known as the Mexican Hat Wavelet) and how the variation of the translation and scale parameters allows various versions of the wavelet to be compared to the signal—in this case a simple sine wave. The result of the integration is a two-dimensional matrix of coefficients (similar to correlation coefficients) as a function of translation and scale. In

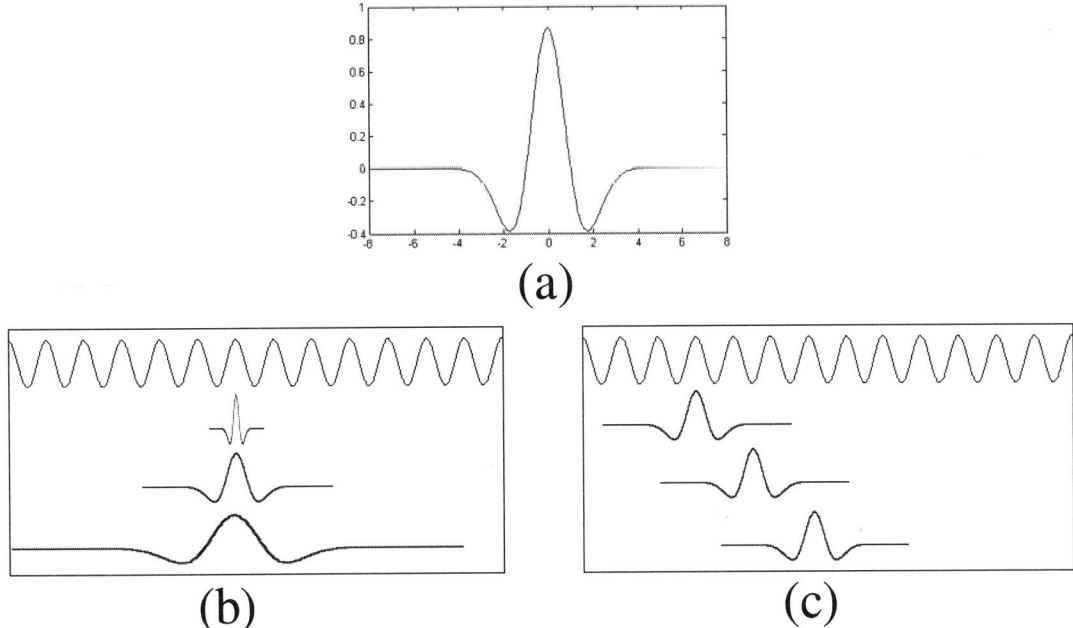

Figure A1. (a) The Second Derivative Gaussian, or Mexican Hat, Wavelet. The effect of (b) scale and (c) translation upon the Mexican Hat Wavelet when compared to a sinusoidal signal.

practice, our analyzed signal is not continuous (as in the discrete values of geomagnetic data) and so a discrete version is adopted where the integral becomes a summation over all the data points.

There are an infinite number of possible wavelets that can be used, the only requirement being that the total area under the curve is zero, or

$$\int_{-\infty}^{\infty} \psi(t)\,dt = 0 \qquad (2)$$

For each wavelet, a specific scale value can generally be correlated to a specific characteristic frequency, or pseudo-frequency. The relationship is inversely proportional and can be obtained by finding the equivalent Fourier period for the wavelet scale [*Meyers et al.*, 1993; *Torrence and Compo*, 1998]. One important aspect of CWT is that it gives good time resolution at high frequencies and good frequency resolution at low frequencies.

Discrete Wavelet Transform (DWT).

This is distinctly different from the discrete version of the CWT. Although visually revealing, the CWT analysis at all scales and locations turns out to be very redundant as far as signal reconstruction is concerned. The analysis can be just as accurate if the signal is analyzed with scale and translation values that are powers of two (dyadic). This process is the DWT and is equivalent to passing the signal through a series of high- and low-pass filters. This decomposition can then be used for reconstruction, compression, filtering, and de-noising of signals. A specialized application of the DWT is de-noising signals using the Stationary Wavelet Transform. The SWT uses the same basic deconstruction of the DWT, but restores translation invariance which is lost under the DWT. The SWT de-noising involves averaging several de-noised signals, and so the issue of cutoff frequencies related to scales becomes somewhat imprecise, especially when one considers the issues of frequency versus pseudo-frequency.

The mathematics and requirements of the DWT are much more rigorous than for the CWT—for example, the wavelet basis must be orthogonal. Chapter 3 of *Chan* [1995] and Chapter 5 of *Boggess and Narcowich* [2001] discuss the mathematics and algorithms of the DWT in great detail.

Acknowledgments. This research was supported by NASA grant NAG5-8227 and NSF grant ATM-0000171 to Utah State University. One of us, W. B. Cade III, is supported through the Air Force Institute of Technology.

REFERENCES

Boggess, A. and F. J. Narcowich, *A First Course in Wavelets With Fourier Analysis*, Prentice Hall, Upper Saddle River, NJ, 2001.

Campbell, W. H., Spatial distribution of geomagnetic spectral composition for disturbed days, *J. Geomag. Geoelectr., 28*, 481, 1976a.

Campbell, W. H., An analysis of the spectra of geomagnetic variations having periods from 5 min to 4 hours, *J. Geomag. Res., 81*, 1369, 1976b.

Campbell, W. H., *Introduction to the Geomagnetic Field*, Cambridge University Press, Cambridge, 1997.

Chan, Y. T., *Wavelet Basics*, Kluwer Academic Publishers, Boston, 1995.

Chapman, S., Earth storms: retrospect and prospect, *J. Phys. Soc. Jpn., 17*, 6, 1962.

Chapman, S. and A. T. Price, The electric and magnetic state of the interior of the earth as inferred from terrestrial magnetic variations, *Phil. Trans. Roy. Soc. London, A229*, 427, 1930.

Chapman, S. and J. Bartels, *Geomagnetism*, Clarendon Press, Oxford, 1940.

Chen, C.-K., R.A. Wolf, M. Harel, and J.L. Karty, Theoretical magnetograms based on quantitative simulation of a magnetospheric substorm, *J. Geophys. Res., 87*, 6137, 1982.

Clauer, C. R. and R. L. McPherron, Mapping the local time-universal time development of magnetospheric substorms using mid-latitude magnetic observations, *J. Geomag. Res., 79*, 2811, 1974.

Crooker, N. U., High-time resolution of the low-latitude asymmetric disturbance in the geomagnetic field, *J. Geomag. Res., 77*, 773, 1972.

Crooker, N. U. and G. L. Siscoe, A study of the geomagnetic disturbance field asymmetry, *Radio Sci., 6*, 495, 1971.

Crooker, N.U. and G.L. Siscoe, Birkeland currents as the cause of the low-latitude asymmetric disturbance field, *J. Geophys. Res., 86*, 11,201, 1981.

De Michelis, P., I.A. Daglis, and G. Consolini, Average terrestrial ring current derived from AMPTE/CCE-CHEM measurements, *J. Geophys. Res., 102*, 14,103, 1997.

Feldstein, Y.I., Modelling of the magnetic field of magnetospheric ring current as a function of interplanetary medium parameters, *Space Sci. Rev., 59*, 83, 1992.

Gedalin, M., J. A. Newbury, and C. T. Russell, Shock profile analysis using wavelet transform, *J. Geophys. Res., 103*, 6503, 1998.

Gonzalez, W.D., J.A. Joselyn, Y. Kamide, H.W. Kroehl, G. Rostoker, B.T. Tsurutani, and V.M. Vasyliunas, What is a geomagnetic storm?, *J. Geophys. Res., 99*, 5771, 1994.

Grafe, A., Y.I. Feldstein, V.I. Pisarski, N.M. Rudneva, P. Ochabova, and A. Prigantsova, The local time asymmetry of the geomagnetic disturbance field at low latitudes and its dependence upon the interplanetary activity, *J. Geomag. Geoelectr., 38*, 1183, 1986.

Graps, A., An introduction to wavelets, *IEEE Comp. Sci. Eng., 2(2)*, 50, 1995.

Harel, M., R.A. Wolf, P.H. Reiff, R.W. Spiro, W.J. Burke, F.J. Rich, and M. Smiddy, Quantitative simulation of a magnetospheric substorm, 1. model logic and overview, *J. Geophys. Res., 86*, 2217, 1981a.

Harel, M., R.A. Wolf, R.W. Spiro, P.H. Reiff, C.-K. Chen, W.J. Burke, F.J. Rich, and M. Smiddy, Quantitative simulation of a magnetospheric substorm, 2. comparison with observations, *J. Geophys. Res., 86*, 2242, 1981b.

Kamide, Y., W. Baumjohann, I.A. Daglis, W.D. Gonzalez, M. Grande, J.A. Joselyn, R.L. McPherron, J.L. Phillips, E.G.D. Reeves, G. Rostoker, A.S. Sharma, H.J. Singer, B.T. Tsurutani and V.M. Vasyliunas, Current understanding of magnetic storms: storm-substorm relationships, *J. Geophys. Res., 103*, 17,705, 1998.

Lagoutte, D., J. C. Cersier, J. L. Plagnaud, J. P. Villain, and B. Forget, High-latitude ionospheric electrostatic turbulence studied by means of wavelet transform, *J. Atmos. Terr. Phys., 54*, 1283, 1992.

Lui, A. T. Y. and A.-H. Najmi, Time-frequency decomposition of signals in a current disruption event, *Geophys. Res. Lett., 24*, 3157, 1997.

McPherron, R. L., The role of substorms in the generation of magnetic storms, in *Magnetic Storms, Geophys. Monogr. Ser.*, vol. 98, ed. by B. T. Tsurutani, W. D. Gonzalez, Y. Kamide, and J. K. Arballo, pp. 131-147, AGU, Washington D.C., 1997.

Meyers, S. D., B. G. Kelly, and J. J. O'Brien, An introduction to wavelet analysis in oceanography and meteorology: with application to the dispersion of yanai waves, *Mon. Wea. Rev., 121*, 2858, 1993.

Nakabe, S., T. Iyemori, M. Sugiura, and J.A. Slavin, A statistical study of the magnetic field structure in the inner magnetosphere, *J. Geophys. Res., 102*, 17,571, 1997.

Sugiura, M. and S. Chapman, The average morphology of geomagnetic storms with sudden commencements, *Abh. Akad. Wiss. Göttingen Math. Phys. Kl., 4*, 1, 1960.

Sun, W., B.-H. Ahn , S.-I. Akasofu and Y. Kamide, A comparison of the observed mid-latitude magnetic disturbance fields with those reproduced from the high-latitude modeling current system, *J. Geophys. Res., 89*, 10,881, 1984.

Takahashi, S., M. Takeda and Y. Yamada, Simulation of storm-time partial ring current and the dawn-dusk asymmetry of geomagnetic variation, *Planet. Space Sci., 39*, 821, 1991.

Torrence, C. and G. P. Compo, A practical guide to wavelet analysis, *Bull. Am. Met. Soc., 79*, 61, 1998.

W. B. Cade III, J. J. Sojka, and L. Zhu, Center for Atmospheric and Space Sciences, Utah State University, 4405 Old Main Hill, Logan, UT 84322-4405.

Y. Kamide, Solar-Terrestrial Environmental Laboratory, Nagoya University, Toyokawa, Japan

Energetics of isolated and stormtime substorms

N. Østgaard[1]

Department of Physics, University of Oslo, Norway

E. Tanskanen[2]

Finnish Meteorological Institute, Helsinki, Finland

A fundamental question in the physics of solar-magnetospheric-ionospheric (SMI) interaction is to understand how the solar wind energy is transferred, stored and distributed in the magnetospheric-ionospheric (MI) system. In this paper we discuss how well parameterized expressions can estimate the energy sources and sinks in the SMI system. We also present some recent results from two energy budget studies. We find that the overall energy coupling efficiency, defined as the ratio of energy dissipated in the MI system to the total available solar wind kinetic energy (U_{SW}) is <1%. We also find that the Akasofu ϵ-parameter, which approximates the solar wind input due to day-side reconnection, does not always provides enough energy to balance the energy sinks in the MI system. Allowing for viscous interaction between the solar wind and the magnetosphere, such a mechanism only need to transfer 0.17% of U_{SW} to balance the energy budget. The dominant energy sink in the MI system is found to be the U_J during both isolated and stormtime substorms. The particle precipitation energy flux (U_A) is about 1/2 of U_J. Several studies indicate that U_R is about the same as U_A even during magnetic storms. Isolated substorms are found to be five times as numerous, but half as intense as stormtime substorms. However, over a 2-year period twice as much energy is dissipated during isolated substorms than during stormtime substorms, indicating that isolated substorms are the main process by which solar wind energy is dissipated in the MI energy sinks.

[1]Now at Space Sciences Laboratory, University of California, Berkeley, California
[2]Now at Goddard Space Flight Center, Greenbelt, Maryland

Disturbances in Geospace: The Storm-Substorm Relationship
Geophysical Monograph 142
Copyright 2003 by the American Geophysical Union
10.1029/142GM15

1. INTRODUCTION

One of the fundamental questions in the physics of solar-magnetospheric-ionospheric (SMI) interaction is to understand and to quantify how the flow of energy from the Sun carried by the solar wind is transferred, stored and distributed in the magnetospheric-ionospheric (MI) system during magnetically disturbed conditions like substorms, convection events and magnetic storms. The growing interest of space weather activities during the last few years has further emphasized the importance of energy budget studies of the SMI system during different levels of magnetic disturbances. It is fairly well accepted that the solar wind, carrying the interplanetary magnetic field (*IMF*), can penetrate the magnetic shielding of the Earth through a reconnection process on the day-side magnetopause and thereby transfer energy and momentum into the magnetosphere, although other mechanisms of energy transfer as viscous interaction between the solar wind and the magnetosphere may be important during specific magnetic conditions. The transferred energy can then either be deposited "directly" into the ionosphere with a typical delay of ~20 min or be stored in the magnetosphere and with a typical delay time of ~60 min be dissipated during substorms [*Bargatze et al.*, 1985; *Tsurutani et al.*, 1985; *Liou et al.*, 1998]. These two categories of energy deposition events are usually referred to as directly driven events and loading-unloading events [*Baker et al.*, 1984; *Rostoker*, 1991; *Elphinstone et al.*, 1996]. The directly (solar wind) driven processes indicate convection activity that is formed as an immediate response to the enhanced energy input [*Baker et al.*, 1997]. The loading-unloading events refer to the storing of magnetic and kinetic energy in the magnetosphere during growth phase and the abrupt substorm breakup with the subsequent global expansion of the auroral oval. The main part of the energy needed to build up substorms is gained during substorm expansion and recovery phases, whereas the growth phase is needed for pre-conditioning the magnetotail to allow the global instability to grow [*Kallio et al.*, 2000]. Magnetospheric substorms and magnetic storms are the main processes by which the solar wind energy is carried to the MI energy sinks. Magnetic storms may include both substorms and convection events and it has been suggested that there are fundamental differences between substorms occurring at storm and non-storm times [*Baumjohann et al.*, 1996].

The purpose of this paper is two-fold. First, in Section 2, we give a tutorial presentation of the most important sources and sinks of energy in the SMI system and how well we can quantify these different energy forms. Secondly, in Sections 3 and 4, we present some recent results from energy budget studies during isolated and stormtime substorms. One study utilizes the imaging instruments on board the Polar satellite to obtain an estimate of the energy flux of precipitating electrons from 0.1 to 100 keV and examines the total energy budget for 7 isolated substorms [*Østgaard et al.*, 2002 A, in press]. Another study focuses on the Joule dissipation for isolated and stormtime substorms, separately, from 1997 and 1999 and gives statistical results in terms of solar wind input, Joule dissipation, intensity, duration, occurrence frequency and time of onset [*Tanskanen et al.*, 2002].

2. ENERGY SOURCES AND SINKS IN THE SMI SYSTEM

The main source of energy for the MI system is the solar wind, although the conductance increase due to solar radiation must be considered when the Joule heating of ionosphere is estimated. The three most important forms of ionospheric and magnetospheric energy dissipation rates are the energy increase of the ring current (U_R), the Joule heating rate of the atmosphere (U_J) and the energy flux by particle precipitation (U_A) [*Akasofu*, 1981; *Weiss et al.*, 1992; *Baker et al.*, 1997; *Knipp et al.*, 1998], although other forms of energy dissipation like the plasma sheet heating, the energy returned to the solar wind by plasmoid ejections from the tail and the production of relativistic electrons might be considered as well [*Baker et al.*, 1997; *Lu et al.*, 1998; *Ieda et al.*, 1998].

2.1. The Solar Wind Energy Source

To quantify the energy transfer from the solar wind into the magnetosphere *Perrault and Akasofu* [1978] and *Akasofu* [1981] examined magnetic storms and substorms to estimate the different forms of ionospheric and magnetospheric energy dissipation. By using the *Dst* index to estimate U_R and the auroral electrojet index *AE* to estimate U_J and U_A in both hemispheres they obtained an estimate of the total energy dissipation rate in the MI system, $U_T = U_R + U_J + U_A$. By comparing U_T with the solar wind parameters they derived the Akasofu's energy input parameter ϵ [*Akasofu*, 1981], here given in SI units,

$$\epsilon [W] = 10^7 \upsilon B^2 \sin^4(\frac{\theta_c}{2}) l_0^2 \qquad (1)$$

where υ is the solar wind bulk speed, B is the magnitude of the interplanetary magnetic field and θ_c is the clock angle of the *IMF* which is defined as the polar angle between the *IMF* as projected into the *Y-Z* plane and the *Z*-axis in GSM coordinates. The parameter l_0 is an empirically determined scale length, sometimes interpreted in terms of the merging region at the sub-solar magnetopause, assumed to be 7 R_E. The ϵ-parameter is a semi-empirical function which has been shown to approximate the solar wind energy input to the magnetosphere due to day-side reconnection. Because of its θ_c dependence ϵ maximizes when *IMF* turns southward and approaches zero for northward *IMF*. Efforts have been made to find other energy transfer parameters that would give better correlation with auroral measurements [Holzer and Slavin, 1982; Gonzales et al., 1994; Liou et al., 1998; Stamper et al., 1999] but they have not provided more accurate estimates of the magnitude of the energy transfer.

The ϵ-parameter is frequently used and many studies show that during mainly southward IMF conditions the parameter gives a reasonable estimate of the total energy transferred into the magnetosphere [Zwickl et al., 1987; Baker et al., 1997; Lu et al., 1998; Liou et al., 1998]. However, by studying time intervals of intense northward IMF events that last for several hours some studies [e.g., Tsurutani and Gonzalez, 1995] have found there is still energy dissipated in the MI system. As the energy transfer due to reconnection during such IMF conditions cannot balance the energy sinks in the MI system, a viscous type of interaction between the solar wind and the magnetosphere has been suggested as an additional energy transfer mechanism. This was first suggested by Axford and Hines [1961] and has later been supported by energy budget studies [Tsurutani and Gonzalez, 1995; Lu et al., 1998; Knipp et al., 1998; Østgaard et al., 2002 A, in press]. As the physical meaning of viscous interaction in a collisionless plasma is not obvious [Parks, 1991] some kind of wave-particle interaction has been suggested to be the coupling mechanism. Farrugia et al. [2001] argue that the Kelvin-Helmholtz instability is the major contributor to viscous coupling, a mechanism that is most efficient during northward IMF conditions. The energy transferred by viscous interaction (U_{VI}) can be expressed as a fraction, σ_v, of the total available kinetic energy of the solar wind (Eq. 2).

$$U_{VI} = \sigma_v U_{SW} = \sigma_v \cdot 1/2 \rho \upsilon^3 A \qquad (2)$$

where ρ is the solar wind mass density (including the Helium content in the solar wind), υ is the solar wind radial speed and A is the magnetopause cross section. An estimate of A is given by Shue et al. [1997, 1998] who found that A depends on the solar wind dynamic pressure and the $IMF\ B_z$ value. The overall coupling efficiency of the solar wind kinetic power to the magnetosphere is thought to be ~1% [Stern, 1984].

2.2. Ring Current

During magnetically disturbed conditions particle injections in the tail will lead to an intensification of the ring current, which can be monitored as geomagnetic disturbances at low latitudes. Four stations near the equator have therefore been selected to provide a global index (*Dst*) for the ring current [Sugiura, 1964]. Variations in the solar wind pressure modulate curents flowing at the day-side magnetopause [Gonzalez et al., 1994] and lead to a compression of the magnetosphere; this will be observed as an increase in the *Dst* index. As this effect is not related to the ring current, the *Dst* index is corrected for this solar pressure effect. An expression for the pressure corrected *Dst** is given by Gonzalez et al. [1994].

$$Dst*[nT] = Dst - 5.10^5 [\frac{nT}{(Jm^{-3})^{1/2}}] p^{1/2} - 20[nT] \qquad (3)$$

where p is the solar wind dynamic pressure. Using the corrected *Dst** index as input, the following equation has been found to give a good estimate of the ring current energy change, U_R [Perrault and Akasofu, 1978; Akasofu, 1981; Zwickl et al., 1987; Gonzalez et al., 1994],

$$U_R[GW] = 4.10^4 (\frac{\partial Dst*}{\partial t} + \frac{Dst*}{\tau}) \qquad (4)$$

where *Dst** is expressed in nT and τ is the ring current lifetime given in seconds. Many studies have used the scaling factor, $4 \cdot 10^4$ [Baker et al., 1997; Lu et al., 1998], which is derived under the assumption of a symmetric ring current in a dipole magnetic field [Akasofu, 1981], while Prigancova and Feldstein [1992] used a slightly lower value $2.7 \cdot 10^4$. The value from [Akasofu, 1981] may be considered as an upper bound as the effect of an asymmetric ring current should imply a lower estimate of U_R for the same amplitude of *Dst*. The ring current lifetime, τ, which depends on loss processes mainly due to charge exchange between ring current ions and exospheric neutrals, is not thought to be constant, but is strongly dependent on the amplitude of *Dst**. The magnitude of U_R is consequently very sensitive to the choice of τ [Zwickl et al., 1987; Prigancova and Feldstein, 1992], especially for large amplitudes of *Dst* during the main phase of magnetic storms [Gonzalez et al., 1994; Valdivia et al., 1996]. Lu et al. [1998] makes a reasonable compromise between the various differentiated τ values presented by Akasofu [1981], Prigancova and Feldstein [1992] and Gonzalez et al. [1994]. Based on nonlinear dynamical methods Vassiliadis et al. [1999] report τ values similar but less differentiated than the values used by Lu et al. [1998]. They also argue that a smaller τ should be applied for small magnetic storms.

2.3. Joule Heating

The Joule heating is controlled by the Pedersen currents and the electric field, which are linked by the Pedersen conductance. As none of these quantities can be monitored directly on a global scale different methods have been developed to estimate at least two of them remotely. Based on radar measurements [*Ahn et al.*, 1983; *Ahn et al.*, 1989], magnetic measurements [*Baumjohann and Kamide*, 1984] or huge data sets as input to the assimilative mapping of ionospheric electrodynamics (AMIE) procedure [*Richmond*, 1990; *Cooper et al.*, 1995; *Lu et al.*, 1995; *Lu et al.*, 1998], the global Joule heating has been found to be closely related to the *AE* index. It is widely known that the auroral electrojet indices show seasonal variations [*Russell and McPherron*, 1973; *Allen and Kroehl*, 1975; *Kamide and Akasofu*, 1983], such that the activity maximizes near equinox and is lower during solstice. Some studies have examined these variations to establish a more accurate relation between U_J and the geomagnetic indices [*Nisbet*, 1982; *Lu et al.*, 1998]. However, most of the relations do not take into account either the neutral wind effect or variations of the large scale electric field, which both can affect the estimate of U_J significantly [*Lu et al.*, 1995; *Emery et al.*, 1999]. A more thorough discussion about the different effects important for calculating the U_J and an evaluation of the different parameterized relations between U_J and *AE* is given by *Østgaard et al.* [2002 A, in press]. A common limitation using global indices to estimate U_J is that local effects are not properly considered.

As the global indices (i.e., *AE* and *AL*) do not always reflect the maximum disturbance due to the ionospheric currents, local indices in limited local time sectors, such as *CU* and *CL* from the CANOPUS chain [*Rostoker et al.*, 1995] or the *IU* and *IL* indices from the IMAGE chain [*Kallio et al.*, 2000; *Tanskanen et al.*, 2002] have been found to give more accurate representation for geomagnetic activity [*Kauristie et al.*, 1996] due to their extended latitudinal coverage. *Kauristie et al.* [1996] found that westward electrojet index from the IMAGE chain catches at least 70% of the AL activity during period 1800–0400 UT (2030–0630 MLT). Especially in the middle of this optimal UT-sector, the IL index usually show stronger activity than the AL index. However, it is true that between 1600–1800 UT (1830 and 2030 MLT) the IMAGE chain can underestimate substorm intensity, because of its limited longitudinal coverage.

2.4. Auroral Particle Precipitation

As particle precipitation through ionization affects the Hall conductance and thereby increases the Hall currents resulting in disturbances of the geomagnetic field, U_A is believed to be related to the *AE* (or *AL*) indices [*Akasofu*, 1981; *Spiro et al.*, 1982; *Ahn et al.*, 1983; *Richmond*, 1990; *Lu et al.*, 1998; *Østgaard et al.*, 2002 B, in press]. Where instantaneous global measurements of electron precipitation are not available, U_A can be derived indirectly from radars or magnetic data or from statistical particle measurements by low-altitude satellites. *Østgaard et al.* [2002 B, in press] used X-ray and UV emissions to derive U_A from 0.1 to 100 keV. By comparing this result with the parameterized methods based on *AE* or *AL* they found that most of these methods underestimate U_A significantly. This may be due to the fact that most of these parameterized methods are developed by using data that only cover electron energies up to 20–30 keV. To examine how much the energetic electrons (> 20 keV) contribute to the total U_A *Østgaard et al.* [2002 A, in press] calculated how much of the U_A derived from UVI and PIXIE that would be estimated if only the UVI is used. The UVI instrument is most sensitive to electrons < 20–30 keV, while the PIXIE instrument is sensitive to electrons up to ~100 keV. It was found that during growth phase the U_A derived from UV emissions only, gives about 90–100% of the total U_A, while the fraction gets closer to 80% during the recovery phase. The energy contribution from the hard tail in the electron spectrum was consequently found to be no more than 20–30% on a global scale and could not explain why some of the parameterized methods only provide 30% of the U_A [*Østgaard et al.*, 2002 B, in press]. Comparing the electron energy flux derived at different local times the fraction was about 90–100% in the day, dusk and midnight sector and closer to 80% in the dawn sector. These results are consistent with the existence of hard electron spectra in the post-midnight to dawn sector and during the recovery phase of substorms. However, for physical processes like Hall conductance increases, cosmic radio wave absorption or wave-particle interaction this tail is the most important part of the energy spectrum.

2.5. Other Energy Sinks

There are other energy sinks in the MI system that are not so straightforward to estimate. In a statistical study of plasmoids *Ieda et al.* [1998] found that on average 1.8 plasmoids were ejected down the tail during substorms and that the energy released by plasmoids was estimated to be roughly 10^{15} J in the course of a substorm. During a magnetic cloud event *Lu et al.* [1998] suggested the plasma sheet heating to be roughly ~100 GW [*Weiss et al.*, 1992] and in the same range as the U_A. They also examined the SAMPEX data from the Polar satellite of MeV electrons and found the energy flux of relativistic electrons to be about 0.5 GW during the event. *Baker et al.* [2001] have examined relativistic electron data from several spacecraft over a 7-year period and found that a rather steady 1% of the solar wind energy transfer estimated by ϵ is converted to relativistic electron energy. Other minor energy sinks are auroral kilo-metric radiation and ultra-low frequency (ULF) magnetic field oscillations which account for $\leq 1\%$ of the magnetospheric dissipation [*Baker et al.*, 1997].

2.6. Energy Budget Studies

A general understanding of the energy flow from the solar wind to the MI system is given by *Stern* [1984]. Magnetic storm and substorm energetics have been studied since the beginning of the 1980's when the role of ring current was thought to be overwhelming relatively to other energy sinks [*Akasofu*, 1981]. Since then several studies have found that during both substorms and magnetic storms the ionospheric energy sinks are dominant [*Weiss et al.*, 1992; *Lu et al.*, 1998; *Knipp et al.*, 1998; *Tanskanen et al.*, 2002; *Turner et al.*, 2001, in press; *Østgaard et al.*, 2002 A, in press]. *Weiss et al.* [1992] concluded that both statistical and case studies indicate that the predominant energy dissipation mechanism is the Joule heating. *Gonzalez et al.* [1994] have questioned this result due to the ring current decay time *Weiss et al.* [1992] used and claimed that U_R is still the dominant energy sink during magnetic storms.

Examining a severe magnetic storm 2–11 November, 1993, *Knipp et al.* [1998] utilized a huge database to estimate both ϵ, U_{SW}, U_R, U_J and U_A and presented a total energy budget for the entire storm period. As they used the ring current decay time suggested by *Akasofu* [1981] their U_R values are probably overestimated rather than underestimated. Only at the very onset of the storm U_R was found to be the dominant energy sink. After 12 hours U_J was found to be the dominant energy sink (55%) and for the entire 10-day period of the storm they found the energy to be distributed by 60%, 23% and 17% to U_J, U_A and U_R, respectively. *Lu et al.* [1998] used a comprehensive set of data to study the total energy budget for the magnetic cloud event on January 10–11, 1997. They used the AMIE procedure supported by auroral electron energy fluxes derived from ultraviolet (UV) images to estimate U_J and U_A. Comparing average energy dissipation they found the energy to be distributed 47.5% to U_J and 30%, 22.5% to U_R and U_A, respectively.

Other studies have focused on more specific aspects of the ionospheric energy dissipation. *Lu et al.* [1995] used the National Center for Atmospheric Research Thermosphere-Ionosphere General Circulation Model (NCAR TIGCM) and the AMIE procedure to examine the effect of neutral winds on the U_J during the Geospace Environment Modeling (GEM) campaign period of March 28–29, 1992. They also calculated the ratio U_J/U_A during that period. *Emery et al.* [1999] examined the same storm period as studied earlier by *Knipp et al.* [1998] to determine the thermospheric neutral response to the magnetic storm. Using the TIGCM and AMIE procedures they estimated the neutral wind effect on the U_J, the seasonal variations of U_J as well as the ratio U_J/U_A. *Liou et al.* [1998] used the UV images from Polar to estimate the energy deposition by precipitation. *Kallio et al.* [2000] used the *IL* index from the International Monitor for Auroral Geomagnetic Effects (IMAGE) as input to the relation found by *Ahn et al.* [1983] to present statistical results of the U_J during isolated substorms compared with the solar wind input estimated by the ϵ-parameter.

3. TOTAL ENERGY BUDGET FOR 7 SUBSTORMS

In this section we will present results from a study where the imagery on board the Polar satellite were utilized to obtain a more accurate estimate of the U_A and where the energy sources and sinks during 7 isolated substorms during 1997 were examined in detail (see *Østgaard et al.* [2002 A, in press] for a more detailed description of these results). Although the solar wind conditions for the 7 substorms diversed, none of the substorms occurred during large magnetic storms, see Table 1.

Solar wind density and velocity data from the solar wind experiment [*Ogilvie et al.*, 1995] and *IMF* measurements by the magnetic field experiment [*Lepping et al.*, 1995] on board the WIND spacecraft were used to estimate the solar wind input due to day-side reconnection (ϵ, Eq. 1) and the total available kinetic energy (U_{sw}, Eq. 2)

To estimate the increase of the ring current the pressure corrected *Dst** (Eq. 3) was used as input for Eq. 4 with the

Table 1. Onset times and geomagnetic conditions for the 7 substorms

Date	Onset	Dst	Kp	AE max
1) 970709	0120	−9	3−	600
2) 970709[a]	0400	−17	3−	250
3) 970724	1400	12	4−	600
4) 970724	1830	8	3−	300
5) 970731[a]	0240	−20	4+	1100
6) 970828	0245	−6	2+	900
7) 970828	0600	−48	4+	500

[a]The stations used to calculate AE_{QL}, AU_{QL} and AL_{QL} are not well located regarding the regions of intense electron precipitation.

differentiated ring current life time (τ) values recommended by *Lu et al.* [1998]. Our choice of τ should be reasonable as none of the substorms occurred during large magnetic storms with large amplitudes of *Dst**. As we used the constant, $4 \cdot 10^4$ [*Akasofu*, 1981], derived under the assumption of an asymmetric ring current the U_R values might be slightly overestimated.

The different parameterized methods to obtain U_J estimates based on the *AE* index were reviewed to see how seasonal and hemispherical effects as well as the number of stations that were used to calculate the auroral electrojet indices could properly be taken into account. The quick look AE_{QL}, AU_{QL} and AL_{QL} indices, at 1 min resolution were obtained from the World Data Center for Geomagnetism, in Kyoto, Japan. During the examined events these indices were based on 6–8 of the 12 standard stations, except for one event (July 31, 1997) where only 4 stations were included. The advantage of choosing the quick look auroral electrojet indices is that they are easily accessible, but it requires that one must make certain that the few stations are well located with regard to the regions of intense electron precipitation. Based on these considerations it was found that U_J in both hemispheres for these substorms during summer time conditions could be estimated by Eq. 5 [*Østgaard et al.*, 2002 A, in press].

$$U_J[GW] = 0.54 AE + 1.8 \quad (5)$$

The neutral wind effect and the variations of the large scale electric field were not taken into account. *Lu et al.* [1995] and *Emery et al.* [1999] found that U_J would be decreased by 10–30% if a complete calculation of U_J including the neutral wind mechanical energy increase, U_{NW}, would be carried out. *Codrescu et al.* [1995] and *Lu et al.* [1998] found that if realistic electric field variability is taking into account, the U_J estimate would be increaed by 10–30%. As these two effects are in the same range and opposite, they may cancel, but these considerations also indicate that the U_J value we use is somewhat uncertain.

To estimate the auroral energy dissipation by particle precipitation, U_A, the combined measurements from the Polar Ionospheric X-ray Imaging Experiment (PIXIE) [*Imhof et al.*, 1995] and the Ultraviolet Imager (UVI) [*Torr et al.*, 1995] were utilized. From these two imagers it is possible to derive 5 min-averages of $U_{A,north}$ in the 0.1–100 keV energy interval. A description on how a four-parameter electron spectra can be derived from PIXIE measurements is given by *Østgaard et al.* [2000, 2001] and a description of the technique used to derive a two-parameter electron spectrum from the UVI measurements is given by *Germany et al.* [1997, 1998a, b]. A validation of combining the two measurements and techniques by comparing with directly measured electron spectra by low-altitude satellites is given by *Østgaard et al.* [2001]. The U_A in both hemispheres was calculated by multiplying the $U_{A,north}$ by a factor 2. This is consistent with the results from *Lu et al.* [1998]; using the AMIE procedure supported with UVI data they found that the energy deposition rates in the two hemispheres were almost similar.

Several criteria had to be met which made the number of available substorms for this study limited. We required that the PIXIE instrument should be fully operating to provide measurements in all energy bands during the entire substorm including both the directly driven growth phase and the loading-unloading substorm. As UVI field of view is sometimes too small to cover the entire global substorm only substorms occurring when Polar was close to apogee could be examined. During 1997, seven substorms were found to meet these criteria.

To present the method of analysis we show in Figure 1 the substorms that occurred on July 24, 1997. Panels a-d show the solar wind parameters and Panels e-f show the ϵ-parameter and U_{SW} derived from these parameters. Panel g-i show the three energy sinks, U_R, U_J and U_A and Panel j shows the *AE* index. The solar wind data is shifted in time due to the radial distance of the WIND spacecraft relative to the subsolar point of the magnetopause. We have also added 5 min to account for the propagation time for a disturbance in the solar wind to affect the inner magnetosphere and thereby be observable in the ionosphere and on the ground [*Kan et al.*, 1991]. The results from *Collier et al.* [1998] have been used to estimate the uncertainty of the time shift. The location of the WIND spacecraft, the time shift and the uncertainty of this time shift are indicated in Panel c. To examine the energy budget the time-integrated values of ϵ and U_{SW} as well as the energy deposition in the MI system were calculated. In order to include the energy that was transferred into the magnetosphere prior to and during the substorms, the *AE* index was examined to find a quiet period before the substorm. This time, t_1, was taken to be the beginning of the growth phase. For all the events this time was found to correspond fairly well with the increase in the e-parameter, indicating that the time shifts of the solar wind data are reasonable. The end of integration time, t_2, was determined by the end of PIXIE and UVI data. For the different forms of energy the time-integrated value $W(U_X)$ from t_1 to t_2

$$W(U_X)[J] = \int_{t_1}^{t_2} U_X(t) dt \quad (6)$$

is calculated (see Panel e-i).

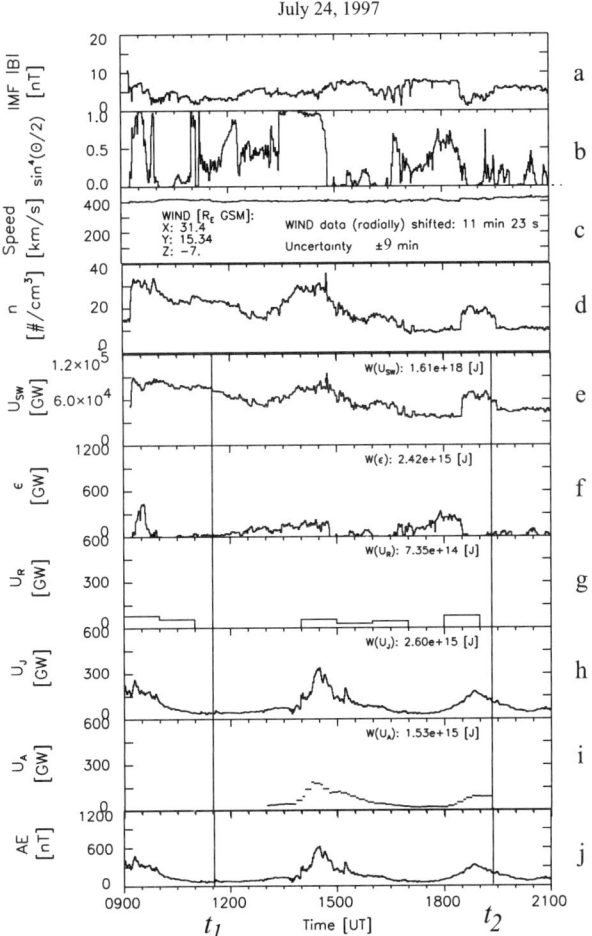

Figure 1. (a-f) Solar wind parameters and energy input measured by WIND. The data are shifted by to the radial distance of the spacecraft and the expected ~5 min propagation time from the subsolar point to the ionosphere [*Kan et al.*, 1991]. (a) The scalar magnitude interplanetary magnetic field. (b) $\sin^4(\frac{\theta_c}{2})$ where θ_c is the clock angle of the interplanetary field. (c) The solar wind bulk speed. (d) Solar wind density. (e) The available solar wind kinetic energy (U_{SW}). (f) ϵ-parameter. (g-i) The various energy dissipations in the MI system. (g) The energy increase of the ring current (U_R). (h) The Joule heating rate in both hemispheres (U_J). (i) The rate of energy deposition by electron precipitation in both hemispheres derived from UV- and X-ray emissions (U_A). (j) The AE quick look index. The solid vertical lines denote the time interval used for integration from t_1 to t_2. The time-integrated energies ($W(U)$) are calculated in panel e-i.

In Table 2 (from *Østgaard et al.* [2002 A, in press]) the time-integrated energy sinks and sources in the SMI system, are listed for all the 7 substorms that occurred during the 4 days in 1997. For $W(U_R)$, $W(U_J)$ and $W(U_A)$ the fractions (in %) of the total energy dissipation ($W(U_T) = W(U_R) + W(U_J) + W(U_A)$) are calculated. Notice that the time of integration is shorter for $W(U_A)$ than for the other energies due to the limited operation time of the imagers on Polar, giving a slight underestimate of $W(U_A)$.

4. STATISTICAL RESULTS FOR ISOLATED AND STORMTIME SUBSTORMS

In this section we present statistical results on isolated and stormtime substorms in terms of solar wind input, Joule dissipation, intensity, duration, occurrence frequency and time of onset (for a more detailed description of these results see *Tanskanen et al.* [2002]). We have examined 839 substorm events in order to study the substorm energy budget. Isolated (IS) and stormtime (SS) substorms were analyzed separately using time integrals of energy input and Joule dissipation as described in the following subsection. A substorm was termed isolated if the Dst index was above −40 nT, otherwise the event was classified as a stormtime substorm.

4.1. Data Set and Method Description

The data set covers two years of IMAGE ground-based magnetic observations at 60-s resolution, combined with simultaneous WIND and ACE solar wind and *IMF* observations. All substorms during 1997 and 1999 in the time sector 16–02 UT (1830–0430 MLT) were analyzed. During this local time period IMAGE is near the mid-night sector, which is the best location to record substorm activity [*Akasofu*, 1964; *Caan et al.*, 1978]. Solar wind magnetic field was measured using WIND Magnetic Field Instrument (60-s time resolution) [*Lepping et al.*, 1995] for 1997 and ACE Magnetic Field Experiment (16-s time resolution) [*Smith et al.*, 1998] for 1999. Plasma velocities were taken from WIND Solar Wind Experiment (92-s time resolution) [*Ogilvie et al.*, 1995] for 1997 and from ACE Solar Wind Electron Proton Alpha Monitor (64-s time resolution) [*McComas et al.*, 1998] for 1999. The *IL* index [*Kallio et al.*, 2000], a proxy of westward electrojet index *AL*, is constructed from 16 (22) magnetometers of the IMAGE magnetometer array [*Syrjäsuo et al.*, 1998] for 1997 (1999) (for more details see *Tanskanen et al.* [2002]). Due to its extended latitudinal coverage, the *IL* index correctly records the occurrence and intensity of substorms both at high and low latitudes, in addition to the typical auroral latitudes. According to *Kauristie et al.* [1996] the maximum electrojet activity derived from the IMAGE chain is equal to or larger than the *AL* index within the local time sector considered in our study.

In selecting substorm events, a minimum *IL* intensity of $|IL| = 100$ nT was used to avoid ambiguous identification

Table 2. Total energy budget for seven selected substorms

Date (hours)[e]	#[a]	$W(U_R)$ [10^{14}J] (%)[f]	$W(U_J)$[b] [10^{14}J] (%)[f]	$W(U_A)$[b] [10^{14}J] (%)[f]	$W(U_T)$[c] [10^{14}J]	$W(\epsilon)$ [10^{14}J]	$W(U_{SW})$ [10^{16}J]	CE[d] [%]
970709 (8.5–4)	2	3.8 (6)	42.2 (69)	15.1 (25)	61.1	47.0	78.5	0.8
970724 (7.8–6.3)	2	7.4 (15)	26.0 (53)	15.3 (32)	48.7	24.2	161.0	0.3
970731 (3.7–2.3)	1	4.6 (11)	23.2 (56)	13.5 (33)	41.3	83.3	78.8	0.5
970828 (10.8–8.3)	2	36.8 (26)	61.4 (44)	41.0 (30)	139.2	135.0	152.0	0.8
Average		(15)	(56)	(29)				

[a]Number of substorms
[b]Both hemispheres
[c]$W(U_T) = W(U_R) + W(U_J) + W(U_A)$
[d]CE: Coupling Efficiency defined as $W(U_T)/W(U_{SW})$
[e]Hours of integration for: $W(U_R)$, $W(U_J)$, $W(\epsilon)$ – and : $W(U_A)$
[f]% of $W(U_T)$

during very low-level activity conditions. The magnetospheric energy input, $W(\epsilon)$, was computed by integrating the ϵ parameter (Eq. 1) as

$$W(\epsilon) = \int \epsilon \, dt. \quad (7)$$

Time integration of the ϵ parameter was started from the beginning of the substorm, defined from the southward turning of $IMF B_Z$, and it was continued until the end of the substorm when the IL index returned to near zero. The ϵ parameter was evaluated at the subsolar magnetopause, obtained by shifting the solar wind time series from the upstream solar wind to the magnetopause at 10 R_E by $\Delta t = \Delta x/v$, where v is the average speed observed around the substorm onset time. Similar time integral for the IL index was used to compute the two-hemisphere Joule dissipation

$$W(U_J) = 2 \int aIL \, dt \quad (8)$$

during the substorm event, where a is a conversion factor from IL in [nT] to power. We have used the conversion factor $a = 3 \cdot 10^8$ given by *Ahn et al.* [1983]. As these substorms occurred during different seasons we have simply multiplied the northern hemispherical value by a factor 2 to get the $W(U_J)$ for both hemispheres. This is slightly different from Eq 5 used for the 7 substorms, where the seasonal effects were taken into account [*Østgaard et al.*, 2002 A, in press].

4.2 Number, Intensity and Frequency of Events

In total, 698 (83%) of the substorms in our data set were isolated with Dst above –40 nT (termed IS for isolated substorms). The intensities of isolated events, as measured by the minimum of the IL index, ranged from –100 nT to –1700 nT, the average being –347 nT. Stormtime substorms (SS) were twice as intense as isolated substorms, but the minimum was around –1700 nT. Thus, there seems to be an upper limit for substorm intensity, which is independent of the simultaneous ring current activity. On the other hand, it seems that the upper limit for the substorm intensity depends on the solar cycle phase, being –1400 nT for the more quiet year 1997 and –1700 nT for 1999 close to solar maximum. Intensity histograms for ISs and SSs, binned every 100 nT, are shown in Figure 2.

Frequency of isolated (stormtime) substorms was computed by dividing the total number of events, n_{IS} (n_{SS}), by the total time during the two years under study when Dst was above and equal to or below –40 nT (T_{IS} and T_{SS}, respectively). Keep in mind that the total time periods are different for different types of events; 7106h for isolated and 563h for stormtime substorms.

$$f_{IS} = \frac{n_{IS}}{T_{IS}} = \frac{698}{7106h} = 0.1 h^{-1}$$
$$f_{SS} = \frac{n_{SS}}{T_{SS}} = \frac{141}{563h} = 0.25 h^{-1} \quad (9)$$

4.3. Input-Joule Heating Analysis

Total magnetospheric energy input, $W(\epsilon)$, and the two-hemisphere Joule dissipation, $W(U_J)$, are illustrated in Figure 3. Energy input and Joule dissipation histograms are shown in the vertical axis, and the number of events in the horizontal axis. Histograms for isolated substorms, binned by every $0.5 \cdot 10^{15}$ J, are shown in the left panel, and histograms for stormtime substorms, binned by every 10^{15} J, are shown in the right panel. The typical value of energy input, $W(\epsilon)$, is $0.1 \cdot 10^{16}$ J for IS while it is $0.3 \cdot 10^{16}$ J for SS, and the histogram shows a much longer tail of $W(\epsilon)$ for SS. For the output, $W(U_J)$, most events occur in the lowest bin, $< 0.1 \cdot 10^{16}$ J, for IS, while there is a maximum at $0.2 \cdot 10^{16}$ J for SS. The ISs are five times as numerous, but only half as energetic as the SSs. The energy input varied from values below 10^{14} J to values larger than $6 \cdot 10^{15}$ J, the median being $1.4 \cdot 10^{15}$ J for IS and $3.5 \cdot 10^{15}$ J for SS. Both input distributions have long tails, with a total of ten events having $W(\epsilon)$ larger than $2 \cdot 10^{16}$ J.

The role of U_J in the total energy budget was further analyzed by computing ratios $W(U_J)/W(\epsilon)$ for each event in our data set. Figure 4 shows these ratios in a histogram format, separately for IS and SS. $W(U_J)$ seems to account for typically (median) more than half of the input, over 60% for IS and about 50% for SS (Table 3). Averages showed even larger ratios for both classes of events (80% and 61%, respectively).

Although single IS on average dissipates less energy than single SS, the yearly $W(U_J)$ during the isolated events exceeds the yearly $W(U_J)$ during stormtime substorms, because ISs are five times more numerous than SSs. Yearly $W(U_J)$ is computed by multiplying the number of events with the average Joule dissipation. In 1997 the yearly $W(U_J)$ through isolated substorms was $2.8 \cdot 10^{17}$ J and in 1999 about $5.0 \cdot 10^{17}$ J. Yearly stormtime $W(U_J)$ was $1.7 \cdot 10^{17}$ J during both years.

We have also examined the duration and the local time of the onset of ISs and SSs. We define the duration of the substorm, Dt, as the time period between the beginning of the substorm growth phase and the end of the recovery phase. The mean duration of all events in our database was 3 hours 50 minutes of which the growth phase took about 60 minutes. Isolated and stormtime substorm did not show large differences in duration (see Table 4). The average local time of onset for ISs was found to be ~2300 MLT. For stormtime substorms the local time of onset was 45 min earlier (~2215 MLT). Table 4 summarizes these results for the entire data set separately for IS and SS.

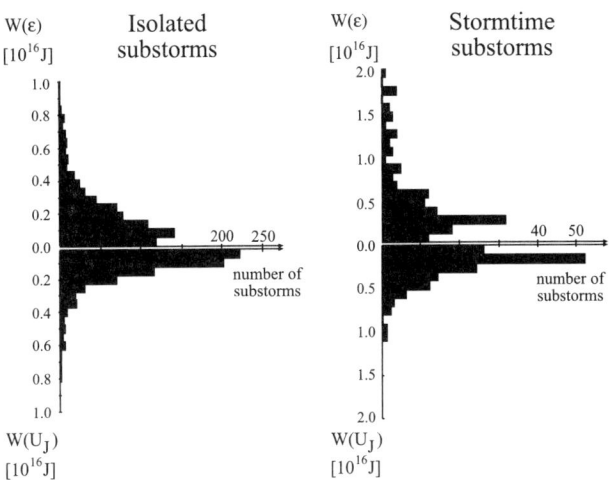

Figure 3. (a) Energy input, $W(\epsilon)$, and two-hemisphere Joule dissipation, $W(U_J)$, histograms for isolated substorms. Both histograms are binned by every $0.5 \cdot 10^{15}$ J. (b) $W(\epsilon)$ and $W(U_J)$ histograms, binned by every 10^{15} J, for stormtime substorms. Note that the scales are different.

5. DISCUSSION

5.1. Total Energy Budget for 7 Substorms

As seen from Table 2, the time integrated total energy dissipation ($W(U_T)$) is distributed as 15%, 56% and 29% to $W(U_R)$, $W(U_J)$ and $W(U_A)$, respectively, which is very close to what *Knipp et al.* [1998] found for the entire 10 days of the magnetic storm period of November, 1993 (17%, 60%, 23%). The main difference seems to be that our U_A is larger than in their study. In the rightmost column of Table 2 we have listed the coupling efficiency, expressed by the ratio $W(U_T)/W(U_{SW})$. The coupling efficiency ranges from 0.3 to 0.8%, which is in reasonable agreement with the 1% that according to *Stern* [1984] is the widely cited order of magnitude estimate of the energy extracted from the solar wind. The higher coupling efficiency found by *Knipp et al.* [1998] (6.9%) and *Lu et al.* [1995] (1.5%) can be explained by the different estimates of the magnetospheric cross section they used. We use the cross section suggested by *Shue et al.* [1997], which depends on the solar wind pressure and $IMF B_Z$ and is usually $\pi \cdot 10^2 R_E^2$. *Knipp et al.* [1998] used $7^2 R_E^2$ and *Lu et al.* [1995] used $15^2 R_E^2$. Using the same cross sec-

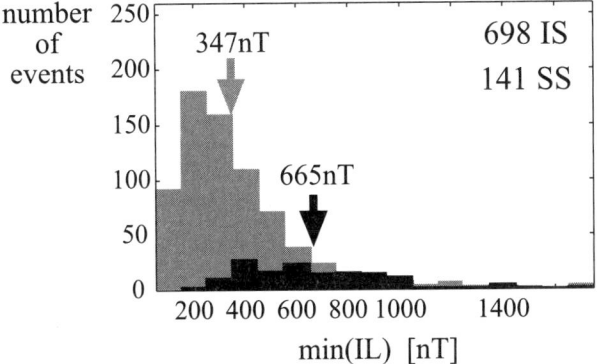

Figure 2. Intensity histograms, binned by every 100 nT, for isolated (IS) and stormtime substorms. Number of the events are marked to the upper right corner.

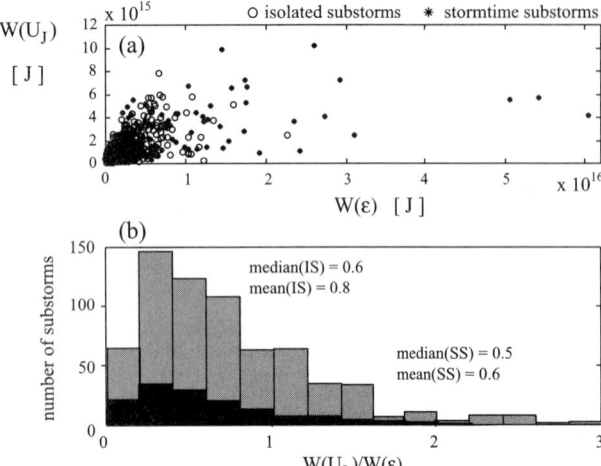

Figure 4. (a) Scatter plot for energy input and two-hemisphere Joule heating. Isolated substorms are marked with circles and stormtime substorms with asterisk. (b) Histograms for input-output ratio, separately for isolated substorms and stormtime substorms. Both means and medians are presented.

tion as we do, their coupling efficiencies would both become 1.1%.

Although none of the substorms occurred during large magnetic storms, the largest value of $W(U_R)$ (26%) on August 28 might be classified as a minor magnetic storm (–48 nT). However, we find that the average of 15% is still similar to the substorms on July 24 where the $W(U_R)$ is fairly small. It should also be noticed that the U_R might has been overestimated by using the scaling factor ($4 \cdot 10^4$) [*Akasofu*, 1981; *Baker et al.*, 1997; *Lu et al.*, 1998] compared to the more conservative value ($2.7 \cdot 10^4$) suggested by *Prigancova and Feldstein* [1992].

The average value found for $W(U_J)$ is 56% which may be a lower estimate as the stations used to calculate the *AE* index were not well located (in longitude) for 2 of the substorms (see Table 1). Another source of underestimation is that during strong events (*AE* >1000 nT) such as the one on July 31, the electrojets may move equatorward (latitudinal shift) of the stations and thus will be poorly detected. The average value of $W(U_J)$ for the substorms is $2.7 \cdot 10^{15}$ J, which is more than 2.5 times the statistical results for isolated substorms (IS) (see Table 3) which means that these are rather intense substorms.

We have also calculated $W(U_A)$ for each substorm, which was found to vary from $6.6 - 24.0 \cdot 10^{14}$ J, with a mean of $1.35 \cdot 10^{15}$ J. This is an order of magnitude larger than suggested by *Akasofu* [1981] ($1.4 \cdot 10^{14}$ J). Part of this may be that our substorms are more intense than the substorms *Akasofu* [1981] examined, but maybe more importantly; the parameterized method he used to estimate U_A only gives 1/3 of the U_A [*Østgaard et al.*, 2002 B, in press]. On average the $W(U_A)$ is ~29% of the total energy dissipation, $W(U_T)$. This should be considered as a lower estimate as the integration times used for $W(U_A)$ are shorter than for the other energies.

The ratio of $W(U_J)$ to $W(U_A)$ was found to be ~2. Other studies have reported different values for the ratio, varying from 1.3 [*Prigancova and Feldstein*, 1992] to 4 [*Lu et al.*, 1995]. However, as discussed by *Østgaard et al.* [2002 A, in press] these variations can be explained by either calculation inaccuracies [*Prigancova and Feldstein*, 1992] or significant underestimation of U_A due to limitations in the data used for the estimate [*Richmond*, 1990; *Lu et al.*, 1995]. Our results are similar to the ratio found by *Lu et al.* [1998], who used the AMIE procedure, supported by global UVI data to obtain U_J and U_A. The ratios from *Knipp et al.* [1998] (2.6) and *Emery et al.* [1999] (2.9) are higher than ours and may be explained by an underestimation of U_A as they do not take into account electrons with energies > 20 keV.

Comparing $W(\epsilon)$ and $W(U_T)$ we see that the ϵ-parameter does not always provide enough energy to balance U_T, which is similar to what *Knipp et al.* [1998] found. As *Akasofu* [1981] originally considered the ionosphere as a minor energy sink, this naturally leads to an underestimation of ϵ when more accurate estimates of U_J and U_A are taken into account [*Pulkkinen and Baker*, 1997]. However, for the July 24 substorms $W(U_T)$ is a factor of 2 larger than $W(\epsilon)$. One might argue that the magnetosphere stores energy from the solar wind and that energy balance only can be required over long time scales, but for the July 24 events it should be noticed that the energy transfer estimated by ϵ during the two substorms is remarkably low, while the $W(U_{SW})$ is large and the $W(U_T)$ is in the same range as the other substorms (Table 2). Even integration of the

Table 3. Comparison of isolated (IS) and stormtime substorms (SS) in terms of energies.

	$W(U_J)^a$	$W(U_J)^b$	$\left(\dfrac{W(U_s)}{W(\epsilon)}\right)^c$	$\Sigma W(U_J)^d$
	[10^{15} J] mean	[10^{15} J] median	[%] median	[10^{17} J]
all	1.3	0.9	60	11.2
IS	1.1	0.8	61	7.8
SS	2.4	1.8	48	3.4

[a]Average of $W(U_J)$
[b]Typical $W(U_J)$, median
[c]Median of ratio $W(U_J)/W(\epsilon)$
[d]Total Joule dissipation during two years under study, $\Sigma W(U_J)$

Table 4. Comparison of IS and SS.

	n^a	Δt^b	f^c [1/h]	t_o^d [MLT]	$min(IL)^e$ [nT]
all	839	3h 50 min	1/9	22:50	−400
IS	698	3h 47 min	1/10	23:01	−347
SS	141	4h 5 min	1/4	22:14	−665

^{a}Number of substorms
^{b}Duration of substorms
^{c}Occurrence frequency
^{d}Magnetic local time of substorm onset
^{e}The minimum of the IL index

ϵ-parameter from an earlier time does not alter this, because another substorm at 0800–1100 UT probably dissipated most of the solar wind energy transferred before and during that event. In Figure 5 (from Østgaard et al. [2002 A, in press]) we show the time-integrated total energy balance between energy input and output as a function of time

$$W(t) = \int_{t_1}^{t} \epsilon(t) - U_T(t) dt \qquad (10)$$

As can be seen from the two upper panels and the bottom panel the ϵ-parameter (dashed line) does not provide enough energy to balance U_T. The overall negative slope of $W(t)$ either indicates that that there has to be some other energy transfer mechanism than day-side reconnection, or that the merging area, l_0^2, in the ϵ-function (see, Eq.1) should be increased, or simply that the magnetosphere is dissipating energy stored much earlier. For the July 24 events the merging area, l_0^2, has to be increased by a factor of 2 to balance the $W(U_T)$. Alternatively, if we allow for solar wind energy to be transferred during northward IMF by viscous interaction (U_{VI}), we find that only $U_{VI} = 0.0017 \, U_{SW}$ has to be added to the ϵ term to balance the energy budget for the July 24 substorms. The solid line shows the result of calculating.

$$W(t) = \int_{t_1}^{t} \epsilon(t) + 0.0017 U_{SW} - U_T(t) dt \qquad (11)$$

For the July 24 events the energy contribution by $W(U_{VI})$ is similar to $W(\epsilon)$, while $W(U_{VI})$ only adds 10 -20% of $W(\epsilon)$ for the other events. In Figure 5 increasing $W(t)$ (solid line) indicates the loading of energy (L) during substorm growth phase and decreasing $W(t)$ indicates the energy dissipation (U, for unloading) from substorm onset through the expansion phase. Onset times were determined from global UV and X-ray imaging. Efforts have been made to quantify the viscous energy transfer efficiency, σ_v. By calculating the U_T during 11 intense northward IMF events that lasted for several hours, where the merging of field lines is almost negligible and hence ϵ would be close to zero, Tsurutani and Gonzalez [1995] found that the transfer efficiency, σ_v, had to be 0.001-0.004 to balance the energy budget. It should be noticed that this study used the very conservative values from Akasofu [1981] to estimate the U_J and U_A, which means that their σ_v should be closer to 0.003-0.012 rather than 0.001-0.004 if more up-date estimates are used. Our suggestion of $\sigma_v = 0.0017$ is therefore in the low side of the σ_v reported by Tsurutani and Gonzalez [1995].

For the July 31 event there is more energy transferred into the magnetosphere than dissipated during the substorm, indicating that there are energy sinks that were not considered. Such energy sinks might be the plasma sheet heating, which Lu et al. [1998] during a magnetic cloud event estimated to be ~100 GW and in the same range as the U_A or the ejection of plasmoids down the tail, which Ieda et al. [1998] estimated to be roughly 10^{15} J in the course of a substorm. However, for the July 31 event including both these energy sinks is still not sufficient to dissipate the stored energy, which means that there may be a net energy gain in the magnetosphere during this event.

5.2. Statistical Results for Isolated and Stormtime Substorms

Energetics of substorms have been studied since 1980's [Akasofu, 1981]. In most studies substorms have been dealt with as a homogeneous group of events. However, there have been suggestions that isolated and stormtime substorms would be different from each other. The results of Baumjohann et al. [1996] seem to show that isolated substorms (which they called non storm substorms) and stormtime substorms behave differently. Hsu and McPherron [2000] concluded that both classes of substorms are caused by the same mechanism. Nevertheless, they found some differences between two classes of events, which were in absolute values and changes of magnitude; the stormtime substorms were larger than isolated substorms. This is consistent with our results showing that on average stormtime substorms (−665 nT) are about twice as intense as isolated substorms (−347 nT).

The intensity of a substorm, measured as a minimum of the IL index, has shown to be a good parameter for measuring the substorm size, but it is not by all means a perfect parameter. This is partly because the auroral electrojet indices are recording both external variations, arising from currents in the ionosphere and magnetosphere, and internal variations, arising from the currents induced in the solid Earth [Viljanen et al., 1995]. Internal part of the IL (and AL

180 ENERGETICS

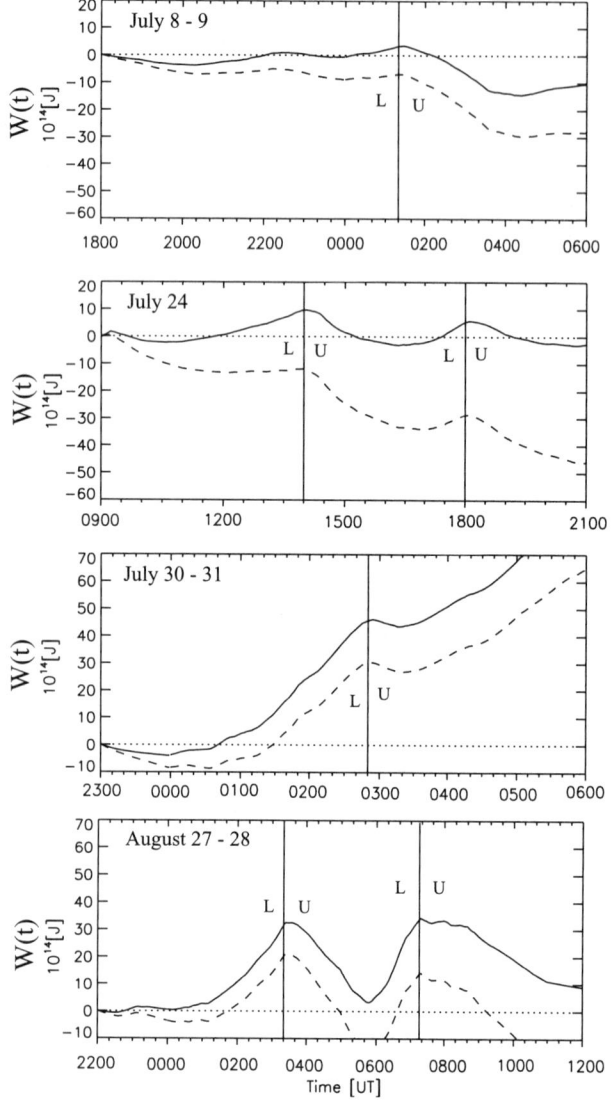

Figure 5. The time-integrated total energy balance between energy input and output ($W(t)$) as a function of time, see Eq. 10. Dashed line: Energy input estimated by ϵ only. Solid line: Energy input is calculated as $\epsilon + 0.0017 U_{SW}$, where U_{SW} is the kinetic solar wind flux given by Eq. 2. We have indicated the times of energy loading (L) and energy unloading (U).

is largest for the rapid time variations, such as substorm onset, when the internal part can be as high as 40% of the *IL* index [*Tanskanen et al.*, 2001]. Internal contribution decreases during expansion and recovery phases to an average of about 15–20% during non-disturbed times [*Tanskanen et al.*, 2001]. Thus, besides using maximal negative variation of *IL* index as a measure of substorm size, the integrated *IL* index, $W(U_J)$, can be used.

$W(U_J)$ during SS is typically two times larger ($1.8 \cdot 10^{15}$ J) than $W(U_J)$ during IS ($0.8 \cdot 10^{15}$ J). However, due to the much larger portion of isolated events the total two-year Joule dissipation was much larger for IS ($7.8 \cdot 10^{17}$ J) than for SS ($3.4 \cdot 10^{17}$ J). This indicates that isolated substorms have a more important role in magnetospheric energetics than previously assumed.

The main difference between IS and SS, besides the differences in *IL* intensity and $W(U_J)$, is the occurrence frequency. The frequency of IS (SS) was computed by dividing the total number of isolated (stormtime) events by the total period of time when the $Dst > -40$ nT ($Dst < -40$ nT). It was found that isolated substorms occurred every ten hours while stormtime substorms occurred every four hours. This method of computing frequencies includes all the moments of the time, regardless of the solar wind conditions and therefore does not provide a consistent comparison between the two classes of substorms. There were several long periods (several days) in the data where no substorms occurred. If we only include moments of the time when the solar wind conditions were favorable to substorm activity, the frequency of isolated substorms would be much larger, even near the frequency of stormtime substorms. *Borovsky et al.* [1993] examined substorms, which occurred during favorable solar wind conditions, meaning that the $IMF\,B_Z$ was southward for an extended period of time. They found the average occurrence frequency of substorms to be 5.74 hours with a significant peak at 2–4 hours. The parameter they studied is comparable to the duration of substorm in our statistics, which was about four hours on average showing a broad peak at 2–5 hours.

Joule heating typically accounts for 50–60% of the total energy input when all substorms are considered. Our data indicate that $W(U_J)$ is slightly lower for ISs than for SSs, which may be due to the larger role of the ring current increase during storm times. Besides of many differences between isolated and stormtime substorms there were also some similarities; for example durations and times of the substorm onsets were quite identical for both classes of substorms. Figure 6 shows a cartoon of typical isolated and stormtime substorms. The figure clearly shows that isolated and stormtime substorms are different from each other, but it is still left an open question whether they can be produced by the same mechanisms.

6. CONCLUSIONS

In this paper we have given a presentation of the energy sources and sinks in the SMI system and discussed how well the different forms of energy can be estimated. We

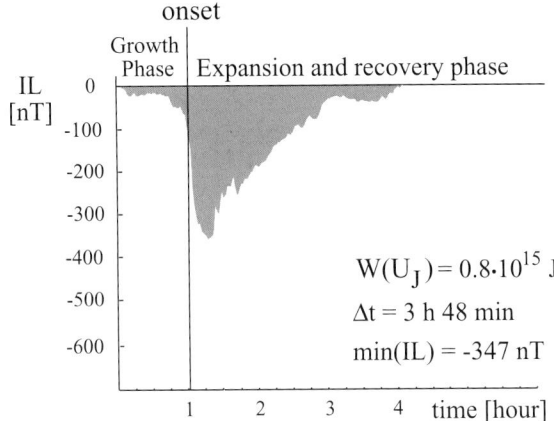

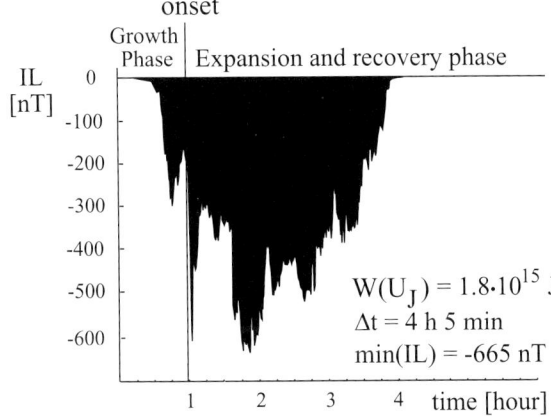

Figure 6. A cartoon of (a) typical isolated substorm and (b) typical stormtime substorm.

have presented results from detailed energy budget analysis of 7 isolated substorms as well as statistical results based on 698 isolated and 141 stormtime substorms. The main results are:

(1) During the 7 isolated substorms the coupling efficiency defined as $W(U_T)/W(U_{SW})$ was found to be 0.3–0.8%, which is in reasonable agreement with the results reported by others [*Stern*, 1984; *Knipp et al.*, 1998; *Lu et al.*, 1995].

(2) For some events the ϵ-parameter does not provide enough energy into the magnetosphere to balance U_T. If we allow for viscous interaction between the solar wind and the magnetosphere, this mechanism only need to transfer 0.17% of U_{SW} to balance the energy budget. This energy transfer would usually represent only a small fraction (10–20%) of what is estimated from the ϵ parameter, but may become important when IMF is northward and $W(U_T)$ is still large. A viscous interaction efficiency of 0.17% is in the low side of what was found by Tsurutani and Gonzales [1995].

(3) Examining 7 isolated substorm the energy transferred from the solar wind to the MI system was distributed with an average of 15% to $W(U_R)$, 56% to $W(U_J)$ and 29% to $W(U_A)$, which is close to what was reported by *Knipp et al.* [1998] during a severe magnetic storm. This distribution of energy is further supported by the statistical results from 698 isolated and 141 stormtime substorms where $W(U_J)$ was found to be slightly less than two thirds of $W(\epsilon)$ during isolated substorms and about half during stormtimes. Although this result emphasizes that the ring current increase is more important during storm times, $W(U_J)$ is the dominant energy sink during both isolated and stormtime substorms. For the 7 isolated substorms $W(U_A)$ was found to be 1/2 of $W(U_J)$ and in fairly good agreement with magnetic cloud and storm studies [*Lu et al.*, 1998; *Knipp et al.*, 1998]. Most energy budget analyses have underestimated the contribution from U_A, as the input data have not covered the entire energy range of precipitating electrons important to calculate U_A.

(4) Stormtime (isolated) substorms appear in every four (ten) hours, when all solar wind conditions were taken into account. If we would examine only moments of time when the solar wind is favorable to substorm occurrence, then the frequency of isolated substorms would be higher.

(5) Isolated substorms are five times as numerous, but half as intense as stormtime substorms. However, over a 2-year period two times more energy is dissipated through Joule heating during isolated substorm ($7.8 \cdot 10^{17}$ J) than during stormtime substorms ($3.4 \cdot 10^{17}$ J). This indicates that the isolated substorm are the main process by which the solar wind energy is carried to the MI energy sinks.

Acknowledgements. The authors wish to thank Richard Vondrak and Tuija Pulkkinen for their valuable comments. We thank R. Lepping for the WIND magnetic field data, A. Lazarus for the WIND solar wind data, C. Smith for the ACE magnetic field data and D. McComas doe the ACE solar wind data. We acknowledge the World Data Center for Geomagnetism (T. Kamei), Kyoto, Japan for providing the preliminary Quick look AE, AL and AU indices, and all the institutes maintaining the IMAGE magnetometer network. The work of E.T. was supported by the Academy of Finland and the work of N. Ø. by the Norwegian Research Council.

N. Ø thanks the staff of the NASA/GSFC Laboratory for Extraterrestrial Physics for their hospitality and support during his stay from June 2000 to September 2001. He also wants to thank the PIXIE and UVI teams for useful collaboration.

REFERENCES

Ahn, B.H., S.-I. Akasofu, and Y. Kamide, The Joule heat production rate and the particle energy injection rate as a function of the geomagnetic indices AE and AL, *J. Geophys. Res.*, *88*, 6275–6287, 1983.

Ahn, B.H., H.W. Kroehl, Y. Kamide, and D.J. Gorney, Estimation of ionospheric electrodynamic prameters using ionospheric conductance deduced from Bremsstrahlung X-ray image data, *J. Geophys. Res.*, *94*, 2565–2586, 1989.

Akasofu, S.-I., The development of the auroral substorm, *Planet. Space Sci.*, *12*, 273–282, 1964.

Akasofu, S.-I., Energy coupling between the solar wind and the magnetosphere, *Space Sci. Rev.*, *28*, 121–190, 1981.

Allen, J.H., and H.W. Kroehl, Spatial and temporal distributions of magnetic effects of auroral electrojets as derived from AE indices, *J. Geophys. Res.*, *80*, 3667–3677, 1975.

Axford, W.I., and C.O. Hines, A unifying theory of high-latitude geophysical phenomena and geomagnetic storms, *Canadian J. Phys.*, *39*, 1433–1464, 1961.

Baker, D.N., S.-I. Akasofu, W. Baumjohann, J.W. Bieber, D.M. Fairfield, E.W. Hones, B. Mauk, R.L. McPherron, and T.E. Moore, Substorms in the magnetosphere, *N A S A Ref. Publ.*, *1120*, 1984.

Baker, D.N., T.I. Pulkkinen, M. Hesse, and R.L. McPherron, A quantitative assessment of energy storage and release in the Earth's magnetotail, *J. Geophys. Res.*, *102*, 7159–7168, 1997.

Baker, D.N., S.G. Kanekal, J.B. Blake, and T.I. Pulkkinen, The global efficiency of relativistic electron production in the Earth's magnetosphere, *J. Geophys. Res.*, *106*, 19,169–19,178, 2001.

Bargatze, L.F., D.N. Baker, R.L. McPherron, and E.W. Hones, Jr., Magnetospheric impulse response for many levels of geomagnetic activity, *J. Geophys. Res.*, *90*, 6387–6394, 1985.

Baumjohann, W., and Y. Kamide, Hemispherical Joule heating and the AE indices, *J. Geophys. Res.*, *89*, 383–388, 1984.

Baumjohann, W., Y. Kamide, and R. Nakamura, Substorms, storms and the near-Earth tail, *J. Geomagn. Geoelectr.*, *48*, 177–185, 1996.

Borovsky, J.E., R.J. Nemzek, and R.D. Belian, The occurrence rate of magnetospheric substorm onsets: Random and periodic substorms, *J. Geophys. Res.*, *98*, 3807–3813, 1993.

Caan, M.N., R.L. McPherron, and C.T. Russell, The statistical magnetic signature of magnetospheric substorms, *Planet. Space Sci.*, *26*, 269, 1978.

Codrescu, M.V., T.J. Fuller-Rowell, and J.C. Foster, On the importance of E-field variability for Joule heating in the high-latitude thermosphere, *Geophys. Res. Lett.*, *22*, 2393–2396, 1995.

Collier, M.R., J.A. Slavin, R.P. Lepping, A. Szabo, and K. Ogilvie, Timing accuracy for the simple planar propgation magnetic field structures in the solar wind, *Geophys. Res. Lett.*, *25*, 2509–2512, 1998.

Cooper, M.L., C.R. Clauer, B.A. Emery, A.D. Richmond, and J.D. Winningham, A storm time assimilative mapping of ionospheric electrodynamics analysis for the severe geomagnetic storm of November 8–9, 1991, *J. Geophys. Res.*, *100*, 19,329–19,342, 1995.

Elphinstone, R.D., J.S. Murphree, and L.L. Cogger, What is a global auroral substorm?, *Rev. Geophys.*, *34*, 169–232, 1996.

Emery, B.A., C. Lathuillere, P.G. Richards, R.G. Roble, M.J. Buonsanto, D.J. Knipp, P. Wilkinson, D.P. Sipler, and R. Niciejewski, Time dependent thermospheric neutral response to the 2–11 November 1993 storm period, *J. Atmos. Terr. Phys.*, *61*, 329–350, 1999.

Farrugia, C.J., F.T. Gratton, and R.B. Torbert, Viscous-type processes in the solar wind-magnetosphere interaction, *Space Sci. Rev.*, *95*, 443–456, 2001.

Germany, G.A., G.K. Parks, M. Brittnacher, J. Cumnock, D. Lummerzheim, J.F. Spann, L. Chen, P.G. Richards, and F.J. Rich, Remote determination of auroral energy characteristics during substorm activity, *Geophys. Res. Lett.*, *24*, 995, 1997.

Germany, G.A., G.K. Parks, M.J. Brittnacher, J.F. Spann, J. Cumnock, D. Lummerzheim, F. Rich, and F.G. Richards, Energy characterization of a dynamic auroral event using GGS UVI images, in *Geospace Mass and Energy Flow: Results From the International Solar-Terrestrial Physic Program*, *Geophys. Monogr. Ser.*, vol. 104, edited by J.L. Horwitz, D.L. Gallagher, and W.K. Peterson, p. 143, AGU, Washington, D.C., 1998a.

Germany, G.A., J.F. Spann, G.K. Parks, M.J. Brittnacher, R. Elsen, L. Chen, D. Lummerzheim, and M.H. Rees, Auroral observations from the POLAR Ultraviolet Imager (UVI), in *Geospace Mass and Energy Flow: Results From the International Solar-Terrestrial Physic Program*, *Geophys. Monogr. Ser.*, vol. 104, edited by J.L. Horwitz, D.L. Gallagher, and W.K. Peterson, p. 149, AGU, Washington, D.C., 1998b.

Gonzalez, W.D., J.A. Joselyn, Y. Kamide, H.W. Kroehl, G. Rostoker, B.T. Tsurutani, and V.M. Vasyliunas, What is a geomagnetic storm?, *J. Geophys. Res.*, *99*, 5771–5792, 1994.

Holzer, R.E., and J.A. Slavin, En evaluation of three predictors of geomagnetic activity, *J. Geophys. Res.*, *87*, 2558, 1982.

Hsu, T.-S., and R. McPherron, The characteristics of storm-time substorms and non-storm substorms, in *Proceedings of ICS-5, Fifth international conference on substorms, ESA SSP-443*, edited by A. Wilson, pp. 439–442, ESA-ESTEC, Noordwijk, Netherlands, 2000.

Ieda, A., S. Machida, T. Mukai, Y. Saito, T. Yamamoto, A. Nishida, T. Terasawa, and S. Kokubun, Statistical analysis of the plasmoid evolution with Geotail observations, *J. Geophys. Res.*, *103*, 4453–4465, 1998.

Imhof, W.L., et al., The Polar Ionospheric X-ray Imaging Experiment (PIXIE), *Space Sci. Rev.*, *71*, 385–408, 1995.

Kallio, E.I., T.I. Pulkkinen, H.E.J. Koskinen, A. Viljanen, J.A. Slavin, and K. Ogilvie, Loading-unloading processes in the nightside ionosphere, *Geophys. Res. Lett.*, *27*, 1627–1630, 2000.

Kamide, Y., and S.-I. Akasofu, Notes on the auroral electrojet indices, *Rev. Geophys. Space Phys.*, *21*, 1647–1656, 1983.

Kan, J.R., L. Zhu, A.T.Y. Lui, and S.-I. Akasofu, A magnetosphere-ionosphere coupling theory of substorms including magnetotail dynamics, in *Auroral Physics*, edited by C.-L. Meng, M.J. Rycroft, and L.A. Frank, pp. 311–321, Cambridge Univ. Press, New York, 1991.

Kauristie, K., T.I. Pulkkinen, R.J. Pellinen, and H.J. Opgenoorth, What can we tell about auroral electrojet activity from a single meridional magnetometer chain, *Ann. Geophys.*, *14*, 1177–1185, 1996.

Knipp, D.J., et al., An overview of the early November 1993 geomagnetic storm, *J. Geophys. Res.*, *103*, 26,197–26,220, 1998.

Lepping, R.P., et al., The WIND magnetic field investigation, *Space Sci. Rev.*, *71*, 207–229, 1995.

Liou, K., P.T. Newell, C.I. Meng, M. Brittnacher, and G. Parks, Characteristics of the solar wind controlled auroral emissions, *J. Geophys. Res.*, *103*, 17,543–17,557, 1998.

Lu, G., D. Richmond, B.A. Emery, and R.G. Roble, Magnetosphere-ionosphere-thermosphere coupling: Effect of neutral winds on energy transfer and field-aligned current, *J. Geophys. Res.*, *100*, 19,643–19,659, 1995.

Lu, G., et al., Global energy deposition during the January 1997 magnetic cloud event, *J. Geophys. Res.*, *103*, 11,685–11,694, 1998.

McComas, D.J., S.J. Bame, P. Barker, W. Feldman, J.L. Phillips, P. Riley, and J.W. Griffee, Solar wind electron proton alpha monitor (SWEPAM) for the advanced composition explorer, *Space Sci. Rev.*, *86*, 563, 1998.

Nisbet, J.S., Relations between the Birkeland currents, the auroral electrojet indices and high latitude Joule heating, *J. Atmos. Terr. Phys.*, *44*, 797–809, 1982.

Ogilvie, K.W., et al., SWE, A comprehensive plasma instrument for the WIND spacecraft, *Space Sci. Rev.*, *71*, 55–77, 1995.

Østgaard, N., J. Stadsnes, J. Bjordal, R.R. Vondrak, S.A. Cummer, D. Chenette, M. Schulz, and J. Pronko, Cause of the localized maximum of X-ray emission in the morning sector: A comparison with electron measurements, *J. Geophys. Res.*, *105*, 20,869, 2000.

Østgaard, N., J. Stadsnes, J. Bjordal, G.A. Germany, G.K. Parks, R.R. Vondrak, S.A. Cummer, D. Chenette, and J. Pronko, Auroral electron distributions derived from combined UV and X-ray emissions, *J. Geophys. Res.*, *106*, 26,081–26,090, 2001.

Østgaard, N., G.A. Germany, J. Stadsnes, and R.R. Vondrak, Energy analysis of substorms based on remote sensing techniques, solar wind measurements and geomagnetic indices, *J. Geophys. Res.*, –, –, 2002 A, in press.

Østgaard, N., R.R. Vondrak, J.W. Gjerloev, and G.A. Germany, A relation between the energy deposition by electron precipitation and geomagnetic indices during substorms, *J. Geophys. Res.*, –, –, 2002 B, in press.

Parks, G.K., *Physics of space plasmas*, The Advanced Book Program, 1 ed., Addison-Wesley Publishing Company, 350 Bridge Parkway, Redwood City, CA 94065, 1991.

Perrault, P., and S.I. Akasofu, A study of magnetic storms, *Geophys. J.R. Astron. Soc.*, *54*, 547–573, 1978.

Prigancova, A., and Y.I. Feldstein, Magnetospheric storm dynamics in terms of energy output rate, *Planet. Space Sci.*, *40*, 581–588, 1992.

Pulkkinen, T.I., and D.N. Baker, Global substorm cycle: What can the models tell us?, in *Surveys in Geophysics*, vol. 18, pp. 1–37, Kluwer Academic Publishers, 1997.

Richmond, A.D., Global measures of ionospheric electrodynamic activity inferred from combined incoherent scatter radar and ground magnetometer observations, *J. Geophys. Res.*, *95*, 1061, 1990.

Rostoker, G., Overviw of observations and models of auroral substorms, in *Auroral Physics*, edited by C.-L. Meng, M.J. Rycroft, and L.A. Frank, pp. 257–272, Cambridge University Press, New York, 1991.

Rostoker, G., J.C. Samson, F. Creutzberg, T.J. Hughes, D.R. McDiarmid, A.G. McNamara, A. Vallance-Jones, D.D. Wallis, and L.L. Cogger, CANOPUS – a ground based instrument array for remote sensing the high latitude ionosphere during the ISTP/GGS program, *Space Sci. Rev.*, *71*, 743–760, 1995.

Russell, C.T., and R.L. McPherron, The magnetotail and substorms, *Space Sci. Rev.*, *15*, 205–266, 1973.

Shue, J.H., J.K. Chao, H.C. Fu, C.T. Russell, P. Song, K.K. Khurana, and H.J. Singer, A new functional form to study the solar wind control of the magnetopause size and shape, *J. Geophys. Res.*, *102*, 9497–9511, 1997.

Shue, J.-H., et al., Magnetopause location under extreme solar wind conditions, *J. Geophys. Res.*, *103*, 17,691–17,700, 1998.

Smith, C.W., J. L'Heureux, N.F. Ness, M. Acuña, L.F. Burlaga, and J. Scheifele, The ACE magnetic fields experiments, *Space Sci. Rev.*, *86*, 613, 1998.

Spiro, R.W., D.H. Reiff, and J.L.J. Mather, Precipitating electron average flux and auroral zone conductance. An empirical model, *J. Geophys. Res.*, *87*, 8215, 1982.

Stamper, R., M. Lockwood, M. Wild, and T.D.G. Clark, Solar causes of the long-term increase in geomagnetic activity, *J. Geophys. Res.*, *104*, 18,325, 1999.

Stern, D.P., Energetics of the magnetosphere, *Space Sci. Rev.*, *39*, 193–213, 1984.

Sugiura, M., Hourly values of equatorial Dst for the IGY, in *Annual International Geophysical Year*, vol. 35, p. 9, Pergamon, New York, 1964.

Syrjäsuo, M., et al., Observations of substorm electrodynamics using the MIRACLE network, in *Substorms-4*, edited by S. Kokubun and Y. Kamide, pp. 111–114, Terra Scientific Publishing Company, Tokyo, 1998.

Tanskanen, E., T.I. Pulkkinen, H.E.J. Koskinen, and J.A. Slavin, Substorm energy budget during low and high solar activity: 1997 and 1999 compared, *J. Geophys. Res.*, *107*, –, 2002.

Tanskanen, E.I., A. Viljanen, T. Pulkkinen, L.H.R. Pirjola, A. Pulkkinen, and O. Amm, At substorm onset, 40% of AL comes from underground, *J. Geophys. Res.*, *106*, 13,119–13,134, 2001.

Torr, M.R., et al., A far ultraviolet imager for the international solar-terrestrial physics mission, *Space Sci. Rev.*, *71*, 329–383, 1995.

Tsurutani, B.T., and W. Gonzalez, The efficiency of "viscous interaction" between the solar wind and the magnetosphere during intense northward IMF events, *Geophys. Res. Lett.*, *22*, 663–666, 1995.

Tsurutani, B.T., J.A. Slavin, Y. Kamide, R.D. Zwickl, J.H. King, and C.T. Russell, Coupling between the solar wind and the magnetosphere: CDAW 6, *J. Geophys. Res.*, *90*, 1191–1199, 1985.

Turner, N.E., D.N. Baker, T.I. Pulkkinen, and G. Lu, Global energy partition during magnetic storms, *J. Geophys. Res.*, 2001, in press.

Valdivia, J.A., A.S. Sharma, and K. Papadopoulos, Prediction of magnetic storms by nonlinear models, *Geophys. Res. Lett., 23*, 2899–2902, 1996.

Vassiliadis, D., A.J. Klimas, J.A. Valdivia, and D.N. Baker, The Dst geomagnetic response as a function of storm phase and amplitude and the solar wind electric field, *J. Geophys. Res., 104*, 24,957–24,976, 1999.

Viljanen, A., K. Kauristie, and K. Pajunpää, On induction effects at EISCAT and IMAGE magnetometer stations, *Geophys. J. Int., 121*, 893, 1995.

Weiss, L.A., P. Reiff, J. Moses, R. Heelis, and B. Moore, Energy dissipation in substorms, in *Proceedings of the International Conference on Substorms ICS–1*, vol. ESA SP-335, pp. 309–317, Eur. Space Agency Spec. Publ., 1992.

Zwickl, R.D., L.F.B.D.N. Baker, C. R. Clauer, and R. L. McPherron, An evaluation of the total magnetospheric energy output parameter, U_T, in *Magnetotail Physics*, edited by A. T. Y. Lui, pp. 155–159, Johns Hopkins Univ. Press, Baltimore, Md., 1987.

N. Østgaard, Space Sciences Laboratory, University of California, Centennial Drive at Grizzly Peak Blvd., Berkeley, CA 94720–7450 (e-mail: nikost@ssl.berkeley.edu)

E. Tanskanen, Laboratory for Extraterrestrial Physics, Code 696, Building 2, NASA/GSFC Greenbelt, MD 20771 (e-mail: etanskanen@lepvax.gsfc.nasa.gov)

Equatorial Ionosphere-Thermosphere System During Geomagnetic Storms

J. Hanumath Sastri

Indian Institute of Astrophysics, Bangalore, India

R. Sridharan and Tarun Kumar Pant

Space Physics Laboratory, V.S.S.C., Trivandrum, India

This paper presents the recent progress in understanding the behavior of the structure and dynamics of the equatorial ionosphere-thermosphere system (EITS) under disturbed geomagnetic conditions. The storm-time modifications in the key parameters and major phenomena of the ionized and neutral domains of the system, which prevail with distinctive time delays with reference to the onset of enhanced geomagnetic activity and excess energy deposition at high latitudes, are traced broadly to fluid dynamical and electrodynamical couplings of the high latitude-low latitude upper atmospheres. Some of the major unsettled issues concerning the storm-time EITS that merit continued and concerted investigations for a comprehensive and quantitative empirical and theoretical understanding are highlighted.

1. INTRODUCTION

The absorption of solar EUV radiation by the earth's upper atmosphere (about 80 km to 400 km) leads to atmospheric heating, photo dissociation of molecular species and creation of free electrons and ions through photo ionization (degree of ionization $\sim 10^{-3}$). The electrons and ions constitute the electrically conducting ionosphere with the neutral atmosphere (thermosphere) dominating the background. The differential heating of the thermosphere on the day and night side hemispheres creates horizontal pressure gradients and the resultant global neutral wind system effectively distributes momentum and energy. The movement of the electrically conducting upper atmosphere across the geomagnetic field by thermospheric zonal winds generates electric fields as in a dynamo, and this is central to the electrodynamics of the Equatorial Ionosphere-Thermosphere System, EITS (Plate 1). The density of ionospheric F region plasma at any given time and location is determined not only by photo ionization and chemical recombination but also by the transport affected by neutral winds, electric fields and field-aligned plasma diffusion. EITS is thus a closely coupled system with linkages to other regions (magnetosphere and lower atmosphere) of the near-Earth space environment as well (Plate 1). At high latitudes, the scenario is complicated due to the highly variable solar wind-magnetosphere interactions and the presence of an additional source of ionization in the form of particle precipitation [*Rishbeth and Garriot*, 1969; *Kelley*, 1989; *Hargreaves*, 1992].

One of the most important features of the ionosphere-thermosphere system is its spatial and temporal variability. The former (as function of latitude, longitude and altitude) is largely due to the coupled physical processes with inherent characteristic time delays. On the other hand, the latter at a given location arises primarily due to external forcing (solar, interplanetary, magnetospheric and lower atmospheric) and manifests on different time scales superposed on the

Disturbances in Geospace: The Storm-Substorm Relationship
Geophysical Monograph 142
Copyright 2003 by the American Geophysical Union
10.1029/142GM16

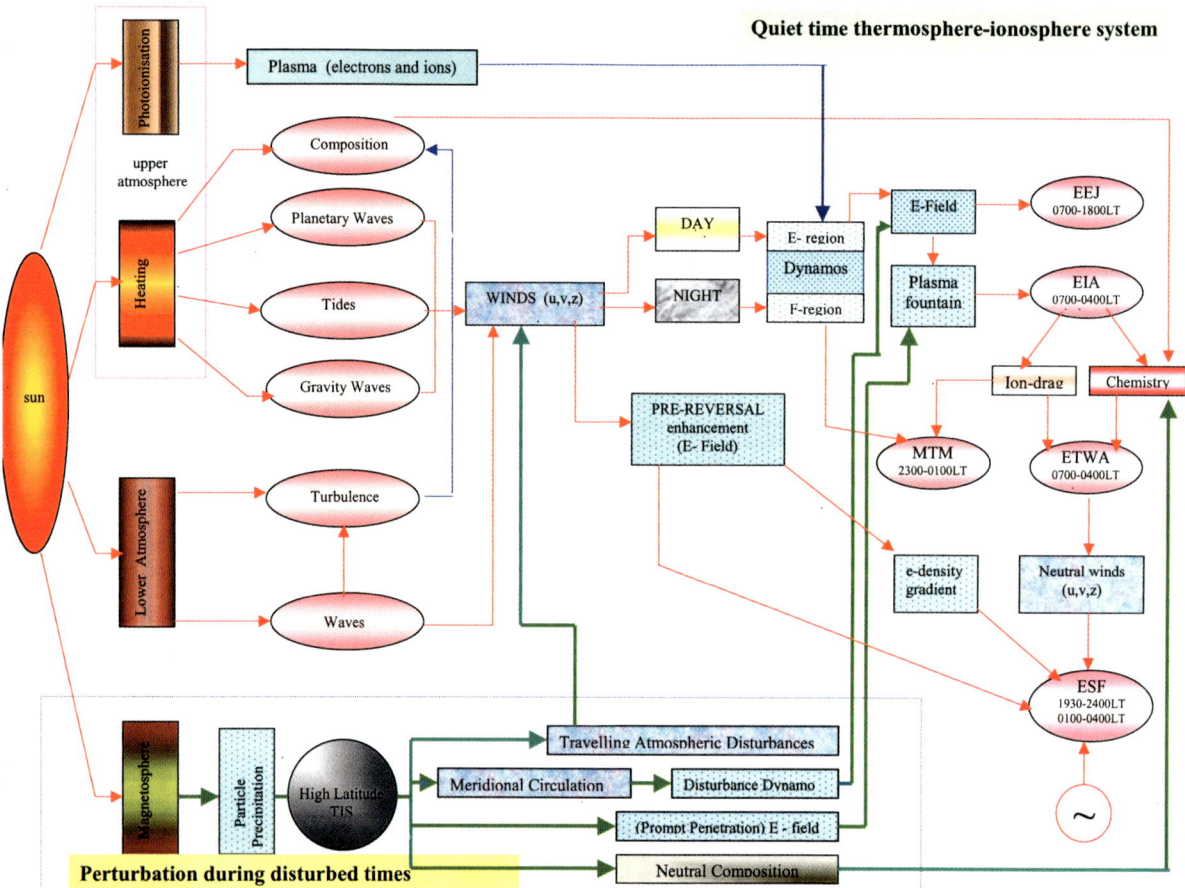

Plate 1. The complex web of interactive processes and the major phenomena of EITS. The major channels/ mechanisms of high latitude-low latitude coupling during disturbed geomagnetic periods are indicated with heavy solid lines. The geomagnetic activity is perceived as perturbations imposed on the quiet time processes operative in the system.

repetitive 24-hr diurnal cycle. The 11-yr solar cycle and 27-day solar rotation induce corresponding temporal modulations. These and other long-term variations (e.g., semi-annual variation) though well known are yet to be understood in quantitative terms.

The focus of this article is on the short-term variability, i.e., on time scales of days, a significant part of which corresponds to periods of active geomagnetic conditions, when complex current systems develop in the near-Earth space environment due to enhanced solar wind-magnetosphere interactions. The most dramatic form of enhanced geomagnetic activity is the magnetic storm, the defining feature of which is the significant development of the ring current encircling the Earth in the westward direction leading to a depression in the H-component of the surface magnetic field at very low latitudes and its slow decay sometimes lasting for several ten's of hours. The storm is thus readily recognizable in the equatorial D_{st} index derived from low latitude magnetic data. Substorm, on the other hand, is a short-lived (2–3 hr) phenomenon at high latitudes on the night side, with characteristic growth, expansion phase onset and recovery phases representing respectively, the accumulation and explosive release of energy from the magnetosphere into the high latitude ionosphere. Though both the phenomena are generally controlled by the same interplanetary condition(s) (southward directed interplanetary magnetic field, for example), the question of storm-substorm relationship, or more precisely, whether the occurrence of intense substorms is necessary for magnetic storms, continues to be unsettled [Kamide et al., 1998 and references therein].

During disturbed geomagnetic condition, the physical and the dynamical state of the ionosphere-thermosphere system deviates substantially from the average quiet-time pattern constituting the 'ionosphere-thermosphere storm' and is a major facet of contemporary research in Solar-Terrestrial Physics [e.g., Buonsanto et al., 1997; Knipp et al., 1998 and references therein]. This could be seen in cooperative International programs like Space Weather, Equatorial Processes Including Coupling (EPIC) and the forthcoming program of Climate and Weather of the Sun-Earth System (CAWSES). The continued interest in 'ionosphere-thermosphere storms' is due to the ever increasing technology-driven capabilities of monitoring the IT system in its entirety, acute awareness of the important roles of plasma-neutral coupling and high latitude-low latitude couplings, and the need to develop predictive capabilities for effective management of technological systems in space and on ground. In this article, we summarize the current understanding of the storm-time behavior of the equatorial ionosphere-thermosphere system (EITS) focusing on its key physical parameters and some of the major phenomena that arise because of the interactive nature of the underlying physical processes.

2. CHARACTERISTICS OF THE EQUATORIAL IONOSPHERE-THERMOSPHERE SYSTEM (EITS)

EITS exhibits several unique phenomena which are the outcome and hence symptomatic of the complex interactive processes that operate in the region and occur with characteristic local time dependencies as shown in Plate 1. At the core of EITS is the ionospheric wind dynamo and the associated global currents and electric fields. At the dip equator the mutually perpendicular electric field (east-west) and magnetic field (north-south) configuration, in combination with the anisotropic ionospheric conductivity, leads to the intense zonal equatorial electrojet (EEJ) current that flows in the altitude range 90–130 km at and around ± 3⁰ of the magnetic equator [Reddy, 1998 and references therein]. The generation of EEJ around 0700 LT marks the beginning of a major phenomena of quiet-time EITS. Thereafter, EEJ grows in strength reaching a maximum just prior to local noon followed by a gradual decay to a low value just prior to sunset and is practically unobservable at night in ground magnetic data due to the much-reduced E-region conductivity although the primary westward electrostatic electric field still exists. EEJ is highly sensitive to electric field disturbances of external (magnetospheric/high latitude) origin because the primary zonal electric field, Ey' associated with them generates a polarization electric field, Ep' larger by a factor of ≈ 25. Studies of EEJ are thus valuable for the characterization and understanding of geomagnetic disturbance-time electric fields and their effects on the other characteristics of EITS.

The E-region east-west and north-south electric fields of ionospheric dynamo origin get mapped to the F region altitudes along the highly conducting magnetic field lines where through interaction with the north-south magnetic field, the former causes vertical E x B plasma drift while the latter induces zonal drift of F-region plasma. Extensive information on equatorial F-region plasma drifts and their dependence on season and solar activity is available [e.g., Sastri, 1995; Fejer et al., 1995; Maynard et al, 1995]. It is established that, the F region plasma drifts are upward by day and downward at night with a conspicuous post sunset prereversal enhancement (PRE). The PRE is the focus of much attention as it significantly affects the properties of both the ionized and neutral species. Theoretical and numerical simulation studies show that the daytime upward drift is due to the E and F region dynamo while the PRE is due to the F region dynamo [*Eccles,* 1998; *Fesen et al.,* 2000 and references therein].

After sunrise, the E region eastward electric field initiates the 'fountain' mechanism wherein the F-region plasma is uplifted over the dip equator only to diffuse along the magnetic field lines aided by the gravitational and pressure gradient forces. This results in depletion (accumulation) of plasma around the dip equator (away from the equator) producing the peculiar double-humped latitudinal distribution known as Equatorial Ionization Anomaly, EIA (Plate 1), which is seen in the peak electron density of the F region, foF$_2$ as well as in the Total Electron Content (TEC). EIA forms generally around 0900 LT with the crests close to the dip equator to start with, and gains in strength (poleward movement and intensification of crests) in the afternoon period. Thereafter the crests retreat equator-wards getting weaker while doing so. The strength of the EIA shows one-to-one correlation with the time integrated strength of the equatorial electrojet confirming that the driving force for both the EIA and the EEJ is the same [*Raghavarao et al.,* 1978]. Though the EIA is present all through the day, the PRE in the F region vertical drift essentially determines the time of disappearance of EIA and hence the total duration of its manifestation. During equinoxes at solar maximum, the large value of PRE rejuvenates the plasma *fountain* with a consequent post sunset strengthening of EIA thereby extending its persistence to early morning hours. Its characteristics, in particular the strength of the crests will also be controlled by the F region meridional neutral winds, because of its ability to move the ionospheric plasma along the magnetic field lines depending on its direction (pole ward/equator ward) and the height-dependent F-region chemical recombination. The net result is an asymmetry in the EIA crest amplitude on either side of the dip equator. The phenomenology of EIA had been extensively investigated and documented in the literature [e.g., *Sastri* 1990; *Bailey et al,* 1997 and references therein].

A further and dramatic indicator of the internal couplings of EITS is the latitudinal distribution of thermospheric parameters. The uplifting of F region plasma around the dip equator depletes the ionization and enhances the zonal neutral winds relatively because of reduced ion drag (frictional force that depends on plasma density and the difference in speed between plasma and neutrals). On the other hand, the ion drag is enhanced at the location of EIA crests resulting in a reduction of neutral zonal flow from the hot dayside to the cold nightside. This leads to the Neutral Anomaly [*Hedin and Mayr,* 1973] with enhanced neutral densities at EIA crest locations and also to a decrease of zonal winds and the associated increase of neutral temperature due to conversion of the kinetic energy to thermal energy referred to as the Equatorial Temperature and Wind Anomaly, ETWA [*Raghavarao et al.,* 1991]. Moreover, substantial vertical neutral winds that are upward (downward) at the EIA crests (trough) location have also been observed [*Raghavarao et al.,* 1993] possibly due to the setting up of a meridional circulation cell in association with the ETWA. Careful evaluation of the local time variability of ETWA highlights the role of both ion drag and the exothermic chemical recombination [*Raghavarao et al.,* 1998; *Fuller-Rowell et al.,* 1997; *Pant and Sridharan,* 2001c].

The phenomenon of much current research interest especially from the point of view of 'space weather' is the destabilized condition of nighttime equatorial F region plasma referred to as Equatorial Spread-F (ESF). ESF is characterized by the presence of a wide spectrum of field-aligned irregularities extending over nearly seven orders of magnitude in the spatial scale which get generated under favorable conditions due to a hierarchy of plasma instabilities and manifest in a variety of forms such as scattered echoes in ionograms, plumes in VHF radar maps, bite-outs in night OI 630 nm airglow intensity, bubbles in *in-situ* satellite measurements and scintillations in VHF/L-band satellite beacon signals (*Kelley,* 1989; *Abdu,* 2001 and references therein). The collisional Rayleigh Taylor Instability is the primary destabilizing mechanism operating in the base of the F-region followed by secondary instabilities, one feeding on the other. The pre-reversal enhancement (PRE) in upward vertical drift over the magnetic equator causes rapid uplifting of the base of F-layer which, in combination with the fast chemical recombination of the dominant molecular ions, results in a steep bottom side gradient making it susceptible to Rayleigh-Taylor instability. This, in turn, results in the upwelling of the low-density plasma of the bottom side F-region to greater heights in the form of a bubble/plume. The importance of ambient ionospheric and thermospheric conditions such as vertical plasma drifts, plasma density gradients on the bottomside and topside, ion composition, meridional and vertical neutral winds, initial seed perturbations .etc in the initiation and non-linear development of ESF and its dynamics has been amply assessed and well recognized [*Kelley et al,* 1981; *Hanson et al.,* 1986; *Sekar and Raghavarao,* 1987; *Sastri,* 1984; *Raghavarao et al.,* 1988; *Sultan,* 1996; *Zalesak et al,* 1982; *Sekar et al.,* 1995, 1997, *Sekar and Raghava Rao,* 1997; *Sekar and Kelley,* 1998; *Basu and Coppi,* 1999; *Viswanathan et al,* 2001]. In-situ measurements by satellites as well as groundbased VHF radar measurements revealed the complex dynamics of ESF bubbles in the form of up drafting and down drafting structures [*Laakso et al.*, 1994; *Patra et al.*, 1997].

The enigmatic aspect of ESF as yet is its unpredictable day-to-day variability. Case studies showed that the time history of vertical plasma drift in the postsunset hours and hence the F region height plays a dominant role in the day-to-day variability of ESF [*Mendillo et al.*, 1992; *Sastri et al.*, 1997a]. Moreover, there is evidence for the occurrence probability of ESF to be very high on days with a strong EIA indicating that the stage is set right during day time itself [*Raghava Rao et al.*, 1988; *Sridharan et al*, 1994, *Mendillo et al.*, 2001]. On the other hand, *Devasia et al* [2002] show that for ESF to occur over the equator when the base height (h´F) is in the range of 270–300km, one should have an equatorward meridional wind, but when h´F is >300km the meridional wind direction does not matter. This result suggests that the observed equatorward winds could be an outcome of the pressure bulges associated with ETWA which in turn is linked to EIA [*Raghavarao et al.*, 1993], thereby providing a plausible explanation for the increasing evidence for a linkage between EIA and ESF.

3. STORM-TIME EITS

It follows from the above discussion that in order to comprehensively understand the major EITS phenomena and be able to now cast/forecast the storm-time changes; one should have a clear understanding of the behavior of the key parameters of EITS. The underlying view point is that the basic physics governing EITS remains the same and it is only the changes in the dynamical forcings brought about by the fluid dynamical and electro dynamical coupling of the low and high latitude upper atmospheres that determine the response of EITS to enhanced geomagnetic activity in general and storms in particular. We shall elaborate this point of view in the next section.

3.1 Thermospheric Structure and Dynamics

One of the very important aspects of the effects of geomagnetic storms is the mechanism(s) by which the copious amount of energy dumped in the high latitude regions gets distributed globally. This is accomplished by equatorward meridional winds and large-scale traveling atmospheric disturbances (TADs) set up by high latitude heating [*Hays and Roble*, 1971; *Hernandez et al.*, 1980, *Sipler and Biondi*, 1978; *Meriwether et al.*, 1986; *Pant and Sridharan*, 1998]. On occasions, even poleward winds have been recorded in the southern hemisphere, presumably due to the prevailing hemispheric asymmetry and the variation of the energy input with time [*Fagundes et al.*, 1995]. Theoretical models typically estimate a delay of 8~10 h for the establishment of the meridional circulation cell from the time of intense high latitude heating. TADs get triggered mainly because of the temporal variability of the energy input to high latitudes and tend to move away from the source regions. They transfer energy to the neutral medium through viscous interactions, heat conduction and frictional loss due to ion drag and this appears as the geomagnetic storm effect at low and equatorial latitudes with a delay of 24 h and more [*Klostermeyer et al.*, 1973; *Prölss*, 1982; *Forbes et al.*, 1987; *Burns and Killeen*, 1992]. The third process identified so far is the release of energy pumped earlier into the ring current during the initial and main phase of a geomagnetic storm and this may last anywhere between 15 to 24 h [*Prölss*, 1982]. The study of the isolated magnetic storm on 29 September 1982 using DE-2 satellite data demonstrates this feature in the latitudinal variation of exospheric temperature at different times during the storm recovery phase (Figure 1). The immediate response of the high latitude regions and the gradual increase of temperature, to start with over mid latitudes by ~4 h and then over equatorial latitudes after >12 h, and the triggering of TADs beyond 37 h are clearly seen. Though high latitude processes are in general responsible for the temperature changes, there are circumstantial evidences for direct overhead deposition of energy at low latitudes even during the initial phase of a storm [*Ranjan Gupta et al.*, 1986]. It is to be noted that though the magnitude of the storm effects on the thermosphere at low and equatorial latitudes is smaller compared to higher latitudes, they do persist for significantly longer periods, i.e., even after the storm subsides [*Fuller-Rowell et al.*, 1994, 1997; *Abdu*, 1997; *Gurubaran et al.* 1995; *Pant and Sridharan*, 1998]. One of the channels by which the stored energy in the ring current gets deposited into the low latitude thermosphere is through isotropic precipitation of energetic neutral atoms (ENA) produced due to resonant charge exchange processes [*Tinsley*, 1981; *Torr et al.*, 1974; *Kozrya et al.*, 1982; *Ishimoto et al.*, 1986]. The images of ENA of ring current origin obtained using satellite data [*Roelef.*, 1987] and the recent results of *Lui et al.* [1996] based on GEOTAIL space craft measurements revealed that the ENA flux and rate of recovery of Dst to be consistent..

The neutral temperature (T_n), which is a measure of energy input into the atmosphere, controls the height distribution of the atmospheric constituents and the consequent thermal gradients establish atmospheric circulation. It is for this reason that atmospheric models aim at making accurate prediction of T_n. Several attempts have been made to formulate working models starting from Jacchia models,

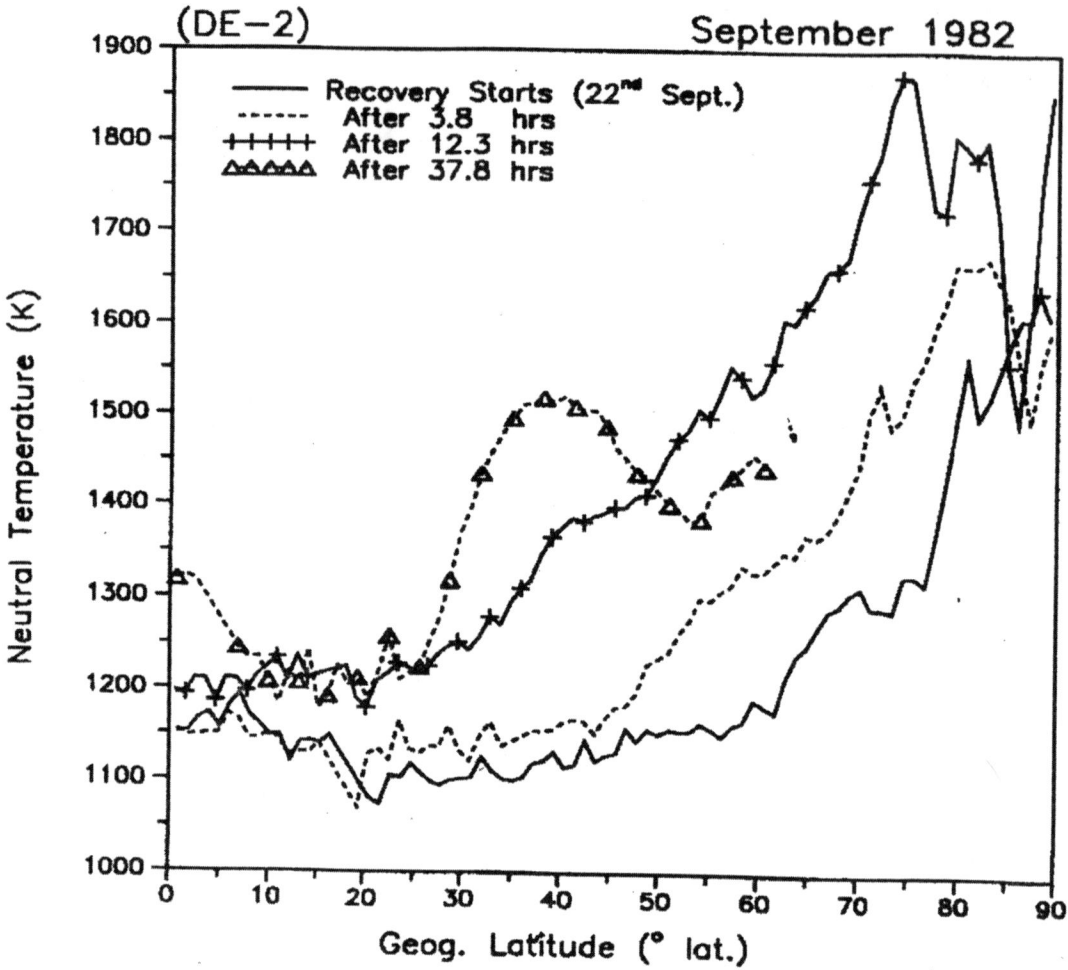

Figure 1. Example of the propagation of energy and momentum from high latitude to low latitude as seen in the DE-2 satellite measured temperatures

MSIS, TIGCM, CITP. ...etc. The most-user friendly and hence widely used model is the MSIS and its upgrades based on Mass Spectrometer and Incoherent Scatter data [*Hedin*, 1983,1987]. More recently the CIRA-86 model has been fused into MSISE-90 model extending its altitude coverage right from ground level up to the exosphere [*Hedin*, 1991]. Since basically the models are climatological in nature, there could be large differences when their predictions are compared with actual measurements during specific storm events. The possibility of several local effects getting integrated out and some of the non-local processes not getting accounted for in the models, could not however be ruled out. This is true especially for low and equatorial latitudes because the database from these regions that has gone into the global models is rather sparse due to the very limited observations made. But there are several unique geophysical processes specific to this latitudinal belt that can alter the thermospheric structure. The particular aspect of the delayed response of the low latitude thermosphere has been found to be of key importance in the explanation of the variations in the thermospheric temperature and its deviations from the model on an event by event basis. Few temperature measurements from low and equatorial regions have revealed enhancements to the tune of 100–400 K [*Hernandez et al.*, 1995; *Biondi and Meriwether* 1985; *Prölss and Roemer*, 1987], the magnitude being highly variable depending upon the intensity of the magnetic storm, its time of occurrence, duration, season, and also on the location of the observing site. Individual case studies, with an aim to parameterize additional factors that have not been taken into account by the present day models would no doubt enhance their forecasting and now casting capabilities. At present the MSIS model is keyed to the geomagnetic indices, K_p and A_p, while the D_{st} index, which represents the 'ring current' could be a better alternative [*Almar et al.*, 1996]. During the main phase of a geomagnetic storm, energy gets pumped into the ring current and gets released later in the storm recovery phase, more favorably at high latitudes. As discussed earlier, advection, diffusion and traveling atmospheric disturbances redistribute the excess energy. There would be a time delay before the low latitude region starts experiencing the storm impact. Pant and *Sridharan* [1998] demonstrated that the rate of change of Dst with time [$d(D_{st})/dt$] could be a good measure of the energy input into the system and showed that the deviations from the model at any instant (in December and February) has a very good correlation with the value of this new parameter 16–18 hours before, thus reproducing all the temporal variability of the measured T_n. Further analysis extended to different months in a year corresponding to both low and high solar activity periods revealed that during the latter periods the time delay varied from about ~ 8 h during May–June to 16–20 h during December–February [*Pant and Sridharan*, 2001a]. Low solar activity epochs also show the local summer minimum and winter maximum but there is an additional uniform delay of ~2h in all the months (Figure 2), clearly highlighting the role of the background

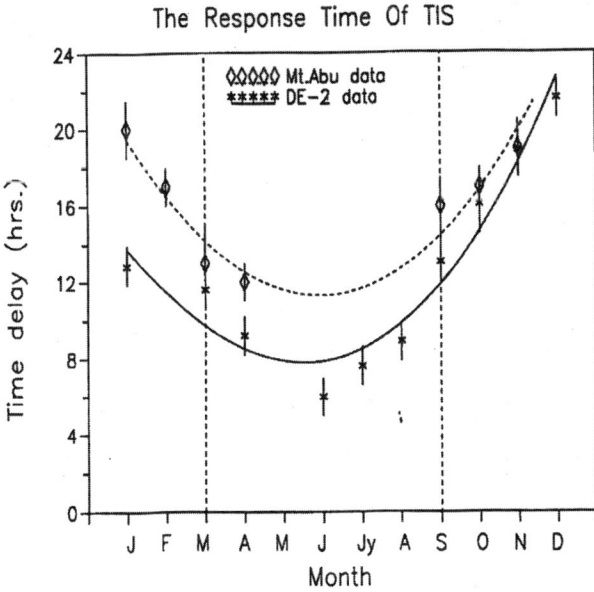

Figure 2. Seasonal dependence of the response-time of the low latitude thermosphere for a geomagnetic storm time energy input. The data base is from DE-2 satellite and also from the ground based spectroscopic measurements of OI630.0nm thermospheric night airglow.

thermospheric conditions in enabling the transport of energy and momentum from high to low latitudes. The seasonal variation was shown to be essentially due to the direct solar forcing rather than due to the energy represented by $d(D_{st})/dt$ [*Pant and Sridharan*, 2001b]. As a logical next step, certain unaccounted local processes like the Equatorial temperature and wind anomaly (ETWA) has also been parameterized by making use of the simultaneous measurements of electron and neutral densities along with the neutral winds, and based on chemical and dynamical processes. It was shown that treating it as an additive input to the model predictions along with the possible effects of $d(D_{st})/dt$, the MSIS model could be made to reproduce actual T_n measurements fairly well. These studies, however, showed that though the variability in the measurements could be reproduced most of the times, some small differences still persisted on occasions, calling for some further fine-tuning of the parameterization process (Figure 3).

In addition to the neutral temperature and neutral composition, the ionospheric F region densities and height too undergo significant changes during geomagnetic storms [*Prölss*, 1995 and references therein]. Analysis of simultaneously measured thermospheric temperatures and meridional winds from Mt. Abu (23.4°N, 73.4°E, 20° diplat.) along with the height of maximum electron density during individual storms have shown that the plasma neutral coupling is still quiet effective though not at the same level as during quiet times. The discrepancies are ascribed to electric field effects not being taken into the model estimates [*Pant and Sridharan*, 1998]. When it comes to the ionospheric domain, in addition to plasma densities the low latitude electric fields also get modified during storms warranting them to be duly accounted for.

3.2 Storm-time Electric Fields

It is established that significant disturbances of ionospheric current and electric field prevail at equatorial latitudes on time scales > 1h during and after geomagnetic storms and substorms. Two major sources of the electric field/current disturbances have been identified as indicated by the heavy lines in Plate 1. The first one is the solar wind-magnetospheric dynamo processes characterized by changes in high latitude plasma convection and polar cap potential distribution controlled by rapid changes in the Bz and By components of the interplanetary magnetic field, through the filed-aligned (Birkeland) currents that couple the high-latitude ionosphere with the magnetosphere. Normally the low latitude ionosphere is effectively insulated from the high latitude electric fields by the shielding ion layer at the inner edge of the

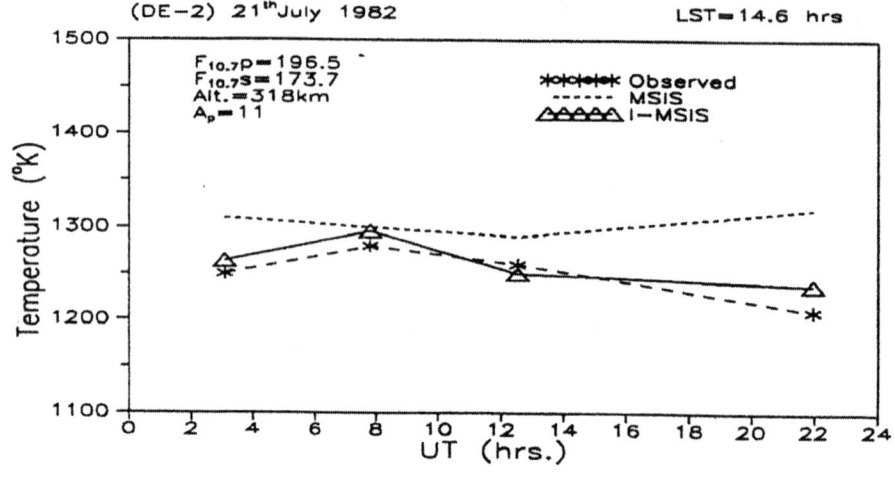

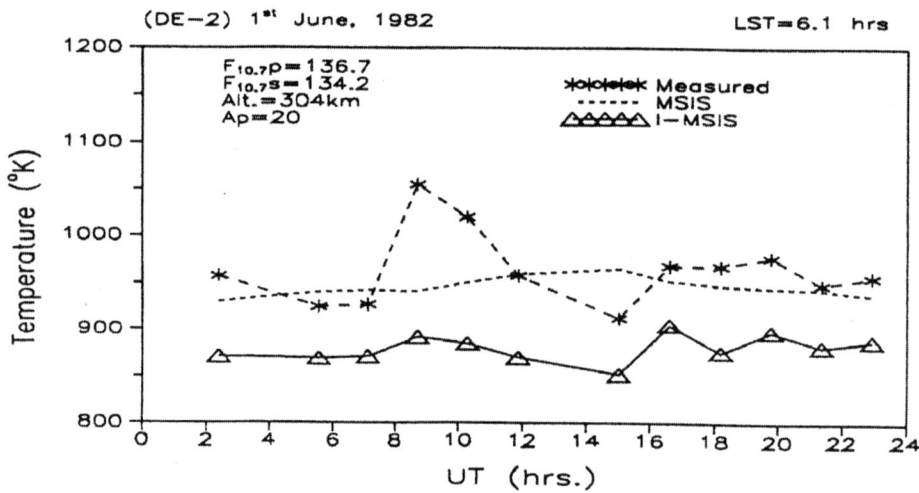

Figure 3. Examples of the effect of the parameterisation of local and nonlocal processes in explaining the deviations in the observed thermospheric temperatures from the model. Note the persistence of some differences on a few occasions although the variability is represented well by the improved MSIS model [after Pant and Sridharan, 2001c].

magnetospheric plasma sheet. But at times of rapid and prominent changes in convection, direct or prompt electric field penetration occurs leading to short-lived (~1h) electric field disturbances in the equatorial regions before shielding reasserts itself. These disturbances get superposed on the normal diurnal pattern of electric fields established by the quiet time ionospheric wind dynamo. A number of theoretical/ numerical global convection models have been developed over the years, enabling characterization of the prompt penetration electric fields as regards the local time dependence of their amplitude and polarity [e.g., *Senior and Blanc*, 1984; *Spiro et al*, 1988; *Fejer et al*, 1990, *Tsunomura*, 1999]. For a sudden increase in polar cap potential (under shielding condition), the models consistently predict the prompt penetration electric fields to be eastward (upward drift) during day and westward (downward drift) at night, i.e., in-phase with the quiet-time pattern with largest amplitudes in the post-midnight period. The polarity pattern of the penetration electric fields gets reversed for a sudden decrease in polar cap potential (over shielding condition). The meridional penetration electric fields, which are southward from midnight to noon and northwards at later times, undergo higher attenuation at equatorial latitudes when compared to the zonal electric fields [*Fejer et al*, 1990].

Substantial evidence for the prevalence of prompt penetration (pp) electric fields in the dip equatorial region due to under shielding and over shielding effects and their agreement with theoretical results have been documented during geomagnetic sudden commencements, sudden transitions in IMF Bz, auroral electrojet activity and symmetric/asymmetric ring current development, all of which in turn are intricately related to one another [*Fejer*, 1991, 1997 and references therein; *Sastri et al*, 1992a,b; 1993; 1997b; *Forbes et al*, 1995; *Fejer and Scherliess*, 1995, *Abdu et al*, 1995, 1997]. The robust result is the common occurrence of large transient eastward electric fields (upward drifts) near the dip equator during the postmidnight hours resulting in an elevated F region which constitutes a necessary (and sometimes sufficient) condition for the onset of equatorial spread F (ESF). It should be noted, however, that experimental evidence for penetration effects in the equatorial vertical electric field (F region zonal plasma drift) has not yet been found, a feature that is consistent with the predictions of the global convection models [*Fejer et al*, 1990].

The second major source of electric field disturbances at low and equatorial latitudes is the ionospheric disturbance dynamo (IDD) wherein the modification in the global thermospheric circulation brought about by storm-time heating at high latitudes generates a large scale system of currents and electric fields [*Blanc and Richmond*, 1980]. This physical process sets up two separate current vortices in the polar and equatorial regions and the current associated with the equatorial disturbance vortex is westward and in opposite direction to that of the quiet-time Sq current. The IDD electric fields are slowly varying and persistent in nature and manifest with a time delay (> 6 hrs) with reference to energy deposition at high latitudes. The time delay corresponds to the growth time of the storm-time wind system at mid latitudes because the wind magnitudes near the equator are insignificant in the *Blanc and Richmond* [1980] model. Credible evidence for IDD electric fields based on analysis of individual events as well as statistical studies is available in the literature [*Fejer et al*, 1983; *Sastri*, 1988; *Mazaudier and Venkateswaran*, 1990; *Fejer and Scherliess*, 1995].

Recent progress in the study of equatorial electric field disturbances has been on (a) Evaluation and comparison with theoretical predictions of the local time dependence of prompt penetration (pp) electric fields and assessment of the longitudinal extent of IDD electric fields, from simultaneous observations from different longitude sectors [*Sastri et al*, 1997b; *Abdu et al*, 1997], (b) increasing awareness of how the simultaneous presence of pp and IDD fields could affect the visibility of pp fields depending on their relative amplitude and polarity [*Abdu et al*, 1997; *Fejer and Scherliess*, 1997; *Sobral et al*, 1997], (c) improving the theoretical models of pp fields taking into account the modifications in ionospheric conductivity near the magnetic equator due to the meridional current sytem driven by the external electric field [*Tsunomura*, 1999] and (d) development of empirical climatological models of both pp and IDD fields from statistical analysis of the extensive vertical plasma drift data of Jicamarca, Peru [*Fejer and Scherliess*, 1997; *Scherliess and Fejer*, 1997]. It has been known for a long time that the response of the equatorial zonal electric field to changes in high latitude convection is asymmetric in that, large and short-lived (life time 2–3 hrs) electric fields commonly manifest only in association with decreases in convection and not with increases in convection [*Fejer*, 1986 and references therein]. The statistical results of *Fejer and Scherliess* [1995] strongly suggest that this asymmetric response may be due to the presence of IDD fields with a polarity opposite to that of penetration electric fields resulting from increases in magnetospheric convection (item b mentioned above). It is to be noted however that transient electric field disturbances with a polarity pattern in agreement with theoretical results do appear sometimes during both the growth and recovery phases of individual isolated auroral electrojet activity [*Sastri et al*, 1992 a, b]. Furthur studies with well identified substorms are necessary to characterize and understand the perturbations in the equatorial ionospheric electric fields/currents during the different substorm phases. This is because of the event-to-event variability of the relative contribution of the directly driven (DD) and unloading (UL) processes to substorm activity in general and substorm phases in particular [*Kamide et al*, 1996; *Kamide and Kokubun*, 1996 and references therein; *Sun et al.*, 2000]. The very recent studies of *Kikuchi et al* [2000] and *Sastri et al* [2001] are cases in point. They have shown that the negative geomagnetic bay in the afternoon sector associated with the expansion phase onset of substorms exhibited a remarkable enhancement of amplitude over the dip equator, strongly suggestive of a contribution of ionospheric currents to the negative bay, in addition to the 3-dimensional current system in the magnetosphere.

3.3 Empirical Models of Electric Field Disturbances

A significant development in recent times is the availability of empirical model of storm-time F region vertical plasma drifts for the 75°E longitude sector [*Fejer and Scherliess*, 1997; *Scherliess and Fejer*, 1997]. The model enables separation of the effects of the magnetospheric dynamo and the ionospheric disturbance dynamo and assessment of their efficiencies and interdependencies by comparison with data for specific events. Figure 4 illustrates the good agreement between the perturbation vertical drifts and zonal electric fields from the empirical model and those from the Rice convection model for an increase in the polar cap potential by 33 kV, i.e., under shielding condition. It needs to be verified whether significant changes in polar cap potential can occur on time scales < 10 minutes as assumed in the simulations. The bottom panel in Figure 4 illustrates the excellent agreement between (i) the disturbance-time vertical drift patterns

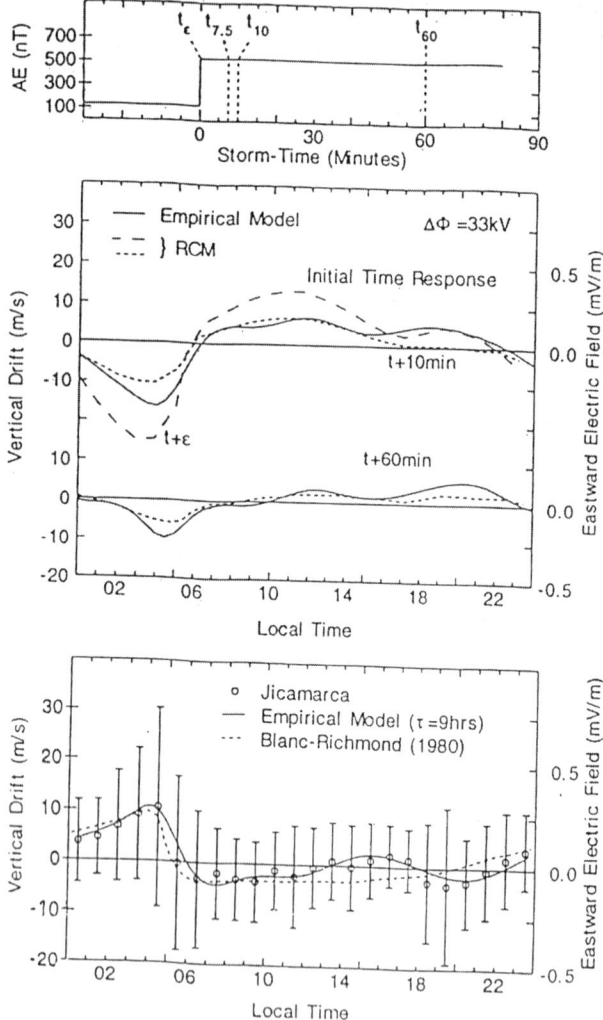

Figure 4. Top two panels: Local time variation of prompt penetration electric fields from the empirical model and from the Rice convection model for an increase in polar cap potential drop by 33 kV [after Scherliess and Fejer, 1997]; bottom panel: Local time variation of the empirical ionospheric disturbance dynamo drifts and that from the Blanc and Richmond (1980) model, both for an increase in AE of about 400 nT over the quiet time level [after Fejer and Scherliess, 1997]

(delay 9 h) from the empirical model, (ii) the ionospheric disturbance dynamo theory of *Blanc and Richmond* [1980] and (iii) the Jicamarca radar observations, for an increase in hemispheric power input corresponding to AE of 400 nT over the quiet time interval. An intriguing result borne out of the study of *Scherliess and Fejer* [1997] is the evidence for electric field disturbances in the postmidnight (00-04 SLT) and forenoon (09-12 SLT) sectors with long delays (22-28 hours). It is interesting to note in this context that the neutral temperature at low latitudes also responds to geomagnetic activity with delays of the same order as mentioned earlier (Figure 2).

Though the empirical models are found to predict reasonably well the pp and IDD electric fields for specific intervals, it is realized that additional factors may have to be taken into account for more accurate predictions. A case study comparing vertical drifts from Jicamarca with model predicted perturbation drifts (pp and IDD) demonstrated as to how a combination of in-phase drifts due to prompt penetration and disturbance dynamo mechanisms can sometimes lead to very large upward drifts/eastward electric fields (60 m/sec or 1.5 mV/m) in the presunrise period (Figure 5). *Maruyama et al* [1998] used the empirical vertical drift model in a theoretical retrospective modeling of the response of the equatorial F region to geomagnetic disturbances. They showed that the increase of F region electron density and decrease in electron temperature at 600 km observed by the HINOTORI satellite in the equatorial dawn sector during disturbed geomagnetic conditions as mainly caused by the substorm-related large upward vertical plasma drifts (eastward electric fields). Development of similar models for other longitudes is warranted and the existing ionospheric and geomagnetic databases could innovatively be used for the purpose [*Sastri*, 1990].

3.4 Equatorial Ionization Anomaly (EIA)

The altitude/latitude structure of the equatorial F region plasma is controlled by electrodynamics and the plasma transport associated with it, with the meridional winds acting as a modulator of field-aligned plasma transport [e.g., *Anderson*, 1981; *Bailey et al.*, 1997 and references therein]. Electric field disturbances are thus the most important contributor to the storm-time behavior of equatorial F region including the EIA. The slowly varying ionospheric disturbance dynamo (IDD) electric fields which are westward at all local times except during 23-06 LT can lead to severe inhibition or weak development of EIA during daytime as well as postsunset hours depending on their strength [*Sastri*, 1988]. On the other hand, the transient prompt penetration electric fields can cause expansion or contraction of EIA

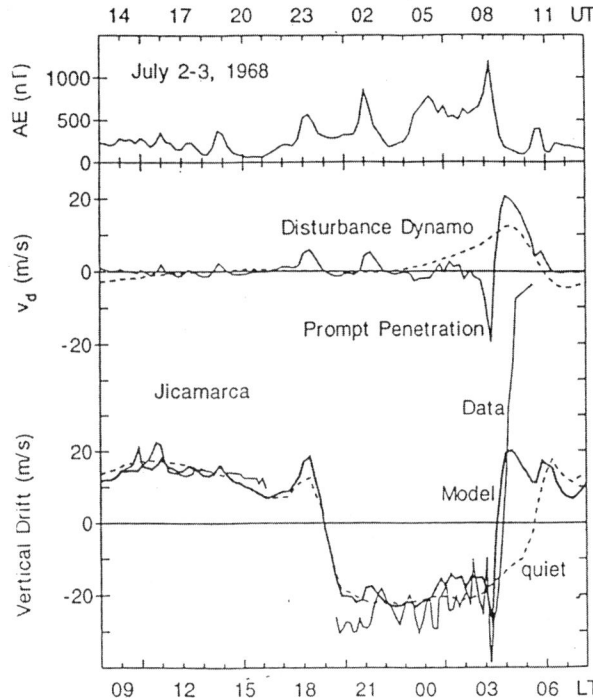

Figure 5. (top panel) Temporal variation during July 2-3, 1968 of the auroral electrojet index, AE; (middle panel) prompt penetration (solid line) and disturbance dynamo (dashed line) vertical plasma drifts calculated from the empirical model; (bottom panel) the average quiet time drifts (dashed line) and the model vertical drifts obtained by adding the quiet time drifts and perturbation drifts. Note the occurrence of very large upward drifts in the post-midnight period due to the combination of penetration and disturbance dynamo electric fields [after Fejer and Scherliess, 1997].

depending on their polarity and amplitude as discussed in detail by *Abdu et al* [1991, 1993].

With regard to the neutral dynamics, the global scale perturbations in thermospheric density, temperature and wind field during geomagnetic storms may some times result in traveling atmospheric disturbances (TAD) that propagate away from the source region. The superposition of TADs on the quiet time neutral circulation due to solar forcing may lead to convergence/ divergence of neutral wind flows over/from the equator and at times even to strong transequatorial winds. These rapid wind variations can significantly influence the EIA [*Fesen et al.*, 1989; *Hajkowicz*, 1991; *Burns and Killeen*, 1992; *Balthazor and Moffett*, 1997; *Knipp et al*, 1998]. The very recent study of the equatorial ionospheric storm of early November 1993 reinforced this view [*Sastri et al.*, 2000]. The prominent morning decrease in the equatorial electrojet strength (Δ SdI) on 4 November 1993 (Ap=77) due to prompt electric field penetration is accompanied by an abnormal positive latitudinal gradient in foF2 from the dip equator to higher latitudes resembling a well developed EIA (Figure 6), instead of the expected delayed onset of the development of EIA due to inhibition of the plasma fountain [*Abdu et al*, 1991, 1993]. This unusual behavior finds explanation in terms of horizontal plasma transport by transe-

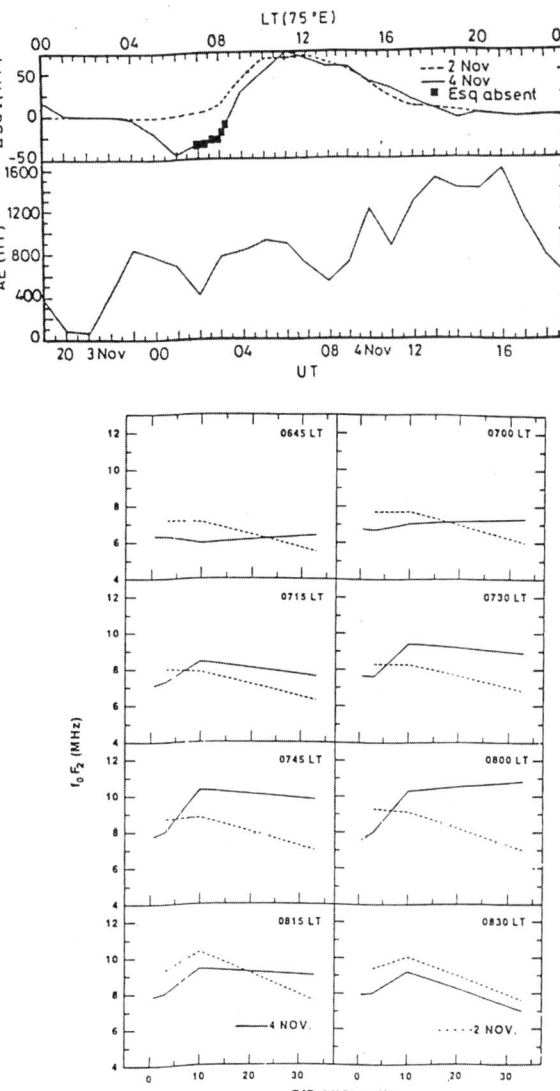

Figure 6. (a) Time variation of the equatorial electrojet index, DSdI in the Indian sector and AE index on 4 November 1993 (Ap=77), and (b) dip angle variation of foF2 over the time interval 0645-0830 LT on November 1993. The corresponding temporal profiles of the parameters on November 2, 1993 (Ap=4), the reference quiet day, are also shown. Note the abnormal positive latitudinal gradient in foF2 during the time of the morning counter electrojet (CEJ) condition on 4 November [after Sastri et al, 2000].

quatorial winds due to TADs predicted by the TIEGCM simulations for this storm event [*Emery et al.,* 1999]. The dominant influence of thermospheric disturbances on the F region parameters continued during the later part of the day as coherent wave-like fluctuations in the height of F layer peak density (hpF2) around the EIA crest location accompanied by a sequence of inhibition and development of EIA. The *inhibition* of EIA over the interval 0900-1100 LT is due to the equatorward wind associated with the TAD opposing the field-aligned plasma diffusion away from the dip equator, while the delayed noontime development of EIA (1115-1300 LT) is due to the *renewal* of the plasma *fountain* due to poleward neutral winds (Figure 7). The telltale signature of the meridional wind induced modulation of EIA is the anti-phase relationship between height and density of F layer peak around the EIA crest location which is precisely what is observed (not shown here).

Though 'positive' ionospheric F region storms are commonly noticed at equatorial latitudes due to the westward IDD electric fields, 'negative' storms do occur on occasions [*Sastri,* 1980; *Huang and Chang,* 1991; *Walker and Wong,* 1993]. The case study of the ionospheric storm of November 4–6, 1958 in the Indian equatorial region indicated that changes in thermospheric neutral composition (enhanced N_2/O density ratio) and not in the zonal electric fields are responsible for the long-duration reductions in daytime foF2 throughout the EIA region, similar to the typical 'negative' storms at mid latitudes [*Sastri,* 1980]. The source of the causative composition disturbance is still unclear.

3.5 Equatorial Spread (ESF)

A number of interactive processes (Plate 1) lead to the post sunset ESF, the basic prerequisite being the pre rever-

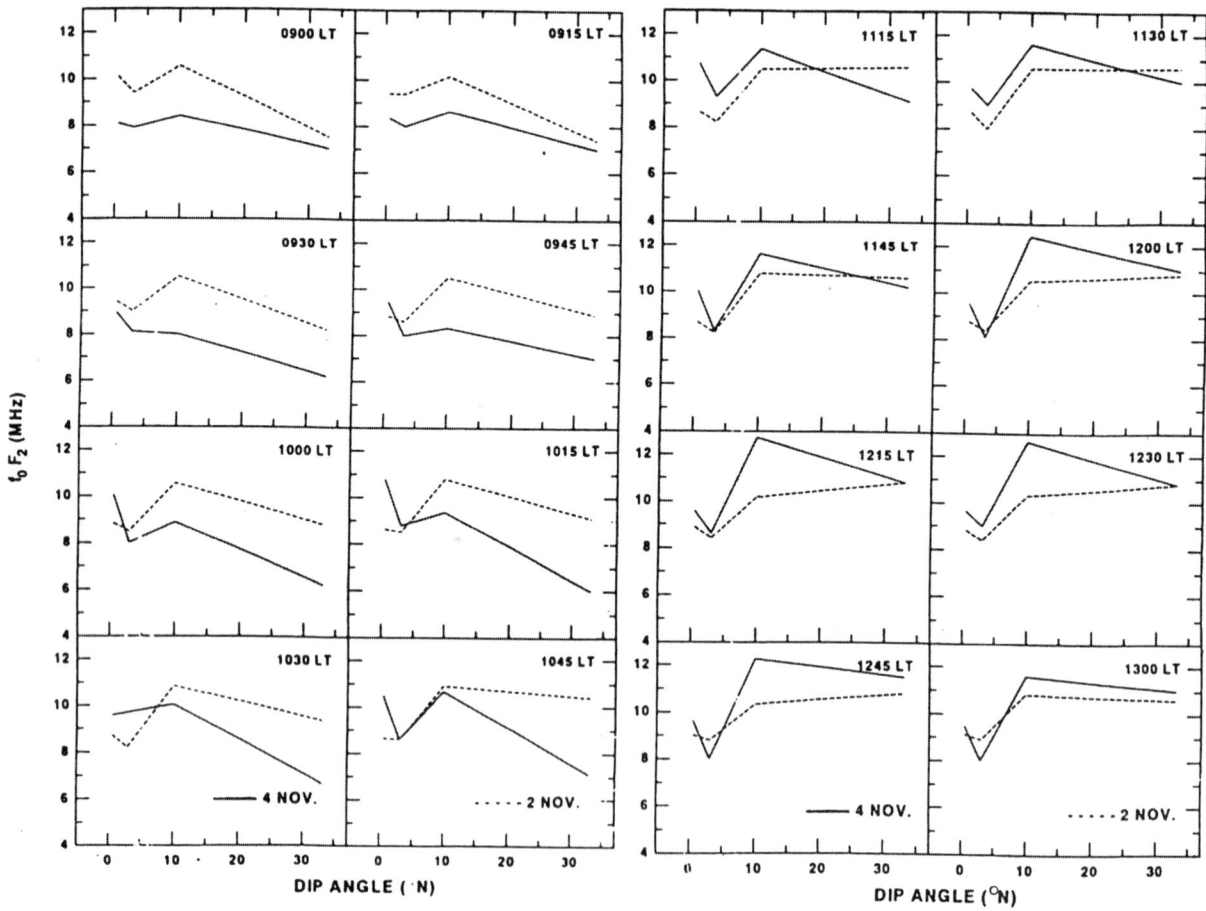

Figure 7. Dip angle variation of f_0F2 on November 4, 1993 in the Indian sector illustrating the inhibition of EIA during the forenoon period (0915-1030 LT) and delayed noon-time development (1115-1300 LT) of EIA [after Sastri et al., 2000].

sal enhancement (PRE) in the vertical plasma drift which, in turn, is controlled by the longitudinal gradient in E region conductivity and zonal winds. Meridional winds, vertical winds and seed perturbations have also been found/proposed to play important roles. Most of these parameters undergo changes during disturbed geomagnetic conditions as discussed earlier. It is therefore natural to expect the normal ESF cycle of postsunset initiation, growth and sustenance during the night to undergo modifications during disturbed conditions, depending on the nature and magnitude of the changes in the controlling parameters. Statistical and case studies of data from a variety of experiments revealed (a) inhibition or dramatic development of ESF in the postsunset hours when ESF usually develops [*Aarons*, 1991; *Abdu et al*, 1995; *Basu et al*, 2001] and (b) the generation of ESF afresh in the postmidnight period, when normally ESF is rare and is not affected by the post sunset processes [*Sastri*, 1979; *Kelley and Maruyama*, 1992; *Hysell and Burcham*, 1998; *Fejer et al*, 1999, *Palmroth et al*, 2000; *Yeh et al*, 2001]. These widely observed patterns are explained as the effects of electric field disturbances due to the ionospheric disturbance dynamo (IDD) and prompt penetration (pp) mechanisms acting independently or in tandem. The local time dependence of the polarity and amplitude of electric field disturbances from these two processes, in fact, determines the favorable or unfavorable conditions for ESF at a given location. As regards post sunset ESF, it is found that if the onset of the geomagnetic activity occurs a few hours prior to local sunset, then, inhibition of ESF takes place due to the superposition of westward electric fields of IDD on the normal prereversal enhancement (PRE) depending on their strength [*Aarons*, 1991; *Abdu et al.*, 1995]. On the other hand, the development in isolation of symmetric/asymmetric ring current or the occurrence of a rapid increase in high latitude convection at or quite close to local sunset, may superimpose a prompt penetration eastward electric field on the normal F layer dynamo-induced PRE and result in a dramatic development of ESF including formation of plasma bubbles [*Abdu et al*, 1995]. A variant of this situation is the simultaneous presence of perturbation electric fields of opposite polarity, i.e., westward IDD fields and eastward penetration fields associated with increases in convection. *Basu et al* [2001] found this to be the case during the main phase of the severe storm of July 15, 2000, when the penetration of an eastward electric field associated with a rapid growth of the ring current as seen in *SYM-H* index (130nT/hr) resulted in a sudden and simultaneous onset of L-band and VHF scintillations at Ascension Island (7.9° S, 14.4° W) (Figure 8). There was however no scintillation activity on this night in the 75° W sector indicating the longitudinal confinement of the ESF irregularities, a feature noted earlier during some other storms. But the noteworthy point here is that the ring current effect on post sunset ESF occurred when the IDD mechanism is active as indicated by large westward plasma drifts seen in ROCSAT-1 measurements (left panel of Figure 8) as well as in zonal drifts of ESF irregularities. The recent study of *Fejer et al* [1999] showed that prolonged geomagnetic activity during daytime, in general, causes a large reduction in the occurrence of post sunset ESF at Jicamarca but only during equinoxes at solar maximum. While the storm-time electric field disturbances can either assist or inhibit post sunset ESF, the disturbances in the transequatorial winds always inhibit especially the formation of plasma bubbles, if not bottomside ESF [*Abdu et al*, 1995, 1997]. During geomagnetic storms, because of the simultaneous heating of both the polar regions, there could be converging winds carrying energy and momentum towards low and equatorial latitudes. It would imply then that, in general, the polarity and magnitude of both zonal electric fields and vertical winds could effectively determine the onset and growth of ESF. Since the EIA responds to changes in electric field or meridional wind with a delay of 2–3 hours, it is necessary to establish through further studies the temporal sequence of the changes in the meridional winds, EIA, vertical winds and their effect on the storm-time behavior of the generation and growth of ESF in the postsunset period. In contrast, geomagnetic activity is commonly found to initiate ESF conditions in the post midnight sector due to anomalous upward reversals in vertical plasma drift under the action of eastward electric fields of IDD and prompt penetration mechanisms [*Kelley and Maruyama*, 1992; *Hysell and Burcham*, 1998]. The geomagnetic activity-related presunrise ESF is recently found to be quite common around solar minimum than during solar maximum suggestive of a control of the downward plasma drifts that usually prevail in this local time sector [*Fejer et al*, 1999]. At solar minimum, when the ambient downward drifts are small in the presunrise hours and therefore the eastward electric field disturbances can easily cause upward reversals in the vertical drift and hence result in ESF, while at solar maximum, relatively large eastward fields would be required to overcome the large ambient downward drifts that usually prevail. This is consistent with the finding of *Sastri* [1979] that the occurrence of anomalous and substantial increase of F layer height in the presunrise hours around solar maximum is infrequent (only 12 events over the four year period, 1957–1960) at Kodaikanal (dip lat.3N). More systematic investigations are called for in delineating the exact role of the ionospheric and thermospheric parameters in the trig-

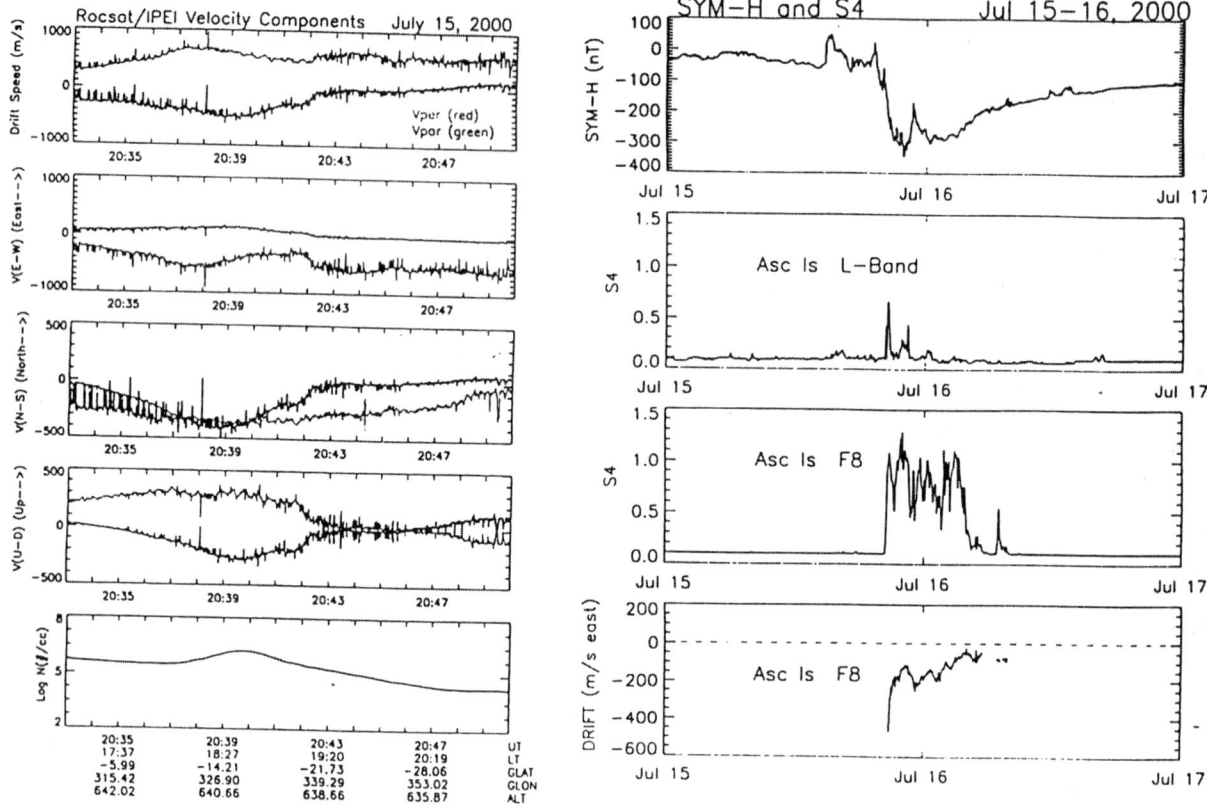

Figure 8. (a) Time histories during July 15-16, 2000 of the SYM-H index and scintillations at 250 MHz and L-band (1.537 GHz) signals from geostationary satellites, F-8 and Inmarsat and irregularity drifts at Ascension Island located under the southern crest of the EIA in the South American sector; (b) ROCSAT-1 satellite measurements on July 15, 2000 of ion drifts parallel (Vpar) and perpendicular (Vper) to the magnetic field and the east-west, north-south and upward-downward components of Vpar and Vper. The bottommost panel shows the ion density variation along the satellite track [after Basu et al., 2001].

gering and sustenance of ESF during post midnight hours of disturbed periods.

4. OUTSTANDING ISSUES AND DIRECTIONS FOR FURTHER WORK

We have presented a review of recent work on geomagnetic disturbance-time behavior of EITS with specific reference to the key parameters of the ionized and neutral domains and the major phenomena (EIA and ESF) of the coupled system. While there is substantial progress, there are still some unresolved problems for realizing reliable forecasting and nowcasting capabilities of storm-time EITS that have immense practical applications. These include the following:

(1) The well-known discrepancy between observations and theoretical models regarding the lifetime of prompt penetration electric fields associated with convection decreases (over-shielding effects) needs to be resolved.

(2) At any given longitude, IDD electric fields they do not follow every episode of strong and persistent auroral electrojet activity. Whether this is because of the longitudinal confinement of the thermospheric circulation disturbance generated by high latitude heating or because of its dependence on the duration and amount of excess energy deposited at high latitudes or both is unclear. The former would trigger off TADs while the latter would alter the global thermospheric circulation [*Fujiwara et al* 1996]. The possible relationship of IDD electric fields with the type of magnetic storm (i.e., Type 1 and 2) therefore merits an assessment.

(3) The question of appearance and non-appearance of prompt penetration electric fields is also relevant. Whether the non-appearance is always because of

masking by IDD fields or due to some other processes needs to be ascertained.

(4) The characteristics of substorm-related electric fields need to be established through analysis of well-identified substorms and contrasted with those of the disturbances associated with other forms of auroral electrojet activity.

(5) As regards the neutral atmospheric response to geomagnetic storm energy input is concerned, identification and quantification of the contributions of the internal and external sources to the overall energetics of the low latitude-equatorial thermosphere should be given top priority.

(6) The changes in the lower thermosphere around the turbopause level during geomagnetic storms, their latitudinal dependencies and consequences have to be systematically studied by coordinated experimental and by theoretical modeling efforts.

(7) The role of meridional winds (converging and diverging) and the gravity wave seed perturbations, their behavior during different phases of geomagnetic storms and possible impact on the evolution and sustenance of ESF is an important aspect that calls for detailed studies.

5. LOOKING INTO FUTURE

Studies of Solar Terrestrial Physics in general and Ionosphere-Thermosphere (IT) interactions and their response to various external forcing in particular, have been receiving a new thrust in recent years due to their practical implications with regard to the performance of technological systems in space and on ground and the ever increasing dependence of modern society (e.g., civil aviation, communications) on these systems. The equatorial IT processes have gained quite a bit of importance due to the realization that our comprehension of them remains incomplete. Coordinated studies initiated by WITS (World Ionosphere Thermosphere Studies) and followed up under STEP (Solar Terrestrial Energy Program) and SRAMP, as also the several adjunct national-level programs like CEDAR (Coupling Energetic Dynamics of the Atmospheric Regions) have enabled identification of specific unresolved problems. Recent project initiatives by SCOSTEP (Scientific Committee for Solar Terrestrial Physics) like the EPIC (Equatorial Process Including Coupling) and the PSMOS (Planetary Scale Mesopause Observing Systems) continue to focus on the complex problems pertaining to the upper atmosphere. Based on the rich experience gained through these earlier concerted efforts, SCOSTEP has come out with a proposal called CAWSES (Climate And Weather of the Sun Earth System) to be taken up in 2003 with the intervening year of 2002 serving as the preparatory period. The scope of CAWSES is enlarged to encompass the climatic aspects of the Sun Earth System and is thus evolving as a logical sequel to STEP and SRAMP initiatives where the emphasis had been on 'Space Weather' processes.

Several coordinated programs and dedicated satellite missions are in the offing/operational. These include the C/NOF Satellite proposed to be launched in 2002 by NRL, USA, the Indian Coherent Radio Beacon Experiment (CRABEX) to be flown on a geostationary satellite (GSAT-II) in 2002, the US geostationary satellite with UV imaging capability, the TIMED satellite mission of NASA (though at a high inclination orbit), the fillip given to the ionospheric tomography, total electron content and VHF and UHF scintillations (because of their direct relevance to satellite based navigation systems) using beacon signals from satellites like GPS/NNSS....etc. With the planned space platforms and the steadily expanding and wide ranging ground-based radio probing and optical remote sensing facilities, near-simultaneous measurements of all the relevant ionospheric-thermospheric parameters on a regular basis is becoming a reality. On the modeling front, significant strides have been made with the development of sophisticated theoretical models like TIEGCM, CTIP and SUPIM including their regular upgrades, and the user-friendly empirical models like IRI and MSISE—90, with reasonable success. With the fresh project initiatives by the international and national bodies and the scientific community, our understanding of EITS is bound to improve significantly in the years to come.

Acknowledgments. The authors are thankful to the Editors (A. S. Sharma, G. S. Lakhina and Y. Kamide) for the kind invitation to contribute the article.

REFERENCES

Aarons, J., The role of the ring current in the generation and inhibition of equatorial F layer irregularities during magnetic storms, *Radio Sci.*, 26, 1131-1149, 1991.

Abdu, M. A., J. H. A. Sobral., E. R. de Paula, and I. S. Batista, Magnetospheric disturbance effects on the equatorial ionization anomaly (EIA): an overview, *J. Atmos. Terr. Phys.*, 53, 757-771, 1991.

Abdu, M. A., G. O. Walker, B. M. Reddy., E. R. de Paula., J. H. A. Sobral., B. G. Fejer and E. P. Szuszczewicz, Global scale equatorial ionization anomaly (EIA) response to magnetospheric disturbances based on the May-June 1987 SUNDIAL—coordinated observations, *Ann. Geophysicae.*, 11, 585-594, 1993.

Abdu, M. A., I. S. Batista., G. O. Walker., J. H. A. Sobral., N. B. Trivedi, and E. R. de Paula, Equatorial ionospheric electric fields during magnetospheric disturbances: local time/longitudinal dependences from recent EITS campaigns, *J. Atmso. Terr. Phys,* 57, 1065-1083, 1995.

Abdu, M. A., Major phenomena of the equatorial ionosphere-thermosphere system under disturbed conditions, *J. Atmos. Solar-Terr. Phys.,* 59(13), 1505-1519, 1997.

Abdu, M. A., J. H. Sastri., J. MacDougall., I. S. Batista and J. H. A. Sobral, Equatorial disturbance dynamo electric fields longitudinal structure and spread-F: a case study from GUARA/EITS campaigns, *Geophys. Res. Lett.,* 24, 1707-1710, 1997.

Abdu, M. A., Outstanding problems in the equatorial ionosphere-thermosphere electrodynamics relevant to spread F, *J. Atmos. Solar. Terr. Phys.,* 63, 869-884, 2001.

Almar, I., Illes-Almar, A. Horvath, Z. Kollath, D. V. Bisikalo, and T. V. Kasimenko, Investigation and modeling of an improved geomagnetic term for the CIRA' 86 model at low latitudes, *Adv. Space. Res.,* 18(9/10), 375-379, 1996.

Anderson, D. N., Modeling the ambient low latitude F-region ionosphere-a review, *J. Atmos. Terr. Phys.,* 43, 753-762, 1981.

Bailey, G. J., N. Balan, and Y. Z. Su, The Sheffield University plasmasphere ionsophere model-a review, *J. Atmos. Solar. Terr. Phys.,* 59, 1541-1552, 1997.

Balthazor, R. L., and R. J. Moffett., A study of atmospheric gravity waves and travelling ionospheric disturbances at equatorial latitudes, *Ann. Geophysicae,* 15, 1048-1056, 1997.

Basu, B., and B. Coppi, Relevance of plasma and neutral wind profiles to the topology and excitation of modes for the onset of spread F, *J. Geophys. Res,* 104, 225-231, 1999.

Basu, S., Su Bau., K. M. Groves., H. C. Yeh., S.-Y.Su., F. J. Rich., P. J. Sultan., and M. J. Keskinen, Response of the equatorial ionosphere in the South Atlantic region to the great magnetic storm of July 15, 2001, *Geophys. Res. Lett.,* 28, 3577-3580, 2001.

Biondi, M. A., and J. W. Meriwether, Jr., Measured response of the equatorial thermospheric temperature to geomagnetic activity and solar flux changes, *Geophys. Res. Lett.,* 12, 267-270, 1985.

Blanc, M., and A. D. Richmond, The ionospheric disturbance dynamo, *J. Geophys. Res.,* 85, 1669-1688, 1980.

Buonsanto, M. J., G. Bust., R. Clark., M. Codrescu., G. Crowley., B. A. Emery., T. J. Fuller-Rowell., G. Hoogeveen., A. Jacobson., D. Knipp., S. Maaurits., P. G. Richards., J. R. Taylor., and W. J. Watkins, Recent results of the CEDAR storm study, *Adv. Space Res.,* 20(9), 1665-1664, 1997.

Burns, A. G., and T. L. Killeen, The equatorial neutral wind response to geomagnetic forcing, *Geophys. Res. Lett.,* 19, 977-980, 1992.

Burnside et al., The neutral thermsophere at Arecibo during geomagnetic storms, *J. Geophys. Res.,* 96, 1289-1301, 1991.

Devasia, C. V., N. Jyoti., K. S. V. Subbrarao., K. S. Viswanathan., Diwakar Tiwari., and R. Sridharan, On the possible linkage of thermospheric meridional winds with the equatorial spread F, *J. Atmos. Solar. Terr. Phys.,* 64, 1-12, 2002.

Eccles, J. V., A modeling investigation of the evening prereversal enhancement of the zonal electric field 1in the equatorial ionosphere, *J. Geophys. Res.,* 103, 26709-26719, 1998.

Emery, B. A., C. Lathuillere., P. G. Richards., R. G. Roble., M. J. Buonsanto and D. J. Knipp,. Time dependent thermospheric neutral response to the 2-11 November 1993 storm period, *J. Atmos. Solar-Terr. Phys.,* 61, 329-350, 1999.

Fagundes, P. R., Y. Sahai., J. A. Bittencourt., and H. Takahashi, Observations of Thermospheric neutral winds and temperatures at Cacheoria Paulista (23° S, 45° W) during a geomagnetic storm, *Adv. Space. Res.,* 16(5), 27-30, 1995.

Fagundes, P. R., Y. Sahai., J. A. Bittencourt., and H. Takahashi, Plasma drifts inferred from thermospheric neutral winds and temperature gradients observed at low latitudes, *J. Atmos. Terr. Phys.,* 58, 1219-1228, 1996.

Fejer, B. G., M. F. Larsen, and D. T. Farley, Equatorial disturbance dynamo electric fields, *J. Geophys. Res.,* 10, 537-540, 1983.

Fejer, B. G., R. W. Spiro, R. A. Wolf., and J. C. Foster, Latitudinal variation of penetration electric fields during magnetically disturbed periods: 1986 SUNDIAL observations and model results, *Ann. Geophysicae,* 8, 441-454, 1990.

Fejer, B. G., Low latitude electrodynamic plasma drifts: a review, *J. Atmos. Terr. Phys.,* 53, 677-693, 1991.

Fejer, B. G., and L. Scherliess, Time dependent response of equatorial ionospheric electric fields to magnetospheric disturbances, *Geophys. Res. Lett.,* 22, 851-854, 1995.

Fejer, B. G., E. R. de Paula., R. A. Heelis., and W. B. Hanson, Global equatorial ionospheric vertical plasma drifts measured by AE-E satellite, *J. Geophys. Res.,* 100, 5769-5776, 1995.

Fejer, B. G., and L. Scherliess, Emperical models of storm time equatorial zonal electric fields, *J. Geophys. Res.,* 102, 24,047-24,056, 1997.

Fejer, B. G., The electrodynamics of the low latitude ionosphere: recent results and future challenges, *J. Atmos. Solar-Terr. Phys.,* 59(13), 1465-1482, 1997.

Fejer, B. G., Effects of the vertical plasma drift velocity on the generation and evolution of equatorial spread F, *J. Geophys. Res.,* 104, 19,859-19,869, 1999.

Fesen, C. G., G. Crowley, and R. G. Roble, Ionospheric effects at low latitudes during the March 22, 1979, geomagnetic storm, *J. Geophys. Res.,* 94, 5405-5417, 1989.

Fesen, C. G., G. Crowley., R. G. Roble, A. D. Richmond, and B. G. Fejer, Simulation of the pre-reversal enhancement of the low latitude vertical ion drifts, *Geophys. Res. Lett.,* 27, 1851-1854, 2000.

Forbes, J. M., R. G. Roble., and F. A. Marcos, Thermospheric dynamics during the March 22, 1979 magnetic storm, 2. Comparisons of model predictions with Observations, *J. Geophys. Res.,* 92, 6069-6081, 1987.

Forbes, J. M., R. G. Roble, and F. A. Marcos, Equatorial penetration of magnetic disturbance effects in the thermosphere and ionosphere, *J. Atmos. Terr. Phys.,* 52, 1085-1092, 1995.

Fujiwara, H., S. Maeda., H. Fukunishi., T. J. Fuller-Rowell., and T. S. Evans., Global variations of thermospheric winds and temperatures caused by substrom energy injection, *J. Geophys. Res.*, 101, 225, 1996.

Fuller-Rowell, T. J., M. V. Codrescu., R. J. Mofett., and S. Quegan, Response of the thermosphere and ionosphere to geomagnetic storms, *J. Geophys. Res.*, 99, 3893-3914, 1994.

Fuller-Rowell, T. J., M. V. Codrescu, B. G. Fejer, W. Borer, F. Marcos, and D. N. Anderson, Dynamics of the low altitude thermosphere: quiet and disturbed conditions, *J. Atmos. Solar-Terr. Phys.*, 59, 1533-1540, 1997.

Gurubaran, S., R. Sridharan., R. Suhasini and K. G. Jani., Variabilities in the thermsopheric temperatures in the region of the crest of the equatorial ionization anomaly-a case study, *J. Atmos. Terr. Phys.*, 57, 695-704, 1995.

Hajkowicz, L. A., Global onset and propagation of large-scale travelling ionospheric disturbances as a result of the great magnetic storm of 13 March 1989, *Planet. Space Sci.*, 39, 583-593, 1991.

Hays, P. B. and R. G. Roble, Direct observations of thermopsheric winds during geomagnetic storms, *J. Geophys. Res.*, 76, 5316-5321, 1971.

Hanson, W. B., B. L. Cragin, and A.Dennis, The effect of vertical drift on equatorial F region stability, *J. Atmos. Terr. Phys.*, 48, 205-22, 1986.

Hargreaves, J. K., The *Solar-Terrestrial Environment—An Introduction to Geospace—The Science of the Terrestrial Upper Atmosphere, Ionosphere, and Magnetosphere*, Cambridge University Press, New York, 1992.

Hedin, A. E., and H. G. Mayr, Mgnetic control of the near equatorial neutral atmosphere, 8, 1669-1672,1973

Hedin, A. E., A revised thermospheric model based on mass spectrometer and incoherent scatter radar data, *J. Geophys. Res.*, 88, 10170-10188, 1983.

Hedin, A. E., MSIS-86 Thermospheric model, *J. Geophys. Res.*, 92, 4649-4662, 1987.

Hedin, A. E., Extension of the MSIS thermosphere model into the middle and lower atmosphere, *J. Geophys. Res.*, 96, 1159, 1991.

Hernadez, G., R. G. Roble., and J. H. Allen, Midlatitude thermospheric winds and temperatures and their relation to the auroral electrojet activity index, *Geophys. Res. Lett.*, 7, 677-700, 1980.

Hernandez, G., T. I. Van Zandt., V. L. Patterson and J. P. Turtle, Comparison of optical and incoherent scatter measurements of nighttime exospheric temperature at the magnetic equator, *J. Geophys. Res.*, 80, 3271-3274, 1995.

Huang, Y.–N. and K. Cheng, Ionospheric disturbances at the equatorial anomaly crest region during the March 1989 magnetic storm, *J. Geophys. Res.*, 96, 13,953- 3,965, 1991.

Hysell, D. L., and J. D. Burcham, JULIA radar studies of equatorial spread F, *J. Geophys. Res.*, 103, 29,155-29,167, 1998.

Ishimoto, M., M. R. Torr, P. G. Richards and D. G. Torr, The role of energetic O^+ precipitation in midlatitude aurora, *J. Geophys. Res.*, 91, 5793-5802, 1986.

Kamide, Y., W. Sun and S.-I. Akasofu, The average ionospheric electrodynamics for the different substorm phases, *J. Geophys. Res.*, 101, 99-109, 1996.

Kamide, Y., and S. Kokubun, Two-component auroral electrojet: Importance for substorm studies, *J. Geophys. Res.*, 101, 13,027-13,046, 1996.

Kamide, Y., W. Baumjohann., I. A. Daglis., W. D. Gonzalez., M. Grande., J. A. Joselyn., R. L. McPherron., J. L. Phillips., E. G. D. Reeves., G. Rsotoker., A. S. Sharma., H. J. Singer., B. T. Tsutani, and V. M. Vasyliunas., Current understanding of storm-substorm relatioships, J. Geophys. Res., 103, 17,705-17,228, 1998.

Kelley, M. C., and M. F. Larsen, C. A. La Hoz, and J. P. McClure, Gravity wave initiation of equatrial spread F: A case study, *J. Geophys. Res.*, 86, 9087-9100, 1981.

Kelley, M. C., The Earth's Ionosphere, Plasma Physics and Electrodynamics, Academic, New York, 1989

Kelley., M. C., and T. Maruyama, A diagnostic model of equatorial spread-F 2. The effect of magnetic activity, *J. Geophys. Res.*, 97, 1271-1277, 1992.

Kikuchi, T., H. Luhr, H. Tachihara, M. Shinohara and T.–I. Kitamura, Penetration of auroral electric field to the equator during a substorm, *J. Geophys. Res.*, 105, 23,251-23,261, 2000.

Klostermeyer, J., Numerical calculation of gravity wave propagation in a realistic thermosphere, *J. Atmos. Terr. Phys.*, 34, 765-775, 1973.

Knipp, D. J., et al., An overview of the early November 1993 geomagnetic storm, *J. Geophys. Res.*, 103, 26,197-26,220, 1998.

Kozrya, J. W., T. E. Cravens and A. G. Nagy, Energetic O^+ precipitation, *J. Geophys. Res.*, 84, 2481-2486, 1982.

Lakkso, K., T. L. Aggson., R. F. Pfaff, and W. B. Hanson, Downdrafting plasma flow in equatorial bubbles, *J. Geophys. Res.*, 99, 11,507-11,515, 1994.

Lui, A. T. Y et al., Current disruption in the earth's magnetosphere: Observations and models, *J. Geophys. Res.*, 101, 13,067-13,088, 1996.

Maruyama, N., S. Watanabe, H. Fukunishi, K. -I. Oyama, B. G. Fejer and L. Scherleiss, *Modeling of the response of the low latitude ionosphere to substorm activities in Substorms-4* (Edited by S. Kokubun and Y. Kamide), pp 115-118, Terra Scientific Publishing Company, Japan, 1998.

Maynard, N. C., T. L. Aggson, F. L. Herrero., M. C. Liebrecht and J. L. Saba, Average equatorial zonal and vertical ion drifts determined from San Marco D electric field measurements, J. Geophys. Res, 100, 17,465-17,479, 1995.

Mazaudier, C. and S. V. Venkateswaran, Delyed ionospheric effects of the geomagnetic storms of March 22, 1979 studied by the sixth co-ordinated data analysis workshop (CDAW-6), *Ann. Gephysicae*, 8, 511-518, 1990.

Mendillo, M., J. Baumgardner, X.-Q. Pi, P. J. Sultan and R. T. Tsunoda, Onset conditions for equatorial spread F, *J. Geophys. Res.*, 97, 13,865-13,876, 1992.

Mendillo, M., J. W. Meriwether, and M. Biondi, Testing the thermospheric neutral wind suppression mechanism for day-to-day variability of equatorial spread F, *J. Geophys. Res.*, 106, 3655-3663, 2001.

Meriwether, J. W., J. W. Moody, M. A. Biondi, and R. G. Roble, Optical interferomteric measurements of nightime equatorial thermsopheric winds at Arequipa, Peru, *J. Geophys. Res.,* 91, 5557-5566, 1986.

Palmroth, M., H. Lakkso, B. G. Fejer and R. F. Pfaff, Jr, DE 2 observations of morningside and eveningside plasma density depletions in the equatorial ionosphere, *J. Geophys. Res.,* 105, 18,429-18,442, 2000.

Pant, T. K., and R. Sridharan, A case study of the low latitude thermsophere during Geomagnetic storms and its new representation by improved MSIS model, *Ann. Geophysicae.,* 16, 1513-1518, 1998.

Pant, T. K., and R. Sridharan, Plausible explanation for the equatorial temperature and wind anomaly (ETWA) based on chemical and dynamical processes, J. *Atmos. Solar Terrs. Phys.,* 63, 885-891, 2001a.

Pant, T. K., and R. Sridharan, Seasonal dependence of the response of the low latitude thermosphere for external forcings, *J. Atmos. Solar Terrs. Phys.,* 63, 987-992, 2001b.

Pant, T. K., and R. Sridharan, Parameterization of the local and non-local processes in the thermospheric energy budget based on DE-2 satellite data, J. *Atmos. Solar Terrs. Phys.,* 63, 1705-1715, 2001c.

Patra, A. K., and P. B. Rao, V. K. Anandan and A. R. Jain, Radar observations of 2.8 m equatorial spread F irregularities, *J. Atmos. Terr. Phys.,* 59, 1663-1641, 1997.

Prolss, G. W., Perturbations of the low latitude upper atmosphere during magnetic substorm activity, *J. Geophys. Res.,* 87, 5260-5266, 1982.

Prolss, G. W., and M. Roemer, Thermospheric storms, *Adv. Sapce. Res.,* 7 (10), 223-229, 1987.

Prolss, G. W., Ionospheric F-region storms, in *Handbook of Atmospheric Dynamics,* Vol 2, (Ed) H. Volland, p 195, CRC Press, Boca Raton, Fla, 1995.

Raghavarao, R., P. Sharma, and M. R. Sivaraman, Correlation of ionization anomaly with the intensity of the electrojet, *Space Res.,* XVIII, 277-280,1978

Raghavarao, R., M. Nageswararao., J. Hanumath Sastri., G. D. Vyas, and M. Sriramarao, Role of equatorial ionization anomaly in the initiation of equatorial spread F, *J. Geophys. Res.,* 93, 5959-5964, 1988.

Raghavarao, R., L. E. Wharton., N. W. Spencer., H. G. Mayr, and L. H. Brace, An equatorial temperature and wind anomaly (ETWA), *Geophys. Res. Lett.,* 18, 1193-1196, 1991.

Raghavarao, R., L. E. Wharton., N. W. Spencer., W. R. Hoegy, Neutral temperature anomaly in the equatorial thermosphere-a source of vertical winds, *Geophys. Res. Lett.,* 20, 1023-1026, 1993.

Raghavarao, R., R. Suhasini, W. R. Hoegy, Spencer, H. G. Mayr, and L. E. Wharton, Local time variation of equatorial temperature and wind anomaly (ETWA), *J. Atmos. Solar-Terr. Phys.,* 60, 1193-1196, 1998.

Ranjan Gupta., J. N. Desai., R. Raghavarao, R. Sekar, R. sridharan and R. Narayan, Excess heating over the equatorial latitudes during storm sudden commencement, *Geophys. Res. Lett.,* 13, 1055-1058, 1986.

Reddy, C. A., Equatorial electrojet: A unique space domain, *PINSA,* 64A, 353-364, 1998.

Rishbeth, H., and O. K. Garriott, Introduction to Ionospheric Physics, Academic Press, New York, 1969.

Roelef, E. C., Energetic neutral atom image of a storm time ring current, *Geophys. Res. Lett.,* 14, 652-655, 1987.

Sastri, J. H., Onset of equatorial spread-F during post-midnight period, *Curr. Sci.,* 48, 12-13, 1979.

Sastri, J. H., Ionosopheric storm of 4-6 December 1958 in the Indian equatorial region, *Indian J. Radio. Space Phys.,* 9, 209-213, 1980.

Sastri, J. H., Duration of equatorial spread F, A*nn. Geophys.,* 2, 353-358, 1984.

Sastri, J. H., Equatorial electric fields of ionospheric disturbance dynamo origin, A*nn. Geophys.,* 6, 635-642, 1988.

Sastri, J. H., Equatorial anomaly in F region- A review, Indian J. Radio. Space Phys., 19, 225-240, 1990.

Sastri, J. H., H. N. R. Rao, and K. B. Ramesh, Transient composite electric field disturbances near the dip equator associated with auroral substorms, *Geophys. Res. Lett.,* 19, 1451-1454, 1992a.

Sastri, J. H., K. B. Ramesh, and D. Karunakaran, On the nature of substorm-related transient electric field disturbances in the equatorial ionosphere, *Planet. Space Sci.,* 40, 95-103, 1992b.

Sastri, J. H., J. V. S. V. Rao, and K. B. Ramesh, Penetration of polar electric fields to the nightside dip equator at times of geomagnetic sudden commencement, *J. Geophys. Res.,* 98, 17517-17523, 1993.

Sastri, J. H., Longitudinal dependence of equatorial F region vertical plasma drifts in the dusk sector, *J. Geophys. Res.,* 101, 2445-2452, 1995.

Sastri, J. H., M. A. Abdu, and J. H. A. Sobral, Onset conditions of equatorial (range) spread F at Fortaleza, Brazil, during the June solstice, *J. Geophys. Res.,* 102, 24,013- 24,021, 1997a.

Sastri, J. H., M. A. Abdu, and J. H. A. Sobral, Response of equatorial ionosphere to episodes of asymmetric ring current activity, *Ann. Geophysicae,* 15, 1316-1323, 1997b.

Sastri, J. H., N. Jyoti., V.V. Somayajulu., H. Chandra and C. V. Devasia, Ionospheric storm of early November 1993 in the Indian equatorial region, *J. Geophys. Res.,* 105, 18,443-18,455, 2000.

Sastri, J. H., J. V. S. Rao, D. R. K .Rao and B. M. Pathan, Daytime equatorial geomagnetic H field response to the growth and expansion phase onset of isolated substorms: Case studies and their implications, *J . Geophys. Res.,* 106, 29,925-29,933, 2001

Scherliess, L., and B. G. Fejer, Storm time dependence of equatorial disturbance dynamo zonal electric fields, *J. Geophys. Res.,* 102, 24,037-24,046, 1997.

Sekar, R., and R. Raghavarao, Role of vertical winds on the Rayleigh-Taylor mode instabilities of the nighttime equatorial ionosphere, *J. Atmos. Terr. Phys.,* 49, 981-986, 1987.

Sekar, R., R. Suhasini, and R. Raghavarao, Effects of vertical winds and electrci fields in the nonlinear evolution of equatorial spread F, *J. Geophys. Res.,* 99, 2205-2213, 1994.

Sekar, R., R. Suhasini, and R. Raghavarao, Evolution of plasma bubbles in the equatorial F region using different seeding conditions, *Geophys. Res. Lett.,* 22, 885-888, 1995.

Sekar, R., and R. Raghavarao, A case study of the evolution of equatroial spread F by a non-linear numerical model using the results from the set of coordinated measurements, *J. Atmos. Solar Terr. Phys.,* 59, 343-350, 1997.

Sekar, R., M. C. Kelley, On the combined effects of vertical shear and zonal electric fields patterns on the non-linear equatorial spread F evolution, *J. Geophys. Res.,* 103, 20,735-20,747, 1998.

Senior, C., and M. Blanc, On the control of magnetospheric convection by the spatial distribution of ionospheric conductivities, *J. Geophys. Res.,* 89, 261-284, 1984.

Sipler, D. P., and M. A. Biondi, Equatorial F region neutral winds from nightglow OI 630nm Doppler shifts, *Geophys. Res. Lett.,* 5, 373-376, 1978.

Sobral, J. H. A., M. A. Abdu, W. D. Gonzalez, B. T. Tsurutani, and I. S. Batista, Effects of storms/substorms on the equatorial ionosphere/thermosphere system in the American sector from groundbased and satellite data, *J. Geophys. Res.,* 102, 14305-14313, 1997.

Spiro, R. W., R. A. Wolfe and B. G. Fejer, Penetration of high latitude electric field effects to low latitude during SUNDIAL 1984, *Ann. Geophyscae.,* 6, 39-53, 1988.

Sridharan, R., D. Pallam Raju, R. Raghavarao and P. V. S. Ramarao, Precursor to equatorial spread F in OI 630 nm dayglow, *Geophys. Res. Lett.,* 21, 2797-2800, 1994.

Sultan, P. J., Linear theory and modelling of the Rayleigh-Taylor instability leading to the occurrence of equatorial spread F, *J. Geophys. Res.,* 101, 26,875-26,891, 1996

Sun, W., W. -Y. Xu., and S.–I. Akasofu, An improved method to deduce the unloading component for magnetospheric substorms, *J. Geophys. Res.,* 105, 13,131-13,140, 2000.

Tinsley, B. A., Neutral atom precipitation-A review, *J. Atmos. Terr. Phys.,* 43, 617-632, 1981.

Torr, M. R., J. G. G. Walker, and D. G. Torr, Escape of the fast oxygen from the atmosphere during the geomagnetic storms, *J. Geophys. Res.,* 79, 5267-5271, 1974.

Tsunomura, S., Numerical analysis of global ionospheric current system including the effect of equatorial enhancement, *Ann. Geophysicae,* 17, 692-697, 1999.

Viswanathan, K. S., D. Tiwari, K. S. V. Subbararo, C. V. Devasia, N. Jyoti and R. Sridharan, First results based on HF radar observations of 8.3 m irregularities of equatorial spread F from the magnetic equator over India, *Radio Sci.,* 2001 (in press).

Walker, G. O., and Y. W. Wong, Ionospheric effects observed throughout East Asia of the large magnetic storm of 13-15 March 1989, *J. Atmos. Terr. Phys.,* 55, 995-1008, 1993.

Yeh, H. C., S.–Y. Su and R. A. Heelis, Storm time plasma irregularities in the predawn hours observed by the low—latitude ROCSAT-1 satellite at 600 km altitude, *Geophys. Res. Lett.,* 28, 685-688, 2001.

Zalesak, S. T., S., L. Ossakow., and P. K. Chaturvedi, Nonlinear equatorial spread F: The effect of neutral wind and background Pedersen conductivity, *J. Geophys. Res.,* 87, 151-166, 1982

J. Hanumath Sastri, Indian Institute of Astrophysics, Bangalore 560 034, India, jhs@iiap.ernet.in; R. Sridharan, Space Physics Laboratory, Vikram Sarabhai Space Centre, Trivandrum 695 022, India, r_sridharan@vssc.org; T. K. Pant, Space Physics Laboratory, Vikram Sarabhai Space Centre, Trivandrum 695 022, India, r_sridharan@vssc.org

Structure of Turbulent Irregularities in High-Latitude Plasma Patches-3D Nonlinear Simulations

N. A. Gondarenko and P. N. Guzdar

Institute for Research in Electronics and Applied Physics, University of Maryland, College Park, Maryland, USA

The high-latitude ionospheric plasma patches have mesoscale irregularities which are believed to be driven by natural instabilities. We present results from three-dimensional nonlinear simulations of the turbulence and sheared flow, generated by the gradient drift and secondary Kelvin-Helmholtz instabilities due to ion inertial effects in high-latitude plasma patches. Numerical simulations, with height-dependent ion-neutral collision frequency show that the turbulence is quasi-two-dimensional with irregularities aligned along the magnetic field direction. The secondary Kelvin-Helmholtz instability of the gradient drift elongated vortices tends to isotropize the spectra. Also, the irregularities penetrate through the entire patch due to the nonlinear development of the instability and do not remain localized on the edges of the plasma patch. The combination of dynamics along the field line direction (3D effects) and nonlinear ion inertial effects with the altitude-dependent ion-neutral collision frequency unify the gradient drift and Kelvin-Helmholtz instabilities. These processes generate the small-scale structures in the polar cap plasma patches, and give rise to irregularity characteristics which are in good agreement with spatial and time scales of the experimentally observed structures.

1. INTRODUCTION

The high-latitude ionospheric plasma, that involves processes coupled to the Earth's magnetosphere, especially when the interplanetary magnetic field turns southward, shows variability and irregularity characteristics, which makes it scientifically fascinating and challenging to interpret. The irregularities of the high-latitude ionospheric plasma have a large dynamic range of spatial scales from meters to hundreds of kilometers [*Tsunoda*, 1988; *Crowley*, 1996; *Basu and Valladares*, 1999] and time scales from seconds to hours, providing a wide range of problems for computer modeling of different regions of the dynamic range, so as to isolate and understand the physics processes playing a dominant role in the range of choice. The large-scale structures ranging from hundreds to thousands of kilometers are referred to as "patches" (in the polar cap) or "blobs" (at auroral latitudes) [*Weber et al.*, 1984, 1986; *Basu et al.*, 1990, 1994]. The density of these structures can be two to ten times the background density. The mesoscale features in the range of 0.1 km to tens of kilometers are believed to arise from naturally occurring instabilities in a two-component plasma. Observations [*Weber et al.*, 1986; *Basu et al.*, 1994] as well as (large-scale) modeling [*Sojka et al.*, 1993] show that the patches convect to distances of the order of 3000 km [*Weber et al.*, 1986] and for periods of a few hours while retaining their integrity. The mesoscale irregularities, observed within these patches throughout the polar cap region [*Weber et al.*, 1984, 1986; *Tsunoda*, 1988; *Basu et al.*, 1990, 1995; *Kivanç and Heelis*, 1997, 1998], occasion-

Disturbances in Geospace: The Storm-Substorm Relationship
Geophysical Monograph 142
Copyright 2003 by the American Geophysical Union
10.1029/142GM17

ally are more intense on the trailing edge than on the leading edge [*Weber et al.*, 1984].

The modeling of the structuring of plasma patches was first done in two-dimensions [*Mitchell et al.*, 1985] transverse to the magnetic field. For the high-latitude plasma patches, the basic natural instability that gives rise to the mesoscale irregularities is the gradient drift instability that can occur in a plasma with a density gradient and a neutral wind and/or cross-field drifts due to electric fields [*Simon*, 1963; *Hoh*, 1963]. The simulations showed that the gradient drift instability grew on the trailing edge of the patch, nonlinearly developed into elongated, penetrating fingers, which fragmented the initial patch into small filamentary structures. However as mentioned earlier, observationally, patches maintain their integrity for four to five hours with mesoscale structures embedded in them. Here, we highlight our recent three-dimensional (3-D) nonlinear simulations to model the mesoscale structures. In the absence of ion inertial effects, the basic structuring occurred transverse to the direction of the magnetic field and to the direction of the ambient density gradient in highly elongated "fingers" [*Guzdar et al.*, 1998]. It was shown, that the inclusion of dynamics along the field line was responsible for slowing down the structuring process due to the stabilizing influence of parallel electric fields, preferentially of the long wavelength modes [*Chaturvedi and Huba*, 1987]. The structuring process for this collision-dominated ion-neutral plasma produced a highly anisotropic spectra of the irregularities in the two directions transverse to the magnetic fields. However for the full three-dimensional case, since the ion-neutral collision varies as a function of altitude, the ion inertial effects need to be included since at the higher altitudes they are larger than the ion-neutral collision effects. Preliminary simulation results, demonstrating the role of the nonlinear ion inertial effects, were presented by *Gondarenko and Guzdar* [1999]. It was shown that the finger-like structures developed in the early phase due to the nonlinear development of the gradient drift instability, are unstable to secondary instabilities. The highly elongated vortices associated with the gradient drift modes are broken up into the smaller ones because of the Kelvin-Helmholtz instability. The small-scale vortices due to the KH instability can further undergo a tertiary instability, which generates a shear flow transverse to the magnetic field. In the more detailed paper by *Gondarenko and Guzdar* [2001] on the subject, we discussed the changes in the character of the irregularities, caused by the inertial terms, and how the ion-neutral collisionality plays a significant role in determining both the level of the fluctuations and the nature of the turbulent spectra of the density and potential irregularities.

However even in those simulations, in order to highlight the role of the ion inertial effects, we assumed the ion-neutral collision frequency to be independent of the altitude. Here we present the complete realistic simulations with both the height-dependence of the ion-neutral collision frequency, and the realistic value of the dimensionless parameter (to be defined later) which accounts for the parallel dynamics. We interpret the results of these simulations based on our understanding derived from our earlier physics-based simulations.

2. BASIC EQUATIONS

The starting equations for the gradient drift instability with ion inertial effects are the electron continuity equation, the momentum transfer equations for electrons and ions [*Drake et al.*, 1988]. The current conservation or "vorticity" equation and the electron continuity equation used in our simulations were derived from equations (1)-(4) of *Gondarenko and Guzdar* [2001]. The "vorticity equation", also known as the current conservation equation ($\nabla \cdot J = 0$), has contributions from the differential collisional motion between ions and electrons (Pedersen drifts), the Hall drifts (ion inertial effects) perpendicular to the direction of the magnetic field, and from the resistive electron motion along the magnetic field. The resulting equations [*Gondarenko and Guzdar*, 1999] are

$$\frac{\partial n}{\partial t} - \frac{c}{B_0}\nabla\phi \times \hat{z} \cdot \nabla n + \frac{\partial}{\partial z}\frac{1}{e\eta_e}\left[\frac{\partial \phi}{\partial z} - \frac{T_e}{ne}\frac{\partial n}{\partial z}\right] = 0 \quad (1)$$

$$\frac{c}{B_0\Omega_i}\nabla_\perp \cdot \left[n\left(\frac{\partial}{\partial t} - \frac{c}{B_0}\nabla\phi \times \hat{z}\cdot\nabla + \nu_{in}\right)\nabla_\perp\phi\right]$$
$$+ \frac{\nu_{in}}{\Omega_i}(\hat{z}\times\vec{V}_n)\cdot\nabla n + \frac{\partial}{\partial z}\frac{1}{e\eta_e}\left[\frac{\partial\phi}{\partial z} - \frac{T_e}{en}\frac{\partial n}{\partial z}\right] = 0 \quad (2)$$

In the above, n, c, and e are the density, the velocity of light, and the electronic charge, respectively. ϕ, T_e, $\vec{V}_n$ refer to electrostatic potential, temperature in energy units, and the neutral-wind velocity, respectively. $\eta_e = m_e\nu_e/ne^2$, $\Omega_q = e_qB_0/m_qc$ ($e_q = \pm e$) is the cyclotron frequency of species q, ν_{qn} is the q species-neutral collision frequency, ν_{ei} is the electron-ion collision frequency, and $q(e, i)$ refers to electrons or ions. For the coordinate system representing the high-latitude ionosphere, the Earth's field lines are nearly vertical and assumed aligned to the z direction, the x direction is sunward, and the y direction is orthogonal to the x and z directions. A plasma patch is assumed to have a finite length along the magnetic field $\vec{B} = B_0\vec{z}$ with the neutral wind velocity $\vec{V}_n = V_n \vec{x}$. The electronic field is $\vec{E} = -\nabla \phi$, where ϕ is the electrostatic potential.

This system of equations can be normalized as follows: $t \to t(V_n/L_n)$, $\phi \to \phi(c/L_0 V_n B_0)$, $n \to n/N_0$, $z \to z/L_{z0}$, $x \to x/L_{\perp 0}$, where $L_{\perp 0}$ is a characteristic length depending on the patch size, N_0 is the undisturbed ambient density of the background ionosphere away from patch/blob (independent of x) and L_{z0} is the characteristic size of the patch along the field line. The normalized electron continuity and vorticity equations have three dimensionless parameters, β, ν, and R defined as

$$\beta \frac{\Omega_e \Omega_i}{\nu_e \nu_{in}} \frac{L_{\perp 0}^2}{L_{z0}^2}, \nu = \nu_{in} L_n/V_0, \text{ and } R = N_{max}/N_0.$$

For the simulations presented in this article, we have chosen $L_{\perp 0} = 8$ km and $L_{z0} = 179$ km. This makes our box size in the midnight-noon (x) direction $L_x = 4\pi L_{\perp 0} \cong 100$ km. In the dawn-dusk direction $L_y = \pi L_{\perp 0}/2 \cong 12.6$ km. The choice of this small section of the patch in the dawn-dusk direction is dictated by the characteristic size of the irregularities in this direction and the availability of computer resources for our computations. Finally, in the z direction $L_z = 2\pi L_{z0} = 1100$ km. For the high-latitude ionosphere parameters $\Omega_e \sim 10^7$ rad/s; $\Omega_i \sim 10^2$ rad/s; $\nu_{ei} \sim 10^3$ s^{-1}; $\nu_{in} \sim 0.1$ s^{-1}, and with the characteristic sizes of the patch in the directions transverse and parallel to the magnetic field, dimensionless parameter β is $\sim 20,000$, and $R = 2$.

In Figure 1a the density and the ion-neutral collision frequency profiles are shown as a function of height, while in Figure 1b the initial density profile along the midnight-noon direction is shown. We only perform simulations for the half of the patch which allows us better resolution of the irregularities. The steep density profile shown is the trailing edge of the initial patch. The peak density of the patch in our simulations corresponds to ~ 400 km in z. The ion-neutral collision frequency profile given in Figure 1a is typical for the period of high sunspot activity. The values of these collisions frequencies were estimated with the approximate formula given in [Kelley, 1989]. This formula can also be used to estimate collisions for solar minimum. For the solar maximum, the $\nu_{in} \sim 0.29$ s^{-1} at the 400 km altitude and for the sunspot minimum it is ~ 0.039 s^{-1} at this height.

The number of grid points used in this explicit finite-difference code are $N_x = 260$, $N_y = 132$ and $N_z = 51$. Density pertubations with an overall amplitude of 0.1% were added to the initial density profile. We introduce seventy harmonics for the perturbed density in the y direction, with a $sin(\pi * x/L_x)$ in the x direction. Along the height (z direction) we do not include any variation. The boundary conditions used in our simulations for the perturbed density and potential ϕ, and vorticity ω are the following. In the x direction $\phi(0, y, z) = \phi(L_x, y, z) = \omega(0, y, z) = \omega(L_x, y, z = 0$. We also use Dirichlet boundary condition for the density. In the z direction we choose the Neumann boundary conditions for the potential and vorticity. The boundary conditions in the y direction are periodic.

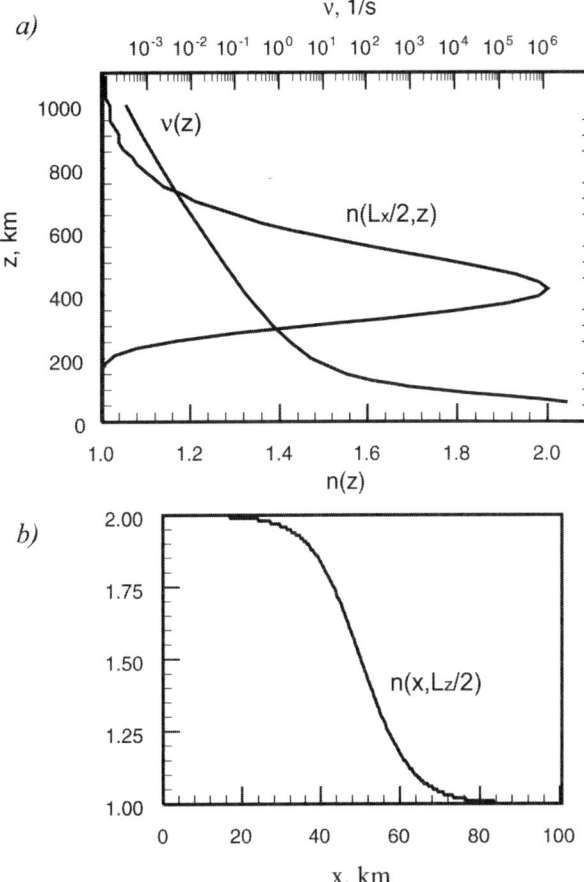

Figure 1. Density and ion-neutral collision frequency profiles (a) versus z and (b) density profile versus x.

3. NUMERICAL RESULTS

In this section we will discuss various results obtained from the nonlinear simulations of the gradient drift instability with inertial effects. The time evolution of the density in the xy plane near the peak of the Chapman function density profile in z are displayed at four different instances of time, $t = 1600, 3200, 4800,$ and 8000 s in Figures 2a-2d, respectively. The self-consistent potential associated with the density irregularities is shown in Figures 2e-2h at the same instants of time. At early time $t = 1600$ s, very fine scale

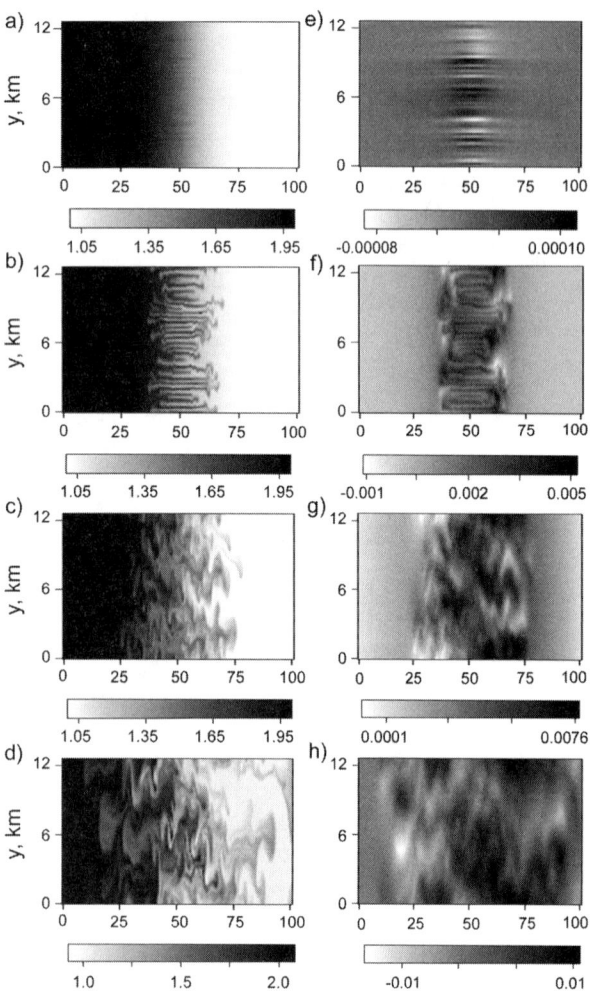

Figure 2. Density and potential contours in the xy plane at $z=2L_z/5$ for (a), (e) t=1600 s; (b), (f) t=3200 s; (c), (g) t=4800 s; (d), (h) t=8000 s.

irregularities, due to the gradient drift instability, begin to develop on the steep gradient part of the initial density profile. The short wavelengths grow preferentially since the inclusion of the dynamics along the fieldline tends to stabilize the long wavelength modes. It is precisely for this reason that the size of the simulation box in the dawn-dusk y direction was chosen to be just ~ 12.6 km. This allowed us to resolve the anticipated high mode-number modes in this direction. The potential contours at $t = 1600$ s in Figure 2e show the elongated vortices in the x direction caused by the primary gradient drift instability. There are ~ 25 fingers and vortices in Figures 2b, 2f. At this instant of time the characteristic size of the vortices in the midnight-noon direction is about 10 km, while the size of the vortices in the dawn-dusk direction is at best 200–300 m. Thus the gradient drift insta-

bility structures are highly anisotropic in the two orthogonal directions perpendicular to the magnetic field. The fingers and vortices propagate further in the x direction up to t ~ 3200 s (Figures 2b, 2f). Till about $t \leq 2800$ s, in the development of the structuring, the spatial characteristics are very similar to that observed in our earlier work [*Guzdar et al.*, 1998] without inertia effects.

The extended vortices in Figure 2f, which can be viewed as alternating shear-layers, become unstable to the Kelvin-Helmholtz instability. This tends to break up the elongated vortex structures in the x direction. This is clearly seen to happen at $t = 3200$ s in both the density (Figure 2b) and potential contours (Figure 2f). Even though collisional dynamics dominates the ion inertial effects at the low altitudes, the height-dependent ion-neutral collisions allows the inertial effects, especially the *nonlinear* ion inertial effects to play a significant role in the subsequent structuring. It allows the secondary Kelvin-Helmholtz instability to occur and influence the nature of the structuring. The time scales of the development of the instabilities with the altitude-dependent ion-neutral collision frequency are longer than the time scales in the simulations [*Gondarenko and Guzdar*, 2001] with constant ion-neutral collision frequency. If the dynamics had been purely collisional, then the extended fingers would bifurcate with primary fingers spawning two fingers which would then extend along the x direction. This would lead to creation of more fingers in the y direction extended along the x direction, thereby accentuating the strong asymmetry in the structuring characteristics between the x and the y directions [*Guzdar et al.*, 1998]. On the other hand, with the height-dependent ion-neutral collision frequency, the secondary Kelvin-Helmholtz instability proceeds towards isotropization of the irregularities scales in the plane perpendicular to the magnetic field by breaking up the long elongated vortices in the x direction. Now as time progresses, the irregularities evolve to larger and larger scale lengths as seen on the density contours at times $t = 4800$ s (Figure 2c) and $t = 8000$ s (Figure 2d). The potential contours at $t = 4800$ s and 8000 s (Figures 2g-2h) show this transition to a larger scale length more clearly. This migration to a longer scale length (in the plane perpendicular to the magnetic field) is the inverse-cascade process, well known in two-dimensional turbulence. Even though our simulations are fully three-dimensional, the dominant convective nonlinearities are intrinsically two-dimensional in nature. It is precisely these nonlinear terms associated with the ion dynamics, and the strong field-aligned nature of the instability, which makes the turbulence quasi-two-dimensional. Also seen in these plots is that as time progresses, the irregularities penetrate into the patch. Interestingly, during the period up to $t = 2800$ s when the collisional gradient-drift dominated the structuring process in elongated fingers, the irregularities "propagated" towards the right boundary rapidly. However, once the secondary K-H instability breaks the fingers up into smaller vortices in the x direction, the "anomalous" transport that rapidly propagated the steep gradient region on the trailing side of the patch is considerably slowed down. The smaller scale length gives rise to a lower value of "anomalous" transport by reducing the effective correlation length. Another important process, namely the generation of sheared flow transverse to the x direction, also inhibits the turbulence from effectively transporting the density.

One of the most interesting aspects of the simulations is the generation of shear flow transverse to the direction of propagation of the patch. Averaging the vorticity equation over y and z and performing an integral over x we can derive the following equation for the average momentum in y [Gondarenko and Guzdar, 1999]

$$\left(\frac{\partial}{\partial t} + v_{in}\right)\langle p_y \rangle - \frac{c}{B_0}\frac{\partial}{\partial x}\left\langle \frac{\partial \tilde{\phi}}{\partial y} \tilde{p}_y \right\rangle = 0 \quad (3)$$

This equation shows that the irregularities (denoted by the $\tilde{p}_y$) are a source for generating self-consistent shear flow. The source term is referred to as the Reynolds stress. The ion-neutral collisions v_{in} try to damp the shear flow and hence provide a threshold for the flow generation instability. The self-consistent generation of shear flow in Rayleigh-Benard convection was first investigated by Howard and Krishnamurti, [1985] and then their instability analysis was improved by Hermiz et al., [1995], where it was shown that vortices driven by the Rayleigh-Benard instability can undergo a secondary instability which can lead to the generation of the shear flow. In Figures 3a-3c are shown the shear flow at $t = 2800, 3200$ and 8000 s respectively. The shear-flow grows only after the initial gradient-drift vortices have developed the extended fingers. The shear layers spreads and systematically gets generated as the irregularities do. The magnitude of the shear velocities at $t = 8000$ s is about 1 m/s. The multi-shear layers strongly inhibit the anomalous transport in the x direction. In fact, shown in Figures 4a-4c are the density lineouts at three different locations in y (at the peak density in z) at $t = 8000$ s. The regions where there is a sharp drop in the density are regions of large transport barriers caused by the self-consistent shear flow. These density lineouts are very similar to the ones reported in the work of Kivanç and Heelis [1997] as seen in their Figure 1c. The DE2 satellite pass for this particular data set was in the midnight-noon direction. The density lineout shows the same characteristics as our simulations, in which there are sharp density drops between regions of lower-amplitude irregularities.

To display the full three-dimensional character of the structuring of the plasma patch, five different isosurfaces equally spaced between the normalized minimum density $N_{min} = 1$ and the maximum density $N_{max} = 2$, initially at $t = 0$ and later at $t = 8000$ s are displayed in Figures 5a and 5b, respectively. There are various interesting features associated with the structuring process. What is clearly evident

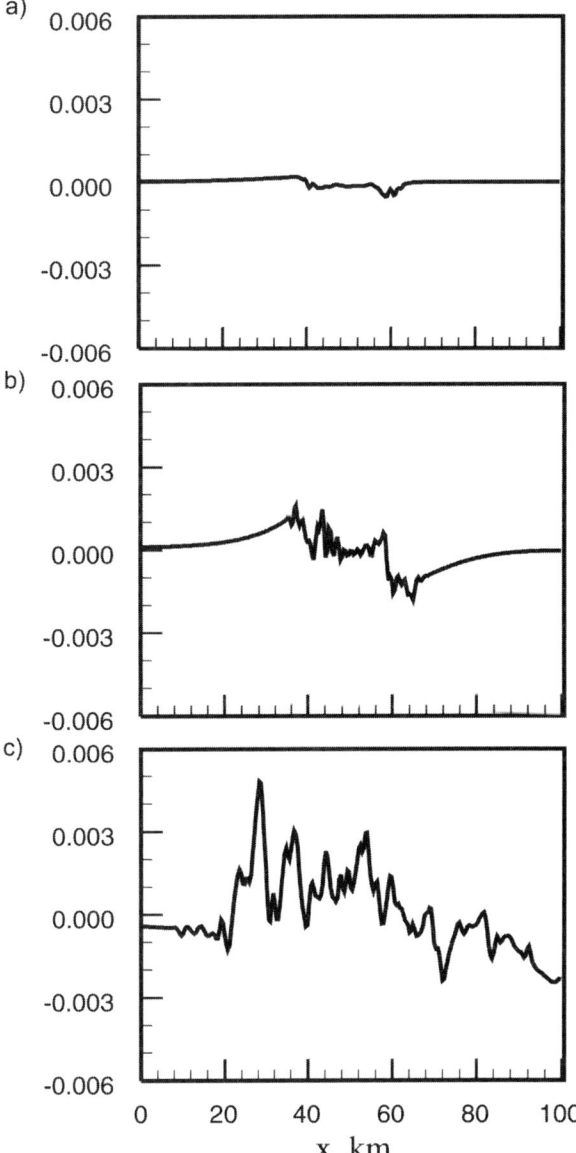

Figure 3. Average v_y velocity as a function of x at t=2800 s, t=3200 s, and t=8000.

the equatorial region, where the initial bottom side Rayleigh-Taylor instability nonlinearly developed into plumes which then extended into the linearly stable topside.

Also it can clearly be seen in Figure 5b that the irregularities are basically field-aligned (along the z direction). Finally, if one compares the extent of the patch along the x direction in Figures 5a and 5b, one sees that the patch has elongated in the x direction (which also is the trailing edge). This extension is due to the "anomalous" transport caused by the irregularities. The third dimension, with the inclusion

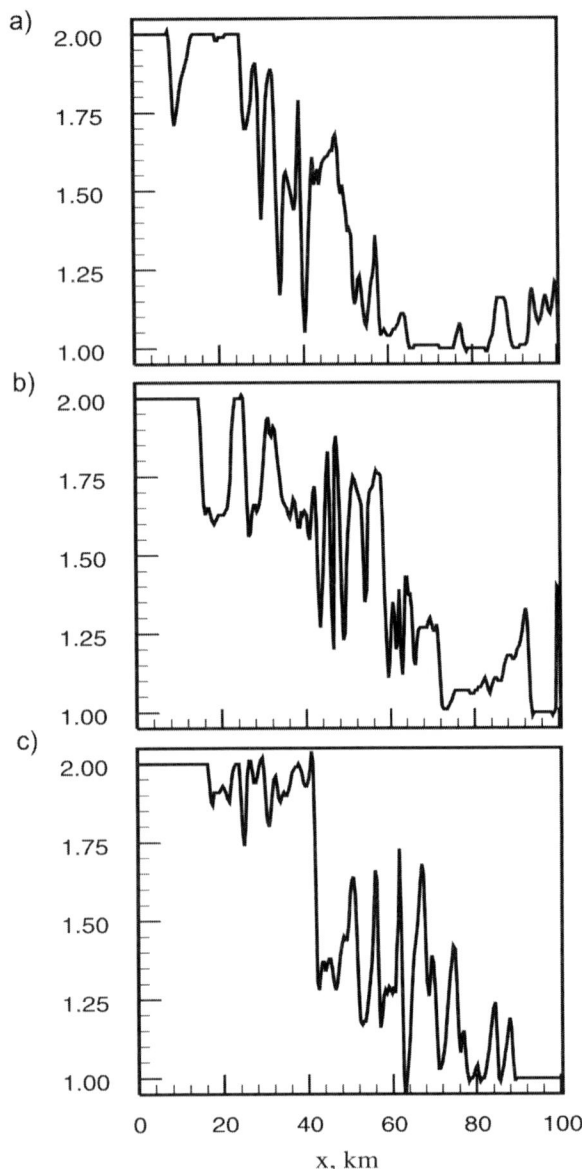

Figure 4. Density versus x at $z=2L_z/5$ and $t=8000$ for (a) $y=L_y/4$, (b) $y=L_y/2$, and (c) $y=3L_y/4$.

is that the structuring occurs throughout the plasma patch. Since the linear theory predicts that the instability should only occur in the region where the density gradient exists, one expects that the instability, even in the nonlinear phase, will remain confined to the edges of the plasma patch. However, the nonlinear development clearly shows that the irregularities permeate the patch and the instability develops on the *nonlinearly* evolving density gradient. This is also consistent with earlier simulations of the collisional Rayleigh-Taylor instability [*Zalesak and Ossakow*, 1980] in

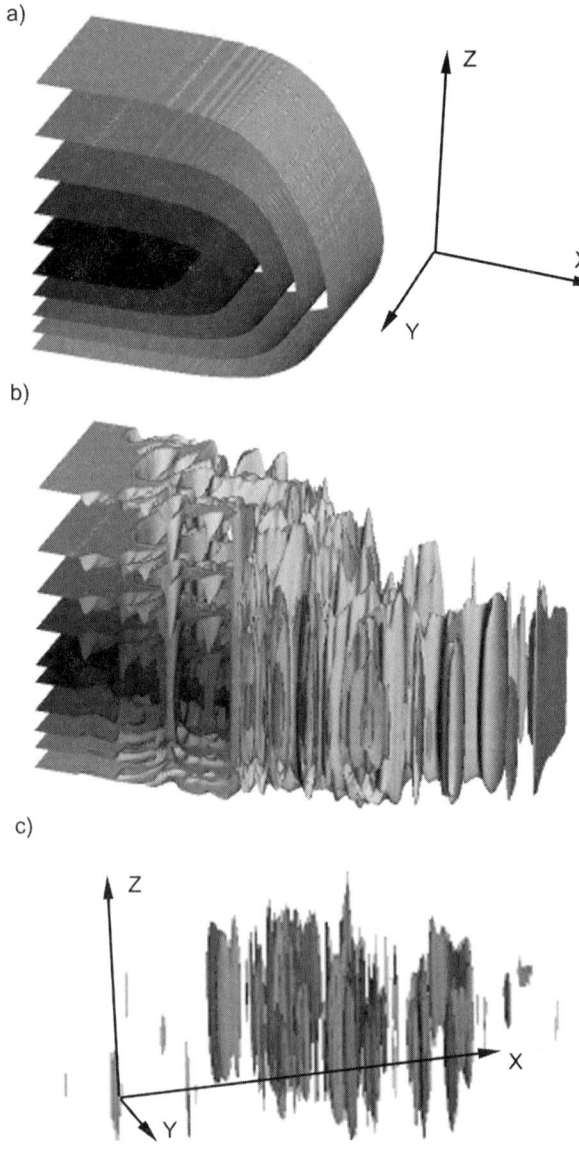

Figure 5. Five equally spaced density isosurfaces at (a) $t=0$ s and (b) $t=8000$ s. (c) Isosurfaces of the parallel electric field at $t=8000$ s.

in the model of the altitude-dependent ion-neutral collision frequency, pushes the instability to high mode numbers in the dawn-dusk (y) direction. The combination of this process with the secondary Kelvin-Helmholtz instability which breaks up the elongated vortices in the midnight-noon (x) direction, and the self-consistent generation of sheared flow in the dawn-dusk direction, keeps the structuring in check. Thereby the patch maintains its robustness and integrity over long periods (4~5 hours) of time.

In Figure 5c we show the isosurfaces of the parallel electric field $E_\parallel$ that was calculated by differentiating the potential over z. The maximum values of the $E_\parallel$ are ~ 0.1 $\mu V/m$. With higher flow convection velocities and larger ratio R of the patch peak density to the background density, these electric field can be of the order ~ 1–10 $\mu V/m$. A combination of ion heating by the transverse fields and these parallel electric field can provide a viable mechanism for generating O^+ outflow from the region of the ionosphere where the dominant ion species is O^+.

The simulation data can now be Fourier-transformed at any specific location to compare with data from satellite passes [*Basu et al.*, 1990; *Kivanç and Heelis*, 1997]. Shown in Figures 6a, 6b are the spectral characteristics of the density in the midnight-noon direction x, and the dawn-dusk direction y, respectively at t = 2800 s (dotted line) and t = 8000 s (solid line). We recall that since the box size is significantly smaller in the dawn-dusk direction compared to the midnight-noon direction, for reasons cited earlier, the long wavelength part of the spectrum in k_y is absent in the dawn-dusk direction. Also, in the simulations the density had a large gradient in the x direction to begin with, while in the y direction the density was initially assumed to be uniform. Thus the long-wavelength part of the simulation may be influenced by our initial conditions. It is however apparent that there is a significant asymmetry at early time t = 2800 s in the xy plane if one compares similar ranges of the high k part of the spectrum. In the range $1 \leq k \leq 10$ for both k_x and k_y, the fluctuation levels are at least two orders of magnitude smaller in k_x compared to k_y. This is due to the gradient-drift instability dominated early phase, where elongated fingers in the x direction have much larger scalelengths than the width of the fingers in the y direction. However if one compares similar ranges of the wave-number spectra in the two directions at the later time t = 8000 s (solid lines), one sees that the spectra are more isotropic. By comparing the spectra at two different instances of time (Figure 6a, 6b) we clearly see the changes in the character of the spectra caused by the onset of the secondary KH and tertiary shear flow instabilities. In the k_x spectra of the density (Figure 6a), the low mode numbers ($k_x \leq 1$) dominates

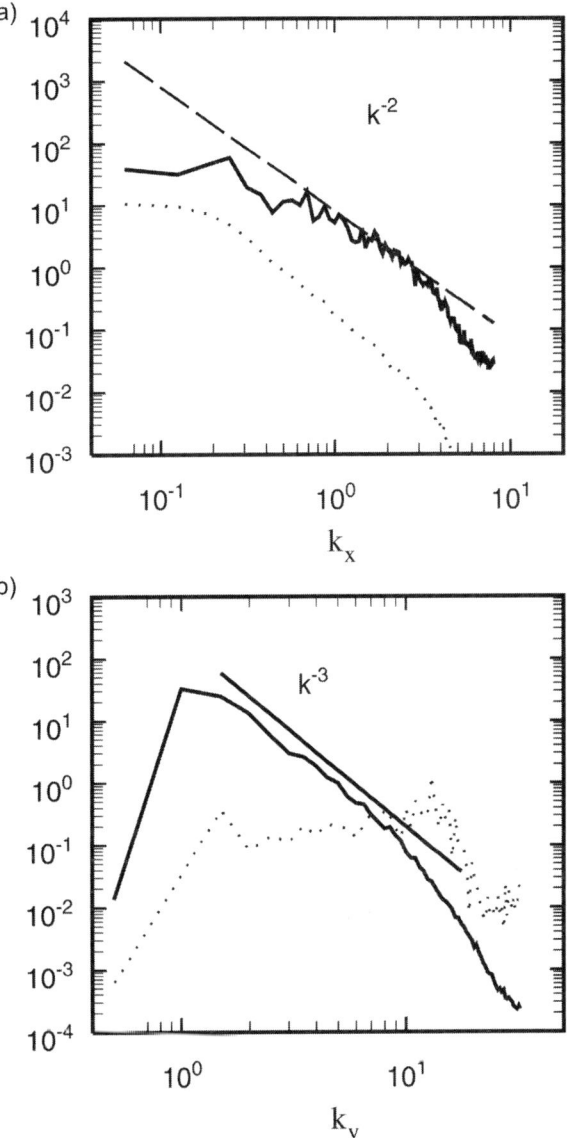

Figure 6. Density power spectra for t=2800 s (the dotted line) and t=8000 s (the solid line), versus (a) k_x and (b) k_y.

the spectra, because of the ambient gradient and the long scale of the gradient drift instability at early time (t = 2800 s). However at the later time (t = 8000 s), the secondary instability KH instability causes a direct cascade to higher mode numbers by breaking up the elongated fingers. On the other hand, in the k_y spectra (Figure 6b), at the early time (t = 2800 s) high k_y dominate the spectrum due to the narrow width of the gradient-drift elongated fingers. These evolve into the lower wavevector region by the inverse cascade process associated with the nonlinear inertial terms.

Thus the cascading of fluctuations scales in the two directions is quite different from isotropic two-dimensional turbulence because of the complex interplay between primary, secondary and tertiary instabilities. The power spectra for the density as a function of k_x for the later time are found to fall off as k_x^{-2} in the interval $k \in [0.5, 4]$. The power law for the density power spectrum in the y direction at the later time is about -3 in the interval $k \in [1.5, 10]$ (Figure 6b). Because the dynamic range of our simulations is limited, the scaling ranges are small and, therefore, the spectral indices are approximate.

Again, if we compare the power spectrum for the density irregularities shown in Figure 6a in the midnight-noon direction with the spectrum displayed in Figure 1c in the work of Kivanç and Heelis [1997], the agreement is reasonably good. Also, the computed power spectrum for the density agrees with that reported in earlier work [*Basu et al.*, 1990]. Now we offer an explanation of the k_x^{-2} fall-off of the density power spectrum. As shown earlier in Figure 4, there are very sharp density drops interspersed in the fluctuations and the over the entire density profile. These density lineouts are also very similar to the ones shown in Figures 2a-2c in simulations with the constant ion-neutral collision frequency [*Gondarenko and Guzdar*, 2001]. If we first obtain a linear fit to the density profile and then subtract this linear fit from the corresponding cut in Figure 4, the fluctuating density then has a series of square wave pulses associated with the sharp drops. The Fourier transform of a square pulse falls off as k_x^{-1}. Thus the power spectrum has the observed spectral index of two.

The issue of isotropy in the plane perpendicular to the magnetic field is further addressed by looking at the velocity v_x and v_y spectra in the midnight-noon direction again for two different instants of time. Displayed in Figures 7a and 7b are the spectra of v_x (solid line) and v_y (dotted line) as a function of k_x (averaged over y) at $t = 2800$ s and $t = 8000$ s, respectively. The spectra of v_x and v_y demonstrate the process of the isotropization of the spectra over the significant range of the k-space. In Figure 7a, the early time strong anisotropy in spectra is clearly seen. The v_x velocity has its largest values at the lower k_x. On the other hand, the v_y velocity peaks at intermediate values of the wave vector and is significantly smaller in magnitude compared to the v_x spectrum. This is consistent with the elongated vortices in x with narrow widths in y. However at the later time, one can see in Figure 7b that the secondary instabilities try to create isotropy of the spectra more strongly at the higher k than at the lower k. At the later time $t=8000$ s (Figure 7b), the power law is -2 in the interval $k \in [0.2, 1]$ while for the higher k the power law is steeper and the spectral index changes to -3.

To further address this issue we show in Figures 8a and 8b the spectra of v_x and v_y as a function of k_y at two different instants of time namely, $t = 2800$ s and $t = 8000$ s, respectively. Here the early time (Figure 8a) anisotropy in the spectra is highly accentuated. The initially unstable modes grow at large k_y ($k_y \sim 10$). This causes the v_x spectrum to be peaked at large mode numbers, in contrast, large scale structures along the x direction make the v_y spectrum peaks at low k. But at later time (Figure 8b) the energy

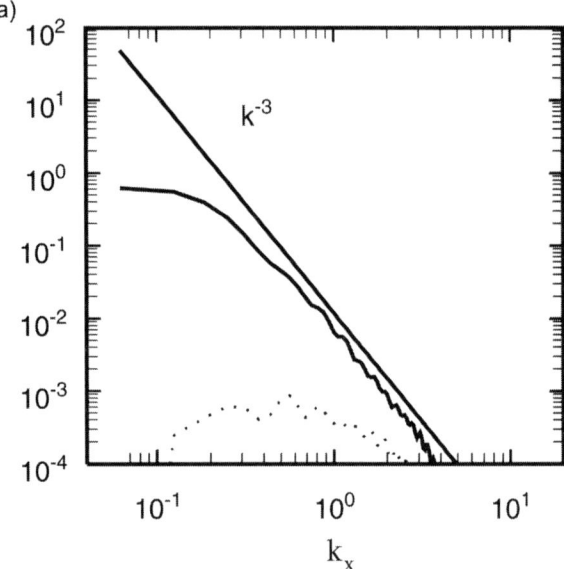

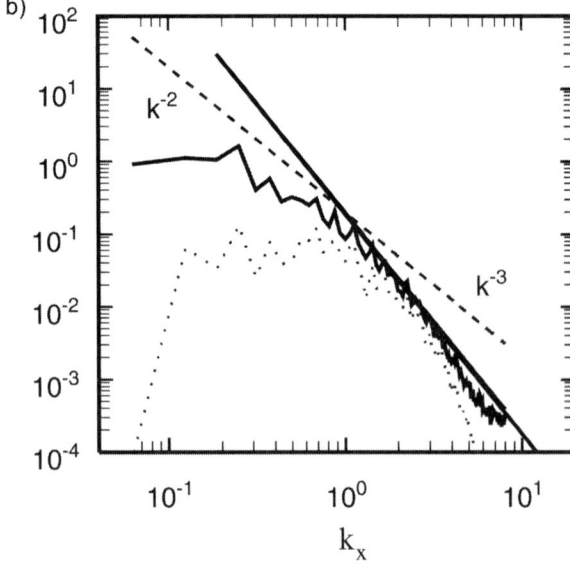

Figure 7. Power spectra of velocities v_x (the solid line) and v_y (the dotted line) versus k_x for (a) t=2800 s and (b) t=8000 s.

transfer to the longer scale lengths by inverse cascade provides the evidence of the quasi-two-dimensional nature of the turbulence. Both spectra in k_x and k_y at early times in Figures 7a and 8a display a strong asymmetry between the v_x and v_y spectra. The character of the instability is determined mostly by the low-altitude ion-neutral collisions. At this time the growth rate for the collisional gradient drift is larger than the inertial gradient drift. At later times the secondary instability and shear-flow generation, which can preferentially occur at the higher altitudes with lower ion-

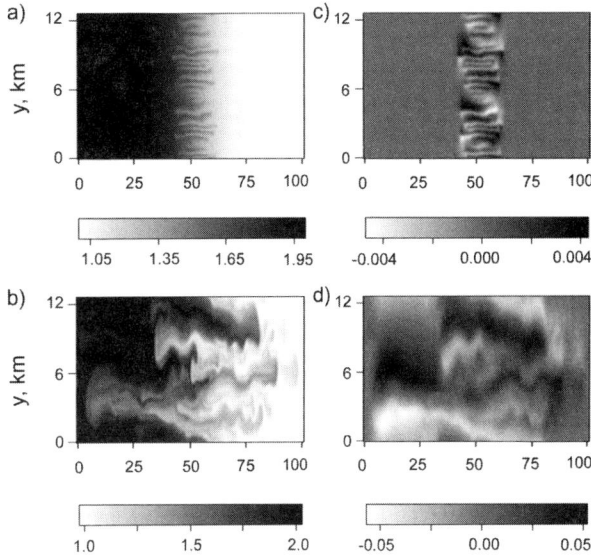

Figure 9. Density and potential contours in the xy plane at $z=2L_z/5$ for (a), (c) t=800 s; (b), (d) t=1800 s with collision frequency v_{in} =0.68 1/s at the density peak.

neutral collision frequencies, start to dominate. Since the v_y velocity spectrum consists of the shear flow, the high k fluctuations associated with the instability suppressed by rapid phase mixing due to the shear flow. At this point the inverse-cascade dominates the evolution of the spectra and they become more isotropic in the plane perpendicular to the magnetic field.

Finally, in Figures 9a-9d we show the time evolution of the structuring of the density and potential in case with the altitude-dependent ion-neutral collision frequency where the collisions dominate at the peak of the patch (v_{in} = 86) and v_{in} is about 0.03 at high altitudes. The effective collision frequencies for the ions O^+ were used from the table of [*Gurevich*, 1978]. Since the collisions dominate over a large range of the altitudes, where the patch has a high density, the gradient drift instability develops very fast (Figures 9b and 9d). On this short time scale the ion inertial effects at higher altitudes were not able to affect the irregularities. The weaker secondary instability failed to generate the shear flow capable of changing the character of the structuring. The shear flow generation was suppressed by the high ion-neutral collisions that dominate in the collision profile.

Thus the full three-dimensional simulations with the altitude-dependent ion-neutral collision frequency show very interesting features which changes the character of the drives as time progresses due to the varying ion-neutral collisionality. Careful analysis of satellite data will therefore need to be performed to investigate the characteristics of the

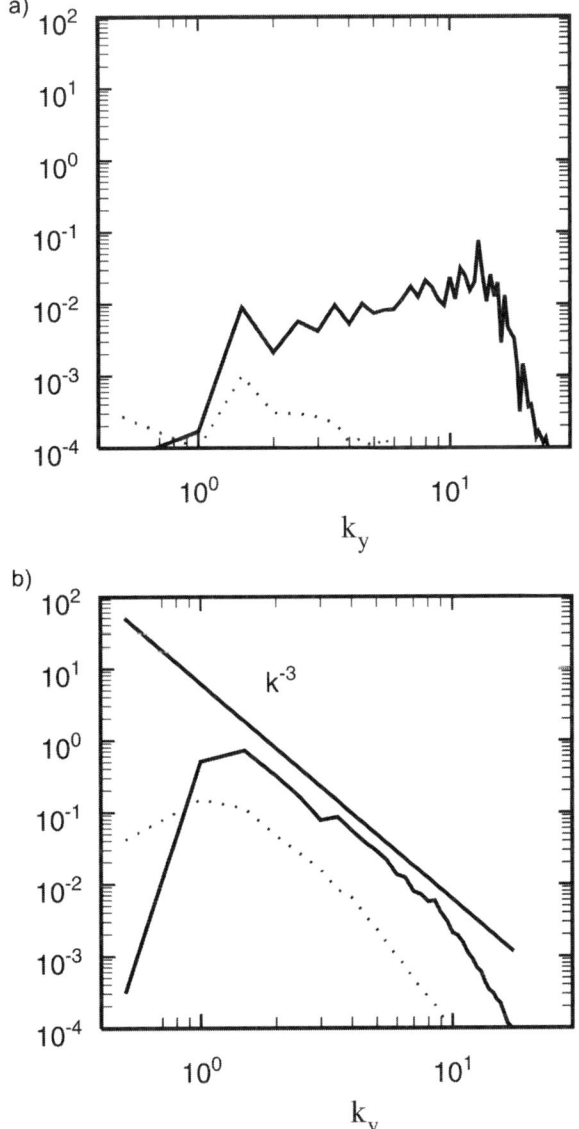

Figure 8. Power spectra of velocities v_x (the solid line) and v_y (the dotted line) versus k_y for (a) t=2800 s and (b) t=8000 s.

evolution of the patch structuring suggested by our simulations.

4. CONCLUSION

We have presented results from fully three-dimensional nonlinear simulations of the structuring of high latitude ionospheric plasma patches with the altitude-dependent ion-neutral collision frequency. The presented results of our simulations demonstrate that the inertial terms combine the gradient drift and Kelvin-Helmholtz instabilities that are the natural instability sources of small-scale structures in the polar cap plasma patches. Our results show that Reynolds stress, the nonlinear convective part of the terms describing the inertial effects, drives the secondary Kelvin-Helmholtz instability, the self-consistent shear flows and the inverse cascade.

The ion-neutral collision frequencies at high altitudes affect the dynamics of the instabilities as time progresses, allowing for the secondary Kelvin-Helmholtz instability to occur and influence the nature of the structuring. The strong coupling of the structures along the fieldline allows for the secondary instabilities to develop even in the lower altitudes, where the dominant drive is the collisional gradient drift instability.

Power spectra of the density and the velocities in the plane transverse to the magnetic field display the isotropization in the turbulence at the late stages of the development of the structuring even though the initial phase is dominated by the gradient drift which displays strong asymmetry. We show the temporal changes in the scale size of the vortices, which reveals the inverse cascade nature of the quasi-two-dimensional turbulence.

Thus in conclusion the combination of three-dimensional effects and ion inertial dynamics produce structuring of patches consistent with many observed features. Furthermore, even though the structuring penetrates through the entire patch, the large scale patch retains its integrity and survives for hours as it convects over the polar cap.

Acknowledgments. We acknowledge useful discussions with Sunanda Basu, Shantimay Basu, Rod Heelis, Todd Pedersen, and Jan Sojka. Rod Heelis' insightful comments have guided us in our modeling effort. This research was supported by the NSF under grant ATM0122874.

REFERENCES

Basu, Su., S. Basu, E. MacKenzie, W.R. Coley, J.R. Sharber, and W.R. Hoegy, Plasma structuring by the gradient-drift instability at high latitudes and comparison with velocity shear driven processes, *J. Geophys. Res.*, 95, 7799, 1990.

Basu, S., Su. Basu, P.K. Chaturvedi, and C.M. Bryant Jr., Irregularity structures in the cusp/cleft and polar cap regions, *Radio Sci.*, 29, 195, 1994.

Basu, S., Su. Basu, J.J. Sojka, R.W. Schunk, and E. MacKenzie, Macroscale modeling and mesoscale observations of plasma density structures in the polar cap, *Geophys. Res. Lett.*, 22, 881, 1995.

Basu, S., and C. Valladares, Global aspects of plasma structures, *J. Atmos. Sol. Terr. Phys.*, 61, 127, 1999.

Chaturvedi, P.K., and J.D. Huba, The interchange instability in high latitude plasma blobs, *J. Geophys. Res.*, 92, 3357, 1987.

Crowley, G., Critical review on ionospheric patches and blobs, in *The Review of Radio Science*, p. 1, Oxford Univ. Press, New York, 1996.

Drake, J.F., M. Mulbrandon, and J.D. Huba, Three-dimensional equilibrium and stability of ionospheric plasma clouds, *Phys. Fluids*, 31, 3412, 1988.

Gondarenko, N.A., and P.N. Guzdar, Gradient-drift instability in high-latitude plasma patches: Ion inertial effects, *Geophys. Res. Lett.*, 26, 3345, 1999.

Gondarenko, N.A., and P.N. Guzdar, Three-dimensional structuring characteristics of high-latitude plasma patches, *J. Geophys. Res.*, 106, 24,611, 2001.

Gurevich, A.V., Nonlinear Phenomena in the Ionosphere, *Springer-Verlag, New York*, 1978.

Guzdar, P.N., N.A. Gondarenko, P.K. Chaturvedi, and S. Basu, Three-dimensional nonlinear simulations of the gradient drift-instability in the high-latitude ionosphere, *Radio Sci.*, 33, 1901, 1998.

Hoh, F.C., Instability of Penning-type discharges, *Phys. Fluids*, 6, 1184, 1963.

Howard, L.N., and R. Krishnamurti, Large scale circulation in turbulent convection: a mathematical model, *J. Fluid Mech.*, 170, 385, 1986.

Hermiz, K.B., P.N. Guzdar, and J.M. Finn, Improved, low-order model for shear flow driven by Rayleigh-Bernard convection, *Phys. Rev. E*, 51, 325, 1995.

Kelley, M.C., The Earth's Ionosphere, Academic, *San Diego, Calif.*, 1989.

Kivanç, Ö., and R.A. Heelis, Spatial distribution of ionospheric plasma and field structures in the high latitude *F* region, *J. Geophys. Res.*, 103, 6955, 1998.

Kivanç, Ö., and R.A. Heelis, Structures in ionospheric number density and velocity associated with polar cap ionization patches, *J. Geophys. Res.*, 102, 307, 1997.

Simon, A., Instability of a partially ionized plasma in crossed electric and magnetic fields, *Phys. Fluids*, 6, 382, 1963.

Sojka, J.J., M.D. Bowline, R.W. Schunk, D.T. Decker, C.E. Valladares, R. Sheehan, D.N. Anderson, and R.A. Heelis, Modeling polar cap *F* region patches using time varying convection, *Geophys. Res. Lett.*, 20, 1783, 1993.

Tsunoda, R.T., High-latitude *F* region irregularities: A review and synthesis, *Rev. Geophys.*, 26, 719, 1988.

Weber, E.J., J. Buchau, J.G. Moore, J.R. Sharber, R.C. Livingston, J.D. Winningham, and B.W. Reinisch, *F* layer ionization patches in the polar cap, *J. Geophys. Res.*, *89*, 1683, 1984.

Weber, E.J., J.A. Klobuchar, J. Buchau, H.C. Carlson Jr., R.C. Livingston, O. de la Beaujardiere, M. McCready, J.G. Moore, and G.J. Bishop, Polar cap *F* layer patches: Structure and dynamics, *J. Geophys. Res.*, *91*, 12,121, 1986.

Zalesak, S.T., and S.L. Ossakow, Nonlinear equatorial spread F: Spatially large bubbles resulting from large horizontal scale initial perturbations, *J. Geophys. Res.*, *85*, 2131, 1980.

N.A. Gondarenko and P.N. Guzdar, IREAP, University of Maryland, College Park, MD 20742. (e-mail: natalia@glue.umd.edu; guzdar@glue.umd.edu)

Relativistic Electron Flux Enhancements During Strong Geomagnetic Activity

D.N. Baker and X. Li

Laboratory for Atmospheric and Space Physics, University of Colorado, Boulder, Colorado

Long-term studies of energetic electron fluxes in the Earth's magnetosphere have revealed many of the temporal occurrence characteristics and their relationships to solar wind drivers. Early work showed the obvious and powerful role played by solar wind speed in producing subsequent highly relativistic electron enhancements. More recent work has also pointed out the key role that the north-south component of the IMF plays: In order to observe relativistic electron enhancement, there must typically be a significant interval of southward IMF along with a period of high ($V_{SW} \geq 500$ km/s) solar wind speed. This has led to the view that enhancement in geomagnetic activity (e.g., magnetospheric substorms) is normally a key first step in the acceleration of magnetospheric electrons to high energies. A second step is suggested to be a period of powerful low-frequency waves that is closely related to high values of V_{SW}. Hence, substorms provide a "seed" population, while high-speed solar wind drives the acceleration to relativistic energies in this two-step geomagnetic storm scenario. This picture seems to apply to most storms examined whether associated with high-speed streams or with CME-related events. In this paper, we also discuss the storm-substorm relationships as they pertain to high-energy electron acceleration and transport. We discuss various models of electron energization that have recently been advanced and we show how such electron energization relates more broadly to overall magnetospheric dynamical processes.

1. INTRODUCTION

The Earth's magnetosphere is an efficient accelerator and effective trapping device for energetic particles. The sources, losses, acceleration mechanisms, and transport processes of energetic particles are primary issues in magnetospheric physics. High energy electrons hold special interest because of their ubiquitous presence in the Earth's magnetosphere and their importance to other scientific and technological issues. Modern instrumentation has given an unprecedented combination of sensitivity, energy resolution, time resolution, and measurement duration that has made it possible to observe hitherto unseen energetic particle phenomena. Long-term measurements have unveiled many significant features of such particles and have also presented a great variety of new challenges in understanding the dynamics of energetic particles in the Earth's magnetosphere.

Important space weather effects relate to large and long-lasting enhancements of radiation belt particle fluxes. Such particles can be damaging to near-Earth spacecraft [*Vampola*, 1987; *Wrenn*, 1995; *Baker*, 2000], as well as to humans in low-Earth orbit [e.g., *Weyland and Golightly*, 2001]. Physical phenomena include dose and bulk charging effects in space vehicles in most near-Earth orbits [*Violet and Frederickson*, 1993; *Koons and Gorney*, 1992; *Gussenhoven*, 1983, 1985]. Recent research has provided a reasonably clear picture of the solar and solar wind drivers of general radiation belt changes [e.g., *Baker et al.* 2001; *Li et al.* 2001a]. Analyses of long-term data sets allow us to characterize the variation of the Earth's outer radiation belt and to describe the most extreme radiation belt conditions over approximately the last sunspot activity cycle.

2. GLOBAL VARIATIONS OF RADIATION BELT ELECTRONS

Records of geostationary-orbit high-energy electron fluxes (Figure 1) for solar cycles in the 1970s and 1980s suggested an 11-year cycle in the outer-belt electron flux [see *Baker et al.*, 1993a]. The long-term geostationary-orbit record indicates that outer radiation belt electrons peak in the declining phase of the sunspot cycle (in association with high-speed solar wind streams and recurrent storms), not at the time of sunspot maximum. The AE-8 model [*Vette*, 1991] flux level estimates are shown by the dashed line in Figure 1. Such comparisons show that the static NASA models give substantially higher flux values at L=6.6 than are actually observed and, furthermore, none of the solar cycle flux variations are captured by AE-8.

There is ample evidence of a "global coherence" of relativistic electron behavior such that electron fluxes vary throughout the entire outer radiation belt with a fair degree of synchronicity [*Kanekal et al.*, 1999; *Baker et al.*, 2001]. Geosynchronous orbit satellite data as shown in Figure 1 provide an important and useful monitoring of outer radiation zone particle fluxes. However, it is reasonably evident that the fluxes of relativistic electrons near L=4 are much higher and more slowly varying than those at L=6.6 (shown

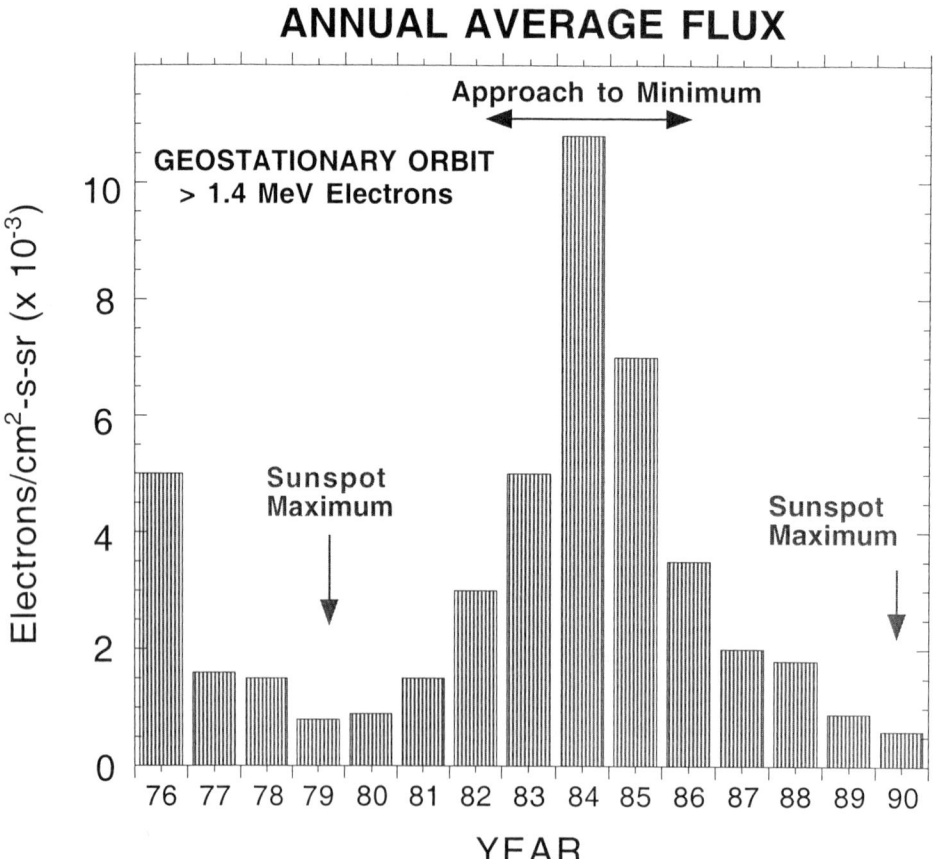

Figure 1. Annual fluxes of electrons with E>1.4 MeV from 1976 through 1990 at geostationary orbit. The NASA AE-8 model flux is indicated by the horizontal dashed line for comparison.

in Plate 1). Thus the electron fluxes are quantitatively, if not qualitatively, different at the heart of the outer zone (L=4) than at its outer fringes (L=6.6).

Plate 1 shows a color-coded representation of the flux of 2–6 MeV electrons (bottom panel) measured by the Solar, Anomalous, and Magnetospheric Particle Explorer (SAMPEX) [see *Baker et al.*, 1993b] from launch in mid-1992 to mid-2001. The upper panel of Plate 1 shows equivalent data (E>2 MeV) from the Comprehensive Energetic Particle Pitch Angle Detector (CEPPAD) experiment [*Blake et al.*, 1995] on the POLAR spacecraft from the time of its launch in February 1996 until mid-2001. The horizontal axis is time in years and the vertical axis in each panel is the McIlwain L-value. The color-coded intensity of electrons for each panel is shown as integral flux [electrons (cm^2-s-sr)$^{-1}$] according to the color bar to the right of the figure. A 27-day running-average smoothing function has been run over the entirety of both the SAMPEX and POLAR data to reduce the high-frequency fluctuations that can be present.

Plate 1 shows many features pertaining to the structure and variability of the outer zone electron population. The high intensity and great breadth of the outer radiation belt in 1993–95 is quite clear. This contrasts with the weaker and narrower outer belt in 1996–97. The resurgence of the electron fluxes (especially in 1998 near the inner portion of the outer belt) after the time of solar minimum is evident. As discussed by previous authors [*Baker et al.*, 1998a, 2001; *Kanekal et al.*, 1999], SAMPEX and POLAR data show many similar features in space and time. However, the POLAR data, being obtained much nearer the magnetic equator, show generally higher absolute electron intensities.

While Plate 1 provides a broad global view of outer zone electron flux variations, a more quantitative view is afforded by taking "cuts" of the data at various L-values. Figure 2 shows such cuts for L=2.0, 3.0, and 4.0. The plots show integral flux on a daily basis, but again with the 27-day smoothing filter applied. Notice the high peak flux values ($\gtrsim 2 \times 10^4$ electrons/cm^2-s-sr) in late 1993 and early 1994 at L=4.0 (Figure 2a). At later times, the L=4 fluxes were much lower on average. Notice also the extended interval in 1996 when the L=4 fluxes were less than 10^1 electrons (cm^2-s-sr)^1. Clearly, the radiation belt electron flux can, even in the heart of the outer zone, exhibit three to four orders of magnitude differences in monthly average intensities over the course of the solar cycle. We show two horizontal lines in Figure 2a, one corresponding to the AE-8 (Max) flux estimate at SAMPEX altitudes and another for AE-8 (Min). Obviously, the model fluxes are drastically different from the observations and the solar cycle trend of measured intensities is opposite to that in the model. Close inspection of Figure 2a shows another important feature not in models, viz., the seasonal (solstice/equinox) effect reported by *Baker et al.* [1999] and recently discussed further by *Li et al.* [2001b]: On average, electron fluxes are much higher around equinox than they are at solstice times.

Figure 2b shows the L=3 electron flux profile near the "slot" region between the inner and outer radiation belts. Notice in 1996–1997 (i.e., around solar minimum conditions) that average flux levels of E>2 MeV electrons never went above 10 electrons/cm^2-s-sr. On the other hand, in 1993–94 and again in 1998–2000 the electron fluxes were quite high, often reaching peak values near 10^4 electrons/cm^2-s-sr. The AE-8 model values shown in the figure are near the lower values actually observed by SAMPEX. During most periods, i.e., during 1992–95 and 1998–2001, the observed fluxes were considerably higher than either AE-8 (Max) or AE-8 (Min).

The L=2.0 profile in Figure 2c shows a behavior similar to L=3.0 in which fluxes were near background levels during sunspot minimum, but were much more elevated in the years away from minimum. The AE-8 models would suggest that there would be no significant flux at L=2.0 for SAMPEX altitudes. It is clear that this is not the case. From Figure 2 we see that spacecraft within the inner magnetosphere (L4) could experience vastly different fluences on a daily and monthly basis depending on which phase of the solar cycle they were actually operating in. Certainly, the static (AE) models are not a good description of observed flux values. More dynamic models [e.g., *Brautigam et al.*, 1992] clearly are needed to characterize observed variability.

3. RECURRENT STORMS AND HIGH-ENERGY ELECTRONS

Examination of the L=4 record in Figure 2a shows that generally the highest electron fluxes were seen in late 1993 and early 1994. More detailed inspection of the unsmoothed SAMPEX data supports this interpretation. Thus the 1993–1994 interval was the most extreme period of E>2 MeV electron radiation in the past solar cycle. We have examined data from SAMPEX electron sensors with thresholds from 0.4 MeV to 6 MeV over the lifetime of the mission in order to judge flux levels. The 1993–1994 interval was also found to be the time of the highest flux values over the entire relativistic energy range.

Plate 2 is a representation of the daily fluxes of 2–6 MeV electrons measured by SAMPEX during 1994. As in Plate 1, the flux values are color-coded and plotted for various L-values versus day of year. Also shown are the white vertical arrows, 27-days apart, along the top of the figure:

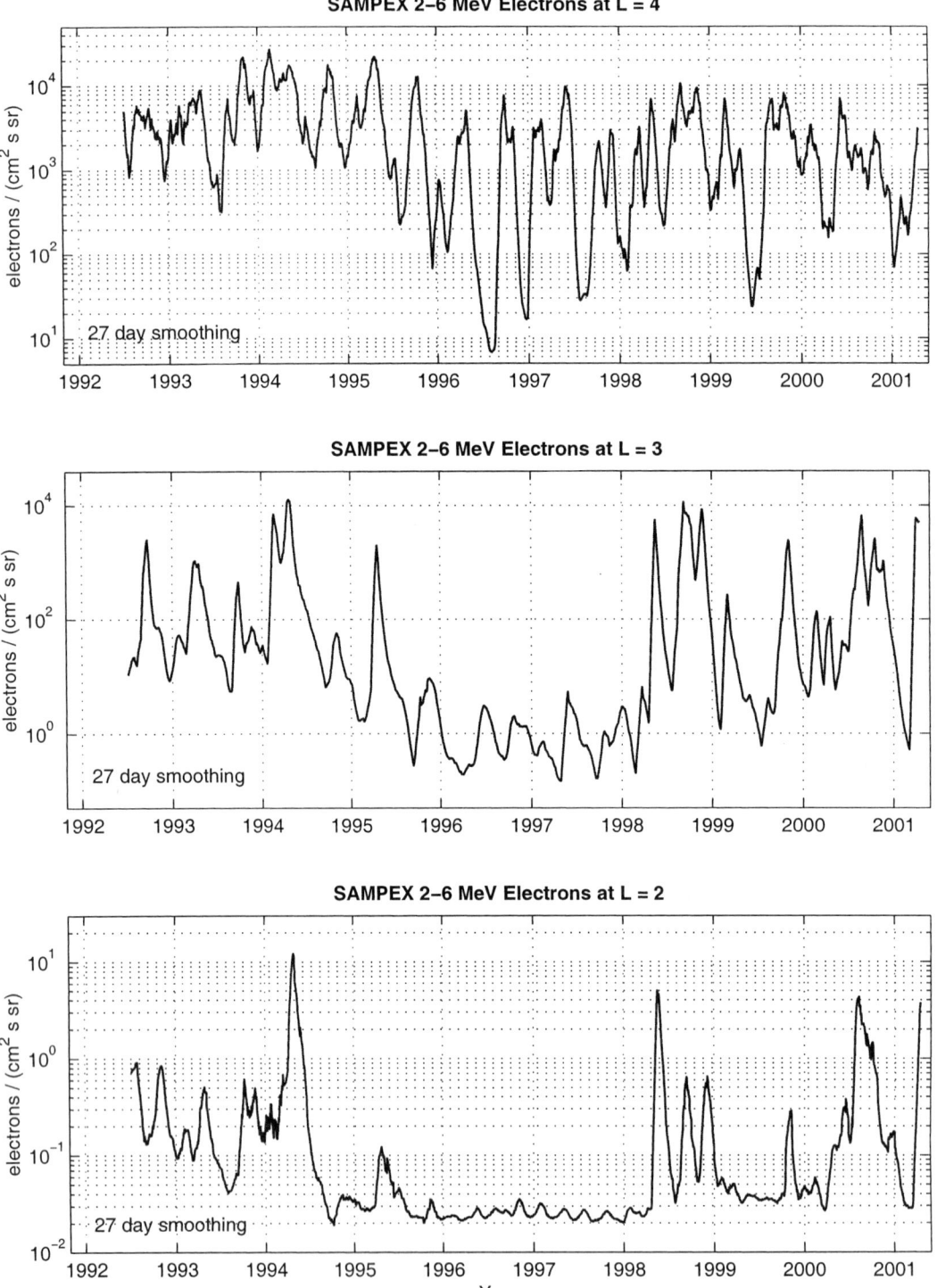

Figure 2. Plots of "cuts" at selected L-values for fluxes of electrons measured by SAMPEX from 1992 to 2001: (a) L=4.0 (b) L=3.0 and (c) L=2.0. The estimated flux levels from the AE-8 (Max/Min) models for L=3 and 4 are also indicated.

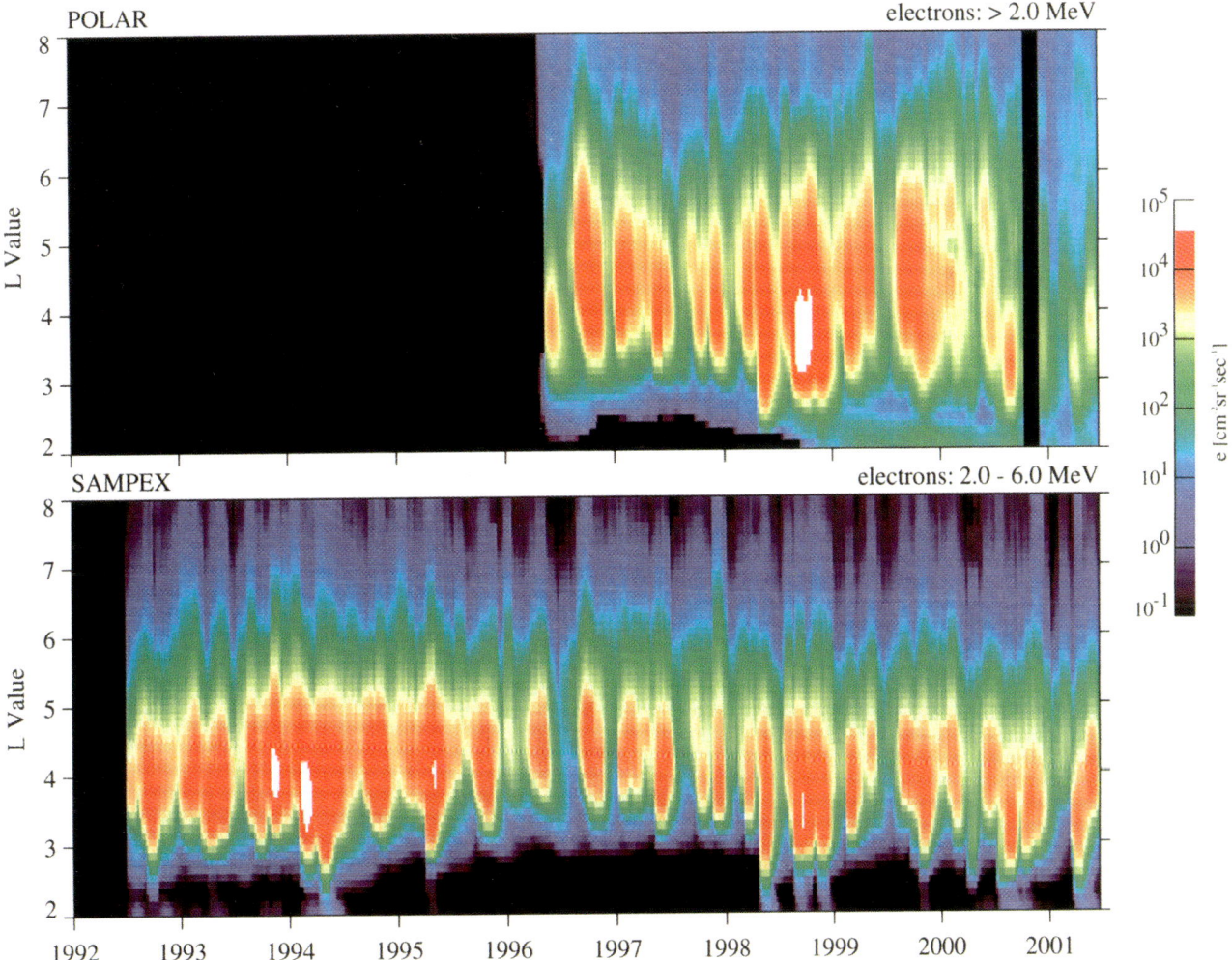

Plate 1. Color-coded intensities of electrons with E>2 MeV (color bar to right of figure). The McIlwain L-parameter is the vertical scale and time (in years) is the horizontal axis. The upper panel shows available POLAR data and the lower panel shows SAMPEX data. A 27-day smoothing filter has been run over all data.

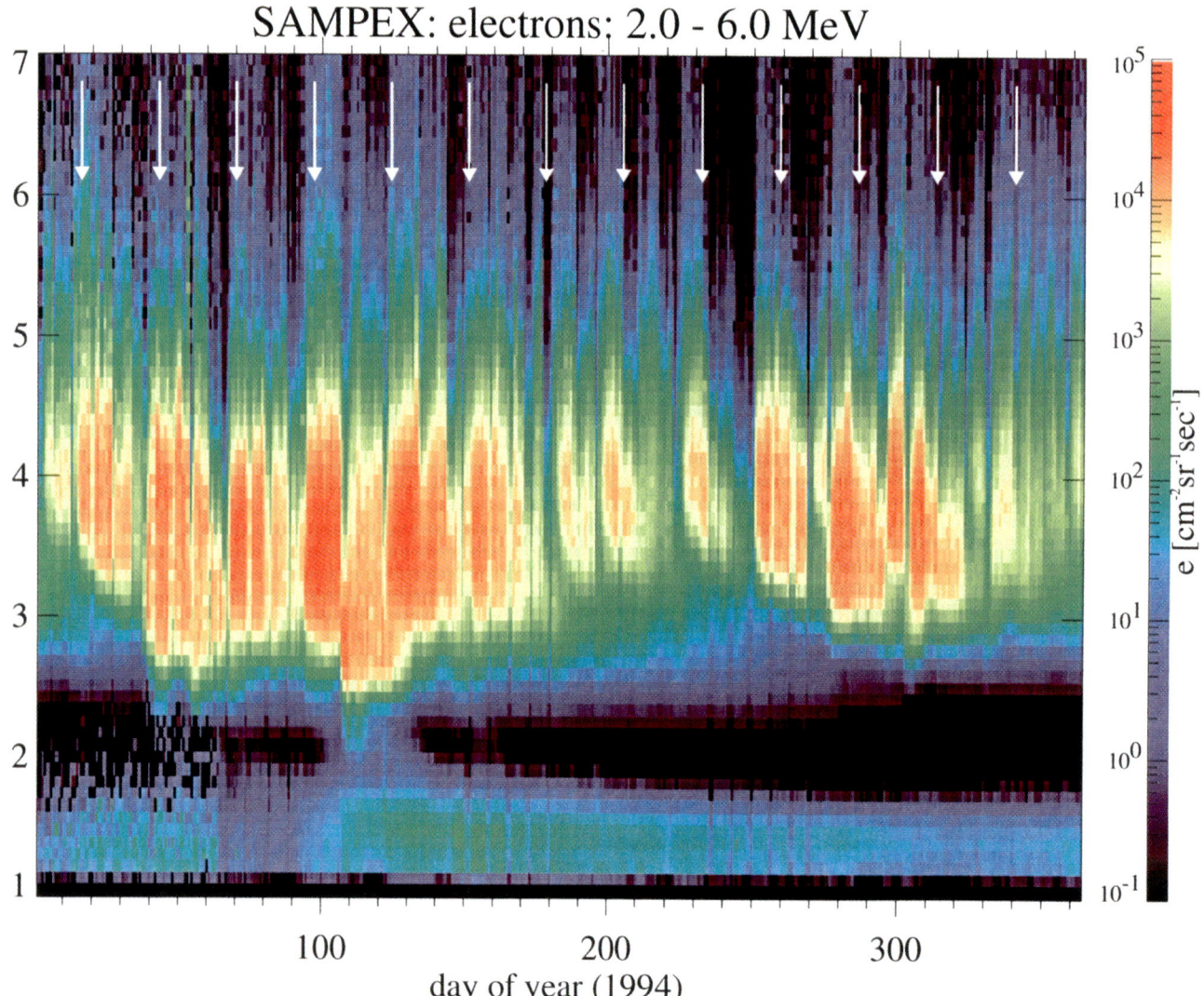

Plate 2. Color-coded intensities of electrons (E=2-6 MeV) measured by SAMPEX during 1994. Fluxes are indicated by color bar. Data are plotted as L-value (vertical axis) versus time. White arrows show 27-day recurrence times.

There was a clear and prominent 27-day periodicity in the electron flux enhancements. This was well associated with solar wind velocity enhancements. Thus, during the approach to sunspot minimum, high-energy electrons are at their highest levels throughout the outer radiation belt and this population is well associated with recurrent geomagnetic storms.

4. CME-DRIVEN STORMS AND ELECTRON INCREASES

The period of time around solar maximum is characterized not by high-speed solar wind streams, but rather by episodic geomagnetic storms driven by coronal mass ejections (CMEs). Powerful CMEs are often preceded by strong interplanetary shock waves which can greatly compress and distort the magnetosphere. Most CME-driven geomagnetic storms give rise to relativistic electron enhancements in the magnetosphere [e.g., *Reeves*, 1998]. Specific CME-related events have been examined in detail in recent papers [e.g., *Baker et al.*, 1998a].

An interesting case occurred in May of 1998. The sun was very active in April and May 1998 with numerous flares and CMEs [*Baker*, 2000]. The result was a major geomagnetic storm with a minimum Dst value of ~ −200 nT (on May 4) and an extended interval of high Kp and AE activity. As can be inferred from careful examination of Plate 1, the period of early May 1998 was a time of very substantial electron flux enhancement throughout the entire outer radiation belt: Electron intensities were evaluated from the "slot" region (L~2.0) all the way out to the vicinity of geostationary orbit (L~6.6).

It is important to note that the May 1998 interval was quite prominent from a space weather standpoint: Several spacecraft anomalies and failures occurred [*Baker*, 2000]. Most of these spacecraft problems occurred near geostationary orbit or elsewhere in the outer magnetosphere. Figure 3 shows a detail of relativistic electron fluxes (2-6 MeV) measured by SAMPEX at L=6.6 for the period April 20, 1998 through May 20, 1998. As can be seen, the electron flux, after being relatively low in late April, went up abruptly to high levels in early May. With some fluctuations in absolute intensity, the fluxes remained elevated until ~May 15. As discussed by *Baker et al.* [1998b], both the Equator-S failure and the Galaxy 4 failure noted in Figure 3 may have been related to deep dielectric charging from the elevated electron flux. The POLAR spacecraft anomaly shown in the figure was probably due to a solar proton event and not to high energy electrons.

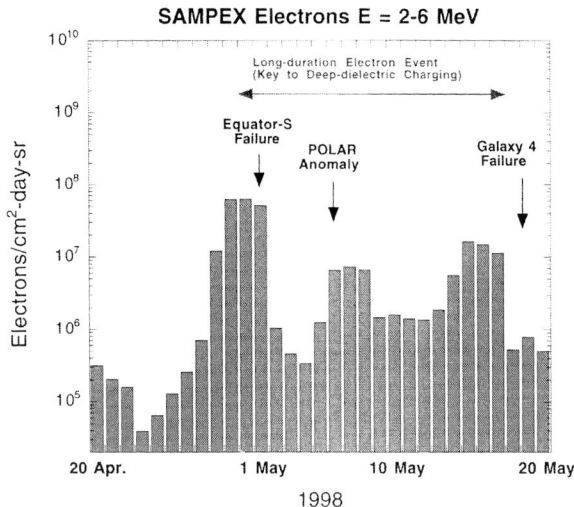

Figure 3. Daily electron fluxes measured at L=6.6 in the energy range $2 \leq E \leq 6$ MeV for the period April 20 to May 20, 1998. The times of several serious spacecraft operational anomalies are indicated by vertical arrows as labeled.

5. ACCELERATION MECHANISMS: ROLE OF SUBSTORMS, RADIAL DIFFUSION, AND WAVE HEATING

Radiation belt electrons are formed by accelerating lower energy electrons. There are two possible sources of lower energy electrons: One source is electrons at larger L that can be energized by being radially transported inward. This is usually called radial diffusion.

Another source is lower energy electrons at the same location that can be energized by wave-particle interactions. Both possible sources usually have a substantially larger phase space density than the radiation belt electrons and thus either of them could be a source of radiation belt electrons. Radial diffusion is usually thought to be the main acceleration mechanism [e.g., *Schulz and Lanzerotti*, 1974].

Recently, a greater emphasis has been placed on in situ heating of electrons by VLF waves on the same L-shell [*Temerin*, 1994; *Summers*, 1998; *Horne and Thorne*, 1998; *Meredith*, 2001; *Meredith*, 2002; *Albert*, 2002]. However, the relative effectiveness of these acceleration mechanisms has not been quantified.

We have previously studied mechanisms that might account for acceleration of electrons to relativistic energies during geomagnetic storms. An important correlation has been found between electron flux enhancements and ULF waved power in the magnetosphere [*Baker et al.*, 1998a]. Figure 4 shows the wave power measured at eight different

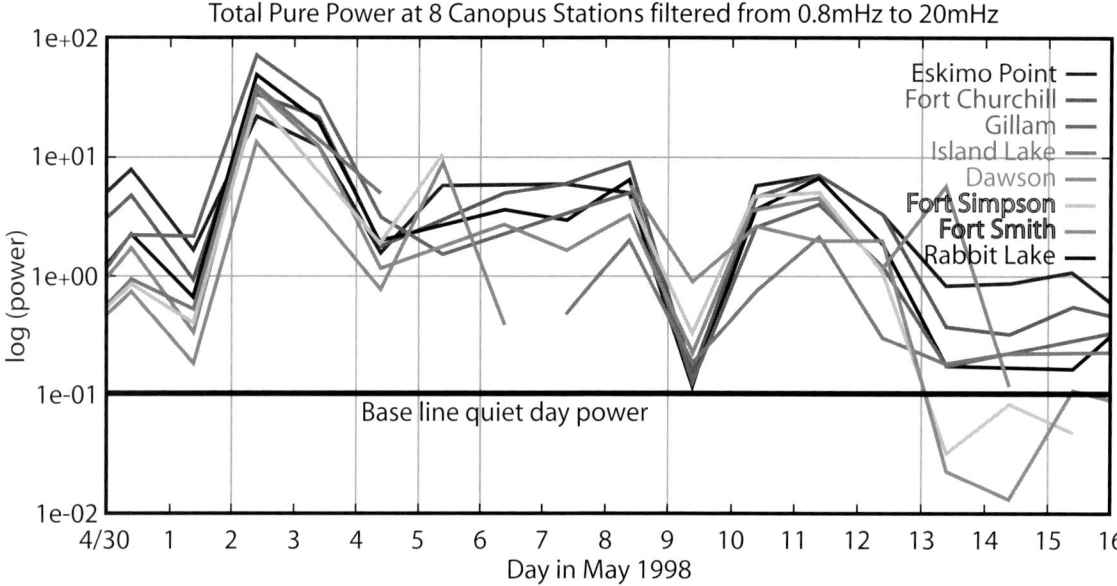

Figure 4. The ULF wave power measured each day during late April and early May of 1998. Several CANOPUS ground stations are represented (as labeled). The wave frequency range is 0.8–20 mHz [Courtesy of G. Rostoker].

ground stations in the Canopus system in Canada for the period April 30 through May 15, 1998. The data show increases from quiet day wave power by as much as a factor of 1000 in the frequency range 0.8 to 20 mHz. It is argued that these ULF waves can play a key role in electron acceleration [e.g., *Rostoker et al.*, 1998; *Hudson et al.*, 2000].

Based on correlations in many cases (as between Figures 3 and 4), we argue that ULF waves play an important role in the acceleration of outer zone electrons to relativistic energies. Another important point, however, is that there needs to be a "seed population" of electrons available on which the ULF waves (or other agents) act [e.g., *Baker et al.*, 1998a]. By analyzing plasma wave and particle data from the CRRES satellite during three case studies, *Meredith et al.* [2002] suggest that the gradual acceleration of electrons to relativistic energies during geomagnetic storms can be effective only when there are periods of prolonged substorm activity following the main phase of the geomagnetic storm. They argue that the prolonged substorm activity provides sustained VLF which in turn accelerate some of substorm injected electrons to higher energies. We feel that magnetospheric substorms are key to providing the seed population [*Baker et al.*, 1998c].

Figure 5 shows a schematic diagram which illustrates energy flow from the solar wind into, and through, the magnetosphere-ionosphere system [adapted from *Baker et al.*, 1997]. The explosive dissipation of stored magnetotail energy that occurs during substorms leads to many forms of energy output including plasmoid formation, ring current injection, plasma sheet heating, and particle acceleration. However, the typical substorm seldom directly accelerates electrons to energies much above 200 to 300 keV [*Baker et al.*, 1997]. A further, second-step acceleration process involving radial diffusion and wave particle interaction is necessary to reach relativistic energies.

The energy that can be gained by radial transport, whether in the form of radial diffusion or fast injections, through the violation of the third adiabatic invariant is limited by the ratio of the magnetic field magnitudes within the region of radial transport. Thus radial transport as an energization mechanism normally requires a substantial source population. For a given value of the first and second adiabatic invariants, the phase space density usually increases with increasing L for L=3-6.6 and beyond [*Selesnick and Blake*, 1997]. Thus there should be a region outside of geosynchronous orbit where the phase space density at constant first and second adiabatic invariants peaks. The central plasma sheet region is a likely source region. Indeed, study of several Wind perigee passes in conjunction with POLAR measurements suggests that the phase space density for given first adiabatic invariant continues to increase toward larger radial distances (~11–14 Re) and precipitously decreases once Wind goes out magnetosphere [*Li, et al.* 1997a,b].

Based on the standard radial diffusion equation [*Schulz and Lanzerotti*, 1974], *Li et al.* [2001a] developed a model

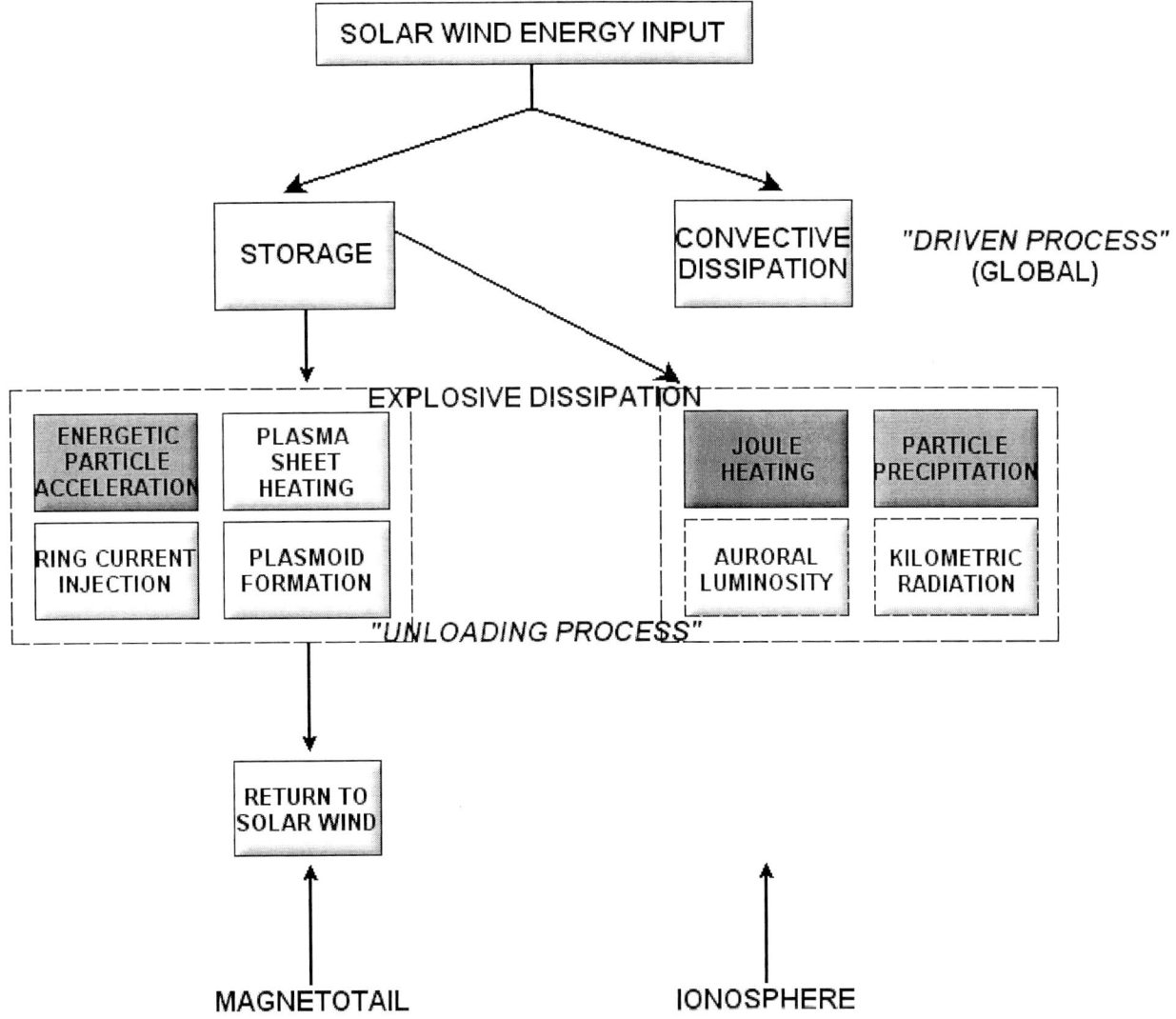

Figure 5. The flow of energy into and through the magnetosphere during periods of enhanced geomagnetic activity [adapted from *Baker et al.*, 1997].

to make quantitative prediction of the intensity of multi-million electron Volt (MeV) electrons at geosynchronous orbit using only measured solar wind parameters. The radial diffusion equation is solved by setting the phase space density larger at the outer boundary than the inner boundary and by making the diffusion coefficient a function of the solar wind parameters. The most important parameters are solar wind velocity and southward component of IMF. Figure 6 displays a comparison of five and one-half years of daily averages of the MeV electron flux measured at geostationary orbit with our prediction based solely on measurements of the solar wind. Both the shorter time scale and the longer seasonal effects, such as the overall reduction in the electron fluxes in the middle of 1996, are reproduced. Furthermore, the model provides a physical explanation for several features of the correlation between the solar wind and the MeV electron flux at geostationary orbit such as the approximate 1–2 day delay between the peak in the solar wind velocity and the peak in the MeV electron flux at geostationary orbit. The delay is simply due to the fact that it takes some time for the electrons to diffuse inward to geostationary orbit in response to changes in the solar wind input, and also some time for such changes to decay [*Li et al.*, 2001a].

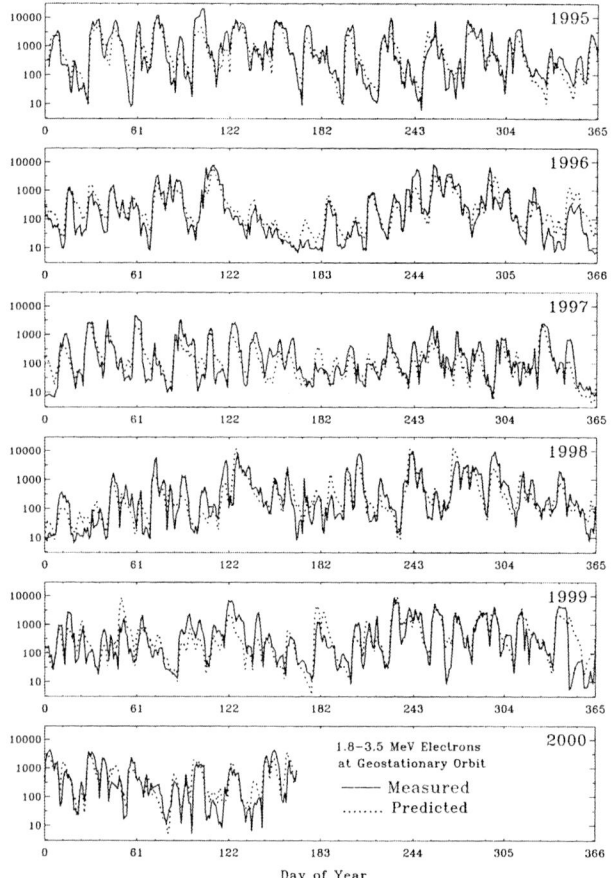

Figure 6. A comparison of five and one-half years of daily averages of the MeV electron flux (#/cm²-s-sr-MeV) measured at geostationary orbit with the predicted results based solely on solar wind measurements. Horizontal axis shows day of year.

6. STORM-SUBSTORM RELATION

We argue here that magnetospheric substorms and geomagnetic storms are closely related to one another when it comes to energetic electron phenomena. It would be remarkable, and quite hard to believe, that a southward turning of the IMF that opens the magnetosphere to energy input would lead to two totally separate and disconnected phenomena. The original view of S. Chapman and many other researchers that storms are just a superposition of substorms was clearly too limited. But at the present time, there is a tendency to let the pendulum swing too far the other way: There is often an assertion that storms could, in principle, occur without substorms. We note that these different points of view are presented in other review papers in this volume.

Our view is in between these two extremes: We believe that substorms are an important, indeed, key, step along the way to geomagnetic storms. The magnetosphere crosses many thresholds in its progression of development and it begins to admit many new forms of energy dissipation (which we might call "branches" or "channels") as it is driven harder and harder by the solar wind (Figure 7). Substorms are an elementary (and essential) component in this progression. Substorms have many important properties like nonlinearity, complexity, self-organization, criticality, and so forth. We very much want to know exactly <u>when</u>, and exactly <u>where</u> a substorm will start. But like a dripping faucet with variable inflow, or like a localized thunderstorm on a hot summer day, there is a quality of unpredictability in substorm physics.

As the magnetosphere progresses toward major storms, however, the external driver (the strong flow of the solar wind energy) overwhelms and drives the magnetosphere into a mode of powerful "direct" response. This is somewhat analogous to a large-scale and highly organized hurricane which moves across sea and land in a dominant and coherent way.

In many ways and for many reasons, we want to know as much as possible about where localized events, i.e., substorms, will strike. But we will always have to accept some probabilistic statements in this set of phenomena. But for the major storm—the cyclonic system or the hurricane of our analogy—we can make much more definitive statements if we track the course of the storm (i.e., observe the

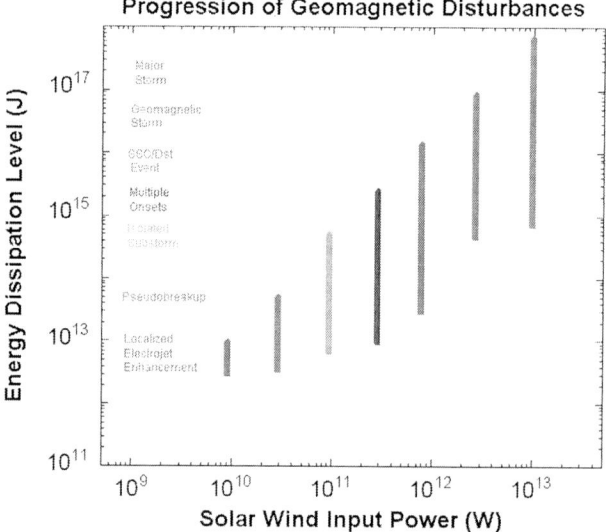

Figure 7. The progression of energy dissipation in the magnetosphere with the increasing solar wind energy input rate. The vertical bars show the range of processes typically occurring during different input power levels [from *Baker et al.*, 2001].

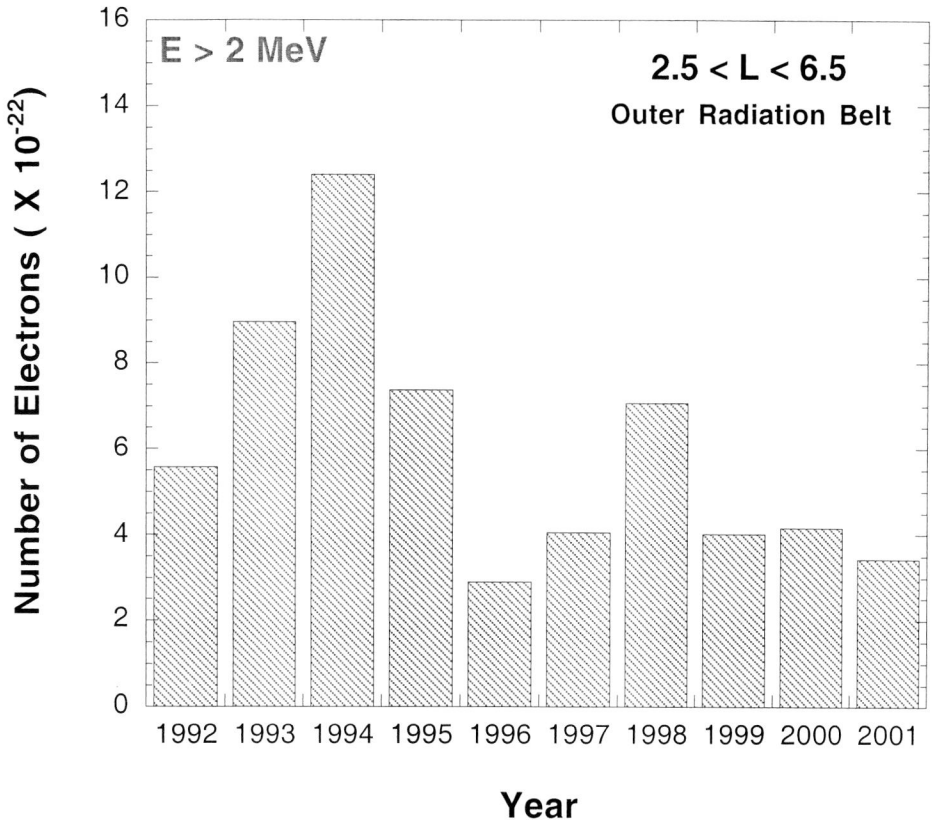

Figure 8. Annual averages of magnetospheric electron content (E>2 MeV) throughout the outer radiation belt (2.5≤L≤6.5) from 1992 to 2001.

external solar wind driver). The largest and most powerful geomagnetic storms involve numerous constituent substorm events and a host of included physical processes including relativistic electron acceleration episodes.

7. LONG-TERM RADIATION BELT BEHAVIOR

SAMPEX, POLAR, and other available data can be used to calculate the total electron content in the entire outer radiation belt [*Baker et al.*, 2001]. The method considers the volume of the outer zone from L=2.5 to L=6.5 and calculates average fluxes (and densities) of electrons above particular energies throughout the volume of interest. Figure 8 shows the calculated number of electrons with E>2 MeV on average for each year from 1992 to 2001. (Note that the 1992 and 2001 estimates are based upon partial-year coverage). The results in Figure 8 clearly make the point in an integrated sense for the entire outer belt that the years corresponding to the approach to solar minimum (i.e., 1993–1995 in this case) greatly exceed solar max years (1999–2001) in average electron content.

Another way of looking at the radiation belts is presented in Plate 3. This shows Northern Hemisphere global maps of the Earth's radiation belts (as measured at 600 km altitude) for three separate years: 1994, 1997, and 2000. The data are shown as color-coded electron fluxes for E=2-6 MeV. All data for each of the respective years have been averaged together. It is seen that 1994 was, on average, a year of very high electron intensity (especially at longitudes west of the South Atlantic Anomaly), while the years 1997 and 2000 were relatively inactive. Note that the electron intensities tend to be high over a broad latitude band (~45° to ~75°) corresponding to the entire outer radiation zone.

8. SUMMARY AND CONCLUSIONS

In this paper we have presented analyses of data from several spacecraft (especially SAMPEX and POLAR) drawn from the last nine years, i.e., 1992 to 2001. Smoothed and/or broadly averaged data were used in order to characterize the outer radiation belt electron environment at different time scales. We find that at any L-value, including

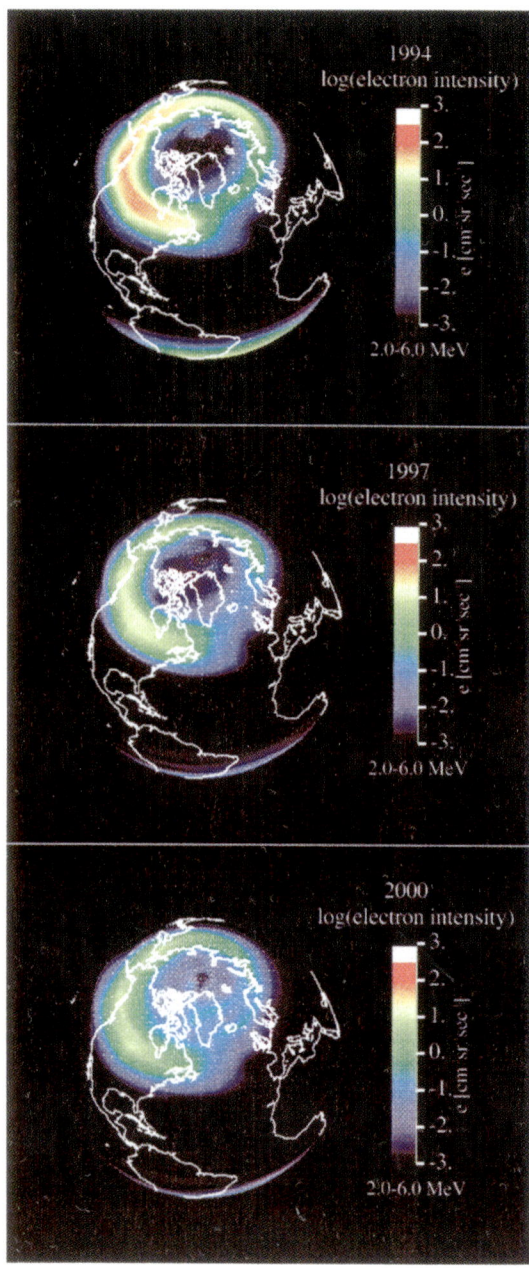

Plate 3. Global radiation belt maps averaged for the years (a) 1994, (b) 1997, and (c) 2000 for electrons with E=2-6 MeV. The year 1994 was by far the most intense in terms of average electron flux.

right at the heart of the outer zone (L=4), there can be tremendous ranges of average particle fluxes. The lowest intensities of electrons were found during summer and winter solstice times around sunspot minimum. The maximum fluxes were found, generally, at equinox times during the declining phase of the sunspot activity cycle. These analyses present quite a different picture of the relativistic electron fluxes in the outer radiation belt as compared to the standard NASA models, [e.g., *Vette*, 1991]. More modern radiation belt models such as those from CRRES [*Brautigam et al*, 1992] give somewhat more realistic descriptions of observed radiation belt behavior than do the essentially static models.

We also discussed the systematic response of MeV electrons at geostationary orbit to the variation of solar wind, demonstrated by the predictability of the daily averaged electron fluxes based on solar wind parameters only. On the other hand, given the solar wind condition, it is still not predictable exactly when and where a substorm will start, an example of chaotic features in magnetospheric physics.

Acknowledgments. The authors thank many colleagues from the SAMPEX, Polar, WIND, ACE, and geostationary orbit satellite teams for use of data and useful conversations. This work was supported by NASA.

REFERENCES

Albert, J. M., Nonlinear interaction of outer zone electrons with VLF waves, *Geophys Res. Lett., 10.1029/2001GL013941*, 30 April, 2002.

Allen, J.H., and D.C. Wilkinson, Solar-Terrestrial Activity Affecting Systems in Space and on Earth, in *Solar-Terrestrial Predictions–IV*: Vol. 1, pp. 75-107, Ottawa, Canada, May 18–22, 1992.

Baker, D.N., The occurrence of operational anomalies in spacecraft and their relationship to space weather, *IEEE Trans. Plasma Sci., 28*, 2007, 2000.

Baker, D.N., R.A. Goldberg, F.A. Herrero, J.B. Blake, and L.B. Callis, Satellite and rocket studies of relativistic electrons and their influence on the middle atmosphere, *J. Atmos. Terr. Phys., 55*, 1619, 1993a.

Baker, D.N., G.M. Mason, O. Figueroa, G. Colon, J. Watzin, and R. Aleman, An overview of the SAMPEX mission, *IEEE Trans. Geosci. Elec., 31*, 531, 1993b.

Baker, D.N., T.I. Pulkkinen, M. Hesse, and R.L. McPherron, A quantitative assessment of energy storage and release in the Earth's magnetotail, *J. Geophys. Res., 201*, 7159, 1997.

Baker, D.N., T.I. Pulkkinen, X. Li, S.G. Kanekal, J.B. Blake, R.S. Selesnick, M.G. Henderson, G.D. Reeves, H.E. Spence, and G. Rostoker, Coronal mass ejections, magnetic clouds, and relativistic magnetospheric electron events: ISTP, *J. Geophys Res., 103*, 17,279, 1998a.

Baker, D. N., X. Li, J. B. Blake, and S. Kanekal, Strong electron acceleration in the Earth's magnetosphere, *Adv. Space Res., 21*, 609, 1998b.

Baker, D.N., S.G. Kanekal, T.I. Pulkkinen, and J.B. Blake, Equinoctial and solstitial averages of magnetospheric relativistic electrons: A strong semiannual modulation, *Geophys. Res. Lett., 26*, No. 20, 3193, 1999.

Baker, D.N., S.G. Kanekal, J.B. Blake, and T.I. Pulkkinen, The global efficiency of relativistic electron production in the Earth's magnetosphere, *J. Geophys. Res., 106*, 19,169, 2001.

Blake, J.B., et al., CEPPAD: Comprehensive energetic particle and pitch angle distribution experiment on POLAR, *Space Sci. Rev., 71*, 531, 1995.

Brautigam, D.H., M.S. Gussenhoven, and E.G. Mullen, Quasistatic model of outer zone electrons, *IEEE Trans. Nucl. Sci., 39*, 1797, 1992.

Gussenhoven, M.S., Geosynchronous environment for severe spacecraft charging, *J. Spacecraft, 20*, 26, 1983.

Gussenhoven, M.S., High level spacecraft charging in the low altitude polar auroral environment, *J. Geophys. Res., 90*, 11,009, 1985.

Horne, R. B., and R. M. Thorne, Potential waves for relativistic electron scattering and stochastic acceleration during magnetic storms, *Geophys. Res. Lett., 25*, 3011, 1998.

Hudson, M. K., S. R. Elkington, J. G. Lyon, and C. C. Goodrich, Increase in relativistic electron flux in the inner magnetosphere: ULF wave move structure, *Adv. Space Res., 25*, 2327, 2000.

Kanekal, S.G., D.N. Baker, J.B. Blake, B. Klecker, R.A. Mewaldt, and G.M. Mason, Magnetospheric response to magnetic cloud (CME) events: Energetic particle observations from SAMPEX and POLAR, *J. Geophys. Res., 104*, 14,885, 1999.

Koons, H.C., and D.J. Gorney, The relationship between electrostatic discharges on spacecraft P78-2 and the electron environment, The Aerospace Corp., El Segundo, *CA, TR-0091 (6940 06)-2*, Mar. 1992.

Li, X., D. N. Baker, M. Temerin, D. Larson, R. P. Lin, G. D. Reeves, M. D. Looper, S. G. Kanekal, and R. A. Mewaldt, Are energetic electrons in the solar wind the source of the outer radiation belt? *Geophys. Res. Lett., 24*, 923, 1997a.

Li, X., D. N. Baker, M. Temerin, D. Larson, R. P. Lin, G. D. Reeves, J. B. Blake, M. Looper, R. Selesnick, R. A. Mewaldt, Source of Relativistic Electrons in the Magnetosphere: Present Knowledge and Remaining Questions, presented at the IAGA, Uppsala, Sweden, 1997b.

Li, X., M. Temerin, D. N. Baker, G. D. Reeves, and D. Larson, Quantitative Prediction of Radiation Belt Electrons at Geostationary Orbit Based on Solar Wind Measurements, *Geophys. Res. Lett., 28*, 1887, 2001a.

Li, X., D. N. Baker, S. G. Kanekal, M. Looper, and M. Temerin, Long Term Measurements of Radiation Belts by SAMPEX and Their Variations, *Geophys. Res. Lett., 28*, 3827, 2001b.

Meredith, N. P., R. B. Horne, and R. R. Anderson, Substorm dependence of chorus amplitudes: Implications for the acceleration of electrons relativistic energies, *J. Geophys. Res., 106,* 13,165, 2001.

Meredith, N. P., R. B. Horne, R. H. Iles, R. M. Thorne, D. Heynderickx, and R. R. Anderson, Outer zone relativistic electron acceleration associated with substorm enhanced whistler mode chorus, *J. Geophys. Res.,* in press, 2002.

Reeves, G. D., Relativistic electrons and magnetic storms: 1992–1995, *Geophys. Res. Lett., 25,* 1817, 1998.

Rostoker, G., S. Skone, and D. N. Baker, On the origin of relativistic electrons in the magnetosphere associated with some geomagnetic storms, *Geophys. Res. Lett., 25,* 3701, 1998.

Schulz, M. and L.J. Lanzerotti, *Particle diffusion in the radiation belts,* Springer-Verlag, New York, 1974.

Selesnick, R. S., and J. B. Blake, Dynamics of the outer radiation belt, *Geophys. Res. Lett., 24,* 1347, 1997.

Summers, D., R. M. Thorne, and F. Xiao, Relativistic theory of wave-particle resonant diffusion with application to electron acceleration in the Magnetosphere, *J. Geophys. Res., 103,* 20,487, 1998.

Temerin, M., I. Roth, M. K. Hudson, J. R. Wygant, New paradigm for the transport and energization of radiation belt particles, *Eos,* AGU, Nov. 1, 1994, page 538.

Vampola, A.L., Thick dielectric charging on high-altitude spacecraft, *J. Electrostatics, 20,* 21, 1987.

Vette, J., The AE-8 trapped electron model environment, *National Space Science Data Center, Report 91-24,* Greenbelt, MD, 1991.

Violet, M.D., and A.R. Fredrickson, Spacecraft anomalies on the CRRES satellite correlated with the environment and insulator samples, *IEEE Trans. Nucl. Sci., 40,* 1512, 1993.

Weyland, M., and M. Golightly, Results of radiation monitoring on the International Space Station and associated biological risks, *Proceedings of International Space Environment Conference,* Queenstown, New Zealand, July 23–27, 2001.

Wrenn, G.L., Conclusive evidence for internal dielectric charging anomalies in geosynchronous communications spacecraft, *J. Spacecraft, 32,* 3, 514, 1995.

D.N. Baker, Laboratory for Atmospheric and Space Physics, University of Colorado, 1234 Innovation Drive, Boulder, CO 80303-7814, USA

X. Li, Laboratory for Atmospheric and Space Physics, University of Colorado, 1234 Innovation Drive, Boulder, CO 80303-7814, USA

Modeling the Magnetosphere Using Time Series Data

A.S. Sharma, A. Y. Ukhorskiy, and M. I. Sitnov

University of Maryland, College Park, Maryland

J. A. Valdivia

Departamento de Fisica, Universidad de Chile, Casilla 653, Santiago, Chile

The solar wind-magnetosphere coupling during storms and substorms leads to many processes with a wide range of space and time scales. These processes are not modeled easily using first principles approaches. The nonlinear dynamical approach provides an alternative and has been used to develop models of magnetospheric dynamics from observational time series data. These models capture the features of the magnetosphere inherent in the data, independent of a priori assumptions of specific physical processes. This modeling approach has given a new description of the global and multi-scale features of the magnetosphere. The global features represent the coherent dynamics and can be described by a low dimensional model, such as local-linear filters, which have provided reliable space weather forecasting tools. The filters are derived from the reconstructed input-output phase space representing the solar wind – magnetosphere system. In these models the input is the solar wind induced electric field VBs and the output is the auroral electrojet index AL for substorms or the Dst index for storms. The local-linear filters represent a mean-field description, obtained by averaging the outputs corresponding to similar states of the system in the reconstructed phase space. These filters can be used to model the multi-scale dynamics but its predictability is not as high as in the case of the global dynamics. A combination of the nonlinear dynamics of the global features with a statistical physics of the multi-scale processes yields a comprehensive model.

1. INTRODUCTION

The magnetosphere is driven by the turbulent solar wind and exhibits complex behavior. During substorms both global and multi-scale features are clearly observed. The global or coherent behavior of the magnetosphere is evident in a variety of large-scale processes such as plasmoid formation and ejection, field line dipolarization, global current systems, etc. At the same time a number of small-scale phenomena observed during substorms, such as MHD turbulence, bursty bulk flows, current disruption, etc. are multi-scale in nature, viz. they have power law spectra over a wide range of perturbation scales. Such a diversity in dynamical properties makes the development of a unified framework for modeling the magnetospheric dynamics during substorms a challenging task. The magnetospheric response during the storm time scale is in some sense less complex, being dominated by directly driven processes. However recent observations by Image spacecraft [*Burch et al.*, 2001] and modeling using ground-based data [*Clauer et al.*, this volume] have shown many new features that need to be included in modeling the magnetosphere during storms.

The early data-derived models of magnetospheric substorms were inspired by the concepts of dynamical chaos [*Vassiliadis et al.*, 1990]. They were based on the assumption that the observed complexity of the magnetospheric dynamics can be attributed to the nonlinear coupling of just a few dominant degrees of freedom [see reviews: *Sharma*, 1995, 1997; *Klimas et al.*, 1996]. Such an approach essentially implies predictability, at least within appropriate time scales, and has led to considerable progress in the development of space weather forecasting tools based on local-linear filters and neural networks [*Price et al.*, 1994; *Vassiliadis et al.*, 1995; *Valdivia et al.*, 1996; *Horton and Doxas*, 1996; *Gleisner and Lundstedt*, 1997].

However many studies have shown that not all aspects of magnetospheric dynamics during substorms conform to the hypothesis of low dimensionality and thus cannot be accounted within the framework of dynamical chaos. For example, the power spectrum of AE index data [*Tsurutani et al.*, 1990] and magnetic field fluctuations in the tail current sheet [*Ohtani et al.*, 1995] have a power law form typical for high dimensional colored noise. Moreover, detailed analyses [*Takalo et al.*, 1993, 1994] have shown that the qualitative properties of the AE time series are more similar to bicolored noise than to low-dimensional chaotic systems.

One of the models used to explain the multi-scale properties of the magnetospheric dynamics is self-organized criticality (SOC) [*Bak et al.*, 1987]. A system in SOC is modeled by a sand-pile or other non-equilibrium cellular automata which evolve to a steady-state critical point due to the fine tuning of the control parameters [*Vespignani and Zapperi*, 1998]. In the vicinity of a critical point the energy transport in the system is carried out by avalanches whose sizes are distributed according to a power law. This feature of SOC has been exploited to account for the power law spectra observed in the magnetosphere during substorms [e.g. *Consolini*, 1997; *Chapman et al.*, 1998; *Takalo et al.*, 1999; *Klimas et al.*, 2000]. However, the fine tuning of parameters required to approach the criticality corresponds to the vanishing values of the input parameter [*Vespignani and Zapperi*, 1998]. This makes the system effectively autonomous and thus questions the applicability of SOC to the magnetosphere, whose dynamics is to a large extent driven by the solar wind input. Moreover, SOC models generally cannot account for the large-scale coherent features of the magnetosphere since in most SOC models the multi-scale properties of the system are essentially independent of the global dynamics. An exception to this is a sand-pile model [*Chapman et al.*, 1998] in which scale-invariant avalanches were found to coexist with system-wide large-scale events.

It is essential to develop a comprehensive model that can reconcile the global and multi-scale features of solar wind–magnetosphere coupling. To this end the nonlinear dynamical approach is combined with the elements of statistical physics to obtain a single model capable of describing both the features. The methods of nonlinear dynamics are used to reconstruct the phase space of the system and to forecast the global constituent of the magnetospheric dynamics using local-linear filters (LLF). However, it has been shown recently that LLF models leave out a significant portion of the time-series and these features correspond to the multi-scale component of the dynamics [*Ukhorskiy et al.*, 2002a]. This multi-scale constituent of the magnetospheric dynamics is high dimensional, similar to the colored noise, and does not allow deterministic predictions, thus imposing limitations on the predicting ability of the dynamical models.

The magnetospheric dynamics during substorms shares a number of properties with non-equilibrium phase transitions [*Sitnov et al.*, 2000, 2001; *Sharma et al.*, 2001]. Using global singular spectrum analysis of VBs-AL data *Sitnov et al.* [2000] have shown that the global magnetospheric dynamics is organized in a manner similar to the first order phase transition. For eaxmple, a phase diagram similar to the "pressure-temperature-density" diagram of the water-steam system was obtained from the *Bargatze et al.* [1985] dataset. Further, a relationship between the magnitude of the largest fluctuations of AL time derivative and the solar wind parameters was obtained [*Sitnov et al.*, 2001]. This is similar to the input-output critical exponent β in second order phase transition. Subsequent analyses [*Ukhorskiy et al.*, 2002a] have shown that dynamical models based on LLF are very similar to the mean-field approach in phase transitions since its output is obtained by an averaging over a chosen range of scales in the reconstructed input-output phase space. Thus, the multi-scale features of the time series not captured by LLFs are essentially the deviations from the mean-field model. According to the phase transition analogy the magnitudes of these fluctuations should be related to the solar wind input in a probabilistic fashion similar to the input-output critical exponent. Such a relation has been established in terms of conditional probability and used to improve space weather forecasting tools. These results were obtained with a new model that describes the global and multi-scale features of the magnetosphere during substorms. This model, referred to as STADY, combines the STAtistical and DYnamical features of the magnetosphere [*Ukhorskiy et al.*, 2002b]. This model yields a dynamical prediction from a low-dimensional model and an estimate of its deviations from the data, computed from the conditional probability of the events.

The data-derived models for substorms are developed using correlated database of solar wind and geomagnetic time series. Among such databases that of *Bargatze et al.* [1985], consisting of solar wind data from IMP 8 spacecraft and the corresponding values of the auroral indices with 2.5 min resolution, is perhaps the most widely used. This database is used in the studies of substorms discussed in the following sections. The database consists of 34 isolated intervals, arranged in order of increasing geomagnetic activity, with each interval representing isolated auroral activity preceded and followed by at least two-hour long quiet periods ($VB_s \approx 0$, AL<50 nT). The solar wind convective electric field VB_s is taken as the input in the model. The magnetospheric response to the solar wind activity is represented by the AL index and is the output of the model. To facilitate the use of VB_s and AL data in a joint input-output phase space, the data sets are normalized by their respective standard deviations.

The solar wind–magnetosphere coupling during magnetic storms is readily described in terms of local-linear filters [*Valdivia et al.*, 1996]. These filters have been obtained from many data sets. The LLFs obtained from the 1 hr averaged OMNI data set have been used for forecasting the Dst index during magnetic storms. Similar filters were obtained using 5 min averaged data for January–June 1979 [*Kamide et al.*, 1998]. The latter contains the magnetic field data from the individual magnetometer stations and have ben used to study the spatio-temporal dynamics during magnetic storms [*Valdivia et al.*, 1999a & 1999b].

2. GLOBAL FEATURES OF THE MAGNETOSPHERE DURING SUBSTORMS: DYNAMICAL MODEL

The magnetosphere exhibits globally coherent dynamical features during substorms [*Baker et al.*, 1990; *Vassiliadis et al.*, 1990; *Siscoe*, 1991; *Sharma et al.*, 1993] and have been modeled using the time-delay embedding technique. This technique was initially developed for autonomous chaotic systems [see review: *Abarbanel et al.*, 1993] and was later extended to input-output systems [*Casdagli*, 1992]. In this approach a scalar time series data of an observable of the system is assumed to be a function of the state of the underlying system and to contain all the information necessary to determine its dynamics. Thus, a space with large enough dimension to unfold the dynamical attractor can be reconstructed from the time series. This reconstructed space contains the information about the dynamical evolution of the states of the system and the time evolution of a specified state can be obtained from the known evolution of similar states.

For driven dynamical systems the states of the system in the reconstructed phase space are given by the input-output delay vectors defined as

$$\left(\vec{I}_n, \vec{O}_n\right)^T = \left(I_n, \ldots, I_{n-(M-1)}, O_{n-1}, \ldots, O_{n-(M-1)}\right) \quad (1)$$

where $I(t_n) = I_n$ and $O(t_n) = O_n$ are the input and the output time series, and M is the total number of delays. A proper reconstruction requires a time delay that is long enough to unfold the underlying dynamics and short enough so that relevant features are not lost. Once such a reconstruction is achieved, the value of the output at the next time step is related to the current state of the system by some nonlinear map F as

$$O_{n+1} = F(\vec{I}_n, \vec{O}_n) \quad (2)$$

The linear expansion of **F** gives the well known expression for the local-linear filters:

$$O_{n+1} = \sum_{i=1}^{M-1} \alpha_i \cdot I_{n-i} + \sum_{j=T_{prd}/dt}^{M-1} \beta_j \cdot O_{n-j} \quad (3)$$

where T_{prd} is the duration for which the prediction is made. The filter coefficients α_i and β_j are computed from the given data set, referred to as the training set. For a given state outside the training set, similar states in the data set are searched and those within a specified range, as measured by the distance in the embedding space using the Euclidean metric, are identified as the nearest neighbors. The evolution of these nearest neighbors determines, on the average, the evolution of the current state under consideration.

To model the solar wind–magnetosphere coupling during substorms the local-linear filters (LLF) defined by (3) with $I=VB_S(t)/\sigma_{VBs}$ and $O=AL(t)/\sigma_{AL}$ are used, where the σ's are the square roots of the respective variances. An example of 1 hour predictions of AL for a high activity period (interval 32 of *Bargatze et al.* [1985] data set) obtained by this model is shown in Fig 1. It can be seen from the figure that LLF reproduces the long-term global variations in data well and this feature is readily used to forecast the average level of substorm activity. This component of AL is dominantly regular and by construction can be described by a low-dimensional model. Thus, the portion of the time series reproduced by LLF corresponds to the global and coherent features of the magnetospheric dynamics. However, it is also evident from Fig. 1 that LLF do not capture the abrupt variations and peaks. This feature of LLF is generic to the magnetosphere and does not depend on the level of substorm activity [*Ukhorskiy et al.*, 2002a]. The inability of LLF to reproduce the sharp variations in data is due to multi-scale constituents of AL time series, which can be described only by a high dimensional model. It should be noted that the properties of AL not captured by LLF dynamical model are different from the multi-scale properties of low-dimensional chaotic systems like Lorenz attractor or Mackey-Glass system, in which the scale-invariance is reconciled with low dimensionality due to the fractal structure of the respective attractors. Both statistical and dynamical properties of this component of AL are closer to those of colored noise.

The analyses of the delay embedding of AL-VB_s time series [*Ukhorskiy et al.*, 2002b] have shown that the averaging over a large number (NN) of nearest neighbors carried out in the embedding space smoothes away the small-scale or high dimensional dynamical constituents. This averaging is what yields the more regular low dimensional component from the time series. The higher the value of NN the wider is the range of scales which are smoothed away and the smaller is the effective dimensionality of the averaged system. It was also found that a single set of parameters (M, NN) corresponding to an unique reconstruction of the dynamical system does not exist. The time delay embedding should be seen rather as a process of striking a balance between the level of "noise" (the range of scales over which the averaging is performed) and the complexity (the effective dimensionality of the averaged system) in a "noisy" dynamical system. Similar conclusions were reached by *Stark* [2001] from the analysis of time delay embedding in stochastic dynamical systems.

The averaging that stabilizes the filter output and helps in extracting the coherent component from AL data is also the cause of information loss, leading to significant limitations on our ability to predict by using the dynamical approach. In particular, due to the averaging the sharpest peaks in data always come out strongly smoothed, and the actual magnitudes of substorms are not predicted well. Thus, to improve the accuracy of the data-derived forecasting tools a new approach beyond the dynamical modeling is needed.

3. MULTI-SCALE DYNAMICS: MODELING BASED ON STATISTICAL PHYSICS

There are many essential features of the magnetosphere beyond the low-dimensional behavior. The power spectrum of the AE index was studied using 5-min averaged data from 1967 to 1980 [*Tsurutani et al.*, 1990]. The power spectrum was found to have a break at about 5 hrs, and at frequencies less than this value the spectrum was close to 1/f and at higher frequencies the spectral index was –2.2, while that of the solar wind VB_s was –1.42. The nonlinear response of the magnetosphere to the solar wind is consistent with the low dimensional behavior [*Sharma*, 1995]. However the power law spectral characteristic was recognized to be a puzzle then. The power spectral nature of the magnetospheric response has been studied using the structure function, which characterizes the fractal nature, and hence the break in the spectrum was interpreted in terms of bi-colored noise [*Takalo et al.*, 1993]. These results have shown that the frac-

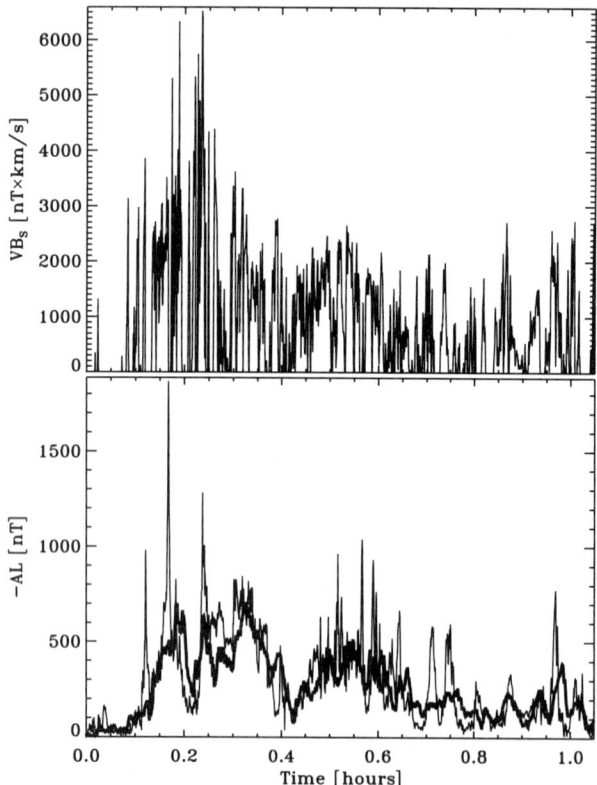

Figure 1. The predictions of AL (thick line, bottom panel) from the solar wind VB_s (top panel) for the high activity interval (32) of *Bargatze et al.* [1985] dataset. The solar wind input data (VB_s) is shown on the upper panel. The prediction misses the sudden changes in the actual AL (thin line, bottom panel).

tal nature, and hence the self—similarity and scale invariance, may not be due to the dynamics of a low dimensional system. Further studies using the structure function characterized the fractal nature of the magnetospheric dynamics, and the break in the spectrum was interpreted in terms of bicolored noise [*Takalo et al.*, 1993]. These results have shown that the fractal nature, and hence the self-similarity and scale invariance, are indications that the dynamics depart significantly from that of a low dimensional system. The multi-scale and intermittent behavior of the magnetosphere has been investigated [*Consolini*, 1996; *Thomas et al.*, 2001] using the multi-fractal approach [*Halsey et al.*, 1986]. These results, based on the probability scale distribution computed from the time series data, further emphasize the presence of the multi-scale behavior in the magnetospheric dynamics.

While many studies have been based on the auroral indices data, many other studies have used the spacecraft data, notably from the fleet of ISTP spacecraft. Studies of the magnetic field fluctuations during the disruption of the magnetotail current have shown power spectrum dependence [*Ohtani et al.*, 1995, 1998]. The plasma flow in the inner plasma sheet measured by Geotail and Wind spacecraft have been used to study the nature of the intermittency in the magnetosphere [*Angelopoulos et al.*, 1999]. The probability density of the magnitudes of the bursty bulk flows show power law dependence in time and their distribution is non-Gaussian. The UVI images from Polar spacecraft was used to study the nature of the dynamics during global auroral energy deposition events [*Lui et al.*, 2000]. In this study using more than 9,000 frames of auroral images the internal scales of the magnetosphere were found to have the same power law in both quiet and active periods. The global energy dissipation during active periods however had a different scale. These features were interpreted as consistent with an avalanching system that exhibits criticality.

The power law nature of the magnetospheric response however may not be due entirely to the internal magnetospheric dynamics. Studies of the solar wind induced electric field VB_S and the energy input into the magnetosphere have recently been found to have power law dependences [*Freeman et al.*, 2000]. This result has interesting implications for magnetospheric dynamics, especially the interpretation in terms of SOC phenomenon. The analysis of the probabilty density functions of the solar wind variables also show power law dependences. The power law form of the inter-burst intervals in the solar wind was found to be distinct from that of ideal SOC but not from SOC-like sandpile models. This result has a wider implication on the signatures of SOC.

The multi-scale behavior can be studied using the local linear filters discussed in the previous section. The dynamical model output is constructed by taking phase space averages of a number of states similar to the current state of the system, viz. the nearest neighbors. The reconstructed phase space is divided into clusters and this yields the probability measure of the events constituting that cluster. The size of the cluster is defined by the radius of a sphere containing the set of nearest neighbors. To obtain the model output the average is taken only over the states within a given cluster while the states outside this cluster are considered to be independent and therefore they do not contribute to the output. Thus, the LLF is similar to the mean-field approach in thermodynamics, and this recognition can be used to build a link between the two components based on dynamics and statistical physics. The output of the dynamical model corresponds to the trajectories in a truncated reconstructed phase space and lie on a low-dimensional surface defined in the mean-field sense. The differences between the model output and the real AL data correspond to the deviations of real trajectories from this mean-field surface due to the high dimensionality of the multi-scale components of AL and due to the dynamic nature of substorms. In the SOC approach these fluctuations are considered autonomous and thus should not depend on the solar wind input. On the other hand, in the models based on phase transition there should be a statistical relationship between the multi-scale fluctuations and the solar wind features. Such a relationship has been obtained in the form of an input-output critical exponent by *Sitnov et al.* [2001]. In this case the probability distribution of substorms is a function of solar wind parameters and should be defined in terms of a conditional probability $P(AL, VB_s)$. The output of the dynamical model (AL_{ARMA}) is to a large extent defined by the solar wind parameters as

$$AL_{ARMA} \approx \int_0^\infty VB_S(t-\tau)\cdot f(\tau)d\tau + \int_{T_{prd}}^\infty AL(t-\tau)\cdot g(\tau)d\tau \quad (4)$$

where T_{prd} is the prediction time scale. Therefore AL_{ARMA} can be used as a measure of the solar wind input, and the conditional probability of AL fluctuations can be calculated in the form of $P(AL, AL_{ARMA})$. As will be shown later the advantage of this particular form of the distribution function is that it can be used directly to yield significant improvements in forecasting AL. The probability distribution function $P(AL, AL_{ARMA})$ is shown on Fig 2. The straight line $AL=AL_{ARMA}$ corresponds to the mean-field surface, i.e. the output of a dynamical model. The distribution of AL about a given value of AL_{ARMA} sets the conditional probability $P(AL, AL_{ARMA})$. As can be seen from the plot there is a

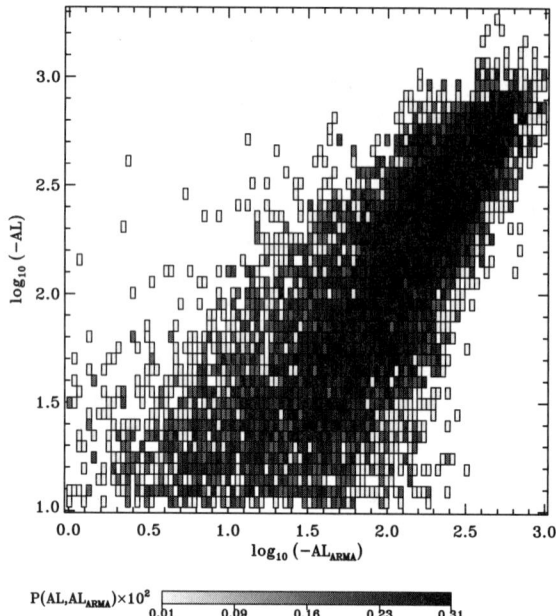

Figure 2. The conditional probability distribution function P(AL, AL_{ARMA}) calculated using *Bargatze et al.* [1985] database.

clear dependence of P(AL, AL_{ARMA}) on AL_{ARMA}. This dependence quantifies the relation between the solar wind input and the fluctuations of AL about the mean-field surface.

4. COMBINED MODEL OF DYNAMICAL AND STATISTICAL DESCRIPTIONS

The conditional probability P(AL, AL_{ARMA}) computed from the training set has been used to construct a new model STADY which combines the DYnamical description with STAtistical description [*Ukhorskiy et al.*, 2002b]. To predict the value of AL at the (n+1) time step $AL_{ARMA}(t_{n+1})$ is first calculated using the dynamical model (3) and the known history of AL for $t<(t_{n+1}-T_{prd})$ and VB_S for $t \leq t_{n+1}$. As discussed above $AL_{ARMA}(t_{n+1})$ should be considered as an estimate of the average level of the substorm activity and the dynamical model often underestimates the AL peaks. This is where the statistical part of the model comes into the play. Knowing $AL_{ARMA}(t_{n+1})$ we can estimate the magnitude of AL deviation from the output of the deterministic model with the use of the distribution function P(AL, $AL_{ARMA}(t_{n+1})$). This function not only specifies the largest possible value of AL for a given AL_{ARMA}, but also ranks it in terms of the probability. An example of AL predictions using STADY is shown in Fig 3 for the data interval shown in Fig 1. As can be seen from this figure STADY yields not only the long term deterministic predictions of the global features of the time series but also yields estimates of the high dimensional multi-scale component of AL in a probabilistic fashion. Thus, for given solar wind conditions STADY can forecast the magnitudes of the substorms and their their associated probabilities, and thus can be used as a practical forecasting tool.

5. MODELING STORM-TIME DYNAMICS

The magnetospheric response on the time scale of magnetic storms is commonly taken to be the disturbance storm time index Dst. This index represents the ring current which is responsible for the variations in the magnetic field measured at mid-latitude stations on the ground. The strong correlation between the Dst and the solar wind variables during storms is well known from earlier studies using linear mod-

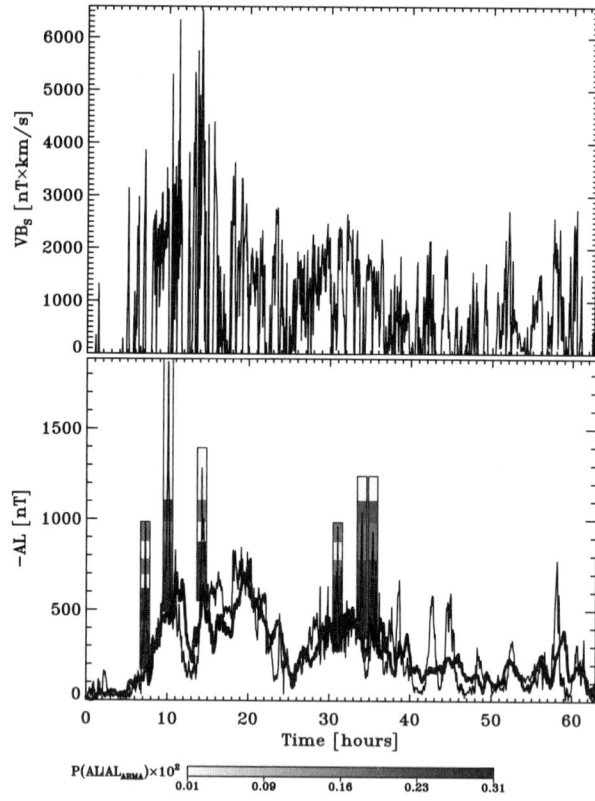

Figure 3. STADY predictions of the high activity Bargatze 32nd interval. The solar wind input data (VB_s) is shown on the upper panel. AL time series are shown on the bottom panel. The dynamical model output is indicated by the bold line. Color bars show the probability of the largest deviations (AL>900 nT) of real data from the dynamical model output calculated using P(AL, AL_{ARMA}).

els [Burton et al., 1975]. A typical storm begins with an increase of the Dst index, marking the onset of the storm or the sudden storm commencement, followed by a strong decrease during the main phase. The Dst minimum marks the beginning of the recovery phase during which the ring current decays. This phase is not directly correlated with the solar wind induced dawn-to-dusk electric field $E_y=VB_z$, given by the product of the solar wind speed V and the north-south component Bz of the IMF. The solar wind induced electric field usually recovers with a faster time scale.

Many studies of the solar wind–Dst relationship have indicated nonlinearity in the evolution of Dst [Burton et al., 1975; Gonzales et al., 1994]. For example, the ring current decay rate is found to be dependent on the Dst value [Gonzales et al., 1994]. The magnetic field measured at low latitudes is affected significantly by the variations of the solar wind ram pressure $p=nmV^2$, which produce changes in the magnetopause currents (n is the solar wind density, and m is the proton mass). Following Burton et al. [1975] a pressure-corrected Dst is defined as $Dst^* = Dst - b\sqrt{p} + c$. For the data set for January-June 1979 with 5 min resolution [Kamide et al., 1998], the best fit values are c=22 nT and b=10.5nT/√nPa, with the dynamic pressure expressed in nano-Pascals.

The time evolution of the ring current, as represented by the Dst index, has been modeled by an input or injection function and a recovery with a characteristic time scale τ [Burton et al., 1975], so that

$$\frac{dD^*_{st}}{dt} = \alpha^* VBs - \frac{D^*_{st}}{\tau} = F[\mathbf{X(t)}] \quad (5)$$

Many forms of the input function have been used [Gonzales et al., 1994], and the most common form is a linear dependence on VBs with the coupling constant α^*. For modeling using the time-delay embedding technique, we define the state vector $\mathbf{X}(t) = [Dst^*, E_w]$, where $E_w = VB_s$ is the solar wind induced electric field. Taking F in Eq. (5) to be a nonlinear polynomial, the coefficients are obtained from the January–June 1979 dataset by a fitting procedure based on a predictor-corrector integration scheme [Valdivia et al., 1996]. In this process the storm under study is excluded so that its features are not built into the model. This yields a global equation for Dst* with a mean absolute fitting error of 4.5 nT. It describes the evolution of Dst*, and thus the interaction of the solar wind and ring current. The coefficient for Dst^* in the global differential equation (5) corresponds to the inverse of the decay time τ and has a value of 12.5 hours, which is consistent with previous studies [Gonzales et al., 1994]. A Dst* dependent decay time can be defined and for Dst* values < –100nT, and the variation of τ due to the nonlinear dependence is significant, with intense storms yielding shorter recovery time scales. [Valdivia et al., 1996].

Using this model one-step predictions are obtained at each time t using the known Dst* and Ew to obtain the predicted values the following hour, t+1. These predictions have a mean absolute error of 3.5 nT, averaged over 50 hours. It may be noted that if we assume persistence the one step mean absolute error is typically about 5 nT. Iterated predictions can be made using the predicted values of Dst* and given values of Ew (t). The iterated prediction for 50 hours has a mean absolute error of 10 nT. The forecasting ability of this model is assessed by obtaining the best model parameters using all 6 month of data except for the storm under study [Valdivia et al., 1996]. In general, these models predict with a good accuracy the development of a storm from the solar wind measurement.

The ring current however displays a clear spatial structure in its evolution and it becomes clear that a more complete understanding of the ring current and magnetic storms will be reached only after a careful analysis of its spatio-temporal multivariate properties, including multiple inputs and outputs. The January–June 1979 dataset [Kamide et al., 1998] contains the measurements of magnetic field variations at 6 low-latitude ground stations with 5 minute resolution. The measured values are normalized by $1/\cos(\chi)$ where χ is the magnetic latitude.

A general description of the solar wind–magnetosphere interaction can be obtained by constructing the phase space of magnetospheric dynamics on the storm time scales using the method of time delay embedding adapted to input-output systems as in the case of the AL index. A spatio-temporal input-output phase space is reconstructed by taking the normalized dusk-dawn electric field Ew as the input and the horizontal component H measued at the magnetometer stations, as the output. The trajectory in the phase space is represented by the time evolution of a state vector defined in terms of H [Valdivia et al., 1999].

In this reconstructed phase space the predicted value $H_i = F[\mathbf{X(t)}]$, where the functional F is obtained by a linear Taylor expansion around $\mathbf{X(t)}$ and the coefficients are computed by a fitting procedure that uses the evolution of the nearest neighbors of the current state [Valdivia et al., 1996]. This model can describe complex functions by adjusting itself to the different conditions of the solar wind–magnetosphere system. The one-step prediction can be used as before to yield an iterated prediction for many time steps by using $H_i(t+1)$ to define $\mathbf{X(t)}$ and to obtain a new set of nearest neighbors, and so on. This local linear procedure is used to make out-of-sample predictions by reconstructing the phase space with all storms except the one to be predicted.

The predictions for the magnetic storm of 24–25 September 1998 from the solar wind input are shown in Plate 1. In this prediction the H component data from six mid-latitude stations, viz. Boulder, Frederickberg Hermanus, Honolulu, Kakioka, and San Juan, have been used. The time series data from these stations for the period January–June 1979 were used to develop linear and nonlinear models. These models are then used to predict the time evolution of the H componets at these stations for the 24–25 September 1998 storm from the solar wind measurements by IMP-8 spacecraft. The predicted Dst, shown in Plate 1, are computed from the predicted values of the H components. The solar wind induced electric field VBs is shown in the top panel of Plate 1 and the lower panel shows the actual (black line) and predicted (blue and red lines) values of Dst. The linear model is similar to the *Burton et al.* [1975] model and the coefiicients are computed from the 5 min. resolution data. The predictions of the nonlinear model (Plate 1, bottom panel–red line) tracks the Dst development better but both the models miss the peak of the storm. Similar trend is seen in the results of the Michigan Ring current Atmosphere interaction Model (RAM) for this storm [*Clauer et al.*, this volume; *Daglis et al.*, 2002].

6. STORM-SUBSTORM RELATIONSHIP

The nonlinear dynamical techniques can be used to study the relationship between storms and substorms. The strong correlation between intense substorms and the main phase of storms has suggested a cause-effect relationship between these two components of geomagnetic activity [*Gonzalez et al.*, 1994]. The scenario that a sequence of substorms can lead to the storm main phase can be studied by using the nonlinear filters in the same way the input-output relationship between the solar wind VBz and AL or Dst has been studied. Linear prediction filters have been used to study such a relationship [*McPherron*, 1997].

Using the local-linear filters obtained from the 5 min. averaged data of AL and Dst for January–June 1979, the Dst was predicted from the AL index for the magnetic storm of 9 March 1979 [*Kamide et al.*, 1998]. Also assuming and an exponential form of the filter the Dst was obtained from AL. In both the cases the Dst is predicted well. However it has been recognized that AL contains may of the features of solar wind and hence it could not be concluded from this result that a sequence of AL intensifications can lead to Dst during a storm main phase. However the nonlinear filters predict the time evolution of Dst from AL better than the linear models and this indicates that the internal dynamics of the magnetosphere plays a significant role in the storm development [*Sharma et al.*, 2001, 2002].

Another important issue in storm-substorm relationship is the relative timing between substorm intensifications and the onset of the storm main phase. The linear prediction filters are an appropriate technique for studying the time delay between the substorms and the storm main phase. The 5 min resolution AL and Dst data [*Kamide et al.*, 1998] were used to obtain data with 10, 20, 30, 40, 50 and 60 min averages and the linear prediction filters computed in each case. In each case the time delay between AL and Dst indices were found to be the resolution of the data. This implies that a time delay between substorms and storm main phase, if there is any, can not be concluded. If the storm main phase is indeed due to a sequence of substorm intensifications, a delay time independent of the resolution of the data is expected. The absence of such a time delay essentially rules out a cause-effect relationship inherent in the substorm injection hypothesis of magnetic storms. This agrees with the recent results from the studies using the ground-based and spacecraft measurements [*Sharma et al.*, 2002]. Recently it has become clear that substorms are not the main processes responsible for the storm main phase but can play an important role during the recovery phase. This consensus (Lonavala consensus [*Sharma et al.*, 2002]) is an important milestone in the understanding of storm-substorm relationship.

7. CONCLUSION

The nonlinear dynamical modeling using time series data has given a number of important results in the understanding of the solar wind–magnetosphere coupling. The first results elucidated the global dynamical behavior as coherent and thus amenable to low dimensional description. This agrees well with the results of numerical simulations using a global MHD code [*Lyon*, 2001]. The important implications of this result is that the magnetospheric dynamics is predictable over the appropriate space and time scales. This recognition has laid the foundation for the development of predictive models of the magnetosphere. The space weather forecasting tools based on the phase space reconstruction techniques now forecast the global magnetospheric conditions near-real time from the solar wind data measurements from ACE sopacecraft.

The multi-scale properties of the magnetosphere is most commonly seen in the power law dependence of the different scale sizes. This feature naturally can not be modeled

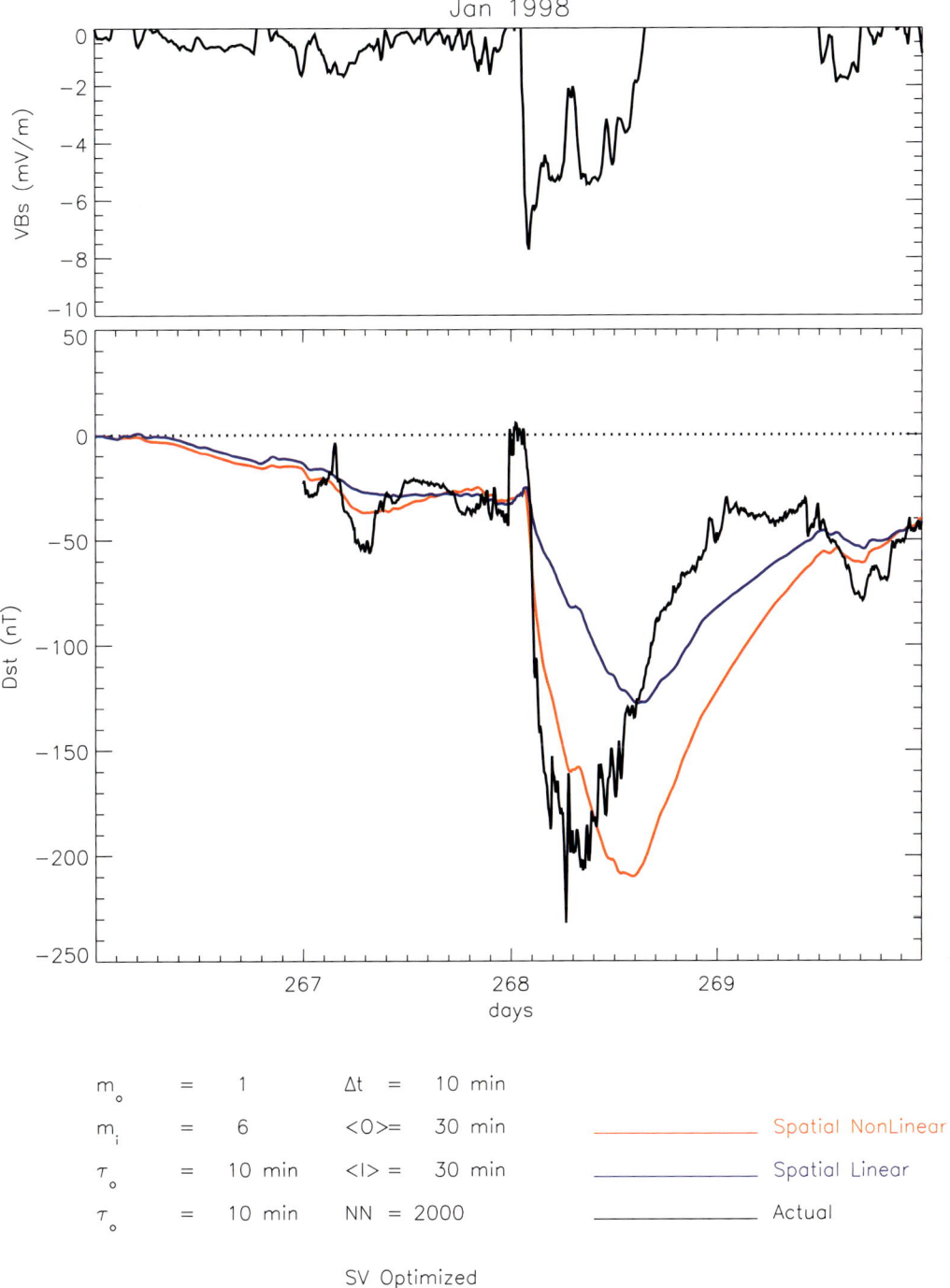

Plate 1. The prediction of the magnetic storm of 28 September 1998. The solar wind induced electris field (top panel) is used as the input and the actual Dst (bottom panel–black line) is compared with predictions using a linear model (blue line) and a nonlinear (red line)

with a low dimensional model and consequently is not amenable to dynamical prediction. The properties of the multi-scale behavior however can be used to forecast the statistical variables from the conditional probabilties. A new approach combining the nonlinear dynamical approach with elements of statistical physics las led to a comprehensive mode; that reconciles the global and multi-scale aspects of the magnetosphere during substorms. The dynamical part of the model leads to the deterministic predictions of the globally coherent components of the time series, while the statistical part yields probabilistic predictions of the multi-scale constituents. This yields significant improvements in the space weather forecasting tools since it yields not only the deterministic predictions of the average level of the magnetospheric activity but also the probabilities of the range of geomagnetic activity for given solar wind conditions.

The global and multi-scale features of the magnetosphere can be understood in the framework of phase transitions. Studies using time series data have shown that the global features correspond to the first order transitions and the multi-scale features to the second order transitions [*Sitnov et al.*, 2000, 2001]. This picture is derived from the phase diagrams and critical exponents computed obtained from the time series data., agrees well with the results from the global MHD simulations of the solar wind–magnetosphere coupling [*Shao et al.*, 2002].

Acknowledgments. The research is supported by NSF grants ATM-0001676 and ATM-0119196, and NASA grant NAG5-1101

REFERENCES

Abarbanel, H. D., R. Brown, J. J. Sidorovich, T. S. Tsimring, The analysis of observed chaotic data in physical systems. *Rev. Mod. Phys., 65,* 1331, 1993.

Angelopoulos V., Mukai T., Kokubun S., Evidence for intermittency in Earth's plasma sheet and implications for self-organized criticality, *Phys. Plasmas, 6,* 4161, 1999.

Bak, P., C. Tang, and K. Wiesenfeld, Self-organized criticality: An explanation of 1/f noise, *Phys. Rev. Lett., 59,* 381, 1987.

Baker, D. N., A. J. Klimas, R. L. McPherron and J. Buechner, The evolution from weak to strong geomagnetic activity:an interpretation in terms of deterministic chaos, *Geophys. Res. Lett., 17,* 41, 1990.

Bargatze, L. F., D. N. Baker, R. L. McPherron and E. W. Hones, Magnetospheric impulse response for many levels of geomagnetic activity, *J. Geophys. Res., 90,* 6387, 1985.

Burch, J. L., S. B. Mende, D. G. Mitchell, T. E. Moore, C. J. Pollock, B. W. Reinisch, B. R. Sandel, S. A. Fuselier, D. L. Gallagher, J. L. Green, J. D. Perez, and P. H. Reiff, Views of Earth's Magnetosphere with the IMAGE Satellite, *Science, 291,* 629, 2001.

Burton, R. K., R. L. McPherron, and C. T. Russell, An empirical relationship between interplanetary conditions and Dst, *J. Geophys. Res., 80,* 4204, 1975.

Casdagli, M., A dynamical system approach to modelling input-output systems, in *Nonlinear Modeling and Forecasting*, edited by M. Casdagli and S. Eubank, vol. XII, of SFI Stud. Sci. Complexity, p. 265, Reading, Mass., Addison-Wesley, 1992.

Chapman, S.C., N.W. Watkins, R.O. Dendy, P. Helander, G. Rowlands, A simple avalanche model as an analogue for magnetospheric activity, *Geophys. Res. Lett., 25,* 2397, 1998.

Clauer, C. R., M. W. Liemohn, J. U. Kozyra, M. L. Reno, The relationship of storms and substorms determined from mid-latitude ground-based magnetic maps, in *Storm-Substorm Relationship, Geophys. Monogr. Series.* edited by A. S. Sharma, G. S. Lakhina, Y. Kamide and K. Papadopoulos, AGU, Washington, D. C., 2002.

Consolini, G., Sandpile cellular automata and magnetospheric dynamics, In: *Proc. 8th GIFCO Conference, Cosmic Physics in the Year 2000: Scientific Perspectives and New Instrumentation*, edited by S. Aiello et al., Soc. Ital. di Fis., Bologna, Italy, 1997.

Daglis, I. A., J. U. Kozyra, Y. Kamide, D. Vassiliadis, A. S. Sharma, M. W. Liemohn, G. Lu, W. D. Gonzalez, B. T. Tsurutani and A. Korth, Intense space storms: 2. Critical issues and open disputes (Invited review), *J. Geophys. Res.*, submitted, 2002.

Freeman M. P., Watkins N. W., Riley D. J., Evidence for a solar wind origin of the power law burst lifetime distribution of the AE indices, *Geophys. Res. Lett, 27,* 1087, 2000.

Gonzalez, W.D., J.A. Joselyn, Y. Kamide, H.W. Kroehl, G. Rostoker, B.T. Tsurutani, and V.M. Vasyliunas, What is a geomagnetic storm?, *J. Geophys. Res., 99,* 5771, 1994.

Gleisner, H., and H. Lundstedt, response of the auroral electrojets to the solar wind modeled with neural networks, *J. Geophys. Res., 102,* 14,269, 1997.

Halsey T., Jensen M., Kadanoff L. et al., 1986, Phys. Rev. A, 33, 1141

Horton, W., and I. Doxas, A low-dimensional energy conserving state space model for substorm dynamics, *J. Geophys. Res., 101,* 27,223, 1996.

Kamide, Y., W. Baumjohann, I. A. Daglis, W. D. Gonzalez, M. Grande, J. A. Joselyn, R. L. McPherron, J. L. Phillips, E. G. D. Reeves, G. Rostoker, A. S. Sharma, H. J. Singer, B. T. Tsurutani and V. M. Vasyliunas, Current Understanding of Magnetic Storms: Storm-substorm Relationships, *J. Geophys. Res., 103,* 17,705, 1998

Klimas, A. J., D. Vassiliadis, D. A. Roberts, and D. N. Baker, The organized nonlinear dynamics of the magnetosphere, *J. Geophys. Res., 101,* 13089, 1996.

Klimas, A. J., J. A. Valdivia, D. Vassiliadis, D. N. Baker, M. Hesse, and J. Takalo, Self-organized criticality in the substorm phenomenon and its relation to localized reconnection in the magnetospheric plasma sheet, *J. Geophys. Res., 105,* 18,765, 2000.

Lui A. T. Y., Chapman S. C., Liou K., P. T. newell, C. I. meng, M. Brittnacher and G. K. Parks, Is the dynamic magnetosphere an avalanching system?, *Geophys. Res. Lett., 27,* 911, 2000.

Lyon, J. G., The solar wind–magnetosphere coupling, *Science, 288,* 1987, 2000.

McPherron, R. L., The role of substorms in the generation of magnetic storms, in *Magnetic Storms, Geophys. Monogr. Ser. Vol. 98*, edited by B. T. Tsurutani, W. D. Gonzalez, Y. Kamide and J. K. Arballo, p. 131, AGU, Washington, D. c., 1997.

Ohtani, S., T. Higuchi, A. T. Lui, and K. Takahashi, Magnetic fluctuations associated with tail current disruption: Fractal analysis, *J. Geophys. Res.*, *100*, 19,135, 1995.

Ohtani, S., T. Higuchi, A. T. Y. Lui, and K. Takahashi, AMPTE/CCE-SCATHA simultaneous observations of substorm associated magnetic fluctuations, J. Geophys. Res., 103, 4671, 1998.

Price, C. P., D. Prichard and J. E. Bischoff, Nonlinear input/output analysis of the auroral electrojet index, *J. Geophys. Res.*, *99*, 13227, 1994.

Shao, X., M. I. Sitnov, A. S. Sharma, K. Papadopoulos, C. C. goodrich, P. N. Guzdar, G. M. Milikh, M. J. Wiltberger and J. G. Lyon, Phase transition-like behavior of magnetospheric substorms: Global MHD simulation results, *J. Geophys. Res.*, in press, 2002.

Sharma, A. S., Assessing the magnetosphere's nonlinear behavior: Its dimension is low, its predictability high (US National Report to IUGG, 1991–1994), *Rev. Geophys. Supple.*, *33*, pp. 645-650, 1995.

Sharma, A. S., Nonlinear dynamical studies of global magnetospheric dynamics, in *Nonlinear Waves and Chaos in Space Plasmas*, edited by T. hada and H. Matsumoto, pp. 359-389, Terra Sci. Pub., Tokyo, 1997.

Sharma, A. S., D. Vassiliadis, and K. Papadopoulos, Reconstruction of low dimensional magnetospheric dynamics by singular spectrum analysis, *Geophys. Res. Lett.*, *20*, 335, 1993.

Sharma, A. S., J. A. Valdivia and Y. Kamide, Dynamical Relationship Between Storms and Substorms, in *Substorms-4: Proc. 4th Internl. Conf. on Substorms*, edited by S. Kokubun and Y. Kamide, Terra Sci., Tokyo, pp. 737-740, 1998.

Sharma, A. S., G. S. Lakhina and Y. Kamide, Storm-substorm relationship, *Current Sci., 81(2)*, 150-152, 2001.

Sharma, A. S., M. I. Sitnov, and K. Papadopoulos, Substorms as nonequilibrium transitions in the magnetosphere, *J. Atmos. Solar Terrest. Physics, 63*, 1399, 2001.

Sharma, A. S., D. N. Baker, M. Grande, Y. Kamide, G. S. Lakhina, R. L. McPherron, G. D. Reeves, G. Rostoker, R. Vondrak and L. M. Zelenyi, Storm-substorm relationship: Current Ynderstanding and outlook, in *Storm-Substorm Relationship, Geophys. Monogr. Series.* edited by A. S. Sharma, G. S. Lakhina, Y. Kamide and K. Papadopoulos, AGU, Washington, D. C., 2002.

Siscoe, G. L., The magnetosphere: A union of interdependent parts, *EOS, Trans. AGU, 72*, 494, 1991.

Sitnov, M. I., A. S. Sharma, K. Papadopoulos, D. Vassiliadis, J. A. Valdivia, A. J. Klimas, and D. N. Baker, Phase transition-like behavior of the magnetosphere during substorms, *J. Geophys. Res., 105*, 12,955, 2000.

Sitnov, M. I., A. S. Sharma, K. Papadopoulos, D. Vassiliadis, Modeling substorm dynamics of the magnetosphere: From self-organization and self-criticality to nonequilibrium phase transitions, *Phys. Rev. E., 65*, 016116, 2001.

Stark, J., Delay reconstruction: dynamics versus statistics, *Nonlinear Dynamics and Statistics*, A.I. Mees, Birkhauser, 2001.

Takalo, J., J. Timonen, H. Koskinen, Correlation dimension and affinity of AE data and bicolored noise, *Geophys. Res. Lett., 20*, 1527, 1993.

Takalo, J., J. Timonen, H. Koskinen, Properties of AE data and bicolored noise, *J. Geophys. Res., 99*, 13239, 1994.

Thomas, B., A. S. Sharma and M. I. Sitnov, Multi-fractal properties of solar wind magnetosphre coupling, *AGU Fall Meeting*, 2002.

Tsurutani, B. T., M. Sugiura, T. Iyemori, B. E. Goldstein, W. D. Gonzalez, S. I. Akasofu, and E. J. Smith, The nonlinear response of AE to the IMF Bs driver: A spectral break at 5 hours, *Geophys. Res. Lett., 17*, 279, 1990.

Ukhorskiy, A. Y., M. I. Sitnov, A. S. Sharma, K. Papadopoulos, Global and multi-scale properties of magnetospheric dynamics in local-linear filter, *J. Geophys. Res.107, in press*, 2002a.

Ukhorskiy, A. Y., M. I. Sitnov, A. S. Sharma, K. Papadopoulos, Modeling and forecasting the multi-scale constituent of magnetospheric dynamics during substorms, *Proc. Int. Conf. Substorms–6*, University of Washington, 2002b.

Valdivia, J. A., A. S. Sharma and K. Papadopoulos, Prediction of magnetic storms with nonlinear dynamical models, *Geophys. Res. Lett., 23*, 2899, 1996.

Valdivia, J. A., D. Vassiliadis, A. J. Klimas, and A. S. Sharma, Modeling the Spatial Structure of the High Latitude Magnetic Perturbations and the Related Current Systems, *Phys. Plasmas, 6*, 4185, 1999.

Valdivia, J. A., D. Vassiliadis, A. J. Klimas, A. S. Sharma and K. Papadopoulos, The Spatio-temporal Structure of Magnetic Storms, *J. Geophys. Res., 104*, 12,239, 1999.

Vassiliadis, D., A. S. Sharma, T. E. Eastman, and K. Papadopoulos, Low-dimensional chaos in magnetospheric activity from time seies, *Geophys. Res. Lett., 17*, 1841, 1990.

Vassiliadis, D., A. J. Klimas, D. B. Baker, and D. A. Roberts, A description of the solar wind–magnetosphere coupling based on nonlinear filters, *J. Geophys. Res., 100*, 3495, 1995.

Vespignani, A., and S. Zapperi, How self-organized criticality works: A unified mean-field picture, *Phys. Rev. E, 57*, 6345, 1998.

A. S. Sharma, Department of Astronomy, University of Maryland, College Park, MD 20742 (ssh@astro.umd.ed)

M. I. Sitnov, Department of Astronomy, University of Maryland, College Park, MD 20742 (sitnov@astro.umd.edu)

A.Y. Ukhorskiy, Department of Astronomy, University of Maryland, College Park, MD 20742 (sasha@astro.umd.edu)

J. A. Valdivia, Departamento de Fisica, Facultad de Ciencias, Universidad de Chile, Casilla 653, Santiago, Chile (alejo@fisica.ciencias.uchile.cl)

Comments on Some Long-Standing Problems in Storm/Substorm Studies

S.-I. Akasofu

International Arctic Research Center, University of Alaska Fairbanks, Fairbanks, AK 99775-7340, USA.

W. Sun

Geophysical Institute, University of Alaska Fairbanks, Fairbanks, AK 99775-7320, USA.

B.-H. Ahn

Department of Earth Sciences, Kyungpook National University, Taegu, Korea.

We identify seven long-standing problems in storm/substorm studies and present our views on them mainly on the basis of earlier literatures, demonstrating that earlier studies should be combined with new results to overcome the difficulties associated with long-standing problems. We introduce a new index AF to monitor the upward field-aligned currents (FAC) and show that it is far better correlated to the Dst index than the AE index.

1. INTRODUCTION

There are several long-standing problems in magnetospheric physics, which have not been solved for several decade or investigators have not yet come to a reasonable consensus. The reasons are complex. Some are no doubt difficult problems to begin with. In some cases, however, we have not carefully synthesized/integrated research results which are well-documented in earlier literatures. In this paper, we examine the second problem by selecting seven long-standing issues and present our views on the basis of results from earlier literatures. The new generation researchers are encouraged to consider both their results and earlier ones for a better understanding of the problems. Each section has suggestions for future studies; they are expressed in the italic form.

2. ASYMMETRIC DEVELOPMENT OF THE MAIN PHASE OF GEOMAGNETIC STORMS

The asymmetric development of the main phase of a geomagnetic storm has long been known, but has not received much attention during the last few decades. It is an aspect of the ring current belt which is so crucial in understanding its formation. Figure 1 shows the main phase field (H component) in a geomagnetic longitude-latitude map for several major storms. It can be seen that the main phase decrease is largest near the magnetic equator and that there is a difference of about 200 nT at the maximum and minimum decreases [*Akasofu* and *Chapman*, 1972; Chapter 8]. Further, the maximum decrease tends to occur in the afternoon-evening sector. *Chapman* [1918] expressed the main phase field D as follows.

$$D(\theta, \Phi, t) = Dst(\theta, t) + DS(\theta, \Phi, t) \quad (1)$$

$$DS(\theta, \Phi, t) = C_n(\theta, t) \sin(n\Phi + \alpha n(\theta, t)) \quad (2)$$

where θ and Φ denote latitude and longitude, respectively [*Akasofu* and *Chapman*, 1972; Chapter 8]. This is a sort of

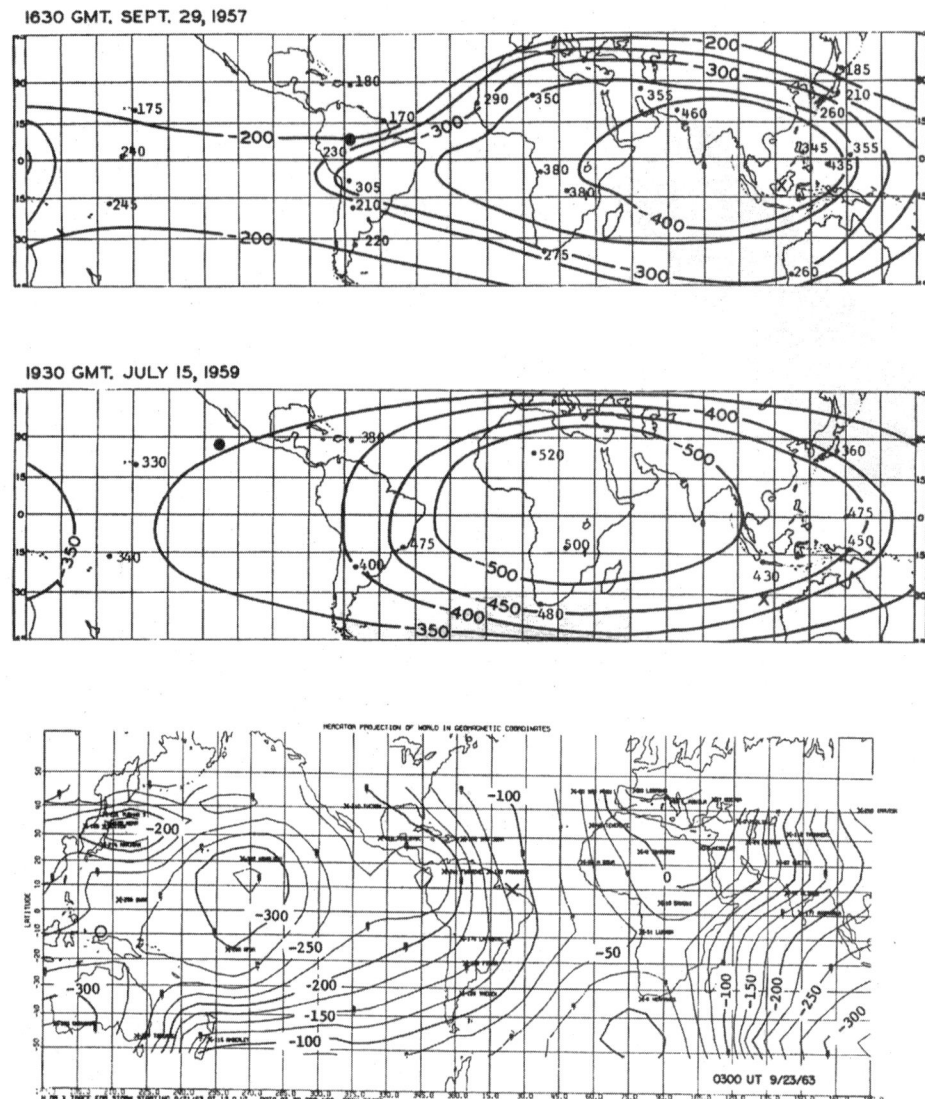

Figure 1. Contours of the H component of geomagnetic field disturbances in the middle and low latitudinal region for the main phase of three huge magnetic storms.

Fourier analysis of the main phase field for a given latitude θ, $D(\Phi,t) = Dst(t)+DS(\Phi, t)$. *Chapman* [1918, 1935] interpreted that Dst represents the ring current field, which is assumed to be symmetric with respect to the geomagnetic axis. The so-called Dst index originates its name from this analysis. The DS term represents the asymmetric component; as a first approximation, it can be expressed simply by $C_1(t)\sin\Phi$. *Chapman* [1935] implied that the DS term was caused by the ionospheric return currents from the eastward and westward electrojet along the auroral zone. On the other hand, Alfven [1950] proposed that the field-aligned currents are involved in the DS field. It is likely that both the 3-D current system associated with the electrojets (Figure 2) and the asymmetric diamagnetic effects are responsible for the asymmetry of the main phase. In order to examine the 3-D current contribution, Figures 3a and 3b show the computed H and D components in middle and low latitudes for weak and intense electrojet. It can be seen that the current system tends to produce a field that resembles the D field in terms of the longitudinal (MLT) distribution. However, a closer examination shows that the computed field is smaller than the observed field and is minimum near the geomagnetic equator, so that the current system cannot reproduce satisfactorily the observed DS field

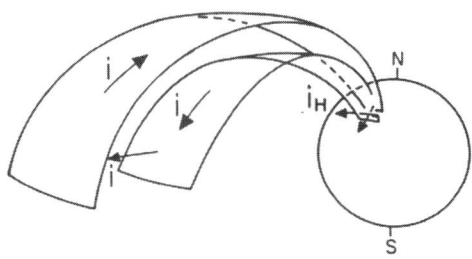

Figure 2. The 3-D field-aligned loop current connect ionospheric currents with equatorial currents suggested by Boström [1964].

Another source of the DS field must exist. It is generally believed that the storm-time ring current belt carries a westward current as a result of the westward drift motion of positive ions. However, as shown by *Akasofu* and *Chapman* [1961; 1962], for an isotropic pitch angle distribution, the westward drift motion does not contribute to the current. The diamagnetism of the trapped positive ions produces an eastward current in the inner half and a westward current in the outer half (see Figure 4). Further, the westward current in the outer part is stronger than the eastward current in the inner part. It is this current system that produces the so-called westward ring current; its field is the largest near the equator. Figure 4 shows the field of the symmetric belt.

As far as we are aware, there has been no quantitative study of the current and field associated with an asymmetric belt even for a simplest case of isotropic pitch-angle distribution. Does the diamagnetic current close itself in the equatorial plan?

Akasofu and *Chapman* [1961] and *Kawasaki* and *Akasofu* [1972] showed also that the asymmetry is prominent during the period when the main phase is developing and tends to decay during the recovery phase. Figure 5 shows clearly this trend. It shows also that substorms activity is intense when the main phase is developing and that the main phase begins to decay as substorm activity subsides.

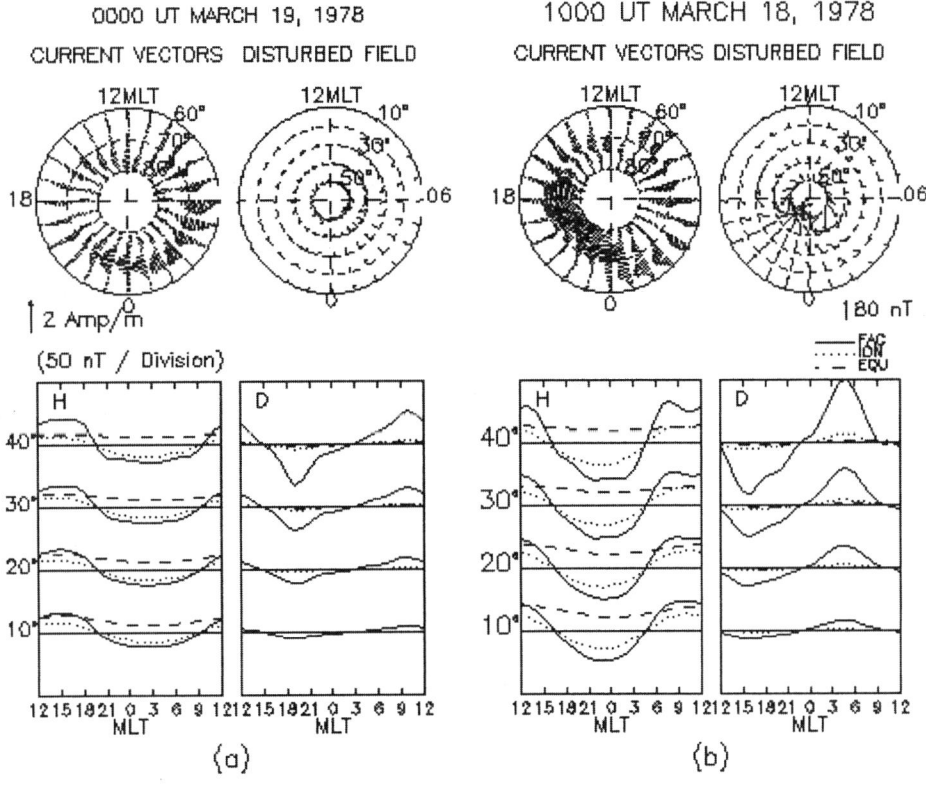

Figure 3(a). Two top panels show the ionospheric current vectors and magnetic field disturbances caused by three-dimensional currents associated with the ionospheric currents at 0000UT on March 19, 1978. The lower two panels show time variations of the H and D components which are produced by three parts of the current loop; field-aligned currents, ionospheric currents and equatorial currents. **3(b)** shows the same as those in 3(a) except for the event at 1000UT on March 18, 1978.

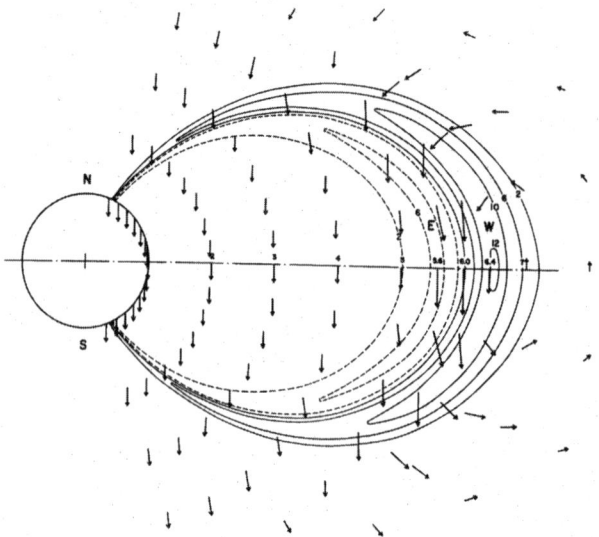

Figure 4. The distribution of electric currents in the ring current and the disturbed magnetic fields (arrows) produced by the ring current; E and W indicate the eastward and westward currents, respectively (Akasofu and Chapman, 1961).

Thus, there is a clear relationship between the development of the main phase and substorm activity. These trends provide crucial hints on the mechanism of the formation of the ring current belt. This is the subject for the next section.

3. STORM - SUBSTORM RELATIONSHIP

Akasofu and *Chapman* [1963] found that the auroral electrojet and auroral activity grow and decay impulsively several times during a single geomagnetic storm. Thus, they suggested that these impulsive phenomena are related to the cause of the main phase. For this reason, they proposed to call the impulsive phenomena *substorms*. The associated phenomena are polar magnetic substorms and auroral substorms. In fact, a Dst decrease can hardly occur without substorms. This can be seen from any AE-Dst relationship (see Figure 6). Subsequently, *McIlwain* [1974] showed that plasma particles are injected from the plasma sheet during individual substorms. Positive ions drift westward and electrons drift eastward around the earth. Further, it was found that O^+ ions flow out from the ionosphere to the magnetotail during substorms [*Shelly et al.* 1976; *Lundin*, et al., 1987; *Hultqvist et al.*, 1988] and can become the dominant positive ions [*Hamilton et al.*, 1988; *Wilken et al.* 1992; *Daglis et al.*, 1992; 1994; 1996; *Daglis*, 1997]. Thus, after flowing out from the ionosphere, O^+ ions must be convected to the ring current belt [*Williams*, 1980; *Gonzalez et al.*, 1994].

It has been established that in a typical situation the so-called "southward turning" of the IMF causes the growth phase and that the growth phase is followed by substorm onset. Thus, it is likely that a chain of processes is needed to produce the ring current belt, namely:

Southward turning of IMF → Enhanced convection → Substorms → O^+ outflow → Injection of O^+ to the ring current belt.

In recent years, there have been several papers in which the authors express doubt about the storm-substorm relationship, stating that the convection alone can cause the main phase. Anyone who doubts this relationship should find storms in which a significant Dst decrease occurred without substorm activity. The present authors cannot find such a case in Figure 6 which shows one full year of the data. In any case, it is important to understand that enhanced convection will inevitably lead to substorm activities, so that it is difficult to understand how the convection alone

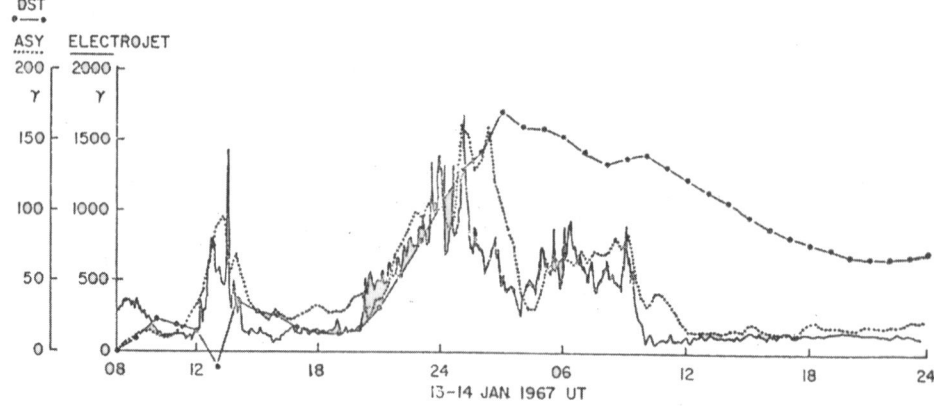

Figure 5. AL, ASY and Dst indices for the January 13–14, 1967, storm (Kawasaki, 1971).

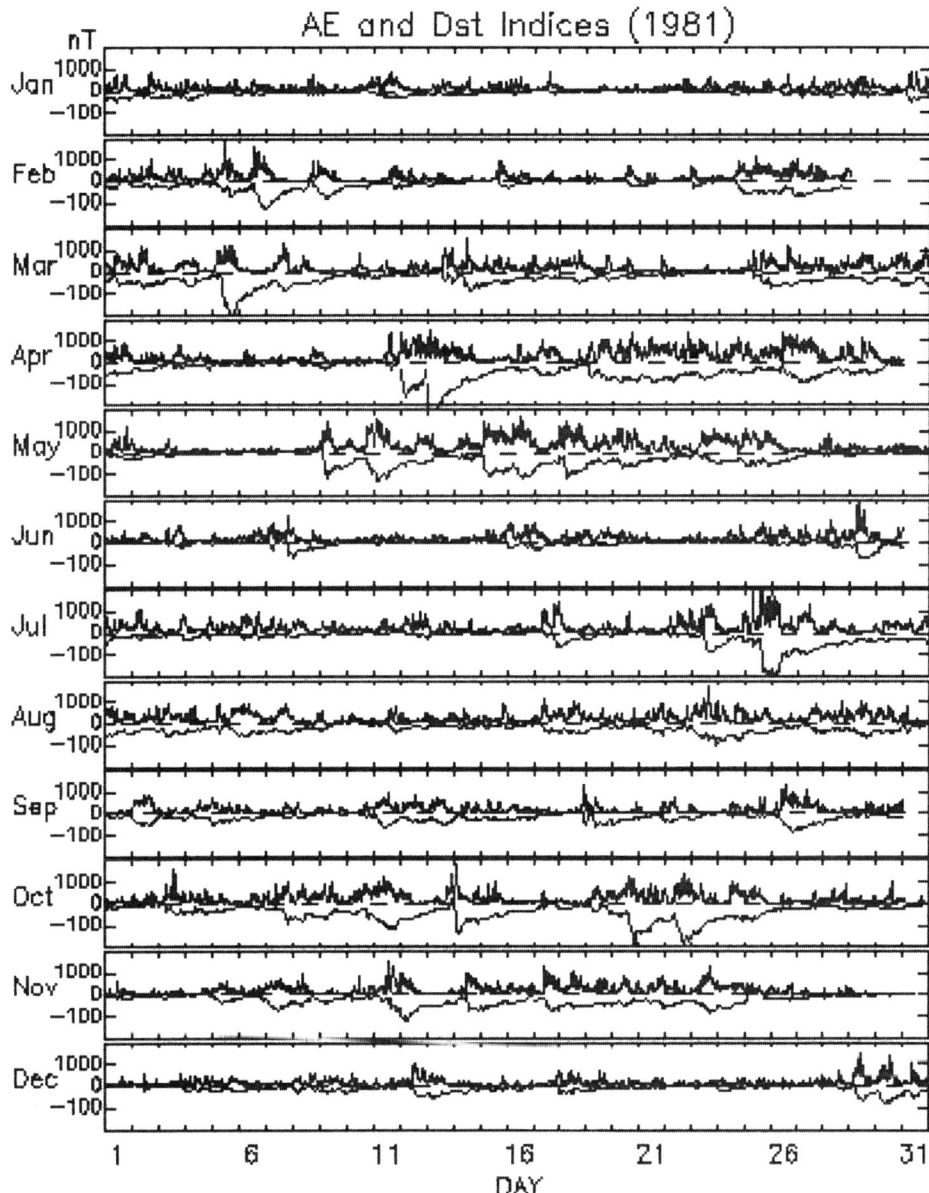

Figure 6. The AE and Dst indices in 1981.

can cause an intense ring current belt without causing substorm activity which supplies O^+ ions.

It is our opinion that the present confusion on the issue arises from the fact that each step in the chain of the above processes is quantitatively complex, so that one may find cases in which the convection alone appears to cause the main phase and also in which substorms do not contribute to the main phase. For example, the injected O^+ ions do not necessarily contribute fully to the ring current belt, depending on their drift path. Thus, if one examines individual cases, there can be a great variety of cases. Another problem is that substorms are episodic phenomena, while the main phase develops gradually, so that a blind comparison of the AE-Dst relationship would lead us to misleading conclusions [*Ahn et al.* 2001].

We must realize that the accuracy of our analysis is determined by the accuracy of the present Dst and AE indices. It is surprising that we still rely on indices that were devised in the 1960s. Both indices were derived in the 1960s [*Akasofu* and *Chapman*, 1961; *Davis* and *Sugiura*, 1966]. Indeed, both the AE and Dst indices are not accurate enough to examine in detail the storm-substorm relationship *quantitatively* for individual cases. For example, it has been claimed that some substorms are associated with a positive change in Dst and thus substorms appear to reduce the ring current. Such a confusion arises from the ignorance that some substorms during major storms are very intense, so that the longitudinal extent of the westward is very large, causing sometimes positive changes at all Dst stations associated with the 3-D current system (see Figure 3 and 7).

In summary, there is no doubt that substorm activity is closely related to the main phase development. Our task should be to examine quantitatively how the injected protons and O^+ from the ionosphere and magnetotail produce magnetic variations on the earth. In some cases, those ions may drift away from the magnetosphere without forming the complete or even a partial ring current. At the same time, it is necessary to improve the accuracy of both Dst and AE indices. Further, we should look for other new indices which might give new light in studying our complex problem. This is the subject of the next section.

4. NEW INDEX "AF"

For the above reasons, we propose here a new geomagnetic index, AF [*Sun* and *Akasofu*, 2002]. In proposing a new index such as AF, it is important that it should be available for a very long period, at least for a few sunspot cycles. Further, the data for the new index should be readily available. Thus, geomagnetic data are most useful for this purpose. Since we have the AE and Dst indices for the electrojets and the ring current, respectively; we propose here a new index AF as a measure of the field-aligned currents.

4.1. Determination of the AF index

It is well known that the Y component variations observed at middle and low latitudes (Y-low) are mainly caused by field-aligned currents [cf. *Sun et al.*, 1984]. Thus, the field-aligned currents can be monitored by using the Y-low observed at eight mid-latitude stations, which are uniformly distributed along longitude as shown in Table 1. In order to find a measure of the total upward field-aligned current, first we normalized the observed Y component variations at eight stations by using an experimental formula (3) that shows the

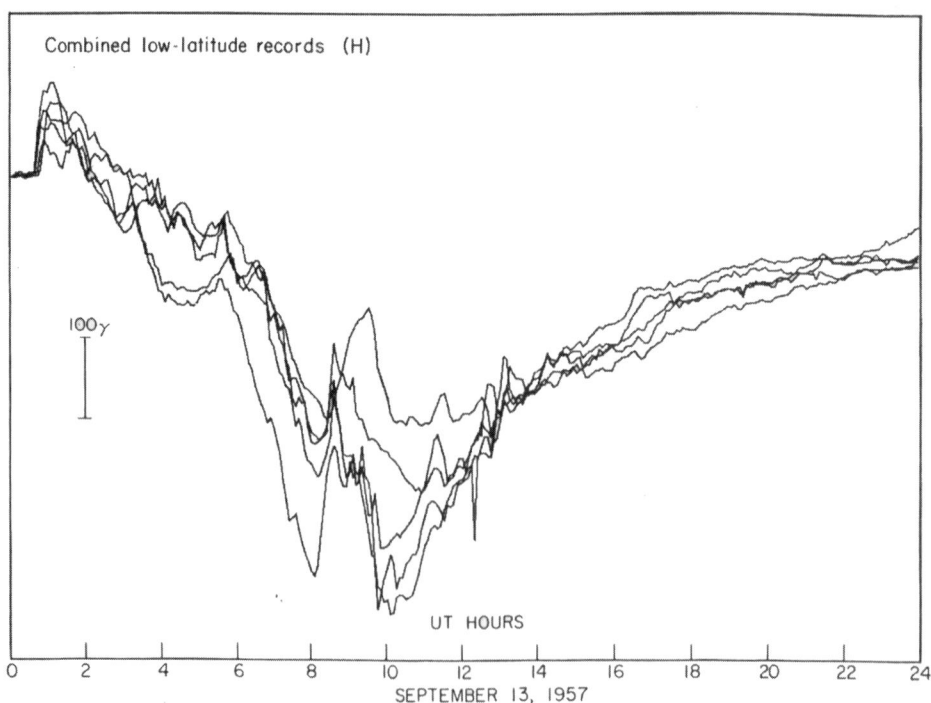

Figure 7. The H component of geomagnetic field disturbances recorded at five low-latitude observatories on September 13, 1957.

Table 1. Codes and coordinates of observatories

CODE	GEOGRAPHIC		ECCENTRIC DIPOLE	
	Longitude	Latitude	Longitude	Latitude
AM	5.53	22.79	67.60	23.68
ABG	72.87	18.64	130.39	8.89
PHU	105.97	21.03	163.03	9.24
KAK	140.18	36.23	195.55	26.97
HON	202.00	21.32	260.36	21.44
TUC	249.17	32.25	307.79	39.80
SJG	293.85	18.12	356.21	27.80
MBO	343.04	14.39	44.61	19.60

dependence of the magnitude of the Y-low on the latitude. The formula (3) was obtained on the basis of magnetometer data observed at 8 stations along 210° magnetic meridian (MM) from July, 1992 to June, 1993 [*Yumoto et al.*, 1992].

$$Y(\theta) = 0.0049\theta^2 + 0.1107\theta + 1.7172 \quad (3)$$

Then, the AF index is determined by the integrated normalized positive Y-low variations in terms of MLT with a resolution of one hour, which can be obtained by interpolating the normalized Y-low variations at eight mid- and low-latitude stations.

4.2. Examples

One should note that geomagnetic indices, including AE, Dst, Kp, ap, etc., are not physical quantities we deal with. Indeed, as we have seen earlier, the Dst index has at least two components. The AE index is not the total intensity of the electrojet. In a similar way, our new index is a quantity which is related to the field-aligned currents, but is not its intensity. In this short paper, we show only two examples in Figure 8a and 8b, demonstrating how useful the AF index is. The figures show that the AF index is far better correlated with the Dst index than the AE index.

The good correlation between the Dst and AF indices indicates that the field-aligned currents (perhaps more specifically O^+) are directly related to the development of the ring current belt.

5. SUBSTORM ONSET

One of the reasons for the lack of progress in substorm research may be that we have not made enough effort to synthesize/integrate observed facts for the purpose of searching for key processes associated with substorms. For

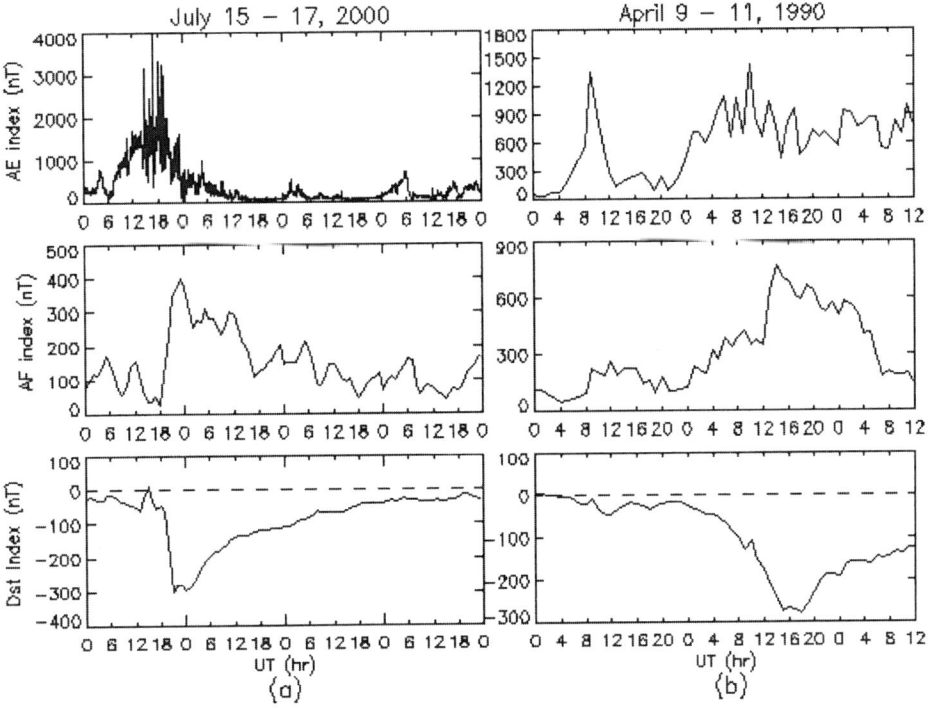

Figure 8(a) and 8(b). From the top to the bottom, variations of the magnitude with time for the AE, AF and Dst indices on October 4, 2000 and April 9–11, 1990, respectively.

this reason, we attempt here to identify essential processes associated with substorm onset by synthesizing/integrating the three well-established observations for decades [*Akasofu*, 1968; 1977]:

(i) A sudden brightening of an auroral arc.
(ii) A sudden growth of the auroral electrojet.
(iii) The so-called "depolarization."

A sudden brightening of an arc located near the equatorward boundary of the auroral oval in the evening and midnight sectors, suggests a sudden increase of the intensity of the upward current there. Thus, the first step in our attempt is to find out how the pair of upward and downward field-aligned currents in the night sector of Boström's current system (Figure 2) constitutes a circuit; the former is located poleward of the latter. One possibility is that they are connected to a southward-directed current in the ionosphere. Figure 9a shows the distribution on the Pedersen current I_p in the ionosphere. One can see that the Pedersen current flows southward in the late evening, midnight, and morning sectors in the location where the pair of field-aligned currents is present [*Iijima and Potemra*, 1978; *Water et al.*, 2001]. Therefore, the downward and upward field-aligned currents and the Pedersen current in the ionosphere must constitute a circuit.

It is quite like that this meridional circuit closes on the equatorial plane. In order to examine such a possibility, the Pedersen current vectors are projected onto the equatorial plane by reversing the direction; see an insert in the left-hand side of Figure 9b [*Akasofu*, 1992]. The projected distribution may be compared with the distribution of the radial component of the I_r on the equatorial plane (right-hand side of Figure 9b), which was obtained on the basis of the AMPTE satellite data [*Iijima, et al.*, 1990]. In spite of the fact that both distributions were obtained by two completely independent methods during different periods, they are remarkably similar, suggesting that the Pedersen current I_p and the equatorial radial current I_r together with the downward and upward field-aligned currents, constitute Boström's current system in Figure 2 [*Boström*, 1964]. Since the aurora must be associated with upward field-aligned currents, auroral arcs must appear along the feet of the upward *sheet* current of Boström's current system.

The electric field observations by the Chatanika incoherent scatter radar shows clearly a southward-directed electric field (E_s) of 20–50 V/km during substorms, but not a westward-directed electric field [*Brekke et al.*, 1974]. Thus, it should be noted that $E_s \cdot I_p > 0$, so that the ionosphere must be the load of the circuit. Therefore, Boström's current system must be driven on the equatorial plane by driving electric field E_d, namely $E_d \cdot I_r < 0$. In the night side of Boström's current circuit, I_r must be directed outward (Figure 2), so that the necessary E_d must be directed inward. If the dynamo process ($V_d \times B$) powers Boström's current system, the necessary azimuthal plasma flow ($V_d = -E_d/B$) must be directed eastward if the frozen-in-field condition would hold there.

It is not difficult to see that the sudden brightening of an auroral arc and the increase of the upward field-aligned current must arise from the fact that Boström's current system is suddenly enhanced. Boström's current system must be driven by a dynamo process associated with an eastward flow of plasma on the equatorial plane.

Indeed, the eastward flow of ionospheric electrons, namely westward current, is a manifestation of this eastward flow of plasma above the ionosphere. The return current from the westward electrojet can reduce or sometimes overwhelm the cross-tail current on the equatorial plane, resulting in the depolarization/overdipolarization/negative depolarization.

In the past, it has been assumed that this phenomenon is associated with a diversion of the cross-tail current into the ionosphere. The diverted current flows into the morning side of the auroral oval and flows along the auroral oval in the night sector and then flows out from the evening sector of the ionosphere back to the magnetotail. If this would indeed be the case, a significant part of the cross-tail potential (namely, the voltage 50 ~ 100 kV) of solar wind-magnetosphere dynamo and thus the westward electric field) must drive the westward electrojet, so that the westward electrojet would have to be the Pedersen current. This is not the case. Further, the diversion process cannot explain any features of auroral arcs during the substorms, including the brightening of a sheet-form over a large local time span. An auroral arc requires a *sheet* of upward field-aligned current, extending in the east-west direction over thousands of kilometers as Figure 2 illustrates.

The intensity of the return current from the westward electrojet can vary, depending on the intensity of the westward electrojet current. The magnitude of the return current can be smaller or larger compared with the westward-directed cross-tail current. A careful examination of this phenomenon clearly shows that the increase of B_z component in the magnetotail can be much greater than what is expected from the simple depolarization [*DeForest et al.*, 1971]. The dipole filed is about Bz ~ 110 nT at the geosynchronous distance. The "depolarization" can cause Bz to become as large as 150 nT. Therefore, the depolariza-

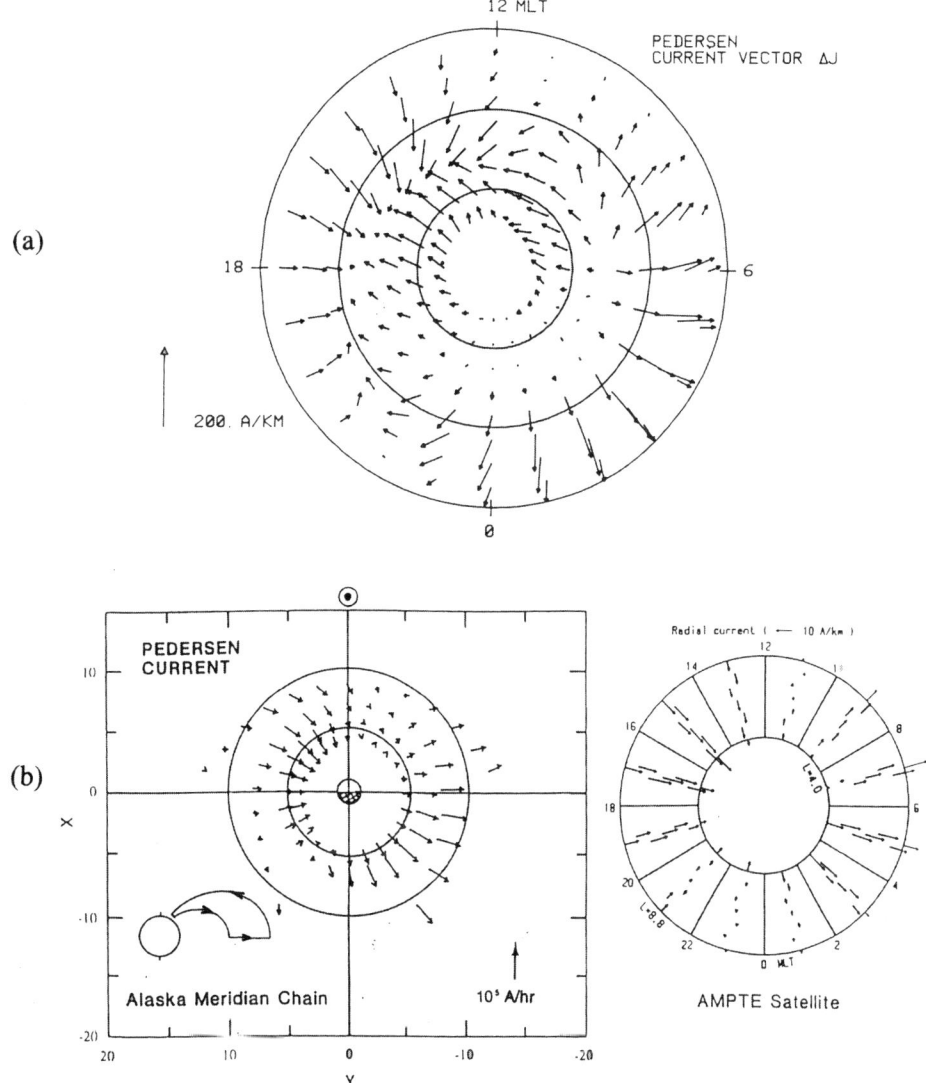

Figure 9(a). The distribution of the Pedersen current in magnetic latitude-MLT coordinates which is obtained by the KRM method on the basis of the Alaska meridian chain of magnetometers. **9(b).** Comparison of the distribution of radial current in the equatorial plane (the left-hand side), deduced from the ionospheric Pedersen current in Figure 9a, and those deduced from AMPTE satellite data (the right-hand side).

tion is not simply the tendency for the stretched field lines to return back to dipole filed lines.

Even the present exercise alone can suggest that the activation of Boström's current system is essential in explaining the three substorm onset phenomena. Thus, the enhancement process of the internal dynamo may be the key in understanding substorm onset and a source of auroral electrons in the night sector. Other source of auroral electrons could be studied a similar way, if the electrons are field-aligned current carriers.

6. RECOVERY PHASE OF GEO MAGNETIC STORMS

As early as 1963, *Akasofu, Chapman* and *Venkatesan* [1963] suggested that the storm recovery phase consists of two components (see Figure 10). The first one decays rapidly. The rapid decay is then followed by a slow decay. This feature may be explained in several ways. The ring current belt may consist of two positive ion species, one decays rapidly and the other slowly. Another interpretation is that the inner part of the ring current decays rapidly because it is imbedded in a dense atmosphere.

Future ENA observations might throw light on this issue. It is important that such a study requires geomagnetic storms which have no major substorm activity after the recovery phase; such substorm activities delay the recovery

7. DIRECTLY DRIVEN AND UNLOADING COMPONENTS IN SUBSTORM ACTIVITIES

Akasofu [1979] proposed that the magnetospheric substorm consists of two processes, the directly driven (DD) and the unloading (UL) process. His concept of the two-component process was recognized by *Rostoker et al.* [1987]. *Sun et al.* [1998; 2000a] separated the directly

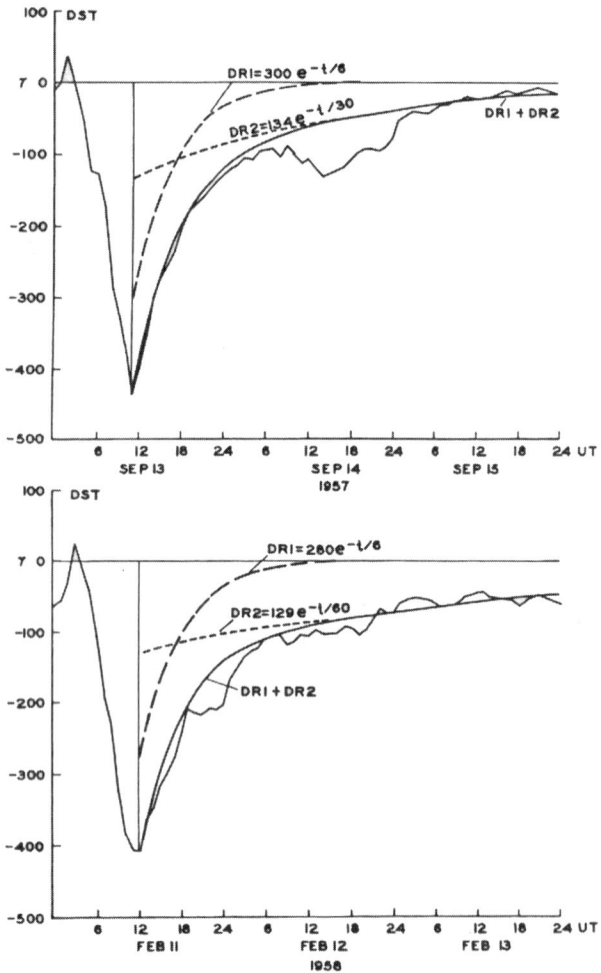

Figure 10. The Dst index for two huge magnetic storms during September 13–15, 1957 and February 11–13, 1958. DR1 and DR2 indicate decay of two components in the ring current belt. (Akasofu et al. 1963)

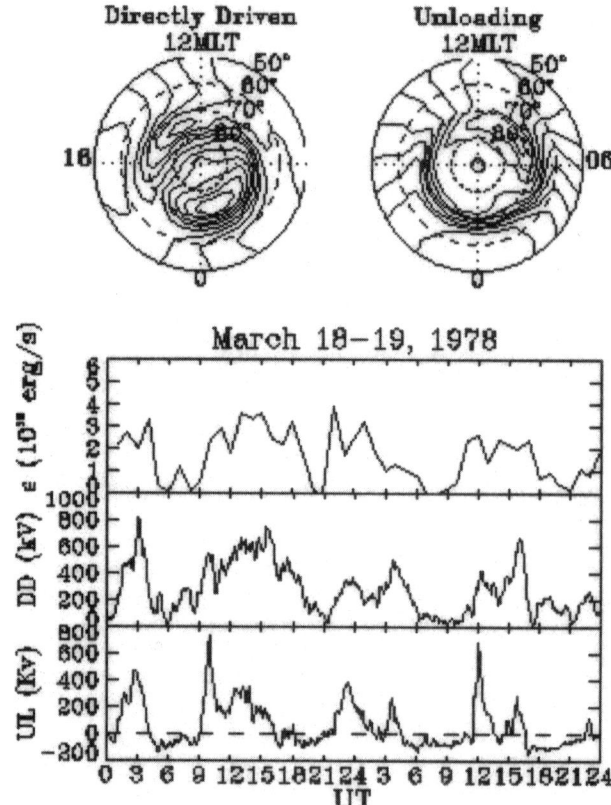

Figure 11. The top two panels show the normalized directly driven and unloading pattern of equivalent current function, respectively. The lower panel shows time variations of Epsilon function, the directly driven and unloading components during March 18–19, 1978.

driven (DD) and unloading (UL) components by using the method of the fundamental orthogonal component (MNOC). The upper two panels of Figure 11 shows the normalized DD and UL patterns of the equivalent current, respectively. The DD and UL patterns apparently resemble the DP2 and DP1 current patterns, respectively. The lower panel of Figure 11 shows time variations of the magnitude of both components during March 180–19, 1978 together with Epsilon function. Further, the DD and UL components for the ionospheric electrodynamics (the electric potential, the ionospheric current and FAC etc.) can be obtained by using the KRM algorithm [Kamide, et al., 1981].

We examine here the correlation of the DD and UL components of the integrated upward filed-aligned current with the corrected Dst index [*Sun et al.* 2000b]. We adopted the main phase during two big storm events during October 21–22, 1990 and November 4–5, 1993. In Figure 12, the left panel shows the correlation between the corrected Dst index

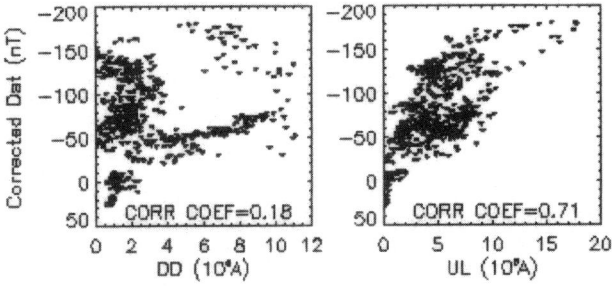

Figure 12. The left panel shows the correlation between the corrected Dst index and the directly driven component of the integrated upward field-aligned currents with correlation coefficient of 0.18. The right panel shows the corrected Dst index and the unloading component of the integrated upward field-aligned currents with correlation coefficient of 0.71.

and the DD component of the upward field-aligned current. The correlation coefficient is 0.18. The right panel shows the correlation between the corrected Dst index and the UL component of the upward field-aligned current. The correlation coefficient is 0.71. In this connection, it may be noted that the correlation coefficient between the standard Dst index and the DD and UL components are 0.26 and 0.70, respectively. The correlation coefficient between the total upward FAC (DD+UL) and the corrected Dst index is 0.70.

These results indicate that the development of the ring current belt during a magnetic storm is closely related to the unloading process during the expansion phase of substorms, but not the directly driven component, which is associated with the convection.

REFERENCES

Ahn, B.-H., G.-H. Moon, W. Sun, S.-I. Akasofu, G.-X. Chen, Y. D. Park, Universal time variations of the Dst index and the relationship between the cumulative AL and Dst indices during geomagnetic storms, *J. Geophys. Res.*, in press, 2002.

Akasofu, S.-I., *Polar and Magnetospheric Substorms*, D. Reidel Pub. Co., Dordrecht-Holland, 1968.

Akasofu, S.-I., *Physics of Magnetospheric Substorms*, D. Reidel Pub. Co., Dordrecht-Holland, 1977.

Akasofu, S.-I., A confirmation of the validity of the electric current distribution determined by a ground-based magnetometer network, *Geophys. J. Inst.*, 109, 1992.

Akasofu, S.-I., What is magnetospheric substorm?, *Dynamics of the magnetosphere*, P. 447, D. Reidel, Norwell, Mass., 1979.

Akasofu, S.-I. and S. Chapman, The ring current geomagnetic disturbances and the Van Allen radiation belts, *J. Geophys. Res.* 66, 1321, 1961.

Akasofu, S.-I., On a self-consistent calculation of the ring current field, *J. Geophys. Res.*, 67, 3617, 1962.

Akasofu, S.-I. and S. Chapman, The ring current and neutral line discharge theory of the aurora polaris, *J. Phys. Soc. Japan*, 17, Suppl. A-I, 169, 1962.

Akasofu, S.-I. and S. Chapman, *Solar Terrestrial Physics*, Clarendon, Oxford, 1972.

Akasofu, S.-I. and S. Chapman, Magnetic storms: the simultaneous development of the main phase (DR) and of polar magnetic substorms (DP), *J. Geophys. Res.*, 68, 3155, 1963.

Akasofu, S.-I., S. Chapman, and D. Venkatesan, The main phase of geomagnetic storms, *J. Geophys. Res.*, 68, 3345, 1963.

Akasofu, S.-I. and J. K. Chao, Interplanetary shock waves and magnetospheric substorms, Planet. Space Sci., 28, 381, 1980.

Akasofu, S.-I. and W. Sun, Advances made in determining magnetospheric current system, *EOS*, 81, pages 361, 365, 2000.

Alfvén, H., *Cosmical Electrodynamics*, Oxford University Press, New York, 1950.

Boström, R., A model of the auroral electrojets, *J. Geophys. Res.*, 69, 4983, 1964.

Brekke, A., J. R. Doupnic, and P. M. Banks, Incoherert scatter measurements of E region conductivities and currents in the auroral zone, *J. Geophys. Res.*, 79, 3773, 1974.

Chapman, S., An outline of a theory of magnetic storms, *Proc. Roy. Soc.*, A95, 61, 1918.

Chapman, S., The electric current-systems of magnetic storms, *Terr. Magn. Atmos. Elect.*, 40, 349, 1935.

Daglis, I. A., The role of magnetosphere-ionosphere coupling in magnetic storm dynamics, in *Magnetic Storms*, Geophys. Monogr., Vol. 98, Edited by B. T. Tsurutani, W. D. Gonzalez, Y. Kamide, and J. K. Arballo, pp. 107-116, AGU, 1997.

Daglis, I. A., W. I. Axford, S. Livi, B. Wilken, M. Grande, and F. Soraas, Auroral ionospheric ion feeding of the inner plasma sheet during substorms, *J. Geomagn. Geoelectr.*, 48, 729, 1996.

Daglis, I. A., S. Livi, E. T. Sarris, and B. Wilken, Energy density of ionospheric and solar wind origin ions in the near-Earth magnetotail during substorms, *J. Geophys. Res.*, 99, 5691, 1994.

Daglis, I. A., E. T. Sarris, G. Kremser, and B. Wilken, On the solar wind-magnetosphere-ionosphere coupling, in *Study of the Solar-Terrestrial System*, Eur. Space Agency Spec. Publ., ESA SP-346, 193, 1992.

Davis, T. N. and M. Sugiura, Auroral electrojet activity index AE and its universal time variations, *J. Geophys. Res.*, 71, 785, 1966.

DeForest, S. E. and C. E. McIlwain, Plasma clouds in the magnetosphere, J. Geophys. Res., 76, 3587, 1971.

Gonzalez, W. D., J. A. Joselyn, Y. Kamide, H. W. Kroehl, G. Rostoker, B. T. Tsurutani, and V. M. Vasyliunas, What is a geomagnetic storm?, *J. Geophys. Res.*, 99, 5771, 1994.

Hamilton, D. C., G. Gloeckler, F. M. Ipavich, W. Studemann, B. Wilken, and G. Kremser, Ring current development during the great geomagnetic storm of February 1986, *J. Geophys. Res.*, 93, 14343, 1988.

Hultqvist, B., R. Lundin, K. Stasiewicz, L. Block, P.-A. Lindquist, G. Gustafsson, H. Koskinen, A. Bahnsen, T. A. Potemra, and L. J. Zanetti, Simultaneous observations of upward moving field-aligned electrons and ions on auroral zone field lines, *J. Geophys. Res.*, 93, 9765, 1988.

Iijima, T. and T. A. Potemra, Large-scale characteristics of field-aligned currents associated with substorms, *J. Geophys. Res.*, 83, 599, 1978.

Iijima, T., T. A. Potemra, and L. J. Zanetti, Large-scale characteristics of magnetospheric equatorial currents, *J. Geophys. Res.*, 95, 991, 1990.

Kamide, Y., A. D. Richmond, and S. Matsushita, Estimation of ionospheric electric field, ionospheric currents, and field-aligned currents from ground magnetic records, *J. Geophys. Res.*, 86, 801, 1981.

Kawasaki, K. and S.-I. Akasofu, The growth and decay of the main phase of the September 21-23, 1963 magnetic storm, *J. Geomag. Geoelec.*, 24, 175, 1972.

Lundin, R., L. Eliasson, and K. Stasiewicz, Plasma energization on auroral field lines as observed by the Viking spacecraft, *Geophys. Res. Lett.*, 14, 443, 1987.

McIlwain, C. E., Substorm injection boundaries, in *Magnetospheric Physics*, edited by McCormac, p 143, D. Reidel, Norwell, Mass., 1974.

Rostoker, G., S.-I. Akasofu, W. Baumjohann, Y. Kamide, and R. L. McPherron, The roles of directly input of energy from the solar wind and unloading of stored magnetotail energy in driving magnetospheric substorms, *Space Sci. Rev.*, 46, 93, 1987.

Shelley, S. F., R. D. Sharp, and R. G. Johnson, Satellite observations of an ionospheric acceleration mechanism, *Geophys. Res. Lett.*, 3, 654, 1976.

Sun, W., B.-H. Ahn, S.-I. Akasofu, and Y. Kamide, A comparison of the observed mid-latitude magnetic disturbance fields with those reproduced from the high-latitude modeling current system, *J. Geophys. Res.*, 89, 10881, 1984.

Sun, W., W.-Y. Xu, and S.-I. Akasofu, Mathematical separation of directly driven and unloading components in the ionospheric equivalents during substorms, *J. Geophys. Res.*, 103, 11695, 1998.

Sun, W., W.-Y. Xu, and S.-I. Akasofu, An improved method to deduce the unloading component for magnetospheric substorms, *J. Geophys. Res.*, 105, 13131, 2000a.

Sun, W. and S.-I. Akasofu, On the formation of the storm-time ring current belt, *J. Geophys. Res.*, 105, 5411, 2000b.

Sun, W. and S.-I. Akasofu, A new field-aligned current index AF and its relation to the storm-time ring current and energetic neutral atorm (ENA) emissions, Geophys. Res. Lett., in press, 2002.

Tsurutani, B. T., B. E. Goldstein, E. J. Smith, W. D. Gonzalez, F. Tang, S.-I. Akasofu, and R. R. Anderson, The interplanetary and solar causes of geomagnetic activity, *Planet. Space Sci.*, 38(1), 109, 1990.

Walters, C. L., B. J. Anderson, and K. Liou, Estimation of global field-aligned currents using the Iridium system magnetometer data, *Geophys. Res. Lett.*, 28, 2165, 2001.

Wilken, B., I. A. Daglis, and S. Livi, Observations of geomagnetic storms by the CRRES satellite, *EOS Trans. AGU*, 73, 457, 1992.

Williams, D. J., Ring current composition and source in *Dynamics of the Magnetosphere*, edited by S.-I. Akasofu, P. 407, D. Reidel, Norwel, Mass., 1980.

Williams, D. J., The Earth's ring current: Causes, generation, and decay, *Space Sci. Rev.*, 34, 223, 1983.

Yumoto, K., and 210° MM Magnetic Observation Group, The STEP 210° magnetic meridian network project, *J. Geomag. Geoelectr.* 48, 1279, 1996.

Measurement Strategies for Future Missions to Understand Geospace Dynamics

R. Vondrak[1], J. Slavin[1], L. Zelenyi[2], M. Guhathakurta[3], S. Curtis[1], B. Tsurutani[4]

Geomagnetic storms and substorms are dramatic reconfigurations of the magnetosphere resulting from fundamental processes that control its dynamics and structure. Improvements in our understanding of these processes require new information on the temporal evolution of the spatially-structured magnetospheric plasma, as well as the variable solar drivers of the connected sun-earth system. A successful approach will consist of an appropriately balanced combination of: a) measurements of the global magnetospheric configuration by widely-distributed spacecraft or by imagers that reveal the connections between different regions; b) comprehensive detailed measurements at important geospace boundaries and regions; c) dense networks of measurements that provide sampling consistent with the coherence length of the plasma; d) measurements of the inner and outer boundaries of the magnetosphere (i.e. the solar wind and the high-latitude ionosphere); e) measurements of the solar plasma and solar processes as a variable driver of geospace dynamics.

Several missions have been strategically designed to supply these needed measurements, some of which are enabled by technology innovations. The measurement capabilities and specific approach of the missions that can resolve fundamental questions of storm-substorm relationships are identified, including some of the new missions planned for the NASA Sun Earth Connection program and for the Russian space research program.

---

[1]NASA Goddard Space Flight Center, Greenbelt, Maryland
[2]Insitute for Space Research, Moscow, Russia
[3]NASA Headquarters, Washington, D.C.
[4]Jet Propulsion Laboratory, Pasadena, California

INTRODUCTION

In recent years there has been much progress in our understanding of the relationships between geomagnetic storms and substorms. We know that both storms and substorms are dramatic reconfigurations of the magnetosphere.

Disturbances in Geospace: The Storm-Substorm Relationship
Geophysical Monograph 142
Copyright 2003 by the American Geophysical Union
10.1029/142GM21

Each is a sequence of events that results in the transport, structuring, and acceleration of plasma. We recognize that while they differ in significant ways, they are sometimes closely related. Although the role of storms and substorms in the basic structure and dynamics of the magnetosphere has been identified, our knowledge and understanding is still incomplete.

Despite remarkable advances in measurements, theory, and modeling, we have been unable to identify clearly their causes or the processes that determine their dynamics and principal effects. Future progress requires simultaneous multispacecraft measurements with a minimum separation determined by the intrinsic nature of the dynamics.

Geomagnetic storms and substorms are characterized by unusually strong and variable magnetospheric electric and magnetic fields that produce charged particle acceleration and transport on a massive scale; energy dissipation rates of hundreds of GW are typically involved. Furthermore, both storms and substorms are closely coupled to their energy source—the solar wind and related solar dynamics. Whenever the interplanetary magnetic field has a southward component there is an enhanced transfer of energy from the solar wind plasma to the magnetosphere. The underlying cause is the dynamo setup along the magnetopause as a result of the interconnection between the IMF and geomagnetic field due to magnetic reconnection. The most intense and longest duration southward interplanetary magnetic fields are those associated with magnetic clouds embedded within interplanetary coronal mass ejections, but large amplitude Alfven waves, the draping and compression of magnetic flux tubes just ahead of coronal ejection events and other types of solar wind events are also important sources of southward IMF [*e.g. Farrugia et al.*, 1997].

Geomagnetic storms are usually the product of long intervals, ~ 3 to 12 hrs or more of intense magnetospheric convection [*Wolf et al.*, 1982; *Chen et al.*, 1994]. These intervals of intense earthward convection "inject" large quantities of ions and electrons into the inner magnetosphere, L ~ 2–7. During their transport from the plasma sheet, X ~ –10 to –15 Re, the ions tend to conserve their first adiabatic invariant and grow in energy from ~ 1–10 keV in the deep tail to ~ 10–100 keV in the "ring current" region at L ~ 3–4. It is the differential azimuthal drift speeds of the ions and electrons in this region that produce the large westward electric current in this region, i.e., the ring current, which gives geomagnetic storms their equatorial ground-based signature: B_H ~ –30 to –300 nT. The time scale for the development of storms is very short, on the order of the MHD wave speed and convection transport delays (~ tens of minutes). However, their decay takes many hours to days and involves various loss processes, especially wave-particle scattering into the loss cone and charge exchange with neutral atoms in the geocorona [*Kozyra et al.*, 1984].

Substorms, while energetically no less intense, are much shorter in duration, ~ 1–2 hrs, and their effects are felt primarily in the high altitude magnetosphere, L > ~ 6, and the high latitude ionosphere. Whereas storms are produced by intense, but relatively steady earthward convection, substorms are characterized by "explosive" energy release in the tail that drives short-lived, 1–10 min, but extremely rapid convection, ~ 10^3 km/s, in the inner plasma sheet [*e.g., Ohtani et al.*, 1999; *Slavin et al.*, 2001] and fast reconfiguration of the tail magnetic fields [*Baker and Pulkkinen*, 1998].

A key unsolved problem in geospace dynamics is the relationship between geomagnetic storms and substorms [*Gonzalez et al.*, 1994; *Kamide et al.*, 1998; *Lui*, 2001]. It seems clear that geomagnetic storms are a consequence of the buildup of an equatorial ring current, resulting from changes in the solar wind that modify the dynamics of the magnetosphere. It is also apparent that substorms occur during magnetic storms. What is not yet understood is whether substorms are elementary components of a storm, such that magnetic storms are the accumulated result of a series of many rapid substorms, or whether both storms and substorms simply occur in association with one another as a result of magnetospheric variations. Important progress in solving this problem of how a substorm is connected to a storm will be made when we can measure unambiguously the entire sequence of events in the initiation and development of storms and substorms.

Although the solar wind is the primary energy source of both storms and substorms, we do not understand how that solar wind energy is coupled to the magnetosphere and how it is dissipated (by means of the ring current, high-latitude Joule Heating and particle energy input, acceleration of energetic trapped particles, magnetopause losses, etc.). Coupling between these regions creates complex feedback that alters the global systems response to external forcing. We need to establish the efficiency of energy transfer from the solar wind into the magnetosphere for a variety of conditions, such as solar wind pressure gradients and interplanetary magnetic field variations. Understanding this transfer will enable predictive models of the response of the magnetosphere to extreme solar wind conditions and allow us to differentiate the dynamic geospace response to steady and nonsteady solar drivers. One of the most fundamental scientific mysteries is the origin of the solar wind. Without the solar wind there would be no magnetosphere, storms, or substorms. Ultimately, we need a predictive capability for space weather that allows us to go from observations of the solar surface to final consequences at Earth, recognizing that the Sun and Earth are a connected system.

This paper describes the strategies for new measurements that can improve our understanding of geospace dynamics, as well as the new technology that can enable them. It also identifies some of the future NASA and Russian missions that can resolve fundamental questions of storm-substorm relationships. This paper does not include missions of other space agencies, such as ESA or ISAS, nor does it include NASA missions in the Explorer program.

STRATEGIES FOR GEOSPACE MISSIONS

While much progress has been realized during the ISTP era, there still remain important open scientific questions, which strongly limit our ability to understand and model geomagnetic storms. These roadblocks to the development of fuller understanding and mature, predictive models include (1) the role of the ionospheric outflow of ions into the magnetosphere, especially O^+, in the development of the ring current [*Delcourt et al.*, 1990; *Daglis and Axford*, 1996], (2) the contribution of substorm-associated reconfigurations of the magnetic field and inductive electric fields to the overall acceleration and transport of energetic charged particles into the inner magnetosphere [*Wolf et al.*, 1997], and (3) the relative roles of charge exchange and wave-particle interactions in the dissipation of the ring current [*Koyzra et al.*, 1984]. Each of these issues are intrinsically complex and key to the understanding of storms. For example, the composition of the inner plasma sheet, $X < \sim -20$ Re, is essential because this is the region where the "seed population" resides for the formation of the ring current during geomagnetic storms [*e.g., Fok et al.*, 1996]. However, the processes controlling the composition of the plasma sheet are, themselves, known to be complicated and linked to conditions in the solar wind. For example, *Moore et al.* (1999) observed large increases in the rate of mass transfer from the ionosphere immediately following global sudden compressions of the magnetosphere.

The quantitative understanding of nearly all aspects of the substorm process is severely limited by the sparseness of the available data sets, especially in the plasma sheet, Furthermore, the sparseness of the existing data makes clear delineation between "cause" and "effect" difficult to impossible. For example, some classes of solar wind events, such as magnetic field discontinuities and pressure pulses, have been observed to "trigger" the onset of substorms [*e.g., Kawasaki et al.*, 1971]. In addition, significant disagreement still exists between substorm theories in which substorms come about as a direct result of magnetic reconnection in the tail as opposed to being initiated by other instabilities occurring much closer to the earth [*e.g., see Slavin et al.*, 2001; *Ohtani et al.*, 1999, *Lui*, 2001]. In summary, substorms exhibit far stronger temporal and spatial gradients in the magnetic fields and plasma populations, and the acceleration of charged particles by "inductive" electric fields is thought to be far more significant than is the case of storms [*e.g., Hesse*, 1995]. These differences can only be resolved by unambiguous space-time resolved measurements on appropriate scales.

Improved measurements are needed to reveal the processes that couple the magnetospheric regions and to establish the relationships between cause and effect. The current data are often ambiguous because they are too sparse, and do not

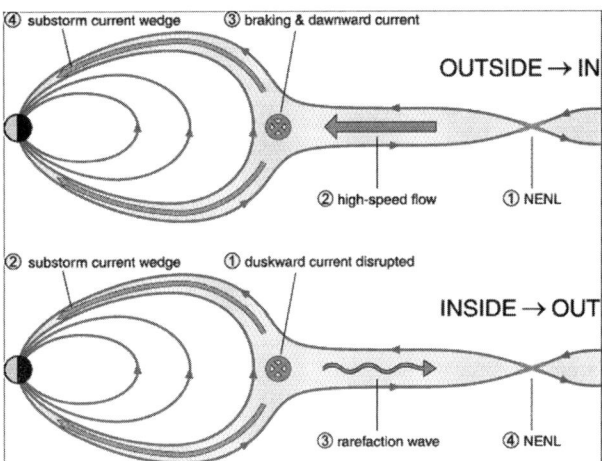

Figure 1. Alternative sequence of events for substorm onset.

have adequate spatial and temporal resolution. As a result the data can be interpreted in different ways, and do not conclusively reveal the critical characteristics and processes. This ambiguity is demonstrated by the fact that reasonable people disagree on the interpretation of observations. For example, as shown in Figure 1, there is an ongoing controversy as to whether substorms are initiated either close to the Earth by a disruption of the cross-tail electric current or further away by magnetic reconnection about 25 Re downtail from the Earth [*Lui*, 2001]. Furthermore, our current local models and simulations have more detail than can be measured, and our global models lack resolution originating in part to a lack of identification of the relevant multiscale physics. We cannot validate some aspects of existing models, nor can we judge which model is correct when their predictions differ. In general, the data sets now available are not adequate to answer the important science questions by providing closure between the theoretical descriptions of magnetospheric dynamics and the observations of these processes.

The main difficulty in the space experiments aimed at studies of these complicated processes is their wide spectrum of space and time scales. This implies the need of multi-point measurements at distances ranging from natural spatial scales (e.g. electron inertia length, Debye length, Larmor radius) to global MHD-like scales (i.e., at least a fraction of the Earth's radius). The small-scale range of magnetospheric plasma phenomena (i.e., effects at ion Larmor radius scales of tens of km), their 2D and 3D characteristics, their spectra and evolution remain largely unexplored experimentally. This complicates a detailed theoretical analysis and modeling of the magnetic field annihilation, or reconnection process. Other processes structure plasma at larger scales, mesoscales. For example, the coherence length for plasma sheet variations is about 1.4 Re. This scale size is much finer than the spacing of most multi-point measurements that are available today. Improvements in our understanding of storms and substorms require new information on the temporal evolution of the spatially-structured magnetospheric plasma over a large region (L = 3 to 30). To attain such a fine resolution over a large volume, we need improved methods such as high-resolution imaging systems, and networks of multi-spacecraft constellations.

A successful approach will consist of a balanced combination of several measurement strategies. These various approaches are summarized in Table 1 and described in this section. Technologies needed to enable them are described in the next section, with specific future missions identified in the final section.

a) Global configuration measurements that reveal the connections between different magnetospheric regions

The International Solar Terrestrial Physics (ISTP) program provided unprecedented, simultaneous measurements of the entire Sun to Earth connected system by observations of the Sun with SOHO and measurements of the geospace measurements with the Global Geospace Science (GGS) program [*Russell*, 1995; *Acuna et al.*, 1995]. GGS accomplished a systems understanding by simultaneous in-situ measurements on strategically-distributed spacecraft and by polar imagers. Such measurements by widely-distributed spacecraft reveal the global configuration and large-scale dynamics, but provide less insight into the mesoscale structures and processes or microphysics.

Another approach to measure the global magnetospheric configuration is global remote sensing with multiple imagers, as is being done by IMAGE [*Burch et al.*, 2001]. IMAGE is able to remotely sense the magnetospheric plasma from its high-altitude vantage point by detection of photons, energetic neutral atoms, and reflected radiowaves. The energetic neutral atom imagers can be used to infer the distribution of ions in the inner magnetosphere and flowing outward from high latitudes, which reveals the patterns of ion injection, energization, and transport. The photon imagers detect atomic emissions originating from the plasma sheet and from proton and electron precipitation into the ionosphere.

b). Comprehensive detailed measurements at important geospace boundaries and regions

Comprehensive detailed measurements by spacecraft clusters can reveal the details of microphysics at important geophysical boundaries (e.g. magnetopause) and regions (e.g. near-Earth neutral line, ring current). The dynamics of the boundaries is controlled by multiscale processes that moderate the flow of mass, energy, and momentum, and can accelerate charged particles. The understanding of the underlying causes of such dynamic processes requires detailed measurements within the current layers that form these boundaries. Processes such as magnetic reconnection that operate on scale lengths as small as tens of kilometers must be accurately measured with sufficient temporal resolution to reveal the underlying physics.

In particular, direct measurements of the current density vector, **J**, are required because of the key role played by the **JXB** force in both the acceleration of plasma and the equilibrium states of the magnetosphere. The need to accurately register these measurements in the rest frame of the plasma is challenging because these magnetospheric

Table 1. Strategies for measurements of geospace dynamics

Mission Type	Measurement Objective	Strategy	Examples
Global configuration	Global dynamics and connections between regions	Distributed spacecraft and/or remote-sensing imagers	ISTP, IMAGE
Microscope	Detailed measurements of plasma dynamics	Tetrahedronal clusters of spacecraft	CLUSTER, MMS
Networks	Mesoscale processes	Dense constellations of spacecraft	ROY/Schwarm, Magnetospheric Constellation
Inner and outer boundaries	Solar wind and high-latitude ionosphere	Upstream monitor and polar imagers	WIND, ACE POLAR, IMAGE
Solar plasma variations	Solar plasma that drives geospace variations	Evolution of plasma from Sun to Earth	SOHO, STEREO

boundaries are generally in motion. The "cluster" technique involving the flight of four spacecraft in a tetrahedronal formation is designed to meet these measurement requirements. Taken together, the four spacecraft function as a "microscope" that directly determines the local current density from the curl of the magnetic field and the orientation and speed of the current layers from the analysis of the arrival times of these discontinuities at the individual spacecraft.

Figure 2 provides an example of these cluster techniques to the Earth's bow shock [*Escourbet et al., 1997*]. As the shock moves over the four spacecraft the gradients in the magnetic field can be directly inverted to obtain the current density vector within this boundary. In this manner the cluster of four spacecraft functions as a "curlometer" that yields current density vectors. An analysis of the magnetic boundary orientation and speed is derived from the time delays between the discontinuity arrival times at the spacecraft and their spatial separation vectors.

The "curlometer" and "discontinuity analyzer" approaches to current layer investigations are now being implemented for the first time by the ESA-NASA Cluster mission [*Paschmann and Daly, 1998*]. This mission is making the first four-spacecraft measurements in the Earth's magnetic cusps, which are known to be the sites for significant mass transfer from the solar wind to the magnetosphere. Techniques learned on this mission will be incorporated into the NASA Magnetospheric MultiScale Mission that will make measurements down to scale lengths below 100 km in the low latitude magnetopause and cross-tail current layer where magnetic reconnection is believed to control the transfer of energy into the mag-

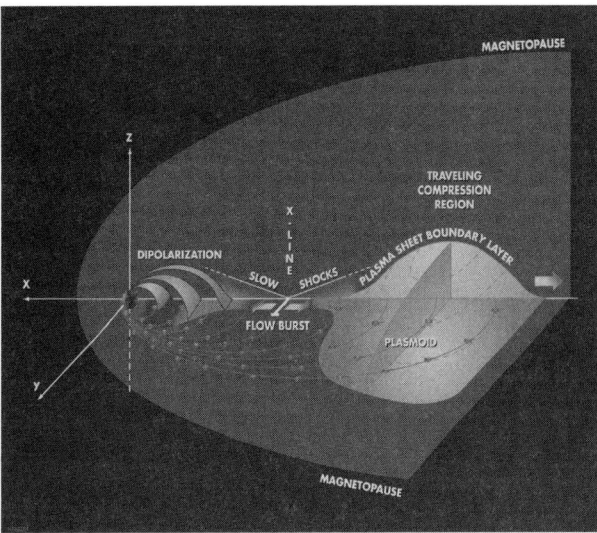

Figure 3. Plasmoid viewed by a magnetospheric constellation mission

netosphere and its later dissipation in geomagnetic storms and substorms.

c). Dense networks of measurements that provide sampling consistent with the coherence length of the plasma

To answer many outstanding questions regarding geospace dynamics, we need knowledge of the instantaneous configuration of the plasma and magnetic field in the magnetosphere. The largest obstacles to achieving this are the vast volume of geospace and the small scale size for magnetospheric structures. For example, Borovsky and coworkers (1998) have made a comprehensive analysis of a large number of plasma sheet flow measurements. They calculate that the mean velocity near the neutral sheet is 60.8 km/s, and the autocorrelation time is 140 s. This implies a coherence length of about 1.4 Earth radii for plasma sheet variations. To fully measure the plasma sheet with a spacing of 1 Earth radius requires hundreds of points to be mapped. An approach targeted at specific science questions can make significant progress with fewer spacecraft deployed in a strategic way.

To provide direct measurements, these satellite constellations will require many multipoint measurements, as shown in Figure 3. Small lightweight sensors will be needed to measure the magnetic field and plasma velocity distribution. Another approach is remote sensing (e.g. active radiowave tomography) to deduce the spatial variation of the plasma density and magnetic field.

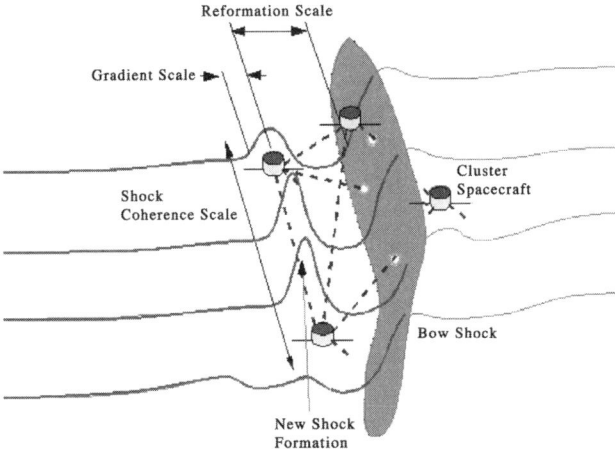

Figure 2. The four CLUSTER spacecraft function as a traveling microscope

d) Measurements of the inner and outer boundaries of the magnetosphere (i.e., the solar wind and the high-latitude magnetosphere).

A full understanding of geospace observations requires simultaneous measurements of the inner and outer boundaries of the magnetosphere (the solar wind and the high-latitude ionosphere). The solar wind measurements can be accomplished by spacecraft at the Sun-Earth Lagrangian point, as has been demonstrated with the ISEE, WIND, and ACE missions. These measurements provide insight into the externally-driven variations, and allow us to differentiate the dynamic geospace response to both steady and non-steady solar drivers, as contrasted with spontaneous dynamics that might be internally driven.

Measurements of the high-latitude ionosphere can be accomplished by polar imagers on high-altitude spacecraft. These measurements provide knowledge of substorm timing, conductivity variations, and energy transfer. They indicate the general state of magnetospheric disturbance during substorms, as well as the timing of substorm onset. As shown in Figure 4 [*Ohtani et al.*, 1999], when combined with measurements at high altitudes in the plasma sheet, they can help resolve the cause-effect sequence of events for substorm triggering processes,

e). Measurements of the solar plasma and dynamics

The ultimate driver of geospace dynamics is the solar wind. Without the solar wind there would be no magnetospheric storms, substorms, or aurora. So, a comprehensive understanding of the Sun-Earth coupled system requires simultaneous measurements of the solar plasma, its evolution during its passage to the Earth, and its structured plasma and field characteristics when it arrives at the Earth.

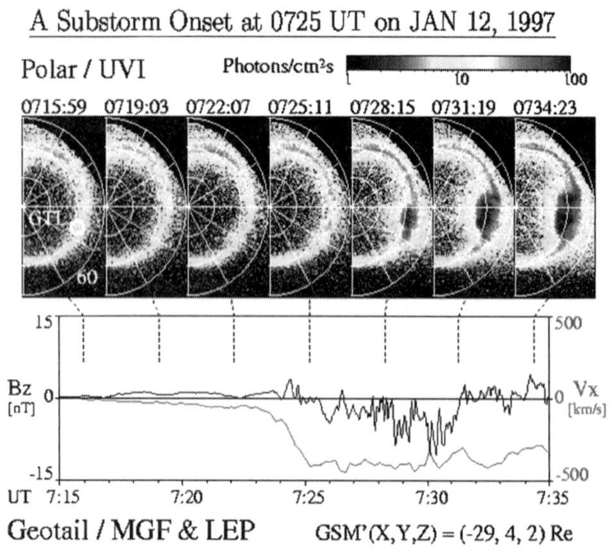

NEW TECHNOLOGY OPPORTUNITIES AND CHALLENGES

Future progress requires both global measurements, as well as measurements that resolve the spatial and temporal variations of the plasma at all scale sizes. Such observations are best accomplished by a combination of remote-sensing imagers and multipoint measurements by distributed spacecraft.

Innovative technology will enable new capabilities for magnetospheric imaging, particularly for tomographic remote sensing. For example, global images of plasma density, convection velocity and magnetic fields can be provided with radio tomography imaging [*Heelis, et al.* 1999: *Ergun*, 1999; *Ganguly et al.*, 1999]. The multiple ray paths between several satellites can be used to reconstruct the density and field distribution with a pixel size of the order of one earth radius in the inner plasma sheet.

Multipoint measurements require fleets of spacecraft whose separation is consistent with the characteristic length and time of the geospace structures and processes. The use of multipoint measurements to resolve space-time ambiguities is not a new strategy and has been used successfully in programs such as DE and ISEE. However, such programs have been constrained by the cost and capabilities of spacecraft to only a few (usually two) measurement platforms that limited the unique space-time separation to be one-dimensional. Fortunately, technology advances have made larger numbers of spacecraft affordable.

To be affordable, each spacecraft for in-situ measurements in a large constellation must be a nano-satellite that weighs no more than approximately 10 kg [*Panetta*, 1998]. A conceptual nano-satellite design is shown in Figure 5. It is a cylindrical spacecraft with a diameter of 30 cm that is spin stabilized and carries a magnetometer and a plasma analyzer. Advanced technology development is required to make these nano-satellites and their on-board instruments compact, lightweight, low power, low cost, highly-autonomous, and survivable in a space radiation environment. Such a small highly-capable spacecraft will be flight validated in the ST-5 mission. ST-5 is the New Millennium Program's (NMP) fifth space technology mission. The purpose of these missions is to "flight vali-

Figure 4. Substorm viewed by Geotail and a POLAR auroral imager

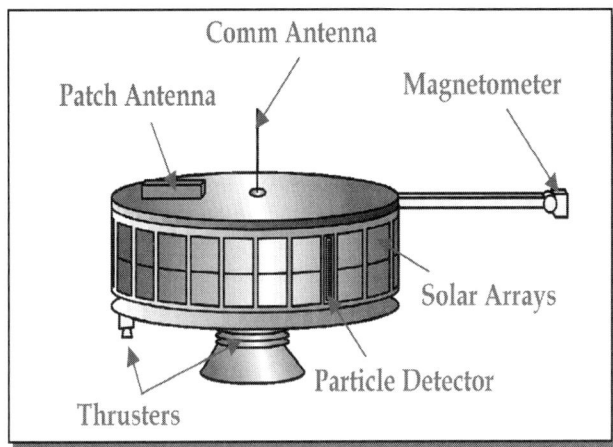

Figure 5. A 10-kg nanosatellite for measurement of geospace dynamics

date" new technologies needed to enable or enhance future science missions, for both the instruments and spacecraft subsystems. ST-5 has three primary mission objectives: 1) design, fabricate, test and fly three 20 kg-class small spacecraft using newly developed miniature technologies; 2) validate in orbit the suitability of these small spacecraft to act as platforms for measurements of the particles and fields required by future missions; and 3) to demonstrate new techniques and technologies for operating the three ST-5 spacecraft as a single "constellation" as opposed to individual spacecraft.

For these purposes, the three small spin-stabilized spacecraft will be launched in geosynchronous transfer orbit (perigee ~ 200 km; apogee ~ 36000 km) as a secondary launch payload. The schedule calls for a launch between mid 2004 and early 2005. The primary mission will be accomplished in 3 months, with the option of extending the operations period if there are sufficient benefits to be gained by doing so. The "science validation instrument" necessary to accomplish the three objectives is a new, miniaturized three-axis fluxgate magnetometer.

SPECIFIC FUTURE MISSIONS

To accomplish the science objectives, new measurements are needed. The NASA strategic missions to accomplish these measurements are two sets of missions referred to as the Solar Terrestrial Probes [*Vondrak*, 1998] and the Living with a Star [*Withbroe*, 2001] programs. In addition, significant missions to understand geospace dynamics are planned by the Russian Space Agency and by other countries.

The Magnetospheric MultiScale (MMS) Mission: Defining the Dynamics of Magnetospheric Boundaries During Storms and Substorms

The Magnetospheric MultiScale (MMS) Mission is a Solar Terrestrial Probe focused on understanding the fundamental plasma processes that operate at space plasma boundaries and within plasma sheets. Broad regions of the magnetosphere are connected by thin boundary layers. At these boundaries, plasmas are transported, accelerated, and energized. The main focus is on how the coupling of processes between different scale sizes influences magnetospheric structure and dynamics. We need to learn how these small-scale processes control large-scale phenomenology, such as magnetotail dynamics, plasma entry into the magnetosphere, and substorm initiation.

Science objectives of the Magnetospheric Multiscale Mission are described in more detail by *Heelis et al.* (1997) and *Burch et al.* (1999). Some of the key questions to be answered are:

What are the processes that permit and control the reconnection of magnetic field lines across collisionless plasma boundaries?

How do energy conversion processes accelerate particles at these boundaries, and what roles do parallel electric fields play?

How are electric currents, which connect distant regions of the magnetosphere, generated, controlled, and disrupted at boundaries?

How do microscale processes near plasma boundaries couple to larger-scale dynamics and structure?

Understanding these fundamental processes requires multipoint measurements that uniquely separate temporal and three-dimensional spatial variations. The spatial and temporal variations need to be measured over scale lengths appropriate to the processes being studied, connecting the small-scale kinetic regime to the larger-scale regimes appropriate for a magnetohydrodynamic description.

The conceptual design [*Burch et al.*, 1999] of Magnetospheric Multiscale employs four spacecraft. The four spacecraft are arranged in an approximately tetrahedronal configuration, with spacing varying between 10 km to several Earth radii. The spacecraft are identically instrumented to measure gradients in the critical in-situ plasma and field parameters. Each spacecraft measures magnetic and electric fields, plasma distributions, energetic particles and plasma waves. The measurements are made with high temporal and spatial resolution. The spacecraft fly in formation to different locations in the magnetosphere, from the dayside magnetopause to the magnetotail. Four orbit phases are planned, with the first three in the equatorial plane and the last in a polar orbit. The total mission duration is planned to be two years.

The Geospace Electrodynamic Connections Mission: Defining the Dynamics of the Interior Boundary of the Magnetosphere During Storms and Substorms

The Global Electrodynamic Connections (GEC) mission [*Sojka and Heelis,* 2001] is a multispacecraft Solar Terrestrial Probe that will allow us to understand the bidirectional electrical coupling between the atmosphere/ionosphere and the magnetosphere. It investigates the low altitude extent of the geospace environment to understand the electrodynamic coupling between the ionosphere and the magnetosphere, and the processes that dissipate energy in the upper atmosphere. The overall focus is to examine the energy exchange processes between the ionized and the neutral components of the upper atmosphere with observations of the global response. Temporal and spatial variations can be resolved with multiple satellites in adjustable orbits, providing comprehensive coverage of ionospheric and atmospheric parameters. Specification of the electrodynamics of the coupled ionosphere/magnetosphere at the appropriate scales and ranges will "close the electrical circuit" for both regions. The scientific goals are to:

Resolve the mechanisms responsible for the complex electrical interaction and transport between the magnetosphere and ionosphere/atmosphere.

Determine the important spatial and temporal scales in energy dissipation, electrodynamics, and transfer processes.

Some of the key questions to be answered are:

How are energy and momentum exchanged between ion and neutral gases in the upper atmosphere?

How does the state of the ionosphere determine its response to a given external forcing?

How are the magnetosphere and ionosphere actively coupled?

What are the roles of turbulence and coherent flows in energy exchange processes?

Four satellites flying in formation in polar orbit are used for GEC. The satellites are in adjustable orbits, with the ability to dip into the atmosphere down to approximately 130 km, where energy and momentum are exchanged between the ionized and neutral gases and where electromagnetic energy from the magnetosphere is ultimately dissipated as heat and light. In-situ sensors will measure all relevant ionospheric parameters and decompose time and space variations. With onboard propulsion these four spacecraft will fly approximately along the same orbit with variable separations that allow spatial and temporal variations to be identified. The GEC mission will quantitatively examine the electrodynamic interactions that control the lower boundary of the magnetosphere and the upper electrical boundary of the Earth's atmosphere.

Innovative technologies used for the GEC mission are advanced propulsion for dipping capability, miniaturized instrumentation, and lightweight booms. Accurate control of the location of the science measurements will be maintained by precision formation flying.

Magnetospheric Constellation Mission: Defining the Mesoscale Magnetospheric Dynamics During Storms and Substorms

The Magnetospheric Constellation Mission [*Spence et al.,* 2001] is an ambitious Solar Terrestrial Probe that will measure with a large fleet of relatively simple spacecraft the finescale structures and the processes that transport energy and momentum in the magnetosphere. Magnetospheric Constellation functions as a network of satellites that images continuously the magnetospheric fields and plasma, with the number of pixels in the magnetospheric image equal to the number of spacecraft in the Magnetospheric Constellation. Methods to accomplish the Magnetospheric Constellation Mission have been described by *Heelis and coworkers* (1997), by *Spence et al.* (2001), and by papers in the volume edited by *Angelopolous and Panetta* (1978).

The primary science objective of the Magnetospheric Constellation is to understand how the magnetotail stores, transports, and releases matter and energy. Some of the key science questions are:

How does solar wind material enter the magnetosphere and where does it go?

How does solar wind structure affect the magnetosphere?

What is the role of large-scale turbulence in the magnetosphere?

What are the mechanisms underlying mass and energy flow in the magnetosphere?

What causes and limits energy surges in the geospace environment?

The measurement approach of the Magnetospheric Constellation science definition team [*Spence et al.,* 2001] is a Magnetotail Constellation Dynamic Response and Coupling Observatory (DRACO) designed to understand the nonlinear dynamics, responses and connections within the Earth's structured magnetotail. It will use a constellation of about 50 to 100 distributed spacecraft, each instrumented to make vector measurements of the magnetic field, plasma flow, and energetic particle fluxes. DRACO will reveal magnetotail processes operating within a domain extending 20 Earth radii (Re) across the tail and 40 Re down the tail. Each spacecraft will have resources of about 10 to 20 kg and 10W. The spacecraft will be in highly elliptical equatorial orbits with common perigees of 3 Re and apogees distributed from 7 to 40 Re, yielding mean interspacecraft separations of 1 to 2 Re.

The Magnetotail Constellation DRACO Mission will reveal how the magnetotail behaves dynamically and will uncover the convection pattern in the magnetotail. The global evolution of small-scale features into larger scale structures is directly related to the response of the system to magnetospheric substorms and the role that small scale turbulence plays in large scale convection. The morphology of the plasma entry at the dayside reconnection sites will be investigated by sequential passes through them. The onset and evolution of substorms in the magnetotail can be investigated, as well as the nature of the feedback of the aurora on the magnetosphere.

ROY (SCHWARM): Exploring Spatial Scales of Magnetospheric Turbulence and its Relationship to Storms and Substorms

The ROY Project is aimed to study previously unexplored small-scale phenomena within the so called "diffusion region" where strong plasma turbulence, and magnetic "reconnection" occur in the critical magnetospheric domains (Figure 6). At the dayside these are mainly the

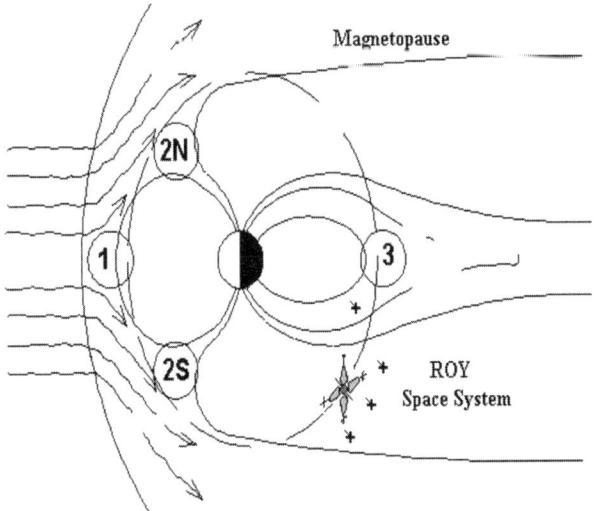

Figure 6. Critical regions of "annihilation" (reconnection) of the magnetic field in magnetosphere: 1 - Subsolar magnetopause, 2 - Northern and Southern Cusp, 3- Substorm generation region

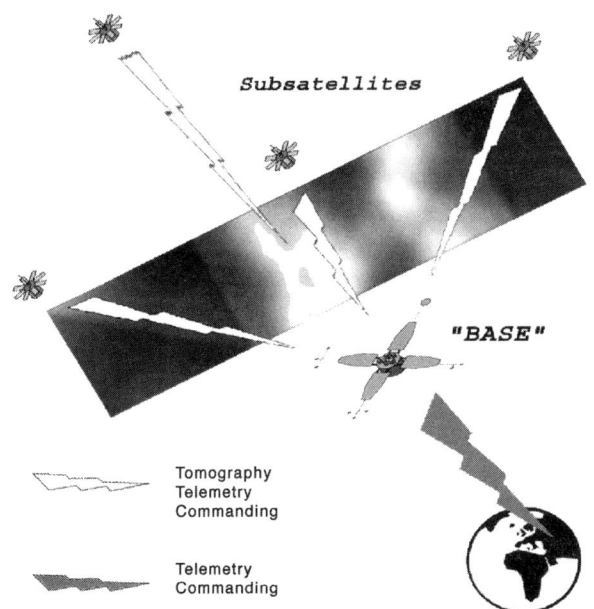

Figure 7. ROY spatial configuration and communication scheme

outer cusps and subsolar magnetopause, but recent findings from the INTERBALL Project show that flanks of the magnetosphere can also be sites of these active plasma processes leading to large scale bulk flows and particle acceleration. On the nightside the most interesting regions are at the earthward edge of the cross-tail current, but their locations are variable.

The main aim of the ROY project is to determine spatial scales of the strong plasma turbulence, characteristic amplitudes and velocities of plasma and magnetic field inhomogeneities, their dynamic spectra, accompanying waves and particles characteristics. An orbit with an apogee ~12–15R_e and a perigee ~5000 km is selected as the optimal for the scientific purposes of the project. The project includes a group of satellites consisting of the main (BASE) satellite and 4 subsatellites at distances 10–300 (1000) km (see Figure 7). The BASE satellite will be equipped with the full set of particle, field and wave plasma diagnostic instrumentation. Its large onboard memory (~10 Gbytes) will allow the storage, partial onboard processing, and compression of data from all the spacecraft and transmission to the Earth using directed antenna. The subsatellites (of the order of 75 kg each) will have thrusters to control orbital configuration and their distance from the BASE determined by GPS/GLONASS navigation technique. They will perform comprehensive in situ plasma measurements and receive coherent radiowaves emitted from the BASE to provide the data for a crude tomography, which will be the main scientific experiment onboard. This medium-wave radio tomography will provide a totally new kind of information about the small-scale variations in the plasma flowing past and between the spacecraft. The most important advantage of the method is the possibility to reconstruct the parameters of the surrounding media not just in the spacecraft vicinity but in the whole region between the satellites as well as to eliminate the Langmuir sheath effects around the spacecraft.

The following scientific experiments will be carried out: angle-energy-mass spectrometry of hot and superthermal plasma; magnetic and electric field measurements; wave spectra registration in a wide range; active probing experiments to periodically check the plasma density and magnetic field measurements by independent techniques.

Currently the ROY project is developed as a joint program with Germany (where it is called SCHWARM). Construction of the BASE satellite will be done in Russia, while the subsatellites are designed in Germany.

The ROY Project differs from previous missions (e.g. ISEE, INTERBALL, and CLUSTER) in that the selected apogee altitude for the polar orbit provides long time measurements inside the critical regions of the magnetic field reconnection. Furthermore, the combined in situ and tomographic measurements significantly improve the quality of the result. Local satellite data can be used as a very good input (initial approximation) for the tomographic reconstruction of 2D and 3D distributions, their motion and evolution.

Project ROY is currently supported by the Russian Ministry of Science and Technology. With the endorsement of the Russian Space Agency the mission could be launched in 2008-2009.

INTERBALL-PROGNOZ: Ionosphere-Magnetosphere Connections During Storms and Substorms

The recently proposed INTERBALL ROGNOZ project is planned to be a segment of the future national Space Weather Program. This multi spacecraft multidisciplinary mission is a joint project of Russian and Ukrainian Space Agencies. It will include the high-apogee spacecraft INTERBALL-3 of the "Prognoz-M1" type (the same as used in the INTERBALL project, provided by Russian collaborators) and three low-apogee micro-satellites launched on the same orbit in the ionosphere (provided by Ukrainian collaborators). INTERBALL-PROGNOZ is designed as the low-cost fast-track mission widely relying on experience and spare hardware of the INTERBALL project to serve the broad range of scientific and technical goals.

The main scientific objectives of the mission are:

Development of the prototype of the national space weather system, including monitoring measurements of solar radiation from hard X-ray to ultraviolet ranges, solar cosmic rays, interplanetary magnetic field and solar wind, onboard facilities for continuous real-time data transfer.

Continuation of the magnetospheric and ionospheric research targeted at studies of the full Sun-ionosphere chain of energy transformation and serving of space weather prognostic and monitoring applications. Measurements will include: local magnetospheric, solar wind and magnetosheath particle, field, and wave diagnostics; ionospheric plasma and wave measurements; monitoring of field-aligned currents, precipitating particles, cross-polar cap potential; active ionospheric experiments.

In order to meet all requirements of the research program, the following orbital scenario (Figure 8) is designed for the INTERBALL-3 spacecraft: (1) Launch in the Sun-Earth L1 point fly-by orbit and return to Earth. Duration of this stage of the flight will be 3-4 months. (2) Transfer to near-Earth elliptic orbit with apogee 350000-400000 km. Other parameters of the orbit will be selected to maximize the spacecraft presence inside the magnetotail. Duration of this stage is planned to be 2-3 years. (3) Transfer to L2 fly-by orbit. This stage is currently considered as optional. The transfer to the L1 orbit will be performed by means of standard MOLNIYA launcher. The transition from this orbit to elliptic orbit with apogee 350000 km (required ΔV is 150 m/s) will be performed by means of additional thrusters to be installed onboard INTERBALL-3. The optional maneuver to the L2 orbit will require the Moon gravity assist.

The low apogee satellites will be placed on the ~700 km high circular sun synchronous orbit oriented along the dusk-dawn meridian. Such an orientation of the orbit is necessary for successful monitoring of the cross-polar cap potential. The spacecraft separation will be adjusted during the mis-

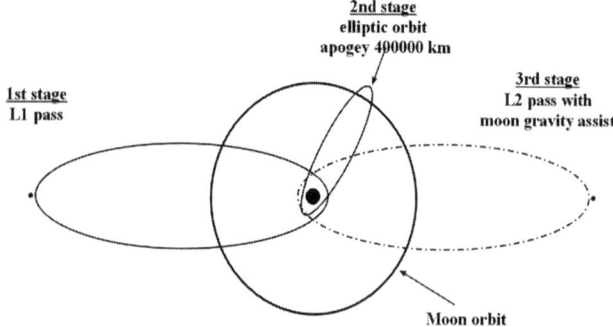

Figure 8. Orbital scenario for INTERBALL-3 high apogee mission

sion in order to carry out a number of active and passive ionospheric experiments. For this project the novel microsat is now being designed by "Yuzhnoe" space center in Dnepropetrovsk, Ukraine. All three satellites will be launched together by TSIKLON or DNEPR launcher.

The project is now under consideration of the Russian Aviation and Space Agency and Ukrainian National Space Agency. If approved, the launch is expected in 2004–2005.

RESONANCE: Magnetosynchronous Observations as Diagnostics for Storms and Substorms

The aim of the RESONANCE mission is to study wave-particle interactions in the inner Earth's magnetosphere. The scientific program of the project consists of two parts. The first, so called "passive part", is directed to the investigation of natural phenomena in the magnetospheric plasma. The main goals of this part are the following:

Dynamics of the magnetospheric cyclotron masers and its long-term evolution;

Ring current formation and global magnetospheric phenomena;

Plasmasphere refilling after magnetic perturbation;

The role of small-scale phenomena in the global dynamics of magnetospheric plasma.

The second part of the Program, the "active part", will focus on the joint experiments of the RESONANCE satellite(s) and a ground-based HF heating facility, as shown in Figure 9. This experiment is based on the assumption that parameters of the natural magnetospheric oscillatory systems could be changed by artificial modifications of the ionosphere. Powerful HF electromagnetic emissions will heat the ionosphere and thus modulate the ionospheric mirrors for cyclotron waves by means of satellite injections or by

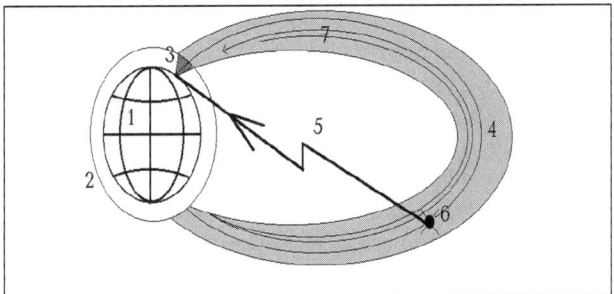

Figure 9. Scheme of the joint experiment with a ground based heating facility (one satellite option) 1 – the Earth, 2 – ionosphere, 3 – region of ionosphere heated by the radiation from the heater, 4 – magnetic flux tube, 5 – TM transmission, 6 – the satellite, 7 – trajectories of particles and guided waves

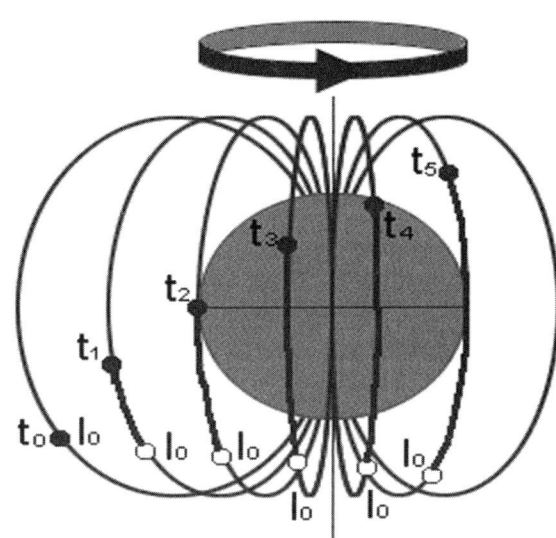

Figure 10. Schematic of magnetosynchronous orbit.

using the existing heating facilities. Natural magnetospheric oscillations will be measured onboard the RESONANCE satellite. Results of these measurements will be transmitted to the receiving station near the heating facility, and will be used to modulate the HF radiation. In this experiment the telemetry line (with a controlled phase shifter), connecting the satellite and the heating facility, will regulate the feedback superimposed on the existing natural feedback of the magnetospheric oscillatory system.

The novel type of a magneto-synchronous satellite orbit (see Figure 10) proposed and designed for this project allows one to conduct the measurements within the same magnetic flux tube for sufficiently long time intervals. Calculations show that the duration of about an hour may be reached if the transverse size of the flux tube at the ionospheric level is about 50–100 km. This duration is significantly longer than the characteristic time scales of the magnetospheric masers and, therefore, such an orbit provides a good opportunity to study these processes. The orbit inclination may be used as an adjustable parameter. The footprint of the flux tube might be at the HAARP and MURMANSK facilities.

Note that the position of the magnetic flux tube should be known *a priori* to adjust the satellite orbit. This is true only for the magnetic flux tubes that co-rotate with the Earth. This imposes an important constraint on the magnetic latitude of the operating orbit, which must be in the inner magnetosphere not far from the plasmapause; otherwise reliable optimization becomes impossible.

Additionally, the second satellite might be launched on a near-equatorial orbit, periodically meeting with the magneto-synchronous satellite in its apogee to make the two-point measurements of a large scale naturally oscillating process. The position of each satellite and the relative distance between them will be determined accurately with onboard navigation receivers (GPS/GLONASS system).

STEREO: Tracking CMEs from Sun to Earth as Drivers of Storms and Substorms

STEREO [Rust et al., 1997], the Solar Terrestrial Relations Observatory, is a Solar Terrestrial Probe whose aim is to increase the understanding of the three-dimensional structure of the Sun's corona. The primary goals are to study the origin of coronal mass ejections (CME), their evolution in the interplanetary medium, and the dynamic coupling between CMEs and the Earth's environment.

STEREO seeks answers to fundamental questions: Are CMEs driven primarily by magnetic or nonmagnetic forces? What initiates CMEs? What is the role of magnetic reconnection? What is the origin of the associated waves, shocks, and particle radiation? These questions cannot be answered conclusively with the single vantage point observations currently available. In order to understand and forecast CMEs, three-dimensional images are needed.

To accomplish the measurement goals, two identical spacecraft will be placed in Sun-centered orbits at 1 AU, with one drifting ahead of the Earth and one lagging behind the Earth. They will drift away from the Earth at about 22 degrees per year. From these vantage points it will be possible to measure stereoscopically the three-dimensional structures in the solar corona. Simultaneous image pairs will be obtained by STEREO at gradually increasing angular separations in the course of the mission. In situ measurements on each spacecraft will provide accurate information about the state of the ambient solar wind and energetic particle populations ahead of CMEs, while also determining the plasma, magnetic field and energetic particle characteristics of the interplanetary disturbances as they pass. These measurements will enable definitive tests of CME and interplanetary shock models.

Space Weather Research Network: Studying the Effects of Storms and Substorms

There are two groups of missions planned for the Living With a Star program [Withbroe, 2001] that will form a Space Weather research network: the solar dynamics elements and the geospace dynamics elements.

The solar dynamics elements consist of the Solar Dynamics Observatory and the Solar Sentinels that observe the Sun and track space disturbances that originate there and are modified by the heliospheric environment. The Solar Dynamics Observatory (SDO) will observe the Sun's dynamics and help us understand the nature and source of variations, from the solar interior into the solar atmosphere. The Solar Sentinels will provide a global view of the Sun and inner heliosphere and describe the origin and evolution of eruptions from the Sun to the Earth.

Geospace dynamics will be measured by spacecraft near the Earth. The Radiation Belt Mappers will study the origin and dynamics of the radiation belts and determine the evolution of penetrating radiation during geomagnetic storms. The Ionospheric Mappers will gather knowledge on how the ionosphere behaves as a system that directly responds to solar photons and energetic particles, as well as solar wind energy stored and then suddenly discharged during geomagnetic storms and substorms.

Solar Probe: Exploring the Inner Boundary of the Sun-Earth System

The mechanisms generating the solar wind, the cause of the formation of planetary magnetospheres, substorms, storms, and aurora, are not known. Solar Probe is a mission designed to determine the ultimate origin of all space weather.

Solar Probe [Tsurutani, 2001] will provide during its near-Sun flyby in-situ measurements of the solar corona and high-resolution images and magnetic measurements of the photosphere and polar atmosphere. The questions to be investigated by Solar Probe are: What heats the corona and accelerates the solar wind? Where in the corona do sources of solar wind originate? What are the roles of waves and turbulence in coronal heating? What processes accelerate, store, and transport energetic particles in the corona?

A full complement of both in-situ and remote sensing instrumentation will be flown, including high-resolution disk imaging. The Solar Probe measurements will provide a "ground truth" for interpreting the many measurements of the Sun and solar activity that have been made from near the Earth. The imaging and in situ instruments will provide the first three-dimensional view of the corona, measurements of the magnetic fields with high spatial and temporal resolution, and helioseismic measurements of the polar regions, as well as local sampling of plasmas and magnetic fields with high spatial resolution at all latitudes.

Solar Probe will arrive at the Sun along a polar trajectory perpendicular to the Sun-Earth line with a perihelion of four solar radii from the Sun's center. Two perihelion passages are planned, so as to measure the solar wind properties in coronal

Table 2. Some future missions to understand geospace dynamics

Mission	Type	Method of implementation	Launch Date
Magnetospheric Multiscale (MMS)	Microscope	4 spacecraft in a tetrahedronal array in variable high-altitude orbits	2008
Global Electrodynamic Constellation (GEC)	Network	4 spacecraft in low polar orbit, with ability to dip to low altitudes	2009
Magnetospheric Constellation	Network	50 to 100 spacecraft distributed between 20 and 40 Re in the magnetotail	2011
ROY/Schwarm	Microscope	Main satellite and four subsatellites with apogee of 12-15 Re in polar orbit	2008-9
Interball-3	Outer boundary and network	One spacecraft in L1 and L2 transfer orbits with 3 other spacecraft in low polar orbit	2004-2005
Resonance	Global configuration	2 satellites in equatorial plane and in magnetosynchronous orbit	2005-6
STEREO	Solar plasma variations	2 spacecraft ahead and behind the Earth	2005
Solar Probe	Solar plasma variations	Near-solar flyby at 4 solar radii	2010

holes near solar minimum and in streamers near solar maximum. To reach the Sun, Solar Probe must first fly to Jupiter and use a gravity assist to lose its angular momentum about the Sun. Unique to Solar Probe is a heat shield that will withstand temperatures as high as 2,300 K and keep instruments and spacecraft electronics at room temperature.

CONCLUSIONS

The existing capabilities provide measurements of the geospace system that have recently produced much scientific progress. However, there are major unanswered questions, because existing measurements are too sparse, have inadequate spatial and temporal resolution, and are unable to unambiguously separate cause and effect. Fortunately, the recent progress provides the scientific basis for new approaches by the next generation of missions that will provide both more complete measurement arrays and increased space-time resolution.

Future progress requires a robust strategy involving several complementary mission types. Fortunately, future science missions in the NASA Solar Terrestrial Probes and the Living with a Star program, as well as international missions, will provide the necessary information at the needed spatial and temporal resolution. Table 2 shows the schedule for the launch of some of these missions.

Acknowledgements. Research for the Resonance project is partially supported by INTAS/ESA 99-1006 grant. Research for the Roy project is partially supported by the INTAS 2000-465 grant.

REFERENCES

Acuna, M., K. Ogilvie, D. Baker, S. Curtis, D. Fairfield, and W. Mish, The Global Geospace Science program and its investigations, *Space Science Rev., 71*, 5-21, 1995.

Angelopolous, V. and P. Panetta, *Science closure and enabling technologies for constellation class missions*, University of California Press, Berkeley, California, 1998.

Baker, D.N., and T.I. Pulkkinen, Large-scale structure of the magnetosphere, *New Perspectives on the Earth's Magnetotail,* ed. A. Nishida, D.N. Baker and S.W.H. Cowley, pp. 21-31, AGU, Washington, D.C. 1998.

Borovsky, J., M. Thomsen, and R. Elphic, The driving of the plasma sheet by the solar wind, *J. Geophys. Res., 103*, 17617-17639, 1998.

Burch, et al. The Magnetospheric Multiscale Mission: Resolving Fundamental processes in Space plasmas, NASA/GSFC, Greenbelt, Maryland, December, 1997.

Burch, J., S. Mende, D. Mitchell. T. Moore, C. Pollock, B. Reinisch, B. Sandel, S. Fuselier, D. Gallagher, J. Green, J. Perez, and P. Reiff, Views of the Earth's magnetosphere with the IMAGE satellite, *Science, 291*, 619-624, 2001.

Chen, M., M. Shultz, and L.R. Lyons, Simulations of phase space distributions of stormtime proton ring current, *J. Geophys. Res., 99*, 5745, 1994.

Daglis, I.A., and W.I. Axford, Fast ionospheric response to enhanced activity in Geospace: Ion feeding of the inner magnetotail, *J. Geophys. Res., 101*, 5047, 1996.

Delcourt, D.C., J.-A. Sauvaud, and T.E. Moore, Cleft contribution to ring current formation, *J. Geophys. Res., 95*, 20937, 1990.

Ergun, R. et al., Feasibility of a multisatellite investigation of the Earth's magnetosphere with radio tomography, *J. Geophys. Res.,105*, 361,2000.

Escourbet, C., R. Schmidt, and M. Goldstein, CLUSTER-Science and Mission Overview, *Sp. Sci. Rev., 79*, 11-32, 1997.

Farrugia, C.J., L.F. Burlaga, and R.P. Lepping, Magnetic clouds and the quiet-storm effect at Earth, *Magnetic Storms*, eds. Tsurutani et al., pp. 91-106, AGU, Washington, D.C., 1997.

Fok, M.-C., T.E. Moore, and M.E. Greenspan, Ring current development during storm main phase, *J. Geophys. Res., 101*, 15311, 1996.

Ganguly, S., G. Van Bavel, and A. Brown, Imaging electron density and magnetic field distributions in the magnetosphere: a new technique, *J. Geophys. Res., 105*, 16063, 2000.

Gonzalez, W. D., J. Joselyn, Y. Kamide, H. Kroehl, G. Rostoker, B. Tsurutani, and V. Vasyliunas, What is a magnetic storm?, *J. Geophys. Res., 99*, 5771, 1994.

Heelis, R. et al., *Geospace Multiprobes: Report from the Science Definition Team*, NASA/GSFC, Greenbelt, Maryland, December, 1997.

Hesse, M. Three-dimensional reconnection in space and astrophysical plasmas and its consequences for particle acceleration, *Rev. Mod. Astr. 8*, 323, 1995.

Kamide, Y., et al., Current understanding of magnetic storms: Storm-substorm relationships, *J. Geophys. Res., 103*, 17,705, 1998.

Kozyra, J.U., et al., Effects of energetic ions on electromagnetic ion cyclotron wave generation in the plasmapause regions, *J. Geophys. Res., 89*, 2217, 1984.

Lui, A., Current controversies in magnetospheric physics, *Rev. Geo., 39 (4)*, 835-864, 2001.

Moore, T., et al., Ionospheric mass ejection in response to a coronal mss ejection, *Geophys. Res. Lettr., 26*, 2339-2342, 1999.

Ohtani, S., et al., Substorm onset timing: The December 31, 1995 event, *J. Geophys. Res., 104 ,22,713*, 1999.

Panetta, P. et al., NASA-GSFC Nano-satellite technology development, 12th AIAA/USU Conference on Small Satellites, Paper SSC98-VI-5, 1998.

Paschmann, G., and P. Daly, *Analysis Methods for Multispacecraft Data*, International Space Science Institute, Bern, 1998.

Russell, C.T., ed., The Global Geospace Science mission, *Space Science Reviews, 71*, Nos. 1-4, 1995

Rust, D. et al., The Sun and heliosphere in three dimensions: the STEREO mission, NASA/GSFC, Greenbelt, MD, 1997.

Slavin, J.A., et al., ISTP Simultaneous Observations of earthward flow bursts and plasmoid ejection during magnetospheric Substorms, *J. Geophys. Res., in press*, 2001.

Sojka, J, R. Heelis, et al., Understanding Plasma Interactions with the Atmosphere: the Geospace Electrodynamic Connections Mission, NASA/GSFC, Greenbelt, Maryland, 2001.

Spence, H., et al., Understanding the Global Dynamics of the Structured Magnetotail: the Magnetospheric Constellation Mission, NASA/GSFC, Greenbelt, Maryland, 2001.

Tsurutani, B., Solar Probe, submitted to *J. Spacecraft and Rockets*, 2001.

Vondrak, R., Multiscale objectives of the Geospace Multiprobes, *Physics of Space Plasmas, 15*, 331-336, 1998.

Withbroe, G., Living with a Star, in *Space Weather, Geophys. Monograph 125*, P. Song, H. Singer, and G. Siscoe, eds., pp.45-41, AGU, Washington, 2001.

Wolf, R.A., et al., Computer simulation of inner magnetospheric dynamics for the magnetic storm of July 29, 1977, *J. Geophys. Res., 87*, 5949, 1982.

Wolf, R.A., et al., Modeling convection effects in magnetic storms, *Magnetic Storms*, eds. Tsurutani et al., pp. 161-172, AGU, Washington, 1997.

Richard R. Vondrak, NASA/Goddard Space Flight Center, Code 690, Greenbelt, Maryland 20771

James A. Slavin, NASA/Goddard Space Flight Center, Code 696, Greenbelt, Maryland 20771

L. Zelenyi, Institute for Space Research, Moscow, Russia 117810

Madhulika Guhuthakurta, NASA Headquarters, Code S, Washington, D.C. 20546

Steven A. Curtis, NASA/Goddard Space Flight Center, Code 695, Greenbelt, Maryland 20771

Bruce Tsurutani, Jet Propulsion Laboratory, Mail Code 169-506, Pasadena, California 91109